Ecological Studies, Vol. 120

Analysis and Synthesis

Edited by

G. Heldmaier, Marburg, FRG
O. L. Lange, Würzburg, FRG
H. A. Mooney, Stanford, USA
U. Sommer, Kiel, FRG

Ecological Studies

Volumes published since 1990 are listed at the end of this book.

Springer-Verlag
Berlin
Heidelberg
GmbH

James F. Reynolds John D. Tenhunen (Eds.)

Landscape Function and Disturbance in Arctic Tundra

With 109 Figures, 11 in color
and 47 Tables

 Springer

Professor Dr. JAMES F. REYNOLDS
Duke University
Department of Botany
Phytotron Building
Box 90340
Durham, NC 27708
USA

Professor Dr. JOHN D. TENHUNEN
Universität Bayreuth
Lehrstuhl für Pflanzenökologie II
Bayreuth Institute for Terrestrial Ecosystem Research (BITÖK)
D-95440 Bayreuth
Germany

ISBN 978-3-662-01147-8   ISBN 978-3-662-01145-4 (eBook)
DOI 10.1007/978-3-662-01145-4

Library of Congress Cataloging-in-Publication Data

Landscape function and disturbance in arctic tundra / James F.
  Reynolds, John D. Tenhunen (eds.).
      p.   cm. — (Ecological studies : v. 120)
  Includes bibliographical references and index.
  ISBN 978-3-662-01147-8
      1. Tundra ecology—Alaska—North Slope. 2. Landscape ecology—Alaska—North Slope.
  3. Man—Influence on nature—Alaska—North Slope. 4. Tundra ecology—Arctic regions—Math-
  ematical models. 5. Landscape ecology—Arctic regions—Mathematical models. 6. Man—Influ-
  ence on nature—Arctic regions—Mathematical models. I. Reynolds. James F., 1946- ,
  II. Tenhunen, John D., 1946- . III. Series.
  QH105.A4L36   1996
  574.5'2644'097987—dc20

© Springer-Verlag Berlin Heidelberg 1996
Originally published by Springer-Verlag Berlin Heidelberg in 1996
Softcover reprint of the hardcover 1st edition 1996

Typesetting: Best-set Typesetter Ltd., Hong Kong
SPIN 10087973   31/3137-5 4 3 2 1 0 – Printed on acid-free paper

Foreword

The plants dominating arctic tundra differ in a fundamental way from almost all other plants on earth. They can carry on metabolic and reproductive processes during the short, cold growing seasons of the far North at air and soil temperatures only slightly above or even below 0°C. These cold-adapted plants are mostly herbaceous perennials or dwarf shrubs that form a mosaic of communities that are subject to large fluctuations in their physical environment, both atmospheric and edaphic. These once pristine and relatively isolated ecosystems are now subject to considerable disturbance, due mostly to increasing human populations and their use of off-road vehicles and industrial equipment.

Arctic tundra is underlain with permafrost, which is as deep as 600 m in places. It has been permanently frozen for thousands of years. Embedded in it are ice wedges and lenses in polygonal patterns that characterize much of the tundra. The permafrost is fundamental to arctic tundra ecosystems—it is the "glue" that holds these ecosystems together. During the short summers, the "active" soil layer above the permafrost table thaws briefly to depths of ca. 20–75 cm and allows root penetration, growth, and nutrient uptake by the tundra vegetation. Without the complementary interactions between permafrost and tundra vegetation, thermokarst, or melting of the ground ice followed by erosion and massive disturbance, would result in transport of the materials sequestered in these ecosystems via the streams to the ocean.

Since more than 25% of the earth's soil carbon is stored in arctic and subarctic permafrost soils, irreversible thermokarsting would release large amounts of carbon dioxide and methane into the atmosphere. Such positive feedback to the atmosphere would stimulate additional temperature increase, resulting in further loss of ecosystem stability. The release of nutrients that have been frozen and dormant for so long in upland tundra soils would stimulate production in river and lake ecosystems downstream, where nutrients have been limiting for centuries or even millennia. If upland ecosystems begin to disappear with the thermokarst, as happened within only 11 years to the Voth Tundra near Barrow, we would face a situation of considerable overall resource loss and environmental impact. If the insulation to permafrost provided by tundra vegetation is damaged or destroyed by *any* disturbance, the underlying permafrost is doomed, as is the tundra.

This volume, which presents research from the US Department of Energy's R4D project on disturbance in arctic ecosystems, is a welcome and thorough contribution to arctic science. It clearly lays out many of the linkages controlling ecosystem development that are shifted under disturbance regimes. This large, multidisciplinary study was based at the Imnavait Creek watershed near Toolik Lake, ca. 250 km south of Prudhoe Bay on the North Slope of Alaska, and was a product of an important US National Research Council Committee Report of 1982. Much of the research presented here was undreamed of 20 years ago when several of the chapter authors and I camped on this watershed in an unusually cold July to formulate research priorities. Often, with snow falling at dawn, we would be awakened by Bob White's bagpipes to enjoy Al Johnson's and Skip Walker's pancakes and coffee and another fine day of field work.

This volume suggests approaches for ecosystem management that are a result of the close integration between fieldwork and computer modeling. The groups working in specific disciplines have provided process summaries that extend our knowledge of the tundra, while the modeling—as well as the overall framework of the book—demonstrate the new promise of integrated landscape approaches for creating management tools. It is essential that we continue in the direction of the research described here, to build our capacity to understand the dimensions of ecosystem impacts that may result due to climate change or man's interference with natural ecosystem behavior. The lessons from R4D as summarized here provide a unique opportunity for truly joining ecosystem science with land and resource management in the Arctic now and in the future.

Winter 1995   W.D. Billings
                                                                  Department of Botany
                                                                       Duke University

Contents

I   **Introduction** 1

1   **Ecosystem Response, Resistance, Resilience, and Recovery in Arctic Landscapes: Introduction**
J. F. Reynolds and J. D. Tenhunen 3

1.1   Introduction 3
1.2   NRC Committee Report 6
1.3   The R4D Program 8
1.3.1   Objectives and Conceptual Framework 8
1.3.2   Program Implementation 9
1.3.3   Landscape Function 12
1.4   Summary 14
References 16

2   **Integrated Ecosystem Research in Northern Alaska, 1947–1994**
G. R. Shaver 19

2.1   Introduction 19
2.2   Early Days at NARL 19
2.3   The U. S. Tundra Biome Program 21
2.4   The Meade River RATE Program 22
2.5   Eagle Creek and Eagle Summit 23
2.6   The Arctic LTER Program at Toolik Lake 25
2.7   Other Studies In Alaska and Elsewhere 26
2.8   Summary and Prospects 27
References 29

3   **Disturbance and Recovery of Arctic Alaskan Vegetation**
D. A. Walker 35

3.1   Introduction 35
3.2   Disturbance and Recovery 35
3.3   Typical Disturbance and Recovery Patterns 39
3.3.1   Small Disturbed Patches 39
3.3.2   Contaminants 43
3.3.2.1 Hydrocarbon Spills 43

3.3.2.2   Seawater and Reserve-Pit Spills   45
3.3.3   Fire   47
3.3.4   Transportation Corridors   48
3.3.4.1   Bulldozed Tundra and Related Disturbances   48
3.3.4.2   Off-Road Vehicle Trails   50
3.3.4.2.1 Summer Travel   50
3.3.4.2.2 Winter Travel   52
3.3.4.3   Permanent Roads and Pads   54
3.3.4.4   Gravel Mines   55
3.3.4.5   Native Species in Revegetation of Gravel Pads and Mines ..   56
3.3.4.6   Road Dust   58
3.3.4.7   Roadside Impoundments   59
3.3.5   Cumulative Impacts   61
3.4   Conclusions   62
References   64

4   Terrain and Vegetation of the Imnavait Creek Watershed
           D. A. WALKER and M. D. WALKER   73

4.1   Introduction   73
4.2   Terrain   73
4.2.1   Glacial Deposits   74
4.2.2   Retransported Hillslope Deposits   75
4.2.3   Colluvial Basin Deposits   79
4.2.4   Floodplain Deposits   79
4.3   Vegetation   79
4.3.1   Flora   79
4.3.2   Vegetation Types   80
4.3.2.1   Lichen-Covered Rocks   85
4.3.2.2   Dry Heath   85
4.3.2.2.1 Exposed Sites   88
4.3.2.2.2 Snowbeds   88
4.3.2.3   Tussock Tundra   89
4.3.2.4   Riparian Areas   90
4.3.2.5   Mires   92
4.3.2.6   Beaded Ponds   93
4.4   West-Facing Toposequence   93
4.5   Terrain Sensitivity to Disturbance   97
4.6   Conclusions   98
Appendix A. List of Plants for Imnavait Creek, Alaska   99
References   106

5   Vegetation Structure and Aboveground Carbon
           and Nutrient Pools in the Imnavait Creek Watershed
           S. C. HAHN, S. F. OBERBAUER, R. GEBAUER, N. E. GRULKE,
           O. L. LANGE, and J. D. TENHUNEN   109

5.1	Introduction	109
5.2	Description of Vegetation	109
5.3	Sampling Methods	112
5.3.1	Cover	112
5.3.2	Biomass and Nutrient Pools	112
5.4	Cover	113
5.5	Aboveground Biomass	114
5.5.1	Live Biomass	114
5.5.2	Photosynthetic Biomass	116
5.5.3	Lichen Biomass	117
5.5.4	Organic Litter	117
5.5.5	Watershed Patterns	118
5.6	Nutrient Pools	118
5.6.1	N and P in Heath Cryptogams	118
5.6.2	N and P in Communities	121
5.7	Discussion and Conclusions	124
References		126

II	Physical Environment, Hydrology, and Transport	129
6	Energy Balance and Hydrological Processes in an Arctic Watershed L. HINZMAN, D. L. KANE, C. S. BENSON, and K. R. EVERETT	131
6.1	Introduction	131
6.2	Radiation and Thermal Regimes	131
6.2.1	Surface Energy Balance	133
6.2.2	Snow Cover and Soil Thermal Regime	137
6.3	Hydrological Processes	140
6.3.1	Snowmelt	141
6.3.2	Plot and Basin Water Balance	142
6.3.3	Runoff and Basin Discharge	143
6.3.4	Precipitation, Evaporation, and Evapotranspiration	145
6.4	Energy Balance and Hydrology Models	146
6.4.1	Simulation of the Thermal Regime	146
6.4.2	Simulation of Snowmelt	148
6.4.3	Simulation of Catchment Runoff	150
6.5	Conclusions	152
References		152

7	Shortwave Reflectance Properties of Arctic Tundra Landscapes A. S. HOPE and D. A. STOW	155
7.1	Introduction	155
7.2	Shortwave Reflectance Studies in Arctic Environments	156

7.2.1   Environmental Considerations   156
7.2.2   Radiometric Data   156
7.2.3   Image Data.....................................   156
7.3   Spectral Reflectance   157
7.3.1   Aboveground Biomass   157
7.3.2   Vegetation Composition   158
7.3.3   Landscape Patterns   158
7.3.4   Effects of Dust Deposition   159
7.4   Albedo   160
7.4.1   Undisturbed Tussock Tundra   160
7.4.2   Effects of Dust Deposition   161
7.5   Conclusions   162
References   163

8   Isotopic Tracers for Investigating Hydrological Processes
         L. W. Cooper, I. L. Larsen, C. Solis, J. M. Grebmeier,
         C. R. Olsen, D. K. Solomon, and R. B. Cook   165
8.1   Introduction   165
8.1.1   Units   166
8.1.2   Conservative vs Nonconservative Isotopes   166
8.2   Nonconservative Tracers   167
8.3   Sulfur-35   167
8.4   Oxygen-18   168
8.4.1   Oxygen-18 Content of Snowpack   169
8.4.2   Oxygen-18 Content of Imnavait Creek   169
8.4.3   Oxygen-18 Content of Soil Moisture   172
8.4.4   Covariance of Oxygen-18 and Deuterium in Watershed
         Compartments   172
8.4.5   Covariance of Oxygen-18 and Deuterium in Plant Water ...   174
8.5   Long-Lived Radioisotopes: Lead-210 and Cesium-137   174
8.5.1   Distribution of ^{137}Cs on Tundra and in Lake Sediments   175
8.5.2   Cycling of ^{137}Cs in Annual Berries   176
8.5.3   Distribution of ^{210}Pb in Tundra   178
8.6   Conclusions   178
References   179

III   Nutrient and Carbon Fluxes   183

9   Surface Water Chemistry and Hydrology
         of a Small Arctic Drainage Basin
         K. R. Everett, D. L. Kane, and L. D. Hinzman   185
9.1   Introduction   185
9.2   Watershed Instrumentation   185
9.3   Snowmelt Period   187
9.3.1   Snowmelt Hydrology.............................   187

9.3.2   Snowmelt Chemistry 189
9.3.2.1   Overland Flow 189
9.3.2.2   Water Track Flow 191
9.3.2.3   Imnavait Creek Flow 192
9.4   Post Snowmelt Period 192
9.4.1   Atmospheric Inputs 192
9.4.1.1   Rainfall 193
9.4.1.2   Dry Deposition 194
9.4.1.3   Rime 195
9.4.2   Water Chemistry 195
9.4.2.1   Overland Flow 195
9.4.2.2   Active Layer Flow 196
9.4.2.3   Imnavait Creek Flow 196
9.5   Conclusions 198
References 200

10   **Nutrient Availability and Uptake by Tundra Plants**
          J. P. SCHIMEL, K. KIELLAND, and F. S. CHAPIN III 203

10.1   Introduction 203
10.2   Controls on Mineralization and Nutrient Supply 203
10.2.1   Patterns of Nutrient Supply in the Soil 203
10.2.2   Patterns of Mineralization 204
10.2.3   Controls on N and P Mineralization 205
10.2.4   Controls on Decomposition and Mineralization 206
10.2.4.1 Temperature 206
10.2.4.1.1 Enzyme Activities 206
10.2.4.1.2 Microbial Activity at Low Temperatures 206
10.2.4.1.3 Freeze–Thaw Events 207
10.2.4.2 Effects of Low Oxygen on Microbial Activity
          and Mineralization 207
10.2.4.3 Substrate Quality 208
10.3   Fate of Available Nutrients 208
10.3.1   Microbial Nutrient Uptake and Competition with Plants ... 209
10.3.2   Plant Uptake 210
10.3.2.1 Soil Factors Controlling Nutrient Absorption 211
10.3.2.2 Rooting Strategies 212
10.3.2.3 Uptake Characteristics of Tundra Plants 213
10.3.2.4 Retranslocation vs Current Uptake 214
10.4   Disturbances 215
10.4.1   Vehicle Tracks 215
10.4.2   Road Dust 216
10.4.3   Gray Water 217
10.4.4   Climate Change 217
References 218

11   **Landscape Patterns of Carbon Dioxide Exchange**
        in Tundra Ecosytems
        S. F. OBERBAUER, W. CHENG, C. T. GILLESPIE,
        B. OSTENDORF, A. SALA, R. GEBAUER, R. A. VIRGINIA,
        and J. D. TENHUNEN 223

11.1   Introduction 223
11.2   Methods 224
11.2.1 Community Types 224
11.2.2 Leaf Photosynthesis 224
11.2.3 Ecosystem Efflux 225
11.2.4 Ecosystem Net CO_2 Exchange 225
11.3 CO_2 Uptake 225
11.3.1 Factors Affecting CO_2 Uptake 226
11.3.1.1 Light 226
11.3.1.2 Temperature 226
11.3.1.3 Phenology 228
11.3.1.4 Water Availability 229
11.3.1.5 Nutrition 229
11.3.2 Landscape Patterns in Leaf Photosynthesis 230
11.4 CO_2 Efflux 233
11.4.1 Factors Affecting CO_2 Efflux 233
11.4.1.1 Live Plant Biomass 234
11.4.1.2 Soil Quality 234
11.4.1.3 Thaw Depth and Depth to Water Table 236
11.4.1.4 Soil Moisture 238
11.4.1.5 Soil Temperature 239
11.4.2 Landscape Patterns of CO_2 Efflux 240
11.4.3 Daily and Seasonal Patterns of CO_2 Efflux 244
11.4.4 Dust Deposition Effects on CO_2 Efflux 245
11.5   Landscape Patterns in Net CO_2 Exchange 247
11.6   Conclusions 250
References 252

12   **Control of Tundra Methane Emission**
        by Microbial Oxidation
        S. C. WHALEN, W. S. REEBURGH, and C. E. REIMERS 257

12.1   Introduction 257
12.2   Sampling Procedure 258
12.3   Results and Discussion 260
12.3.1 Methane Flux and Environmental Variables in Tundra
        and Taiga 260
12.3.2 Physiology, Controls, and Potential for Microbial CH_4
        Oxidation 265

12.3.3   Methane Oxidation by Tundra Soils in a Warmer
          Climate 270
12.4   Conclusions 271
References 272

13   **Dynamics of Dissolved and Particulate Carbon
          in an Arctic Stream**
          M. W. Oswood, J. G. Irons III, and D. M. Schell 275

13.1   Introduction 275
13.2   Site Description 276
13.2.1   Imnavait Creek Watershed 276
13.2.2   Description of Imnavait Creek 276
13.3   Field and Laboratory Procedures 277
13.4   Physical Regime 279
13.5   Carbon in Imnavait Creek 280
13.5.1   Concentrations 280
13.5.2   Transport 282
13.5.3   Spatial Variability 283
13.5.4   Seasonal Dynamics 283
13.6   Conclusions 285
References 287

IV   **Modeling Landscape Function** 291

14   **Patch and Landscape Models of Arctic Tundra:
          Potentials and Limitations**
          J. F. Reynolds, J. D. Tenhunen, P. W. Leadley, H. Li,
          D. L. Moorhead, B. Ostendorf, and F. S. Chapin III ... 293

14.1   Introduction 293
14.2   Modeling Framework 294
14.2.1   Spatial Simulation Units 294
14.2.2   Types of Models 296
14.3   Bottom-Up Models 298
14.3.1   Ecosystem Gas Exchange 298
14.3.1.1 Motivation 298
14.3.1.2 Description 298
14.3.1.3 Potentials and Limitations 301
14.3.2   Plant Growth 302
14.3.2.1 Motivation 302
14.3.2.2 Description 303
14.3.2.3 Potentials and Limitations 304
14.3.3   Nitrogen Uptake 306
14.3.3.1 Motivation 306
14.3.3.2 Description 306

14.3.3.3   Potentials and Limitations 307
14.3.4   Decomposition 307
14.3.4.1   Motivation 307
14.3.4.2   Description 308
14.3.4.3   Potentials and Limitations 309
14.4   Top-Down Models 311
14.4.1   Hydrologic Transport 311
14.4.1.1   Motivation 311
14.4.1.2   Description 312
14.4.1.3   Potentials and Limitations 313
14.4.2   Topographically Derived Vegetation Model 315
14.4.2.1   Motivation 315
14.4.2.2   Description 315
14.4.2.3   Potentials and Limitations 316
14.5   Conclusions 318
References 319

15   Modeling Dry Deposition of Dust Along
            the Dalton Highway
            R. LAMPRECHT and W. GRABER 325

15.1   Introduction 325
15.2   Model Fundamentals 325
15.3   Modeling Heavy Particle Dispersion 329
15.4   Estimation of Atmospheric Boundary Layer Parameters ... 330
15.5   Dust Characterization and Mass Transfer Through
            the Atmosphere 333
15.6   Theory of Particle Dry Deposition into Vegetation 335
15.7   Numerical Results 339
15.8   Conclusions 343
References 343

16   Modeling Decomposition in Arctic Ecosystems
            D. L. MOORHEAD and J. F. REYNOLDS 347

16.1   Introduction 347
16.2   Controls on Decomposition 347
16.3   Arctic Decomposition Models 350
16.3.1   ABISKO 351
16.3.2   ARTUS 352
16.3.3   BARK 353
16.3.4   GENDEC 354
16.3.4.1   General Description 354
16.3.4.2   Validation 355
16.4   Model Comparisons 357
16.5   Effects of Environmental Changes 359

16.5.1   Climate Change   359
16.5.2   Effects of Elevated CO_2   360
16.5.3   Impacts of Road Dust Deposition   360
16.5.4   Tussock Phosphorus Dynamics   361
16.6   Conclusions   362
References   363

17   **Hydrological Controls on Ecosystem Gas Exchange**
          in an Arctic Landscape
          B. OSTENDORF, P. QUINN, K. BEVEN,
          and J. D. TENHUNEN   369

17.1   Introduction   369
17.2   Description of Models   370
17.2.1   Community Gas Exchange   370
17.2.2   Spatial Variation in Water Availability   371
17.3   Coupling of Hydrology and Ecosystem Gas Exchange   374
17.3.1   Vegetation Distribution   374
17.3.2   Spatial Variation in Water Table   375
17.4   Water Balance and Seasonal Changes in Water Fluxes   376
17.4.1   Evapotranspiration   376
17.4.2   Discharge   377
17.4.3   Interception and Surface Water Retention   379
17.5   Carbon Balance and Seasonal Changes in Carbon Fluxes ...   379
17.5.1   Predicted Water Table and Soil Respiration   379
17.5.2   Predicted Watershed Level Net CO_2 Balance   380
17.6   Discussion and Conclusions   382
References   384

18   **Road-Related Disturbances in an Arctic Watershed:**
          Analyses by a Spatially Explicit Model of Vegetation
          and Ecosystem Processes
          P. W. LEADLEY, H. LI, B. OSTENDORF,
          and J. F. REYNOLDS   387

18.1   Introduction   387
18.2   Environmental Gradients and Vegetation Distribution   388
18.2.1   Vegetation and Topography   388
18.2.2   Role of Water and Light   389
18.2.3   Role of Nutrients   389
18.3   Description of Model   390
18.3.1   Overview   390
18.3.1.1 T-HYDRO   390
18.3.1.2 T-VEG   393
18.3.1.3 T-NUT   393
18.3.1.4 T-PLT   394

18.3.2   Disturbance Scenarios 396
18.3.2.1   Effects of Altering Discharge 396
18.3.2.2   Effects of Road Dust 397
18.3.3   Model Validation and Limitations 398
18.4   Model Predictions for Undisturbed Watershed 398
18.4.1   Vegetation 398
18.4.2   Discharge 400
18.4.3   N Availability and NPP 400
18.4.4   Model Evaluation 401
18.5   Model Predictions for Disturbed Watershed 402
18.5.1   Discharge Disturbance 402
18.5.1.1   Road #1 402
18.5.1.2   Road #2 404
18.5.1.3   Roads #3 and #4 406
18.5.2   Dust and Discharge Disturbance 406
18.5.3   Effect of Disturbance on Spatial Patterns 407
18.6   Discussion 408
18.6.1   Model Comparisons 408
18.6.2   Patterns of N Availability 409
18.6.3   Extrapolation Potential: Some Cautionary Notes 411
18.7   Conclusions 412
References 413

V   Summary 417

19   Ecosystem Response, Resistance, Resilience, and Recovery
           in Arctic Landscapes: Progress and Prospects
           J. D. TENHUNEN and J. F. REYNOLDS 419

19.1   The NRC Tasks and R4D Accomplishments 419
19.2   Conclusion 425
References 426

Subject Index 429

Contributors

Benson, C. S., Geophysical Institute, University of Alaska, Fairbanks, Alaska 99775, USA

Beven, K., Institute of Environmental and Biological Sciences, Lancaster University, Lancaster LA1 4YQ, United Kingdom

Chapin III, F. S., Department of Integrative Biology, University of California, Berkeley, California 94720 USA

Cheng, W., Desert Research Institute, P.O. Box 60220, Reno, Nevada 89506, USA

Cook, R. B., Environmental Sciences Division, Oak Ridge National Laboratory, P.O. Box 2008, Oak Ridge, Tennessee 37831-6038, USA

Cooper, L. W., Environmental Sciences Division, Oak Ridge National Laboratory, P.O. Box 2008, Oak Ridge, Tennessee 37831-6038, USA

Everett, K. R., Byrd Polar Research Center and Department of Agronomy, Ohio State University, Columbus, Ohio 43210, USA (deceased)

Gebauer, R., Department of Botany, Phytotron Bldg, Box 90340, Duke University, Durham, North Carolina 27708-0340, USA

Gillespie, C. T., CEG/DEV Vandenburg Air Force Base, California 93437, USA

Graber, W., Paul Scherrer Institut, Würenlingen and Villigen, 5232 Villigen PSI, Switzerland

Grebmeier, J. M., Department of Ecology and Evolutionary Biology, University of Tennessee, Knoxville, Tennessee 37996, USA

Grulke, N. E., USFS Pacific Northwest Forest and Range Research Station, Forestry Sciences Lab, 3200 Jefferson Way, Corvallis Oregon 97331, USA

Hahn, S. C., Bayreuth Institute of Terrestrial Ecosystem Research (BITÖK), University of Bayreuth, 95440 Bayreuth, Germany

Hinzman, L. D., Water Research Center, Institute of Northern Engineering, University of Alaska, Fairbanks, Alaska 99775, USA

Hope, A. S., Geography Department, San Diego State University, San Diego, California 92182, USA

Irons III, J. G., Institute of Northern Forestry, USDA Forest Service, 308 Tanana Drive, Fairbanks, Alaska 99775-5500, USA

Kane, D. L., Water Research Center, Institute of Northern Engineering, University of Alaska, Fairbanks, Alaska 99775, USA

Kielland, K., Institute of Arctic Biology, University of Alaska, Fairbanks, Alaska 88775, USA

Lange, O. L., Lehrstuhl für Botanik II, Universität Würzburg, 97082 Würzburg, Germany

Lamprecht, R., Paul Scherrer Insitut, Würenlingen and Villigen, 5232 Villigen PSI, Switzerland

Larsen, I. L., Instituto de Física, Universidad Nacional Autónoma de México, Ap. Postal 20-364, México, 01000 D.F.

Leadley, P. W., Botanisches Institut, Universität Basel, Schönbeinstr. 6, 4056 Basel, Switzerland

Li, H., Department of Botany, Phytotron Building, Duke University, Durham, North Carolina 27708, USA

Moorhead, D. L., Department of Biological Sciences, Texas Tech University, Box 4149, Lubbock, Texas 79409, USA

Oberbauer, S. F., Department of Biological Sciences, Florida International University, Miami, Florida 33199, USA

Olsen, C. R., Environmental Sciences Division, ER-74, Office of Health and Environmental Research, Department of Energy, Washington, District of Columbia 20585, USA

Ostendorf, B., Bagrenth Institute for Terrestrial Ecosystem Research (BITÖK), University of Bayreuth, 95440 Bayreuth, Germany

Oswood, M. W., Institute of Arctic Biology, University of Alaska Fairbanks, Fairbanks, Alaska 99775, USA

Quinn, P., Department of Environmental Sciences, Stirling University, Scotland FK9 4LA, United Kingdom

Reeburgh, W. S., Earth System Science Program, University of California, Irvine, California 92717, USA

Reimers, C. E., Institute of Marine and Coastal Science, Rutgers University, New Brunswick, New Jersey 08903, USA

Reynolds, J. F., Department of Botany, Phytotron Building, Duke University, Durham, North Carolina 27708-0340, USA

Sala, A., Department of Biological Sciences, University of Montana, Missoula, Montana 59851, USA

Schell, D. M., Water Research Center, University of Alaska Fairbanks, Fairbanks, Alaska 99775, USA

Schimel, J. P., Department of Biology, University of California, Santa Barbara, California 93106, USA

Shaver, G. R., The Ecosystems Center, Marine Biological Laboratory, Woods Hole, Massachusetts 02543, USA

Solis, C., Instituto de Física, Universidad Nacional Autónoma de México, Ap. Postal 20-364, México, 01000 D.F.

Solomon, D. K., Department of Geology and Geophysics, University of Utah, Salt Lake City, UT 84112 USA

Stow, D. A., Geography Department. San Diego State University, San Diego, California 92182, USA

Tenhunen, J. D., Bayreuth Institute for Terrestrial Ecosystem Research (BITÖK), University of Bayreuth, 95440 Bayreuth, Germany

Virginia, R. A., Environmental Studies Program, Dartmouth College, Hanover, New Hamphire 03755-3577, USA

Walker, D. A., Joint Facility for Regional Ecosystem Analysis, Institute of Arctic and Alpine Research, and Department of Environmental, Population and Organismic Biology, University of Colorado, Boulder, Colorado 80309-0450, USA

Walker, M. D., Joint Facility for Regional Ecosystem Analysis, Institute of

Arctic and Alpine Research, and Department of Environmental, Population and Organismic Biology, University of Colorado, Boulder, Colorado 80309-0450, USA

Whalen, S. C., Department Environmental Science and Engineering, University of North Carolina, Chapel Hill, North Carolina 27599, USA

I Introduction

1 Ecosystem Response, Resistance, Resilience, and Recovery in Arctic Landscapes: Introduction

J. F. REYNOLDS and J. D. TENHUNEN

1.1 Introduction

The Arctic, which includes most of Alaska, the Yukon and the Northwest Territories, Greenland, northern Scandinavia, Siberia, and the Arctic Ocean, is a fragile component of the global earth system. Although arctic systems have been relatively stable for thousands of years, they are easily altered by anthropogenic disturbance (lves 1970; NAS 1982; Billings and Peterson 1992). This susceptibility is due to a number of factors, including a short growing season, low temperatures, low primary productivity, the presence of permafrost, and the extreme sensitivity of the vegetative and organic surface layers to any disruption of their physical integrity and thermal regime. The latter leads to themokarst – a localized thawing of ground ice – and severe erosion (Billings 1973). In the Arctic, even subtle variations in air temperature, radiation, atmospheric and oceanic chemistry, and ocean heat transport are likely to have large effects on sea ice, snow, glaciers, permafrost, and tundra ecosystems; in turn, this may alter crucial air/ice/ocean/land feedbacks controlling regional and global climate, ocean circulation patterns, and atmospheric concentrations of trace gases (OIES 1988; Weller 1995). Motivated largely by predictions that global warming is expected to be most extreme in the polar regions of the globe (Mitchell et al. 1990; Maxwell 1992), a tremendous surge in international research on arctic systems science has occurred in recent years (e.g., OIES 1988; IGBP 1990; ARCUS 1991; M. D. Walker et al. 1994a; Chap. 2, this Vol.).

The fragile nature of the Arctic also generated unprecedented public and scientific attention in the 1970s and 1980s in the United States because of the potential impacts of energy development on the North Slope of Alaska. The most controversial and highly visible issue was the development of the Prudhoe Bay and Kuparuk oil fields and the construction of the ca. 1300-km trans-Alaskan pipeline, which runs north to south from Prudhoe Bay on the Beaufort Sea to the terminal ports at Valdez (Fig. 1.1). The pipeline necessitated construction of a gravel access road, known as the Dalton Highway (or TAPS – Trans Atlantic Pipeline System – "Haul Road"), which runs along approximately half of the length of the pipeline in the north (Cohen 1988). Although the initial construction of the pipeline and road was deemed an engineering and environmental "success story" (Alexander and Van Cleve

J.F. Reynolds and J.D. Tenhunen (Eds.)
Ecological Studies, Vol. 120
© Springer-Verlag Berlin Heidelberg 1996

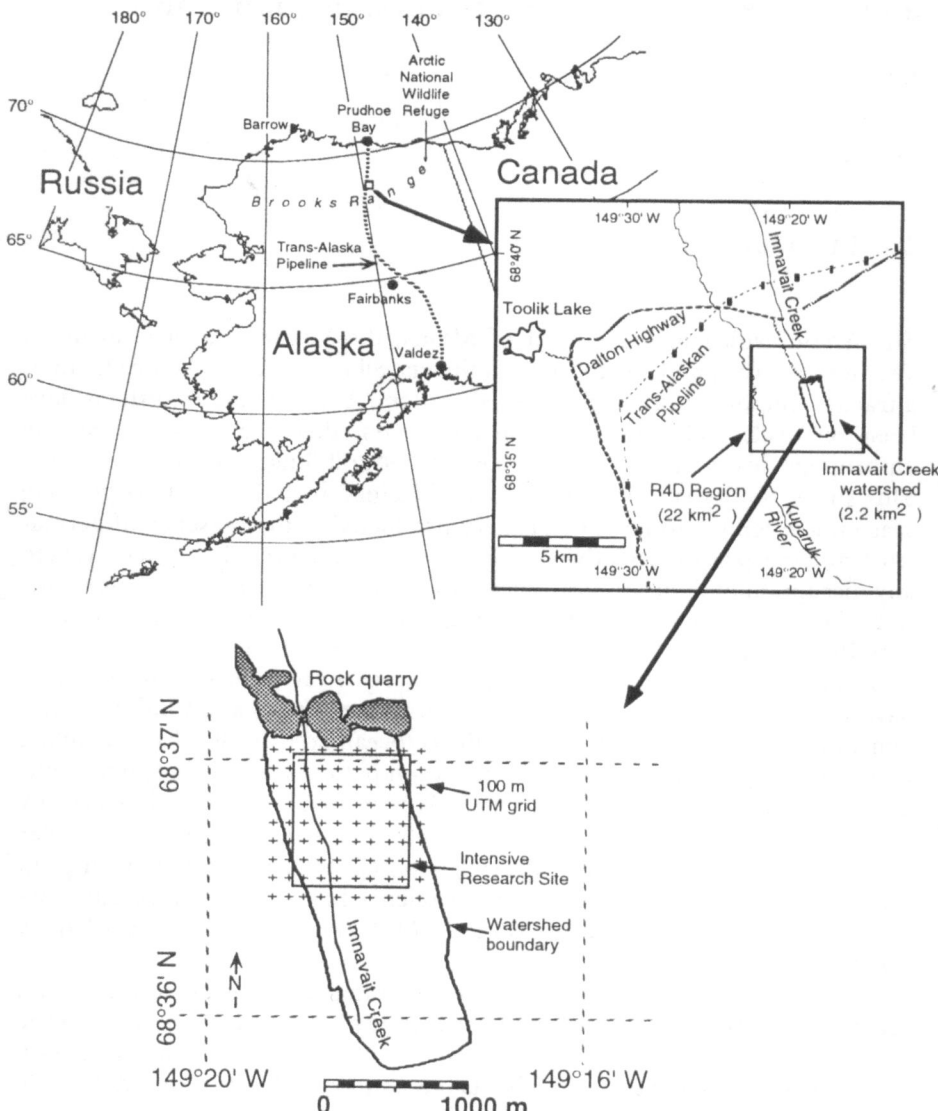

Fig. 1.1. Map of Alaska depicting location of trans-Alaskan pipeline, Arctic National Wildlife Refuge, and R4D study area. *Inserts* Imnavait Creek watershed, R4D region, UTM grid, and intensive research site. See Fig. 4.1 (this Vol.) for location of meteorological instruments and stream gauge

1983), it was recognized that energy development in the Arctic would have major, long-term, cumulative impacts on the environment (Table 1.1; also see D. A. Walker et al. 1987; Chap. 3, this Vol.). The Dalton Highway opened up vast areas of northern Alaska that were previously inaccessible to tourists,

Table 1.1. Potential impacts of energy development in Alaska. (NAS 1982; ARC 1986)

Disturbance of rivers and riparian vegetation by gravel mining and crossings. Roads and pipelines follow river valleys and intersect streams. Drainage and erosion patterns are changed and the growth of fish and their food organisms affected.

Removal and modification of vegetation. Roadsides, drilling pads, and ground denuded by road construction and mining revegetate very slowly in the Arctic and Subarctic. Tractors and other terrain vehicles have long-term effects on vegetation and nutrient cycling.

Acid precipitation and air pollution from local and global sources. Arctic "haze" (see Heintzenberg 1989) suggests atmospheric pollutants are being transported to the arctic from Europe and North America. It is necessary to separate local and long-distance pollutants and their effects on the biota.

Snow and ice roads on tundra. Much of the winter transportation for the exploration of petroleum reserves is on ice roads (i.e., snow is compacted and water is then added to form a road), which affect vegetation.

Drainage blockage in permafrost areas. Gravel roads in the Arctic and some in the Subarctic develop permafrost in their base, which then serve as dams to surface drainage.

Coal mining waste-revegetation. Surface of disturbed areas regain stability only when an insulating layer of vegetation becomes reestablished that allows permafrost to establish close to the surface.

Fire in northern forests. The frequency of man-induced fires can be expected to increase as development spreads across the Arctic.

Dust loads on vegetation. Oil drilling operations, mining activities, and truck traffic generate large dust plumes during the snow-free months, with numerous effects.

Climate change. Changes in atmospheric carbon dioxide concentrations and other greenhouse gases may have major effects on temperatures in the Arctic.

Other impacts. Such as, oil spills, road salting, surface mining, spoil siting, thermal erosion, and waste disposal.

hunters, and explorers. Furthermore, because the oil, gas, coal, and mineral resources of arctic Alaska comprise a significant proportion of US reserves (Washburn and Weller 1986), the Arctic will undoubtably experience anthropogenic disturbance and modification into the next century. Presently, there are renewed efforts in the US Congress to open over 1.5 million acres of the Arctic National Wildlife Refuge (Fig. 1.1) for oil exploration as the reserves of the Prudhoe Bay and Kuparuk oil fields decline.

Do we presently have enough information to implement ecosystem-based management practices in the Arctic? This question raises a plethora of issues that challenge our basic understanding of arctic ecosystems. For example, do arctic terrestrial ecosystems respond in predictable ways to specific types of disturbance? Will short-term disturbances have long-term ecological consequences? Will localized effects be transferred to adjacent systems, e.g., terrestrial to aquatic? Is it possible to generalize the effects of impacts on one type of landscape unit to another? Is knowledge gained from temperate ecosystems applicable to the arctic? This volume describes results of studies by a multidisciplinary research team that was charged with seeking answers to these questions, especially in relation to disturbance due to energy development in tundra regions of northern Alaska. In this chapter, a short background of the events that led to this integrated research effort (known as the R4D) and the general objectives and goals of this program are described. In Chapter 19 (this Vol.), we summarize our present understanding and speculate on our ability to implement ecosystem-based management practices in arctic tundra landscapes.

1.2 NRC Committee Report

For the past three decades, the Department of Energy (DOE) has been the lead US government agency supporting basic ecological research in Alaska on energy-related impacts. The DOE Organization Act of 1977 specifically directed DOE to conduct research on the environmental effects of energy-related development in order to advance the goals of "restoring, protecting, and enhancing environmental quality" (NAS 1982). These activities fit into a more general "national needs" research agenda outlined by the US Arctic Research Commission, which address environmental concerns tied to energy development in Alaska (ARC 1986):

> In accordance with national needs, emphasis should be on gaining knowledge to advance . . . (a) discovery and development, with minimal adverse environmental impact, of terrestrial nonrenewable resources, especially oil, gas, mineral, and hydrologic resources; *and* (b) prediction of ecosystem reactions, particularly the effects of natural and human-induced disturbances on renewable resources . . .

In 1980 a National Research Council (NRC) committee was established[1] to evaluate the terrestrial environmental research programs of the DOE. The findings of the committee were published as a report (NAS 1982). In general, the committee felt that DOE could meet its goals in the Arctic if it had a research program that:

1. focused on problems of environmental quality resulting from energy technologies;
2. utilized experimental manipulation to gain the capability to predict the impacts of energy development on environmental quality;
3. incorporated knowledge from prior arctic ecosystem research to bear on the problem of maintenance of environmental quality;

---

[1] Formally known as the Committee to Evaluate the Department of Energy's Arctic Terrestrial Environmental Research Programs.

4. produced scientific results that were applicable to a number of sites and types of landscapes; and
5. addressed both short- and long-term environmental effects of disturbances so that better management decisions could be made with respect to development, protection, and restoration.

The committee, composed of a number of experienced arctic researchers[2], had strong opinions as to how DOE should carry out its research mandate (NAS 1982) (italics have been added for emphasis):

> The Committee recommends that a large part of the future research supported by DOE concentrate on *long-term intensive studies* of the environmental effects of energy related developments in the arctic region. These ... studies should be centered initially in *one landscape*, either in the tundra or in the taiga zone of Alaska. A number of research projects ... should work within this landscape unit; research sites should include a stream and several vegetation types such as tussock and shrub tundra plus riparian vegetation of tundra and boreal forest. These studies *should emphasize the effects of disturbance on species and on ecosystem processes*. To carry this out, a series of manipulations of the environment should be incorporated into the landscape study. ... The separate research projects will be tied together by virtue of their investigation of a single landscape, through their common need for information about all parts of the system, and through ecosystem-level measurements (e.g., total energy available from photosynthesis) that would be needed by all projects.

The NRC committee gave the following reasons in support of this approach:

1. The results can be *generalized* to other arctic sites. This is possible because basic processes of the natural environment are studied. Thus, an understanding is gained both of the expected results of an impact and also of the *causal mechanisms*. If the same species or processes are present at a different site, then the impact can be predicted and perhaps experimentally validated.
2. The *whole landscape* is examined for the selected impacts. A number of scientists will examine several effects on various processes and species; this approach will give an intensive study of a type of impact that would not occur if a single scientist examined a specific impact. Also, by focusing on a single landscape, unexpected effects at interfaces of landscape units will be discovered.
3. The research will produce information on *long-term* effects and recovery. Particularly in the terrestrial system, effects of disturbances may take years to become obvious. It is also important to look at recovery from impacts both from a need to understand the system and a managerial need to

---

[2] Members were Albert Johnson (chair), John Andrews, Kaye Everett, Wyatt Gilbert, John Hobbie, Philip Johnson, Harold Mooney, Larry Tieszen, Robert Weeden, and George West.

predict whether an impact will be short-term or permanent and whether the impact extends to other systems, e.g., via transport and deposition processes.

4. *Basic* scientific information will be used from the proposed study and from the scientific literature to solve applied problems. The emphasis is on how species and ecosystem processes respond to environmental impacts. Information is especially needed from arctic environments.

5. The concentration at *a single site is cost-effective* for northern research. Arctic and subarctic research is costly because of logistics and these costs can be greatly reduced by having projects at one or two sites instead of eight or ten.

6. This approach provides a *framework for making judgments about priorities* for research and performance of research under way. Once the site and impacts to be tested are chosen, the research priorities can be assigned according to how well the projects will answer questions about how these impacts are operating on the landscape and on the ecological processes.

7. The *predictions* will be made on the basis of knowledge of species, processes, and controlled experiments using the expected impacts and will permit better planning to maintain environmental integrity.

8. It will enhance the training of arctic scientists who by virtue of their experience in tundra and taiga environments will be able to predict the impact of environmental disturbance.

The committee acknowledged that their recommendation for a single, intensive site had some disadvantages. For example, no single site would likely include the range of environmental conditions under which energy development will occur. Furthermore, findings from a primary site could not be easily generalized to other types of areas without some independent tests of their validity. In view of this, they suggested that secondary (extensive) sites also be established.

1.3 The R4D Program

1.3.1 Objectives and Conceptual Framework

The NRC committee's report had a major impact at DOE. In 1983–1984, a new program was established that was coined R4D (Response, Resistance, Resilience, and Recovery of arctic ecosystems to Disturbance). Underlying the conceptual framework for the R4D program was the assumption that arctic ecosystems are in a "stable" state, i.e., in equilibrium or steady state, and that they will remain so unless disturbed. Ecosystems are considered stable if it is likely that all state variables and processes return to this equilibrium state following disturbance (DeAngelis and Waterhouse 1987). Disturbance is considered as an agent or event that forces an ecosystem away from its stable state (Rykiel 1985). A disturbance results in a *response*, which can be described in terms of the attributes persistence and *resistance*. Persistence is the length of time after impact that the ecosystem remains unaltered, whereas resistance is inversely related to the degree of subsequent change from the equilibrium state (Pimm 1984). *Recovery* from disturbance is measured by ecosystem *resilience*, which is proportional to the time required for a system to return to its predisturbed state (Patten and Witkamp 1967; Holling 1973; Webster et al. 1974; DeAngelis 1980; Pimm 1984; Gerritsen and Patten 1985; Carpenter et al. 1992). Depending on the nature of the disturbance, various structural and functional components of a ecosystem may exhibit widely differing responses and recoveries, which complicates our ability to determine total system behavior. Also, large differences in persistence and resistance may be expected depending on the specific type of disturbance (Pimm 1984). Thus, associated with all quantitative measures of response and recovery are large problems related to variability, which has many potential causes in both time and space (Pickett and White 1985; Cairns 1995).

The concepts of stability, resistance, and resilience are conceptual simplifications that have great heuristic value but are difficult to quantify for real ecosystems. Perhaps the main impediment to our understanding of ecosystem response to disturbance is the lack of consensus as to which ecosystem properties should be measured, e.g., species physiology, food webs, nutrient turnover rates, landscape heterogeneity, etc., and our inability to interpret the significance of changes in the context of long-term ecosystem development and return to the "stable" state (Kimball and Levin 1985; Kelly and Harwell 1990; but see Odum 1985 and Rapport 1985). If a few key components or processes could faithfully be related to the state of an entire ecosystem, then the task of predicting the response and recovery of ecosystems to disturbance would be greatly simplified (Kelly and Harwell 1990). Unfortunately, most theoretical and empirical work on disturbance to date has focused on community structure rather than on community processes or ecosystem function (Pickett et al. 1989). In recent years – roughly coincident with the start of the R4D program – there has been a growing interest in the interpretation of disturbance processes at the landscape scale (e.g., Turner 1987, 1989; Costanza et al. 1990; Turner and Dale 1991; DeAngelis and White 1994; Wu and Levin 1994). Thus, the challenge offered by the NRC committee, emphasing landscapes, was timely.

1.3.2 Program Implementation

Implementation of the R4D program initially involved soliciting proposals from the scientific community. Following peer review, individual scientists from a number of universities and research institutes were selected to participate; various disciplines were represented, e.g., ecology, physiology, geogra-

Fig. 1.2a. Areal view of Foothills Region of Brooks Range, looking south. Three rivers are visible: Kuparuk (*upper right*), Imnavait Creek (*lower right*), and Toolik (*left*). Rock quarry of Material Site 117 and access road visible in forground of Imnavait Creek watershed. (Photo courtesy of D. A. Walker)

Fig. 1.2b. Infrared color aerial photo of the R4D Imnavait Creek watershed and vicinity. (Courtesy of C. Benson)

phy, soil science, and hydrology. The researchers had specialized interests ranging from soil-plant interactions, biogeochemical cycling, hydrological transport, to remote sensing. Next, a series of directives from DOE eventually led to the formation of a research team and a management structure that, in turn, set research objectives and selected a site. These DOE directives (some of which are described in NAS 1982) were intended to ensure that the committee's recommendations were realized, viz., that an integrated, systems-level study, relying on basic research from an single intensive site, was formed. The R4D program was supported by DOE for approximately 8 years.

The intensive site selected was the Imnavait Creek watershed, a 2.2 km² watershed located near Toolik Lake in the southern part of the northern foothills of the Brooks Range (68° 37′ N, 148° 18′ W) (Fig. 1.1). Imnavait Creek is a small beaded-stream tributary of the Kuparuk River (see Chap. 13, this Vol.). This site is 250 km south of Prudhoe Bay and 3 km south of the Dalton Highway and pipeline corridor between the Kuparuk and Toolik rivers. Elevation ranges between ca. 770–980 m and the landscape is characterized by gently rolling hills, rising less than 100 m from the valley bottoms to the ridge crests (Fig. 1.2). During the period 1985–1993, the mean annual temperature was −7.5 °C and the average precipitation was ca. 35 cm (Chap. 6, this Vol.). The northern part of the watershed is bounded by Material Site 117, a rock quarry used for construction of the Dalton Highway; the eastern and western boundaries are formed by NNW to SSE trending ridges; and the southern boundary is a flat colluvial basin. The Imnavait Creek watershed is contained within a larger, 22 km², extensive area called the R4D *region*. Further details of the terrain and vegetation of the Imnavait Creek watershed and R4D region are provided in Chapter 4 (this Vol.). The R4D study area is located within a broad band of tussock tundra vegetation. Tussock tundra occupies over 220 000 km², covering 80% of the Alaskan arctic region, including the majority of the area being proposed for drilling in the Arctic National Wildlife Refuge (D. A. Walker et al. 1982).

1.3.3 Landscape Function

The long-term objectives of the R4D program were: (1) to document and to model the ecosystem effects of disturbances associated with energy development (see Table 1.1), so that appropriate, cost-effective management procedures for minimizing harmful effects of disturbances could be developed; and (2) to utilize these management tools for environmental impact assessments in arctic and alpine areas likely to be impacted by future energy development (McCammon and Hendrey 1987). In support of these general objectives, a wide range of specific hypotheses were tested as illustrated in the various chapters of this volume. The background to many of these hypotheses are questions that concern the role of water as both a direct and indirect agent in determining the structure and functioning of arctic ecosystems and, hence, how they will respond to the effects of disturbance. Research was conducted on a complex mosaic of different soil types, in different topographical positions, and in different vegetation types within the Imnavait Creek watershed.

While these landscape approaches add much complexity, landscape heterogeneity cannot be ignored if we hope to understand and model disturbance effects (Turner and Dale 1991; Hannsson et al. 1994). At the same time, there are few guidelines indicating which components of heterogeneity to include in our models. As an example, tussock tundra appears homogeneous over large areas in the vicinity of the Imnavait Creek site. Yet the landscape age at the Imnavait Creek research site is much older than that found near Toolik Lake (12 km distant; Fig. 1.1). The Imnavait Creek site is on old Sagavanirktok (mid-Pleistocene) glacial drift, whereas Toolik Lake is on younger Itkillik I-II (late-Pleistocene) glacial drift (Hamilton 1986; M. D. Walker et al. 1994b; Chap. 4, this Vol.). D. A. Walker et al. (1995) report that the younger arctic landscapes tend to have lower amounts of standing biomass and a preponderance of drier heath vegetation types (see Chap. 5, this Vol.). In fact, upon close inspection, we found that the spatial patterning of 'tussocks' (refers to the physiognomy of the tussock-forming sedge *Eriophorum vaginatum*; see Fig. 4.7, this Vol.) at the Imnavait Creek and Toolik Lake sites is very different (Fig. 1.3). How do we realistically incorporate such sources of variation into models and are the

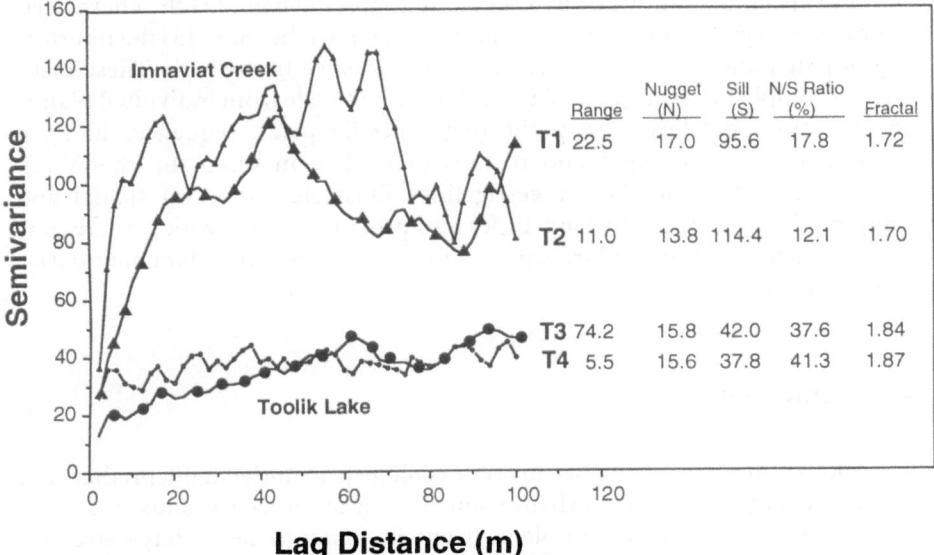

Fig. 1.3. Semivariograms based on tussock density of *Eriophorum vaginatum* obtained from the Imnavait Creek and Toolik Lake areas. Transects were place along similar contour lines at each site. Total length of transect was 256 m, where each was divided into 127 2 × 3 m quadrats. Parameters were estimated by fitting the spherical model of Webster (1985)

models useful if we do not? Since these tussocks provide structure that influences microclimate, water transport, and ecosystem processes (Chap. 14, this Vol.), should we attempt to explicitly include this level of spatial heterogeneity in our models? Difficult and critical questions of this type must be answered during the construction of landscape models. Many compromises between simplicity and unwieldy complexity must be made that ultimately affect predictions.

Most of the field studies and models described in this volume focus on topographic gradients or toposequences (see Fig. 4.9, this Vol.) as a means of examining relationships among landscape units. In the Imnavait Creek watershed, distinct patterns of vegetation, soils, and hydrology occur along a gradient from the dry ridges (dominated by mosses and lichens), through mid-slope tussock tundra (dominated by *E. vaginatum*), down to the beaded stream at the valley bottom (see Chaps. 4 and 5, this Vol.). Many arctic studies have demonstrated strong correlations between soil moisture gradients, the distribution and productivity of tundra vegetation, and an array of chemical, physical, and biological variables, e.g., thaw depth, soil development, soil aeration, soil temperature regime, pH, etc. (Webber 1978; French 1981; D. A. Walker et al. 1989; Chaps. 4, 5, 11, 14, 17, 18, this Vol.). Additionally, topography and hydrology exert strong effects on the belowground environment (Miller 1982; Miller et al. 1982; Cheng et al. 1996; Chaps. 10, 11, this Vol.). In view of this, three distinct tasks are addressed by the field and modeling research efforts reported in this volume: (1) the description of processes that occur at specific locations or *patches* along toposequences; (2) the characterization of energy and materials *transport* between patches; and (3) the interfacing of patch and transport processes in a *landscape* framework. These tasks were accomplished using spatial units that vary in scale from individual plants, to vegetation patches, flow paths (e.g., specific patch sequences along a toposequence), landscapes and regions (described in Chap. 14, this Vol.). The models based on these investigations illustrate important spatial and temporal linkages in arctic landscapes and provide a framework for integrating structure, function, and transport within and across ecosystem boundaries of the watershed.

1.4 Summary

The NRC committee's challenge to arctic ecologists to understand, predict, and manage energy-related disturbance impacts on arctic ecosystems is a good example of a "real world" problem. It entails many issues of large size and scope, high spatial and temporal variability, and general uncontrollability with respect to experimentation. This is typical of problems related to important societal needs, e.g., cumulative impacts in wetlands (Baskerville 1986) or the consequences of global change (Mooney and Chapin 1994). The aspects of this "real world" problem actually examined by the R4D effort at the Imnavait Creek watershed were simpler and much narrower in scope than proposed by the NRC. As a result, the research reported in this volume represents a preliminary – but essential – step in our efforts to develop a capability for addressing the real world problems of disturbance impacts in the Arctic. Three general areas of concern remain to be addressed in future studies: explicit definition of an ecosystem treatment that best summarizes disturbance effects in tundra, development of quantitative measures for response and recovery of tundra ecosystems, and identification (and quantification, where possible) of uncertainties in subsequent predictions (Table 1.2).

We are grateful to the many people that contributed to the completion of this book. A number of technical staff helped edit and type the manuscripts, including Terry Scott, Wendy Shotwell, Phyllis Mills, Ellen Chu, and Stephanie Canon. Dave Tremmel provided valuable computer expertise and did many of the technical drawings. A number of scientists generously provided independent reviews of the manuscripts. We are grateful to Donald Short and Albert

---

Table 1.2. General areas of concern to address in advancing our understanding of real-world impacts of disturbance on arctic landscapes. (Modified from Rykiel et al. 1988; Kelly and Harwell 1990)

Definition of disturbance (anthropogenic stress)

Type: What is it?
Initial predominant effect: What does it do?
Mode of action: How does it do it?
Scope: How big is it? What spatial and temporal scales are involved?
Duration: How long does it last?
Intensity: How much damage does it do?
Timing and frequency relations: When and how often does it occur?
Reliability: Does it always occur within a particular interval?

Response of landscapes to disturbance

What components are affected?
What processes are affected?
What are the relevant spatial and temporal scales to characterize response?

Recovery of landscapes to disturbance

Do indicators exist for components and processes?
Can resilience be quantified?
Has irreparable harm been done?
What are the long-term vs short-term effects?

Uncertainties

Variability in component and process exposure
Variability in component and process response
Variability in component and process recovery
Long-term vs short-term effects
Extrapolation across different types of disturbances
Extrapolation across different types of arctic landscapes

Johnson of San Diego State University who were extremely supportive of R4D research while we were in tenure there. We acknowledge the support at DOE from Helen McCammon and from the DOE program managers, George Hendrey and Edward Rykiel. We thank Otto L. Lange for his editorial guidance. Lastly, we greatly appreciate the insight and perceptions of the NRC committee; K. R. Everett for his broad knowledge of Alaskan arctic tundra, contributions to site selection, and fostering of tundra hydrology and biogeochemistry; P. J. Webber for promoting tundra landscape approaches; D. W. Billings for his pioneering research in tundra ecosystem physiology; and to P. C. Miller for his efforts at integrating process information via simulation modeling. All have contributed significantly to the progress documented in this volume.

References

Alexander V, Van Cleve K (1983) The Alaska pipeline: a success story. Annu Rev Ecol Syst 4: 1443–1463

ARC (Arctic Research Commission) (1986) National needs and arctic research: a framework for action. A working paper. US Arctic Res Comm, Los Angeles

ARCUS (Arctic Research Consortium of the United States) (1991) Arctic systems science: a plan for integration. ARCUS, Fairbanks, Alaska, 34 pp

Baskerville G (1986) Some scientific issues in cumulative environmental impact assessment. In: Beanlands GE, Erckmann WJ, Orians GH, O'Riordan J, Policansky D, Sadar MH, Sadler B (eds) Cumulative environmental effects: a binational perspective. Can Environ Assessment Res Council and US Natl Res Council, Ottawa, Ontario, and Washington, DC, pp 9–14

Billings WD (1973) Arctic and alpine vegetations: similarities, differences, and susceptibility to disturbance. BioSci 23: 697–704

Billings WD, Peterson KM (1992) Some possible effects of climatic warming on arctic tundra ecosystems of the Alaskan North Slope. In: Peters RL, Lovejoy TE (eds) Global warming and biological diversity. Yale University Press, New Haven, pp 233–243

Cairns J Jr (ed) (1995) Rehabilitating damaged ecosystems, 2nd edn. Lewis, Boca Raton

Carpenter SR, Kraft CE, Wright R, He X, Soranno PA, Hodgson JR (1992) Resilience and resistance of a lake phosphorus cycle before and after food web manipulation. Am Nat 140: 781–798

Chapin FS III, Shaver GR (1985). Arctic. In: Chabot BF, Mooney MA (eds) Physiological ecology of North American plant communities. Chapman and Hall, New York, pp 16–40

Cheng W, Virginia RA, Gillespie CT, Oberbauer SF, Tenhunen JD, Reynolds JF (1996) Spatial and temporal variation in soil nitrogen, microbial biomass, and respiration along an arctic toposequence. (submitted)

Cohen S (1988) The great Alaska pipeline. Pictorial Histories, Missoula, Montana

Costanza R, Sklar FH, White ML (1990) Modeling coastal landscape dynamics. BioScience 40: 91–107

DeAngelis DL (1980) Energy flow, nutrient cycling, and ecosystem resilience. Ecology 61: 764–771

DeAngelis DL, Waterhouse JC (1987) Equilibrium and nonequilibrium concepts in ecological models. Ecol Monogr 57: 1–21

DeAngelis DL, White PS (1994) Ecosystems as products of spatially and temporally varying forces, ecological processes, and landscapes: a theoretical perspective. In: Davis SM, Ogden JC (eds) Everglades: the ecosystem and its restoration. St Lucie Press, Delray Beach, pp 9–27

DeAngelis DL, Waterhouse JC, Post WM, O'Neill RV (1985) Ecological modelling and disturbance evaluation. Ecol Modeling 29: 399–419

French DD (1981) Multivariate comparisons of IBP Tundra Biome site characteristics. In: Bliss LC, Heal OW, Moore JJ (eds) Tundra ecosystems: a comparative analysis. Cambridge Univ Press, London, pp 47–75

Gerritsen J, Patten BC (1985) System theory formulation of ecological disturbance. Ecol Modeling 29: 383–397

Hamilton TD (1986) Late Cenozoic glaciation of the Central Brooks Range. In: Hamilton TD, Reed KM, Thorson RM (eds) Glaciation in Alaska: the geologic record. Alaska Geological Society, Fairbanks, Alaska, pp 9–49

Hannsson L, Fahrig L, Merriam G, Noss RF (eds) (1994) Mosaic landscapes and ecological processes. Chapman and Hall, London

Heintzenberg J (1989) Arctic haze: air pollution in polar regions. Ambio 18: 50–55

Holling CS (1973) Resilience and stability of ecological systems. Annu Rev Ecol Syst 4: 1–24

IGBP (International Geosphere-Biosphere Programme) (1990) A study of global change: the initial core projects. Report 12, International Council of Scientific Unions, Stockholm

Ives JD (1970) Arctic tundra: how fragile? A geomorphologist's point of view. Trans R Soc Can, 4th Ser, VII: 39–42

Kelly JR, Harwell MA (1990) Indicators of ecosystem recovery. Environ Manage 14: 527–545

Kimball KD, Levin SA (1985) Limitations of laboratory bioassays: the need for ecosystem-level testing. Bioscience 35: 165–171

Maxwell B (1992) Arctic climate: potential for change under global warming. In: Chapin FS III, Jefferies R, Reynolds JF, Shaver G, Svoboda J (eds) Arctic ecosystems in a changing climate. Academic Press, San Diego, pp 11–34

McCammon H, Hendrey G (1987) Response, resistance and resilience to, and recovery from, disturbance in arctic ecosystems: executive summary. Unpubl Internal Rep, US Dep Energy, Washington, DC

Miller PC, (1982) Environmental and vegetational variation across a snow accumulation area in montane tundra in central Alaska. Holarct Ecol 5: 85–98

Miller PC, Mangan R, Kummerow J (1982) Vertical distribution of organic matter in eight vegetation types near Eagle Summit, Alaska. Holarct Ecol 5: 117–124

Mitchell JFB, Manabe S, Tokioka T, Meleshko V (1990) Equilibrium climate change. In: Houghton JT, Jenkins GJ, Ephraums JJ (eds) Climate change: the IPCC scientific assessment. Cambridge Univ Press, Cambridge, pp 131–172

Mooney HA, Chapin FS III (1994) Future directions of global change research in terrestrial ecosystems. Trends Ecol Evol 9: 371–372

NAS (National Academy of Science) (1982) Arctic terrestrial environmental research programs of the Office of Energy Research, Department of Energy: evaluation and recommendations. Natl Acad Sci Press, Washington, DC

Odum HT (1985) Trends expected in stressed ecosystems. BioSci 35: 419–422

OIES (Office for Interdisciplinary Earth Studies) (1988) Arctic interactions. Rep OIES-4, Boulder, Colorado

Patten BC, Witkamp M (1967) Systems analysis of ^{134}Cs kinetics in terrestrial microcosms. Ecology 48: 813–824

Pickett STA, White P (1985) The ecology of natural disturbance and patch dynamics. Academic Press, New York

Pickett STA, Kolasa J, Armesto JJ, Collins SL (1989) The ecological concept of disturbance and its expression at various hierarchical levels. Oikos 54: 129–136

Pimm SL (1984) The complexity and stability of ecosystems. Nature 307: 321–326

Rapport DJ, Regier HA, Hutchinson TC (1985) Ecosystem behavior under stress. Amer Nat 125: 617–640

Rykiel EJ (1985) Toward a definition of ecological disturbance. Aust J Ecol 10: 361–365

Rykiel EJ, Coulson RN, Sharpe PJH, Allen TFH, Flamm RO (1988) Disturbance propagation by bark beetles as an episodic landscape phenomenon. Landscape Ecol 1(3): 129–140

Turner M (ed) (1987) Landscape heterogeneity and disturbance. Springer, Berlin Heidelberg New York

Turner M (1989) Landscape ecology: the effect of pattern on process. Annu Rev Ecol Syst 20: 171–197

Turner MG, Dale VH (1991) Modeling landscape disturbance. In: Turner MG, Gardner RH (eds) Quantitative methods in landscape ecology, Ecological Studies 82, Springer, Berlin Heidelberg New York, pp 323–351

Walker DA, Acevedo DA, Everett KR, Gaydoes KR, Brown J, Webber PJ (1982) Landsat-assisted environmental mapping in the Arctic National Wildlife Refuge, Alaska. CRREL Rep 82–27, US Army Cold Regions Res Eng Lab, Hanover, New Hampshire

Walker DA, Webber PJ, Binnian EF, Everett KR, Lederer ND, Nordstrand EA, Walker MD (1987) Cumulative impacts of oil field on northern Alaskan landscapes. Science 238: 757–761

Walker DA, Binnian E, Evans BE, Lederer ND, Nordstrand E, Webber PJ (1989) Terrain, vegetation, and landscape evolution of the R4D research site, Brooks Range Foothills, Alaska. Holarct Ecol 12: 238–261

Walker DA, Auerback NA, Shippert MM (1995) NDVI, biomass, and landscape evolution of glaciated terrain in northern Alaska. Polar Res (in press)

Walker MD, Daniëls FJA, van der Maarel E (1994a) Circumpolar arctic vegetation: introduction and perspectives. J Veg Sci 5: 758–764

Walker MD, Walker DA, Auerback NA (1994b) Plant communities of a tussock tundra landscape in the Brooks Range Foothills, Alaska. J Veg Sci 5: 843–866

Washburn AL, Weller G (1986) Arctic research in the national interest. Science 233: 633–639

Webber PJ (1978) Spatial and temporal variation of the vegetation and its production, Barrow, Alaska. In: Tieszen LL (ed) Vegetation and production ecology of an Alaskan arctic tundra. Springer, Berlin Heidelberg New York, pp 37–112

Webster JR, Waide JB, Patten BC (1974) Nutrient cycling and the stability of ecosystems. In: Howell FG, Gentry JB, Smith MH (eds) Mineral cycling in southeastern ecosystems. Energy Dev Admin (ERDA) Symp Ser, Tech Info Center, Washington, DC, pp 1–27

Webster R (1985) Quantitative spatial analysis of soil in the field. Adv Soil Sci 3: 1–70

Weller G (1995) Global pollution and its effect on the climate of the Arctic. Sci Total Environ 160/161: 19–24

Wu J, Levin SA (1994) A spatial patch dynamic modeling approach to pattern and process in an annual grassland. Ecol Monogr 64: 447–464

2 Integrated Ecosystem Research in Northern Alaska, 1947–1994

G. R. SHAVER

2.1 Introduction

Current research on human impacts in tundra ecosystems and landscapes builds upon a rich history of basic ecological studies carried out in Alaska since World War II. This chapter traces that history from 1947 to 1994, showing how the major ecological ideas and research themes have evolved within the community of arctic ecologists. The central hypotheses of the research described in the rest of this book are in many ways a product of this evolution, and this book represents the most recent synthesis of our understanding.

We start in 1947 because that was the year the Arctic Research Laboratory, later renamed the Naval Arctic Research Laboratory (NARL), was established at Point Barrow (Reed 1969; Britton 1973). It was at NARL where, for the first time in Alaska, groups of ecologists began working on interrelated projects, each focusing on different aspects of the same ecosystem type and often working on the same sites or experimental plots. The stable presence of NARL allowed the kind of long-term field studies that are important in separating signal from noise in ecological processes and patterns (Pitelka 1973). This tradition of integrated, or at least complementary, ecosystem research continued, culminating in two research programs based at Toolik Lake, Alaska: the Department of Energy-sponsored R4D program (see Chap. 1, this Vol.) and the National Science Foundation (NSF)-sponsored Arctic Long-Term Ecological Research (LTER) program, which includes lake and stream research as well as terrestrial research. Ecological research in arctic Alaska before 1947 has been reviewed by Hultén (1940), Wilimovsky (1966), Fuller (1977), and Everett (1980a).

2.2 Early Days at NARL

The role of disturbance in arctic ecosystems, and especially the impact of disturbance on soils and nutrient cycling, was among the first themes studied at NARL. Early research focused on naturally occurring, rather than anthropogenic, disturbances. Two hypotheses were particularly influential: the nutrient recovery hypothesis of Pitelka and Schultz (1964; Schultz 1969), and

J.F. Reynolds and J.D. Tenhunen (Eds.)
Ecological Studies, Vol. 120
© Springer-Verlag Berlin Heidelberg 1996

the thaw lake cycle hypothesis first described by Britton (1957). Additional early work on plants and soils is described by Johnson (1969), Johnson and Tieszen (1973), and Tedrow (1973). Early limnological research is reviewed by Hobbie (1973, 1980).

The nutrient recovery hypothesis was one of the first attempts in any ecosystem to link plants, soils, and animal grazing by quantifying element budgets. The hypothesis was developed to explain the relatively regular, but dramatic, fluctuations in the abundance of brown lemmings in the tundra near Barrow (Pitelka 1973). Pitelka and Schultz (1964) suggested that these fluctuations, which occurred every 3–4 years over a period of almost 20 years, were correlated with fluctuations in plant productivity, plant nutrition, and forage quality, and also in decomposition, nutrient availability, and thaw depth.

The basic idea behind the nutrient recovery hypothesis was that only a limited amount of Ca, N, and especially P was actively cycled in the wet tundra at Barrow, and that these elements were alternately accumulated in either plants and soils, or in lemmings. Peaks in lemming populations could theoretically only occur when sufficient amounts of these elements had accumulated in the vegetation to support the lemmings' nutritional and reproductive requirements. Plant productivity, on the other hand, should be greatest just after the lemming populations had peaked, when grazing by the lemmings dramatically increased element availability by releasing elements that had been sequestered in plant biomass. Another important ecosystem interaction studied was the effect of grazing and biomass removal on soil temperature and depth of thaw, where biomass removal increased heat transfer into the soil by reducing the insulating effect of the plant canopy and litter layer.

The nutrient recovery hypothesis was important to later research in arctic Alaska, because it helped change our focus from individual organism-environment interactions to an integrated approach encompassing interactions among trophic levels as well as ecosystem-level element budgets. It also stimulated research on plant mineral nutrition and decomposition. Although the nutrient recovery hypothesis was later found to be simplistic and wrong in many details (Batzli et al. 1980), it provided a starting point and was of heuristic value.

The thaw lake cycle was proposed by Britton (1957) to describe landscape-level, long-term, cyclic successional processes acting over most of the Alaskan arctic coastal plain including the surroundings at Barrow. Britton hypothesized that much of the diversity in soils and vegetation over this area was determined by succession occurring in the bottoms of shallow thaw lakes that had been naturally drained. These lakes, both drained and undrained, cover most of the land area of Alaska's northern coastal plain. They are typically oriented with their long axis perpendicular to the prevailing wind direction. The natural drainage of the lakes results from erosion of their shorelines during the summer thaw period. The lakes gradually grow longer downwind until they drain as they merge with a stream, another lake, or the seacoast. Britton proposed that frost action, especially ice-wedge polygon development, strongly mediates succession in the bottoms of these drained lakes. A result of this polygonization was the formation, growth, and coalescence of small thaw ponds, which eventually led to the reformation of large thaw lakes. Research on various aspects of the thaw lake cycle continued through the 1970s and 1980s (Webber 1978; Billings and Peterson 1980; Webber et al. 1980; Bliss and Peterson 1992), and the basic sequence described by Britton is now generally accepted. Other important early work on vegetation, succession, and disturbance include Benninghoff (1952), Sigafoos (1952), Bliss and Cantlon (1957), and Churchill and Hanson (1958), and the reviews by Johnson (1969) and Johnson and Tieszen (1973).

Both the thaw lake cycle and the nutrient recovery hypotheses were important to later research, because clear testable predictions about the role of natural change and disturbance in an arctic landscape at both long (thaw lake cycle) and short (nutrient recovery hypothesis) time scales resulted. These ideas were particularly useful in the following years for evaluating the potential and time scale for recovery from human impacts associated with natural resource development on the arctic Coastal Plain (e.g., Walker et al. 1987a,b; Walker and Walker 1991; Chap. 3, this Vol.).

2.3 The U. S. Tundra Biome Program

The Tundra Biome program, based in coastal wet sedge tundra at Barrow from 1970 to 1974, was the first multi-investigator ecological research project in northern Alaska to be designed as an integrated, ecosystem-level program (Blair 1977; Brown et al. 1980). The research was divided into terrestrial studies (Tieszen 1978; Brown et al. 1980) and aquatic studies (Hobbie 1980). All of the major ecosystem components – primary producers, decomposers, herbivores, predators, climate and microclimate, and soils – were studied on the same sites. In addition, comparative studies were carried out at Prudhoe Bay (Brown 1975); at Eagle Summit, Alaska; and at Niwot Ridge, Colorado. The U. S. Tundra Biome program was also part of the International Tundra Biome program, which included 14 other study sites in alpine, antarctic, and arctic ecosystems (Wielgolaski and Rosswall 1972; Bliss and Wielgolaski 1973; Rosswall and Heal 1975; Bliss et al. 1981). Several of these national programs produced individual volumes detailing their results (e.g., Wielgolaski 1975a,b; Bliss 1977; Heal and Perkins 1978).

The Tundra Biome program had three broad objectives (Brown et al. 1980): (1) "to develop a predictive understanding of how the tundra system operates," focusing on wet coastal tundra as an example; (2) "to obtain ... [a] data base from a variety of cold-dominated ecosystems ... in the U. S." for the purposes of modeling, simulation, and comparison with other circumpolar arctic sites; and (3) "to bring basic environmental knowledge to bear on problems of degradation, maintenance, and restoration of the temperature-sensitive and cold-dominated tundra and taiga ecosystems."

Computer simulation modeling (at that time a relatively new approach) played a major role in meeting these objectives (see Reynolds and Leadley 1992). The aims of the modeling were, firstly, to provide a means of integrating and comparing individual process-level studies, and secondly, to predict ecosystem-level process rates and responses. Detailed models were developed relating processes at small scales, as in the case of photosynthesis in individual leaves, to ecosystem-level processes such as the productivity of vegetation canopies (Miller et al. 1980). Developing these models depended on a very detailed understanding of underlying adaptations and process controls (Tieszen 1978).

The Tundra Biome modeling efforts were only partially successful. The models were never widely used to predict ecosystem characteristics such as primary production, lemming population cycles, or response to disturbance. Nevertheless, model development forced investigators to work together in new ways and catalyzed many new hypotheses about interactions between ecosystem components (Tiwari et al. 1980) and provided a foundation for future efforts (See Chap. 14, this Vol.). In the Alaskan research programs that followed the Tundra Biome, models and their data requirements were often used to help set priorities.

The ecosystem focus of the Tundra Biome program, combined with the use of quantitative ecosystem-level models, led to further emphasis on nutrient cycling and soil microbial processes as important regulators of ecosystem structure and function (Chapin et al. 1980a). Some arctic ecologists began to view the effect of low soil temperatures on decomposition and mineralization along with the presence of permafrost and poor soil drainage, as having greater importance in regulation of plant growth and element uptake than the direct effects of low air and soil temperatures per se (Chapin 1983). This interest in nutrient cycling had been foreshadowed, of course, by the early work of Pitelka and Shultz (1964). The importance of soil characteristics in determining the distribution of individual plant species and vegetation types was also apparent in wet tundra ecosystems at Barrow (Webber 1978; Webber et al. 1980). In many instances distribution patterns could be explained in terms of the thaw lake cycle, and new hypotheses related long-term community succession with changes in the cycles of C, N, and P (Chapin et al. 1980a).

2.4 The Meade River RATE Program

The Research on Arctic Tundra Ecosystems (RATE) program followed up on the Tundra Biome program from 1975 to 1977 at Atkasook, on the Meade River about 100 km south of Barrow. The Meade River site was chosen because it was more ecologically diverse, including vegetation typical for the entire Alaskan North Slope (Batzli 1980). The vegetation at Meade River included moist *Eriophorum vaginatum* tussock tundra, deciduous and evergreen shrub-dominated communities, and wet sedge tundra (Komarkova and Webber 1980).

The principal theme of the Meade River studies was the role of herbivory in tundra ecosystems (Batzli 1980). The impact of lemmings on wet sedge tundra at Barrow had been well documented by Pitelka, Schulz, the Tundra Biome workers, and others (Batzli et al. 1980), but relatively little was known about other species of herbivores and the effects of herbivory on other plant forms and tundra types. To place the work on herbivory in an ecosystem context, basic research on soils, vegetation, and landscape patterns in community distribution were also undertaken (Everett 1980b; Komarkova and Webber 1980), along with descriptive studies of the natural disturbance regime and long-term successional patterns in relation to disturbances such as the thaw lake cycle, river meandering, and wind erosion (Billings and Peterson 1980; Peterson and Billings 1980).

The work at Meade River showed that a strong relationship existed between plant growth form and herbivory in arctic tundra, with plant secondary compounds as well as energy or nutrient content determining both the level of herbivory and the herbivore species (Batzli and Jung 1980; Batzli and Sobasky 1980; White and Trudell 1980). Plants at Meade River responded to herbivory, depending on their basic carbon and nutrient allocation patterns, and to the location and vulnerability of major storage tissues (Archer and Tieszen 1980; Chapin 1980; Chapin et al. 1980b). The deciduous shrub species were subject to higher levels of herbivory, but were also less damaged by herbivory than evergreen species because of higher rates of nutrient uptake and the ability to replace nutrients lost to herbivores. Grasses and sedges also recovered well from herbivory, because they store large amounts of carbon and nutrients in belowground tissues.

Herbivores (especially mammals) had important impacts on overall nutrient cycling at Meade River, but these impacts seemed to operate over a longer time scale that at Barrow. At Meade River the herbivore effects on nutrient cycles were due more to changes in spatial distribution of nutrients and to long-term controls on species composition of the vegetation (Batzli 1980; McKendrick et al. 1980), whereas at Barrow the lemmings seemed to affect the overall rate of nutrient turnover directly on a 3- to 4-year cycle.

2.5 Eagle Creek and Eagle Summit

During the late 1970s, a second group of ecologists focused their research on a montane tundra site in central Alaska, at Eagle Creek (Miller 1982a). The primary research areas included tussock tundra at Eagle Creek and a series of vegetation types associated with a late-melting snowbank at nearby Eagle Summit. Both areas had been studied previously by a number of investigators (Wein and Bliss 1974; Chapin et al. 1979; Brown et al. 1980). Several comparative investigations were conducted at Meade River, Toolik Lake, and Cape Thompson, Alaska.

The Eagle Creek study focused on plant production processes and the regulation of vegetation composition, distribution, and productivity. The study included both detailed ecophysiological investigations of plant processes such as photosynthesis (e.g., Limbach et al. 1982) as well as extensive comparisons of vegetation composition and structure in contrasting habitats or microenvironments (e.g., Oberbauer and Miller 1979; Miller 1982b). As in the Meade River RATE research, comparisons of different plant growth forms were emphasized to test the generality of knowledge gained working with sedges and grasses at Barrow. The importance of soil nutrient availability to plant growth was also emphasized (Shaver and Chapin 1980), developing further earlier work at the same site, which had indicated strong nutrient limitations (Wein and Bliss 1974; Chapin et al. 1979), as well as work at Meade River on nutritional differences among plant growth forms (Chapin et al. 1980b) and their response to fertilizer applications (McKendrick et al. 1980).

In the work at Eagle Creek, an effort was made to integrate results and observations by developing a single, highly detailed simulation model, the Arctic Tundra Simulator (ARTUS; Miller et al. 1984). This modeling emphasis was justified by the perceived need to predict the effects on arctic ecosystems of climate change and disturbances related to energy development (Miller 1981; Miller et al. 1984). It was argued that such predictions required a detailed understanding of basic physiological and biophysical processes. Although much information was available from earlier work on natural disturbances, such as the thaw lake cycle (Britton 1957; Billings and Peterson 1980), river meandering (Bliss and Cantlon 1957), and other frost phenomena (Churchill and Hanson 1958; Shaver et al. 1979), the similarities and differences between natural and anthropogenic disturbances were unclear.

The ARTUS model did aid in setting research priorities and in integrating the separate, process-level studies carried out at Eagle Creek. ARTUS also helped define the relative sensitivities of primary production by various species to changes in temperature, nutrients, and other environmental factors. ARTUS was less successful, however, in making large-scale predictions of response to climate change or disturbance, largely because it was narrowly focused on detailed knowledge of a few dominant plant species (Reynolds and Leadley 1992). A more general and less detailed model, the Northern Ecosystems Simulator (NECS; Miller et al. 1983), was more successful at predicting regional responses to climate change, but work on this model was discontinued after Phil Miller's death.

Perhaps the most important conclusion of the Eagle Creek study was that the variability in species and growth form composition among study sites was due to "different levels of nitrogen and phosphorus availability [rather than] length of the snowfree season, water availability, and soil pH" (Miller 1982b). This conclusion was a major departure from early studies in alpine as well as arctic tundras (e.g., Bliss 1956; Johnson and Billings 1962; Billings and Mooney

1968; Billings 1973), which had largely focused on temperature, moisture, and snow cover as controlling factors, but it was consistent with work at Barrow and Meade River. As in the Meade River study, the Eagle Creek researchers found it useful to group plant species by growth form when predicting response to environmental factors or resource availability. The relationship to growth form is the result of underlying patterns of nutrient use and nutrient turnover (Miller et al. 1984).

Controls on soil nutrient availability and cycling, as well as the effects of nutrients on tundra plant growth, were heavily emphasized in later research at Toolik Lake. Studies of natural and human disturbances at other sites in Alaska and Canada also demonstrated that most disturbances to tundra led to dramatic increases in nutrient availability, nutrient turnover, and productivity (Bliss and Wein 1972; Hernandez 1973; Wein and Bliss 1973; Babb and Bliss 1974; Babb 1977; Chapin and Shaver 1981; Fetcher and Shaver 1983).

2.6 The Arctic LTER Program at Toolik Lake

The Toolik Lake research camp was set up in 1975 to support aquatic ecologists conducting an independent part of the RATE program. Terrestrial researchers used the camp as a convenient stopover point on the Dalton Highway (Brown and Berg 1980; Shaver and Chapin 1986, 1991; Shaver et al. 1986a; Walker et al. 1987a). Beginning in 1979, terrestrial ecosystem studies were based there. Research on both terrestrial and aquatic ecosystems continued at Toolik Lake throughout the 1970s and 1980s with support from NSF and other agencies. The Arctic LTER (Long-Term Ecological Research) program at Toolik Lake was established in 1988 based on a collaboration among NSF-supported investigators (cf. Callahan 1984; Franklin et al. 1990). The Toolik Lake LTER program includes research on lakes, streams, and terrestrial ecosystems, and it promotes long-term observations and whole-system experiments.

It seemed evident from earlier work at Barrow, Meade River, and Eagle Creek that 2- to 5-year studies might not aid in predicting long-term change, e.g., in response to climate change or physical disturbance. The long-term monitoring program of the Arctic LTER has helped to establish correlations between climate variables and important ecosystem properties, to relate measurements made in a particular site and year to longer-term and larger-area patterns, and to separate signal from noise in a heterogeneous and variable system. For example, it has been demonstrated that average growth over the long term is correlated with long-term average temperatures, whereas interannual variations in temperature and growth are not at all correlated (Shaver et al. 1986b). Multiyear studies of primary production by individual species (Shaver et al. 1986a) and whole communities (Chapin and Shaver 1985) have shown that annual variation in productivity in response to climate is strongly and differentially affected by carbon and nutrient storage. Surveys within aquatic ecosystems have shown the manner in which predation by fish determines community structure and trophic interactions in lakes (O'Brien et al. 1979) and have revealed consistent patterns in excess dissolved CO_2 in lake and stream waters that implicate transport in soil water as a major carbon cycling pathway (Kling et al. 1991; Chap. 13, this Vol.).

Whole-system experiments were developed within the LTER program to understand interactions among ecosystem processes and to develop and test predictions of whole-system models. The experiments are maintained over many years to determine whether initial short-term responses are sustained in the long run, or whether other sets of responses and controls act over longer time scales. Changes in species composition and trophic structure may occur only several years after a disturbance, and productivity of the new community may be regulated by environmental factors very different from those operating in the initial community.

The experimental approach has produced several important successes including a demonstration that productivity of tundra vegetation is more strongly limited in the long-term by nutrient availability than by carbon supply or temperature (Shaver et al. 1986a, 1992), and that changes in species composition of the vegetation may lag, rather than lead, changes in productivity and biomass (Chapin and Shaver, unpubl. data). This understanding has been particularly useful in developing simulation models of the long-term effects of climate change and disturbance on tundra ecosystems, showing how potential changes in tundra carbon cycling are constrained by carbon–nutrient interactions (Rastetter et al. 1991, 1992; Leadley and Reynolds 1992; Chap. 18, this Vol.).

As in the case of studies described in this book, a long-term goal of the LTER research is to "scale up" to the landscape and watershed level by increasing understanding of interactions between neighboring tundra communities and between terrestrial and aquatic ecosystems. To this end, primary foci are on the transport of C, N, and P in soil water and controls on the amounts and chemical forms of these elements. Transport in the soil water in addition to erosion of particulate organic matter provide the energy and nutrients driving biological response in arctic lake and stream ecosystems (Peterson et al. 1985, 1993). Transported nutrients and carbon also have large effects on landscape-level carbon budgets (Kling et al. 1991; Chap. 13, this Vol.). Both the amounts and the chemistry of elements in soil water are strongly influenced by spatial variation in vegetation and soils (Shaver et al. 1990; Giblin et al. 1991).

2.7 Other Studies in Alaska and Elsewhere

Research at a number of other sites in Alaska has focused on the role of natural disturbances in northern ecosystems, the responses of these ecosystems to human disturbance, or both. The largest and most complete of these studies was carried out under US Department of Energy sponsorship at Cape Thompson and Amchitka Island (Wilimovsky and Wolfe 1966; Merritt and Fuller 1977). These two studies were designed primarily as environmental impact surveys in preparation for a harbor excavation with nuclear explosives (Cape Thompson) and a nuclear weapons test (Amchitka). The two resulting volumes provide much useful and detailed information including observations on the role, spatial pattern, and effects of both natural and human disturbance (chap. 3, this Vol.). Perhaps the first comprehensive environmental impact survey ever completed - the Cape Thompson volume (Wilimovsky and Wolfe 1966) - contributed significantly to the decision not to complete the excavation project.

The extensive literature on effects of development-related disturbances (e.g., roads, pipelines, and oil spills) has been reviewed by several authors (Johnson 1981; Walker et al. 1987a; Walker and Walker 1991) and in symposia (Atlas and Brown 1978). The reviews by Walker and Walker (1991) and in Chapter 3 (this Vol.) include especially useful comparisons of natural and anthropogenic disturbances in terms of their spatial and temporal scales.

The Arctic Systems Science (ARCSS) program (ARCUS 1991, IASC 1994) was recently initiated by the NSF. This is a major integrated study of the physical, chemical, biological, and social systems that make up the arctic and how they will affect and be affected by natural and anthropogenic changes. One component of ARCSS concerns the interactions of terrestrial, freshwater, atmospheric, and ice processes. Among its major objectives are to understand how the arctic may change in the future, particularly in terms of hydrological and biogeochemical processes and their interactions with plant and animal communities. The ARCSS program provides a unique opportunity for ecologists to link ecosystem-level studies with complementary work in a wide range of disciplines.

Finally, it is important to remember that progress in ecosystem-level research in Alaska has been accompanied by progress in other arctic regions (Bliss et al. 1981) and in subarctic areas with arctic-like ecosystems (e.g., Persson 1980; Sonesson 1980; Van Cleve et al. 1986). These important studies provide even richer opportunities for comparison, prediction, and development of a general understanding of the role played by disturbance and its interaction with nutrient cycling in the Arctic. International comparisons include those by Wielgolaski and Rosswall (1972), Bliss and Wielgolaski (1973), Rosswall and Heal (1975), and Bliss et al. (1981); the most recent syntheses are by Chapin et al. (1992), Chapin and Körner (1995), and Oechel (in prep.).

2.8 Summary and Prospects

Since the Arctic Research Laboratory opened at Barrow in 1947, ecosystem-level research has repeatedly shown the importance of nutrient limitation to arctic vegetation development and the strong linkage between disturbance and nutrient cycling. The nutrient recovery hypothesis of Pitelka and Schultz (1964) was the first to point out the potential importance of disturbance by herbivory as a means of increasing nutrient cycling in tundra, with a consequent increase in primary production. The thaw lake cycle proposed by Britton (1957) and other early work on natural disturbances (e.g., Sigafoos 1952; Churchill and Hanson 1958) provided a broad perspective on natural rates and patterns of disturbance, which were later shown to be strongly correlated with both vegetation and cycling patterns of soil nutrients.

The models developed during the Tundra Biome program provided important tests of these early ideas. These models were among the first to quantitatively demonstrate the interactions among climatic factors, and to distinguish between direct and indirect effects of climate on vegetation. The models of the Tundra Biome were further elaborated in the studies at Eagle Creek and Meade River especially through the development of predictions about how growth form composition of the vegetation was regulated. The model predictions were based primarily on new knowledge about differences among growth forms in their nutrient turnover and allocation patterns. The Meade River study added further insights relating herbivory to plant nutrient allocation patterns, soil nutrient distribution, and overall nutrient cycling.

Research of the Arctic LTER program based at Toolik Lake has further extended the conclusions of the earlier studies, both in time and over a broader array of climatic regimes and ecosystem types (including stream and lake ecosystems), by combining long-term, whole-system experiments with even longer-term monitoring of control sites. Finally, later work at Toolik Lake, along with studies discussed in this book, has shown the importance of transport processes between communities in an arctic landscape.

The central hypotheses of the R4D Program build on this firm foundation of basic research, applying what we know about natural disturbances and processes controlling vegetation response to make useful and testable predictions about the factors controlling response to human disturbances. For the future many opportunities remain to be explored through integrated ecosystem research in the Arctic particularly when intensive, descriptive studies are combined with long-term monitoring, whole-system experiments, and comparisons of studies at multiple sites. Further development of the research described in this book on system-level controls over ecosystem stability, resistance, and resilience is an important long-term goal. We might build on our present solid understanding of arctic ecosystems to ask, "What are the general characteristics of terrestrial ecosystems that confer stability or lack of stability in response to various kinds of disturbances?" Other opportunities include "scaling up" from internally homogeneous, individual ecosystem types to complex, heterogeneous landscapes, watersheds, and large regions, and understanding the role of the Arctic as a whole in global climate, hydrology, and biogeochemical cycles.

References

Archer S, Tieszen LL (1980) Growth and physiological responses of tundra plants to defoliation. Arct Alp Res 12: 531–552

ARCUS (1991) Arctic system science. A plan for integration. ARCUS, Fairbanks, Alaska, 34 pp

Atlas RM, Brown J (eds) (1978) Effects of hydrocarbon spills in Alaska. Arctic (Spec Symp issue) 31: 155–411

Babb TA (1977) High arctic disturbance studies associated with the Devon Island Project. In: Bliss LC (ed) Truelove Lowland, Devon Island, Canada: a high arctic ecosystem. Univ Alberta Press, Edmonton, pp 647–654

Babb TA, Bliss LC (1974) Effects of physical disturbance on arctic vegetation in the Queen Elizabeth Islands. Appl Ecol 11: 549–562

Batzli GO (ed) (1980) Patterns of vegetation and herbivory in arctic tundra: results from the Research on Arctic Tundra Environments (RATE) Program. Arct Alp Res 12 (4): 401–518

Batzli GO, Jung HG (1980) Nutritional ecology of microtine rodents: resource utilization near Atkasook, Alaska. Arct Alp Res 12: 483–499

Batzli GO, Sobasky ST (1980) Distribution, abundance, and foraging patterns of ground squirrels near Atkasook, Alaska. Arct Alp Res 12: 501–510

Batzli GO, White RG, MacLean SF Jr, Pitelka FA, Collier BD (1980) The herbivore-based trophic system. In: Brown J, Miller PC, Tieszen LL, Bunnell FL (eds) An arctic ecosystem: the coastal tundra at Barrow, Alaska. Dowden, Hutchinson and Ross, Stroudsburg, pp 335–410

Benninghoff WS (1952) Interaction of vegetation and soil frost phenomena. Arctic 5: 34–44

Billings WD (1973) Arctic and alpine vegetations: similarities, differences, and susceptibility to disturbance. BioScience 23: 697–704

Billings WD, Mooney HA (1968) The ecology of arctic and alpine plants. Biol Rev 43: 481–530

Billings WD, Peterson KM (1980) Vegetational change and ice-wedge polygons through the thaw-lake cycle in arctic Alaska. Arct Alp Res 12: 413–432

Blair WF (1977) Big biology: The US IBP. US IBP Synth Ser 7: 261, Dowden, Hutchinson and Ross, Stroudsburg

Bliss LC (1956) A comparison of plant development in microenvironments of arctic and alpine plants. Ecol Monogr 26: 303–337

Bliss LC (ed) (1977) Truelove Lowland, Devon Island, Canada: a high arctic ecosystem. Univ Alberta Press, Edmonton, p 714

Bliss LC, Cantlon JE (1957) Succession in river alluvium in northern Alaska. Am Midl Nat 58: 452–469

Bliss LC, Peterson KM (1992) Plant succession, competition, and physiological constraints of species in the Arctic. In: Chapin FS III, Jefferies R, Reynolds J, Shaver G, Svoboda J (eds) Arctic ecosystems in a changing climate: an ecophysiological perspective. Academic Press, New York, pp 111–138

Bliss LC, Wein RW (1972) Plant community responses to disturbances in the western Canadian Arctic. Can Bot 50: 1097–1109

Bliss LC, Wielgolaski FE (eds) (1973) Primary production and production processes, Tundra Biome. Tundra Biome Steering Comm, Stockholm, p 256

Bliss LC, Heal OW, Moore JJ (eds) (1981) Tundra ecosystems: a comparative analysis. Cambridge Univ Press, Cambridge, p 813

Britton ME (1957) Vegetation of the arctic tundra. In: Hanson HP (ed) Arctic biology. Oregon State Univ Press, Corvallis, pp 26–61

Britton ME (ed) (1973) Alaskan arctic tundra. Arct Inst N Am Tech Pap 25: 224

Brown J (1975) Ecological investigations of the tundra biome in the Prudhoe Bay Region, Alaska. Biol Pap Univ Alaska, Spec Rep 2: 215

Brown J, Berg RL (eds) (1980) Environmental engineering and ecological baseline investigations along the Yukon River–Prudhoe Bay Haul Road. US Army CRREL Rep 80-19: 187

Brown J, Miller PC, Tieszen LL, Bunnell FL (eds) (1980) An arctic ecosystem: the coastal tundra at Barrow, Alaska. Dowden, Hutchinson and Ross, Stroudsburg, p 571

Callahan JT (1984) Long-term ecological research. BioScience 34: 363–367

Chapin FS (1980) Nutrient allocation and responses to defoliation in tundra plants. Arct Alp Res 12: 553–563

Chapin FS (1983) Direct and indirect effects of temperature on arctic plants. Pol Biol 2: 47–52

Chapin FS III, Körner C (eds) (1995) Arctic and alpine biodiversity. Ecologial Studies 113, Springer, Berlin Heidelberg New York

Chapin FS III, Shaver GR (1981) Changes in soil properties and vegetation following disturbance of Alaskan arctic tundra. Appl Ecol 18: 605–617

Chapin FS III, Shaver GR (1985) Individualistic growth response of tundra plant species to manipulation of light, temperature and nutrients in a field experiment. Ecology 66: 564–576

Chapin FS III, Van Cleve K, Chapin MC (1979) Soil temperature and nutrient cycling in the tussock growth form of *Eriophorum vaginatum* L. J Ecol 67: 169–189

Chapin FS III, Miller PC, Billings WD, Coyne PI (1980a) Carbon and nutrient budgets and their control in coastal tundra. In: Brown J, Miller PC, Tieszen L, Bunnell FL (eds) An arctic ecosystem: the coastal tundra at Barrow, Alaska. Dowden, Hutchinson and Ross, Stroudsburg, pp 458–484

Chapin FS III, Johnson DA, McKendrick JD (1980b) Seasonal nutrient movements in various plant growth forms in an Alaskan tundra: implications for herbivory. Ecology 68: 189–210

Chapin FS III, Jefferies R, Reynolds J, Shaver G, Svoboda J (eds) (1992) Arctic ecosystems in a changing climate: an ecophysiological perspective. Academic Press, New York

Churchill EC, Hanson HC (1958) The concept of climax in arctic and alpine vegetation. Biol Rev 24: 127–191

Everett KR (1980a) Scientific history. In: Walker DA, Everett KR, Webber PJ, Brown J (eds) Geobotanical atlas of the Prudhoe Bay Region, Alaska. US Army CRREL Rep 80–14: 69

Everett KR (1980b) Distribution and variability of soils near Atkasook, Alaska. Arct Alp Res 12: 433–446

Fetcher N, Shaver GR (1983) Life histories of tillers of *Eriophorum vaginatum* in relation to tundra disturbance. J Ecol 71: 131–148

Franklin JF, Bledsoe CS, Callahan JT (1990) Contributions of the Long-Term Ecological Research Program. BioScience 40: 509–523

Fuller RG (1977) Previous scientific investigations, 1867–1967. In: Merritt ML, Fuller RG (eds) The environment of Amchitka Island, Alaska. US Energy Res Dev Administration, Publ TID-26712. US Natl Tech Inf Serv, p 682

Giblin AE, Nadelhoffer KJ, Shaver GR, Laundre JA, McKerrow AJ (1991) Biogeochemical diversity along a riverside toposequence in arctic Alaska. Ecol Monogr 61: 415–435

Heal OW, Perkins DF (1978) Production ecology of British moors and montane grasslands. Springer, Berlin Heidelberg New York, 426 pp

Hernandez H (1973) Natural plant recolonization of surficial disturbances, Tuktoyaktuk Peninsula Region, Northwest Territories. Can Bot 51: 2177–2196

Hobbie JE (1973) Arctic limnology: a review. In: Britton ME (ed) Alaskan Arctic Tundra. Arct Inst N Am Tech Pap 25: 127–168

Hobbie JE (ed) (1980) Limnology of tundra ponds, Barrow, Alaska. US IBP Synth Ser 13, Dowden, Hutchinson and Ross, Stroudsburg, p 51

Hultén E (1940) History of botanical exploration in Alaska and Yukon territories from the time of the discovery to 1940. Bot Notes 1940: 89–346

IASC (1994) Scientific plan for a regional research programme in the arctic on global change. Natl Acad Press, Washington, DC, 106 pp

Johnson LA (1981) Revegetation and selected terrain disturbances along the trans-Alaska pipeline, 1975–1978. US Army CRREL Rep 81–12

Johnson PL (1969) Arctic plants, ecosystems, and strategies. Arctic 22: 341–355

Johnson PL, Billings WD (1962) The alpine vegetation of the Beartooth Plateau in relation to cryopedogenic processes and patterns. Ecol Monogr 32: 105–135

Johnson PL, Tieszen LL (1973) Vegetative research in arctic Alaska. In: Britton ME (ed) Alaskan Arctic Tundra. Arct Inst N Am Tech Pap 25: 169–198

Kling GW, Kipphut GW, Miller, MC (1991) Arctic lakes and streams as gas conduits to the atmosphere: implications for tundra carbon budgets. Science 251: 298–301

Komarkova V, Webber PJ (1980) Two low arctic vegetation maps near Atkasook, Alaska. Arct Alp Res 12: 447–472

Leadley PW, Reynolds JF (1992) Long-term response of an arctic sedge to climate change: a simulation study. Ecol Appl 2: 323–340

Limbach WE, Oechel WC, Lowell W (1982) Photosynthetic and respiratory responses to temperature and light of three Alaskan tundra growth forms. Holarct Ecol 5: 150–157

McKendrick JD, Batzli GO, Everett KR, Swanson JC (1980) Some effects of mammalian herbivores and fertilization on tundra soils and vegetation. Arct Alp Res 12: 565–578

Merritt ML, Fuller RG (eds) (1977) The Environment of Amchitka Island, Alaska. US Energy Res Dev Administration, Publ TID-26712. US Natl Tech Inf Serv, p 682

Miller PC (1981) Carbon balance in northern ecosystems and the potential effects of carbon dioxide-induced climatic change. Rep Worksh, San Diego, California, 7–9 March 1980. US Dep Energy, Washington, DC

Miller PC(ed) (1982a) The availability and utilization of resources in tundra ecosystems. Holarct Ecol 5: 81–220

Miller PC (1982b) Environmental and vegetational variation across a snow accumulation area in montane tundra in central Alaska. Holarct Ecol 5: 85–98

Miller PC, Webber PJ, Oechel WC, Tieszen LL (1980) Biophysical processes and primary production. In: Brown J, Miller PC, Tieszen LL, Bunnell FL (eds) An arctic ecosystem: the coastal tundra at Barrow, Alaska. Dowden, Hutchinson and Ross, Stroudsburg, pp 66–101

Miller PC, Kendall R, Oechel WC (1983) Simulating carbon accumulation in northern ecosystems. Simulation 40: 119–131

Miller PC, Miller PM, Blake-Jacobsen M, Chapin FS III, Everett KR, Hilbert DW, Kummerow J, Linkins AE, Marion GM, Oechel WC, Roberts SW, Stuart L (1984) Plant-soil processes in *Eriophorum vaginatum* tussock tundra in Alaska: a systems modeling approach. Ecol Monogr 54: 361–405

Oberbauer S, Miller PC (1979) Plant water relations in montane and tussock tundra vegetation types in Alaska. Arct Alp Res 11: 69–81

O'Brien WJ (ed) (1992) Toolik Lake: ecology of an aquatic ecosystem in arctic Alaska. Hydrobiologia 240: 1–269

O'Brien WJ, Buchanan C, Haney J (1979) Arctic zooplankton community structure: exceptions to some general rules. Arctic 32: 237–247

Oechel WC (ed) Global change and arctic terrestrial ecosystems. Ecological Studies, Springer, Berlin Heidelberg New York (in preparation)

Persson T (ed) (1980) Structure and function of northern coniferous forests. Ecol Bull 32: 609, Swed Nat Sci Res Counc Stockholm

Peterson BJ, Hobbie JE, Hershey AE, Lock MA, Ford TE, Vestal JR, McKinley VL, Hullar MAJ, Miller MC, Ventullo RM, Volk GS (1985) Transformation of a tundra river from heterotrophy to autotrophy by addition of phosphorus. Science 229: 1383–1386

Peterson BJ, Deegan L, Helfrich J, Hobbie JE, Hullar MAJ, Moller B, Ford TE, Hershey AE, Hiltner A, Kipphut G, Lock MA, Feibig DM, McKinley VL, Miller MC, Vestal JR, Ventullo RM, Volk GS (1993) Biological responses of a tundra river to fertilization. Ecology 74: 653–672

Peterson KM, Billings WD (1980) Tundra vegetational patterns and succession in relation to microtopography near Atkasook, Alaska. Arct Alp Res 12: 437–482

Pitelka FA (1973) Cyclic pattern in lemming populations near Barrow, Alaska. In: Britton ME (ed) Alaskan Arctic Tundra. Arct Inst N Am Tech Pap 25: 199–216

Pitelka FA, Schultz AM (1964) The nutrient-recovery hypothesis for arctic microtine cycles. In: Crisp DJ (ed) Grazing in terrestrial and marine environments. Blackwell, Oxford, pp 55–70

Rastetter EB, Ryan MG, Shaver GR, Melillo JM, Nadelhoffer KJ, Hobbie JE, Aber JD (1991) A general biogeochemical model describing the responses of the C and N cycles in terrestrial ecosystems to changes in CO_2, climate, and N deposition. Tree Physiol 9: 101–126

Rastetter EB, McKane RB, Shaver GR, Melillo JM (1992) Changes in C storage by terrestrial ecosystems: how C–N interactions restrict responses to CO_2 and temperature. Water Air Soil Pollut 64: 327–344

Reed JC (1969) The story of the Naval Arctic Research Laboratory. Arctic 22: 177–184

Reynolds JF, Leadley PW (1992) Modeling the response of arctic plants to changing climate. In: Chapin FS III, Jefferies R, Reynolds J, Shaver G, Svoboda J (eds) Arctic ecosystems in a changing climate: an ecophysiological perspective. Academic Press, New York, pp 413–440

Rosswall T, Heal OW (eds) (1975) Structure and function of tundra ecosystems. Swed Nat Sci Res Counc Stockholm. Ecol Bull 20: 450

Schultz AM (1969) A study of an ecosystem: the arctic tundra. In: Van Dyne GM (ed) The ecosystem concept in natural resource management. Academic Press, New York, pp 77–93

Shaver GR, Chapin FS III (1980) Response to fertilization by various plant growth forms in an Alaskan tundra: nutrient accumulation and growth. Ecology 61 (3): 662–675

Shaver GR, Chapin FS III (1986) Effect of NPK fertilization on production and biomass of Alaskan tussock tundra. Arct Alp Res 18: 261–268

Shaver GR, Chapin FS III (1991) Production/biomass relationships and element cycling in contrasting arctic vegetation types. Ecol Monogr 61: 1–31

Shaver GR, Chapin FS III, Billings WD (1979) Ecotypic differentiation in *Carex aquatilus* as related to ice wedge polygonization in the Alaskan coastal tundra. J Ecol 67: 1025–1046

Shaver GR, Chapin FS III, Gartner BL (1986a) Factors limiting growth and biomass accumulation in *Eriophorum vaginatum* L. in Alaskan tussock tundra. J Ecol 74: 257–278

Shaver GR, Fetcher N, Chapin FS III (1986b) Growth and flowering in *Eriophorum vaginatum*: annual and latitudinal variation. Ecology 67: 1524–1525

Shaver GR, Nadelhoffer KJ, Giblin AE (1990) Biogeochemical diversity and element transport in a heterogeneous landscape, the north slope of Alaska. In: Turner MG, Gardner RH (eds) Quantitative methods in landscape ecology. Springer, Berlin Heidelberg New York, pp 105–126

Shaver GR, Billings WD, Chapin FS III, Giblin AE, Nadelhoffer KJ, Oechel WC, Rastetter EB (1992) Global change and the carbon balance of arctic ecosystems. BioScience 42: 433–441

Sigafoos RS (1952) Frost action as a primary physical factor in tundra plant communities. Ecology 33: 480–487

Sonesson M (ed) (1980) Ecology of a subarctic mire. Ecol Bull 30: 313, Swed Nat Sci Res Counc, Stockholm

Tiwari JL, Daley RJ, Hobbie JE, Miller MC, Stanley DW, Reed JP (1980) Modeling. In: Hobbie JE (ed) Limnology of tundra ponds, Barrow, Alaska. US IBP Synth Ser 13. Dowden, Hutchinson and Ross, Stroudsburg, pp 407–457

Tedrow JCF (1973) Pedologic investigations in northern Alaska. In: Britton ME (ed) Alaskan Arctic Tundra. Arct Inst N Am Tech Pap 25: 93–108

Tieszen LL (ed) (1978) Vegetation and production ecology of an Alaskan arctic tundra. Springer, Berlin Heidelberg New York, 686 pp

Van Cleve K, Chapin FS III, Flanagan PW, Vierek LA, Dyrness CT (eds) (1986) Forest ecosystems in the Alaskan Taiga: a synthesis of structure and function. Springer, Berlin Heidelberg New York, 230 pp

Walker DA, Walker MD (1991) History and pattern of disturbance in Alaskan arctic terrestrial ecosystems: a hierarchical approach to analyzing landscape change. J Appl Ecol 28: 244–276

Walker DA, Cate D, Brown J, Racine C (1987a) Disturbance and recovery of arctic Alaskan tundra terrain. US Army CRREL Rep 87-11, pp: 1–63

Walker DA, Webber PJ, Binnian EF, Everett KR, Lederer ND, Nordstrand EA, Walker MD (1987b) Cumulative impacts of oil fields on northern Alaskan landscapes. Science 238: 757–761

Webber PJ (1978) Spatial and temporal variation of the vegetation and its productivity, Barrow, Alaska. In: Tieszen LL (ed) Vegetation and production ecology of an Alaskan arctic tundra. Springer, Berlin Heidelberg New York, pp 37–112

Webber PJ, Miller PC, Chapin FS III, McCown BH (1980) The vegetation: pattern and succession. In: Brown J, Miller PC, Tieszen LL, Bunnell FL (eds) An arctic ecosystem: the coastal tundra at Barrow, Alaska. Dowden, Hutchinson and Ross, Stroudsburg, pp 186–218

Wein RW, Bliss LC (1973) Changes in *Eriophorum* tussock communities following fire. Ecology 54: 845–852

Wein RW, Bliss LC (1974) Primary production in arctic cottongrass tussock tundra communities. Arct Alp Res 6: 261–274

White RG, Trudell J (1980) Habitat preference and forage consumption by reindeer and caribou near Atkasook, Alaska. Arct Alp Res 12: 511–529

Wielgolaski FE (ed) (1975a) Fennoscandian tundra ecosystems. Part 1: plants and microorganisms. Springer, Berlin Heidelberg New York, p 366

Wielgolaski FE (ed) (1975b) Fennoscandian tundra ecosystems. Part 2. Animals and systems analysis. Springer, Berlin Heidelberg New York

Wielgolaski FE, Rosswall Th (eds) (1972) Proc IV Int Meet Biol Productivity of Tundra, Leningrad, USSR. Tundra Biome Steering Comm, Stockholm, p 320

Wilimovsky NJ (1966) Synopsis of previous scientific explorations. In: Wilimovsky NJ, Wolfe JN (eds) Environment of the Cape Thompson Region, Alaska. US Atomic Energy Comm, PNE-481. US Natl Tech Inf Serv, pp 1–8

Wilimovsky NJ, Wolfe JN (eds) (1966) Environment of the Cape Thompson Region, Alaska. US Atomic Energy Comm PNE-481: 1250, US Natl Tech Inf Serv, p 1250

3 Disturbance and Recovery of Arctic Alaskan Vegetation

D. A. WALKER

3.1 Introduction

The discovery of oil at Prudhoe Bay in 1968 prompted considerable interest and funding for disturbance research in arctic ecosystems. Most of the early studies focused on small-scale disturbances relating to oil spills, off-road vehicle trails, roadside disturbances, and old oil-well sites. During the late 1980s and 1990s, scientific interest turned to the broader-scale issues relating to the basic ecosystem processes involved in disturbance and recovery (Oechel 1989; Chaps. 1 and 2, this Vol.), cumulative impacts of large oil-field developments (Walker et al. 1987a), restoration of affected areas (Wyant and Knapp 1992), effects of contamination of the arctic atmosphere from sources at lower latitudes (Landers et al. 1992), and issues relating to climate change (Chapin et al. 1992).

This chapter discusses most of the common disturbances to northern arctic Alaskan vegetation, their patterns of recovery, and recent techniques used for rehabilitating the disturbed areas. Much of the literature consists of unpublished agency and industry reports from the Prudhoe Bay region, but the findings apply to most of the circumpolar Low Arctic – the area of continuously vegetated tundra between the arctic tree line to the south and the intermittently vegetated High Arctic to the north (Bliss and Matveyeva 1992).

3.2 Disturbance and Recovery

Disturbance is a change in vegetation or underlying substrate caused by some external factor (White and Pickett 1985; Pickett et al. 1989). Disturbance can result in altered thermal, hydrological, or nutrient regimes, as well as in changes in the species composition, vegetation structure, or primary production. Disturbance is a natural part of all ecosystems (Chap. 1, this Vol.). It plays a central role in the evolution of ecosystems and is essential to maintaining characteristic diversity and productivity that we associate with a given vegetation type. Anthropogenic disturbances often differ in scale from their natural analogs, but the processes of recovery are often similar (Table 3.1). The final stage of recovery is a healthy functioning ecosystem that can maintain a steady-state equilibrium over a few decades.

J.F. Reynolds and J.D. Tenhunen (Eds.)
Ecological Studies, Vol. 120
© Springer-Verlag Berlin Heidelberg 1996

Table 3.1. Natural analogs of anthropogenic disturbances

Anthropogenic disturbance	Natural analog
Microsite disturbances	
Trash and solid waste	Ice-pushed boulders, driftwood from storm surges, floods
Small barren patches	Frost scars, blow outs
Berms of bladed trails	Ice-pushed turf, debris flows, animal dens
Diesel or gasoline spills	None
Mesosite to macrosite disturbances	
Thermokarst and thermal erosion	Natural thermokarst and thermal erosion
Crude oil spills	Oil seeps
Seawater and brine spills	Salt kill from ocean storm surges
Snow drifts from roads, buildings, snow fences	Natural snow drifts
Impoundments	Thaw lakes
Fire	Fire
Off-road vehicle trails	Caribou trails, natural thermokarst, thermal erosion
Ice roads, pads	River icings
Roads, pads, borrow pits	Gravel bars in river floodplains, talus
Offshore gravel drilling islands, causeways	Barrier islands, spits
Regional disturbances	
Dust from roads	Loess from rivers
Acid rain or increased sulfates, NO$_x$ industrial pollution	Fallout from fires
Cumulative impacts of road networks	None
Climate change	
Temperature	Coastal temperature gradient
Precipitation	Elevation gradient, redistribution of snow caused by topographic effects

Recovery is thus a pragmatic term, useful in terms of human life spans. The original native vegetation is the standard against which recovery is measured (Fig. 3.1). A vegetation type may have no possibility of returning to its original state, because the prevailing climate has changed since the original community formed, or because the set of conditions leading to the original community cannot be repeated or the disturbance has completely changed the substrate character (Komárková and Webber 1978; Webber and Ives 1978). Disturbance often affects the thermal and hydrological properties of the soil surface (Hinzman et al. 1991; Chap. 6, this Vol.). For example, complete recovery to original species composition is unlikely on heavily thermokarsted sites because of altered microtopography, hydrology, and thermal regimes, but an ecosystem with greater productivity often occurs because of enhanced decom-

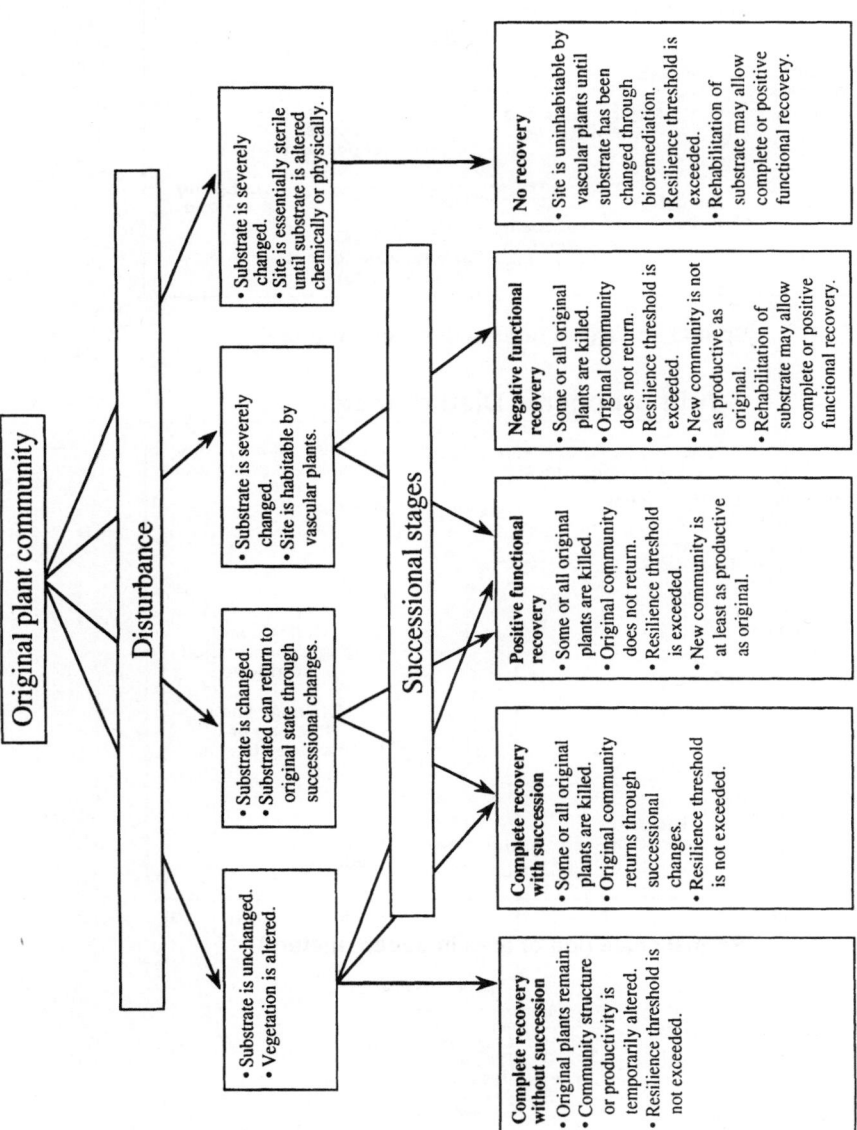

Fig. 3.1. Pathways of recovery following disturbance, without rehabilitation of the site. Substrate rehabilitation on severely disturbed sites can lead to complete or positive functional recovery. (Modified from Walker et al. 1987a)

Natural Disturbances

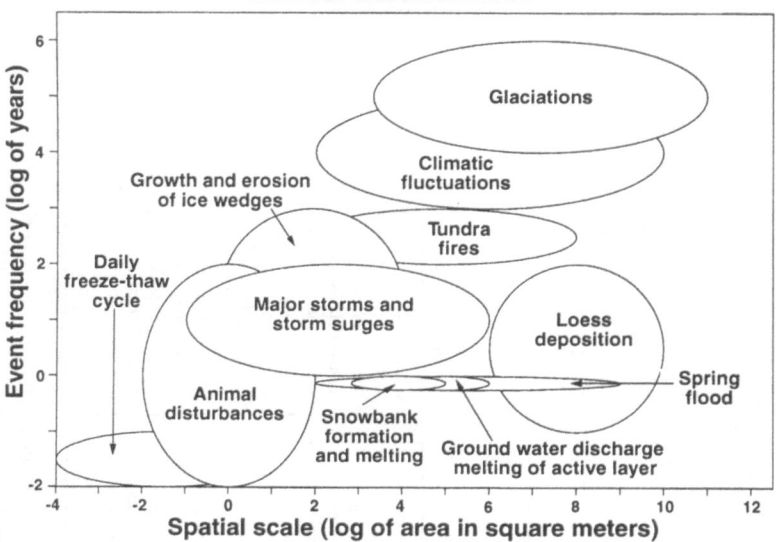

Anthropogenic Disturbances

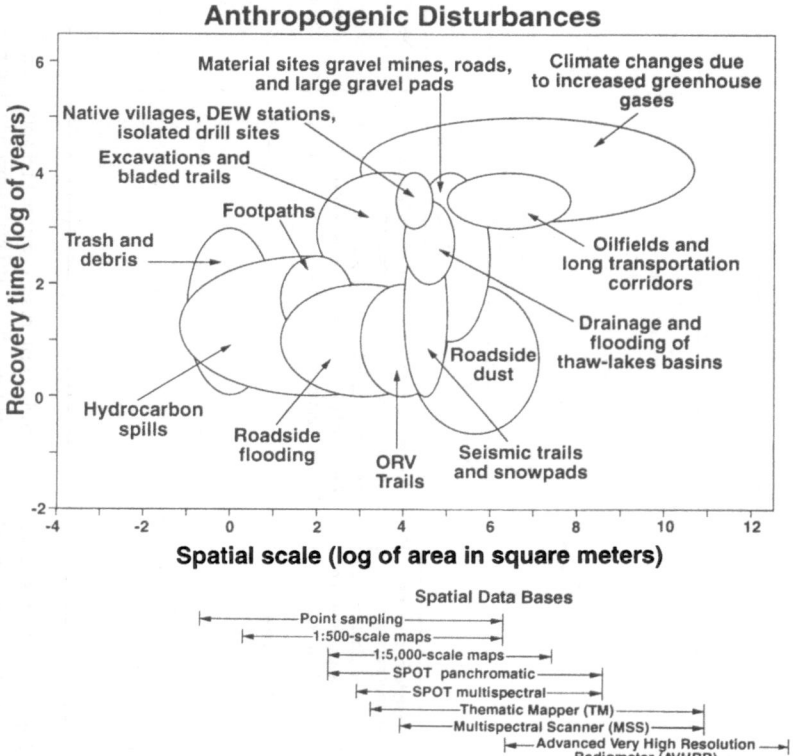

Fig. 3.2. Spatial and temporal domains of natural and anthropogenic disturbances in the arctic and the scales of spatial databases appropriate for monitoring change. *ORV*, off-road vehicle. (Modified from Walker and Walker 1991)

position and higher soil temperatures (positive functional recovery; Ebersole and Webber 1983).

Anthropogenic disturbances to arctic tundra span a wide range of spatial scales (Fig. 3.2). Aggregated disturbances caused by activities both inside and outside the Arctic have impacts on arctic vegetation that are quite different from effects studied at the plot level (Walker and Walker 1991). A hierarchic framework is useful for thinking about disturbances at widely divergent scales and their natural analogs, and for choosing appropriate remote-sensing tools to monitor changes (Fig. 3.2; Chap. 7, this Vol.). This chapter reviews the common disturbances along a spatial gradient from bits of trash ($10^{-1}\,m^{-2}$) to landscapes affected by large oil fields ($10^8\,m^2$).

3.3 Typical Disturbance and Recovery Patterns

3.3.1 Small Disturbed Patches

The most detailed studies of fine-scale arctic disturbances come from Fish Creek and East Oumalik, Alaska. At both sites, short-term recovery was studied on small, newly bared surfaces where debris and solid-waste were removed 30 years after the 1944–1953 exploration of Naval Petroleum Reserve Number 4 (renamed the National Petroleum Reserve in Alaska [NPR-A] in 1976; Komárková 1985; Ebersole 1987). The waste was composed of steel drums, pilings, boardwalks, and remains of camp buildings and equipment. Cleanup operations sponsored by the United States Navy and Department of the Interior removed these materials in 1979 (Gryc 1985). A few similar drill sites outside of NPR-A were not cleared of these materials and remain as monuments to the early days of oil exploration (Fig. 3.3a).

At Oumalik short-term recovery was studied on 73 newly bared sites ranging in size from about $0.1-1.5\,m^2$ (Komárková 1985; Ebersole 1987). After 4 years most of the plant cover was due to vegetative colonization by rhizomatous species. Forty taxa, representing adjacent communities, were present (e.g., *Equisetum arvense*, *Vaccinium vitis-idaea*, *Ledum palustre*, *Lupinus arcticus*, *Arctagrostis latifolia*, and *Eriophorum angustifolium* in moist plots, and *Carex aquatilis* and *Eriophorum angustifolium* in wet plots). On larger plots, seedling establishment was responsible for most of the cover. Seeds in the seed bank apparently did not survive 30 years of burial by the debris; nevertheless, seedlings were abundant (except on dry sites, aquatic sites, and intact organic mats), apparently dispersed from the surrounding vegetation. The vegetation type of the surrounding community was the most important single factor affecting colonization. A few forbs that dominated in the early phases (e.g., *Senecio congestus*, *Epilobium palustre*, and *Saxifraga cernua*) were uncommon on the surrounding 30-year-old disturbed sites, indicating that these species were eliminated during later phases of succession.

Fig. 3.3a–c.

Fig. 3.3d–f. (*continued.*)

g

h

Fig. 3.3a–h. Common anthropogenic disturbances in northern Alaska. **a** Debris from 1960s drilling operation about 60 km south of Prudhoe Bay near the Kuparuk River. Note the wooden derrick in the background, thermokarst, and lush vegetation caused by altered hydrology. **b** Erosion channel along a bulldozed trail to Prudhoe Bay constructed in the 1960s, near Sagwon, approximately 110 km south of Prudhoe Bay. Massive ice in the uplands melted, eroding more than 3 m of soil. The resulting gully is now protected from winter winds and has enhanced snow cover, permitting willows up to 2 m tall. Willow growth is enhanced by the warmer soils, deeper thaw, high decomposition rates and richer nutrient regimes. **c** Peat road, now 15 years old, near the Putuligayuk River in the Prudhoe Bay Oil Field. The road was formed by scraping material from both sides to form a raised surface. Impoundments have developed in the scraped areas, eroding the road surface and making it impassable. These peat surfaces are very slow to revegetate. **d** Winter seismic trail created in 1984 near the Hulahula River in the Arctic National Wildlife Refuge. Multiple tracks created trails wider than 10 m. Trains of vehicles included tracked vehicles mounted with vibrators or drills, small personal carriers, geophone trucks, a recording vehicle, and D-7 Caterpillar tractors pulling ski-mounted trailers for camps. Altogether 2000 km of trails were arranged on a 5- × 10-km grid. **e** Gravel mine occupying the entire floodplain of the Putuligayuk River in the Prudhoe Bay oil field. The mine has altered the river's

Long-term (>30 years) recovery on small disturbance patches was also studied at Fish Creek (Komárková 1985), Oumalik (Ebersole 1985), Cape Thompson (Everett et al. 1985), and along the Canol pipeline in Canada (Kershaw and Kershaw 1986). Recovery of nondegradable and degradable solid-waste surfaces, such as barrels, boards, and other debris, is extremely slow in the arctic climate. For low pieces of debris, e.g., boards and cans, found in moist and wet tundra, vegetation mats eventually envelop them. After 30 years at Fish Creek, only a small percentage of larger solid-waste surfaces, such as barrels and pilings, were colonized by bryophytes, lichens, or vascular plants (Komárková 1985).

3.3.2 Contaminants

Contaminants are sometimes spilled during exploration and development. These include drilling mud, waste water, used crankcase oil, dust-control chemicals, reserve pit fluids, diesel fuel, glycol (antifreeze), crude oil, and salt water. These substances contain a wide variety of chemicals that are toxic to plants.

3.3.2.1 Hydrocarbon Spills

During the early exploration of NPR-A, and construction of the Dalton Highway and trans-Alaska pipeline and operation of the North Slope oil fields, numerous spills of crude and refined petroleum products occurred. At the Fish Creek site spills of crude oil, crankcase oil, and diesel fuel occurred between 1948 and 1949. From 1974 to 1977, when the trans-Alaska pipeline was first built and operated, more than 16 000 oil spills – totaling more than 2 650 000 l – occurred along the pipeline route (Johnson 1981). During 1985–1986, 952 oil spills were reported on the North slope, totaling 731 800 l (Speer and Libenson 1988). Most of these spills consisted of refined petroleum products. Although most took place in water or were confined to gravel pads, some occurred on terrestrial vegetation.

◄───

channel and eliminated riparian and bluff habitat on both sides of the river. f Roadside environment along the Prudhoe Bay Spine Road. Note the barren areas caused by dust and thermokarst on the ice-wedge polygons. This site was an area of low-centered polygons that has been converted to high-centered polygons by erosion of the polygon troughs after the road was constructed in the early 1970s. g Flooded drained-lake basin along a road in the Prudhoe Bay region. The photo was taken in early summer before ice in the culverts thawed. By late summer the large oval impoundment drained, but the vegetation remained changed. Notice the lack of elevated microsites for bird-nest sites in the flooded area compared with the other side of the road. Also note the other impoundment along the road (*arrow*), which does not drain all summer. h A portion of the Prudhoe Bay Oil Field showing a complex of pipelines, roads, and gravel pads in the Deadhorse service area. The large *rectangular gravel pad* on the left side is a drill site with 14 well heads and two reserve pits. Complex cumulative impacts are associated with such large developments. (Photos **a–d**, **f**, and **h** by D. A. Walker; photo **g** by L. F. Klinger; photo **e** courtesy of US Fish and Wildlife Service)

Large spills of crude oil from the trans-Alaska pipeline are rare, but five occurred between 1974 and 1977 (Johnson 1981). On the North Slope, the largest tundra area affected by a spill was a Check Valve 7, north of Franklin Bluffs, where more than 1900 bbl of crude oil were sprayed over approximately 8 ha of primarily wet sedge tundra (Walker et al. 1978; Johnson 1981). Other large documented spills occurred at Check Valve 23 (Brendel and Eschenbach 1985) and at oil-well sites at Prudhoe Bay (Pope et al. 1982; Jorgenson et al. 1991a; Jorgenson and Cater 1991).

Numerous studies have examined the recovery of soils and vegetation after small oil spills (e.g., McCown et al. 1973a; Wein and Bliss 1973b; Deneke et al. 1975; Mackay and Mohtadi 1975; Freedman and Hutchinson 1976; Johnson et al. 1980; Linkins et al. 1984). Most of these addressed general recovery trends related to moisture at the site. For example, at Prudhoe Bay, the recovery from crude oil and diesel spills was generally poor in dry and moist sites after 7 years, although a few taxa colonized the spills. In aquatic tundra sites, sedges and mosses recovered well on oil spills, but virtually not at all on diesel spills (Walker et al. 1985a). Spills on saturated soils or areas with shallow standing water disperse quickly as a film from which volatiles are lost. Wet sites may recover completely within a year or two after light spills. Oil does not penetrate deeply into saturated soils, but spills on dry sites are absorbed by mosses and the underlying organic material and mineral (Linkins and Fetcher 1983). Toxic volatiles are released very slowly in these spills, and the microbial decomposition may take decades without bioremediation. Some components of the vegetation are more susceptible to oil-spill damage; for example, mosses and lichens are easily killed. These forms may also be the first to recolonize the spill area, but colonizing moss taxa generally differ from those in unaffected areas (Walker et al. 1985a). Sedges and willows are the first vascular plants to reappear following an oil spill. At Prudhoe Bay regrowth of some sedges had occurred 1 year after an oil spill of 12 l m^{-2}; but 7 years after the spill, very little vascular plant cover had grown in either the moist or dry spill areas (Walker et al. 1985a). At a more southern site near Fairbanks, cottongrass (*Eriophorum vaginatum*) was successfully established, despite oil-saturated soils in the root zone (Johnson et al. 1980).

Areas receiving a heavier application of oil (as much as 60 l m^{-2}), which is typical of many oil spills, have soil-oil concentrations greater than 100 000 ppm, and are much more difficult to revegetate. The most effective means of revegetating these sites is an initial combination of tilling, to increase soil aeration, and a heavy application of high nitrogen and phosphorus fertilizer (Westlake et al. 1978; Atlas 1985; Brendel and Eschenbach 1985; Jorgenson and Cater 1991; Jorgenson et al. 1991a). These techniques increase the microbial action in the soil. Once the soil-oil content is reduced, then the area can be seeded with grasses. Recent advances in oil-spill cleanup techniques were used on two heavily oiled sites at Prudhoe Bay (Pope et al. 1982; Jorgenson et al. 1991a,b,c). These techniques include:

1. Initial containment with sandbags and absorbent material, surface flushing with warm and cold water, removal of oil and water with a vacuum truck, physical removal of contaminated debris by raking and swabbing with absorbent material, and use of plank walkways to limit trampling
2. Removal of oil-contaminated snow during winter
3. Cross-ripping, bulldozing, and scraping of frozen contaminated soils
4. Incineration of heavily contaminated soil in special furnaces
5. Bioremediation, consisting of deep tilling, fertilization, and control of site hydrology to promote microbial degradation of oil
6. Tundra rehabilitation with commercially available native grass-seed mixes and locally obtained mixes of sedges and legume seeds (Jorgenson et al. 1991a).

The effects of oil spills on permafrost are highly variable. Several investigators have reported that the depth of seasonal thaw in areas of oil spills differs little from unaffected areas (Freedman and Hutchinson 1976). At the Franklin Bluffs site, however, there was a marked difference in thaw 6 years after the spill. Within 80 m of the spill point, thaw deepened to more than 100 cm, whereas thaw depth in undisturbed sites was about 50 cm. Subsidence was apparent in a few areas, but was not severe, mainly because the entire spill area was underlaid by thick alluvial gravels (Walker et al. 1985a).

Recovery from diesel spills is extremely slow. At Fish Creek, 28-year-old diesel spills showed little vegetation recovery, significant depression of the permafrost, and strong diesel odors to a depth of at least 40 cm. Gas chromatography showed that a toxic component persisted in the soil after 28 years (Everett 1978). In contrast, spills of crankcase and crude oil showed good recovery after 28 years, except in the areas of heaviest impact.

Natural oil seeps offer analogues to anthropogenic oil spills. Studies of seeps at Cape Simpson (McCown et al. 1973b) found increased thaw and thermokarst in association with oiled areas. Lush patches of *Eriophorum scheuchzeri* and *Carex aquatilis* were found in association with the seeps. The warmer microclimates near the edge of dark oil pools apparently produced larger plants, advanced phenology, and more abundant fruiting in *Carex aquatilis*. Although the studies noted succession of plant communities in association with active seeps, they gave no details of the species involved. Studies of seeps in a variety of climate regimes and vegetation types could provide useful background information regarding long-term exposure to petroleum.

3.3.2.2 Seawater and Reserve-Pit Spills

Secondary recovery of oil at Prudhoe Bay involves transporting large quantities of seawater in elevated pipelines across the tundra for injection into wells to maintain pressure within oil-bearing strata (Simmons et al. 1983; Jorgenson et al. 1987). Small (2000 l) experimental seawater spills in dry, moist, and wet tundra sites at Prudhoe Bay showed an inverse relationship between soil moisture regime and absorption, and retention of salts (Simmons et al. 1983). In wet sites conductivities approached prespill levels within 30 days; salt water was quickly diluted and flushed from the soil. In contrast, dry sites tended to retain the salts, concentrating them at or near the seasonal thaw line.

ARCO Alaska Inc. conducted studies of a large (ca. 63 600 l) accidental spill in the Kuparuk oil field (Barker 1985; Jorgenson et al. 1987). The spill, which originally covered about 0.3 ha before freezing in December 1982, spread downhill, eventually affecting about 4.5 ha. All vegetation within 90 to 140 m of the spill point was killed (Barker 1985). After 3 years of recovery, *Eriophorum angustifolium* and *Dupontia fisheri* were the most common colonizers on areas of high impact. The only forbs returning in these areas were *Cochlearia officinalis* and *Stellaria laeta*. In moderately affected tussock tundra sites, *Eriophorum vaginatum* recovered well. Thaw depths were significantly greater in the brine-affected areas than in the control areas. Woody shrubs (*Dryas integrifolia* and *Salix* spp.) were strongly affected in all spill areas. Sedges, particularly *E. angustifolium*, showed the most resistance to impact. After 3 years, salts appeared to have leached from the high-impact zone and accumulated in lower soil layers downslope of the original spill. For the four common vegetation types at the Kuparuk spill, vegetation cover was reduced 20–80% where electrical conductivities were about 2–3 mmhos cm^{-1}, and cover was reduced more than 80% where conductivities rose above approximately 8 mmhos cm^{-1} (Jorgenson et al. 1987).

Natural analogues of seawater spills can be found along the coast in areas affected by storm surges. These areas are often colonized quickly by salt-tolerant taxa such as *Braya purpurascens, Cochlearia officinalis, Puccinellia andersonii, Salix ovalifolia, Stellaria humifusa, Fulgensia bracteata,* and *Thamnolia subuliformis*.

Salts are also major constituents of old reserve-pit fluids (French 1985). The primary cation in reserve-pit fluids at Prudhoe Bay is sodium, and the major anions are sulfate, chloride, and bicarbonate. The toxicity of these fluids varies considerably during the summer and among reserve pits (Myers and Barker 1984), which tend to decrease with the age of the pit because they become diluted by snow. At Prudhoe Bay the plant species most affected by reserve-pit fluids were the same as those plant species affected by salt-water spills (Simmons et al. 1983; Myers and Barker 1984). All sedges and most forbs examined were little affected, whereas woody plants, most noticeably *Salix reticulata* were most sensitive to elevated salt levels. High soil salinities also pose a problem for revegetating disturbed sites in the Prudhoe Bay area, where soil calcium carbonate salt levels are naturally high and summer precipitation is low. Initial investigations suggest that pore-water conductivities above 10 mmhos cm^{-1} severely restrict plant growth on all substrates (Jorgenson 1988).

The effects of summer and winter seawater spills can be considerably reduced by flushing with freshwater. For example, one spill that occurred in

December 1989 had salt-contaminated snow scooped from the site. During the following spring and summer, the site was repeatedly flushed with fresh water, resulting in negligible damage (M. T. Jorgenson, pers. comm.).

3.3.3 Fire

Lightning-caused tundra fires in the Alaskan Arctic have been described from the Yukon–Tanana uplands (Wein and Bliss 1973a; Wein 1975), the Seward Peninsula (Racine 1981), the Noatak River area (Racine et al. 1985), and the Arctic Foothills–Kokolik River area (Hall et al. 1978; Johnson and Viereck 1983). The size and importance of tundra fires (based on 25 years of records) varies widely within the Arctic. Tundra fires are rare on the Arctic Coastal Plain and in the Arctic Foothills and central and eastern Brooks Range, but are more common in northwestern Alaska (Gabriel and Tande 1983; Racine et al. 1985).

Recovery patterns vary considerably with local climate, vegetation type, and soil pH. Along a south–north gradient of decreasing temperature and lightning frequency, vascular plant cover plays less of a role in the recovery, and bryophytes increase in importance in both tussock tundra and low-shrub tundra (Racine et al. 1987). Within tussock tundra plant cover increases rapidly following light and moderate fires (Wein and Bliss 1973a; Fetcher et al. 1984; Racine et al. 1987). In one study plant regrowth produced 10–30% cover of vascular vegetation by the end of the first summer following a fire, 50% cover by the fifth year, and 100% cover by the tenth year (Racine et al. 1983, 1987). The recovery was rapid because the leaves of fire-resistant cottongrass (*Eriophorum vaginatum*) tussocks regrow quickly. Low shrubs resprouted more slowly than sedges and contributed less to early recovery. When severely burned, both cottongrass tussocks and shrubs may be killed, and recovery of the vegetation is slower (Johnson and Viereck 1983). Lichen-dominated tundra is predisposed to burning because of the high surface-to-volume ratio of lichens and rapid desiccation because of lack of roots. Lichens are slow to regrow in comparison with most vascular taxa and may require 50–80 years to redevelop a continuous cover (Auclair 1983).

Species involved in recovery vary with substrate pH. For example, *Rubus chamaemorus*, an acidophilous taxon, plays an important role in recovery on acidic sites on the Seward Peninsula, whereas willows are much more important at alkaline sites. A greater diversity of vascular plant species are involved in postfire succession on alkaline sites than on acidic sites including *Potentilla fruticosa, Gentiana propinqua, Equisetum scirpoides*, and *Thalictrum alpinum* (Racine et al. 1987).

Subsidence and thermal erosion following a fire in tussock tundra is usually minimal, because this tundra type generally occupies gentle slopes. Thaw stabilizes within 10 years; erosion and thermokarsting did occur at two areas on the Kokolik River burn site where massive ice was exposed as a result of fire (Johnson and Viereck 1983).

Seeding by vascular plants is an important revegetation mechanism at most burned sites. On the Seward Peninsula, sedges (*Carex* spp.) established from seed covered 50–100% of better-drained sites within 6 years after a fire (Racine et al. 1983). On other well-drained, willow-dominated sites on the Noatak River and Seward Peninsula, dense stands of fireweed (*Epilobium angustifolium*) developed within 2 years following a fire and have persisted for at least 10 years. Grasses (e.g., *Poa* spp., *Calamagrostis* spp., *Arctagrostis latifolia*) are locally important following fires in tussock tundra. Total primary production recovers quickly after most fires, but restoration of the original abundance of species is still not complete after 13 years at a burn site in interior Alaska (Fetcher et al. 1984). Further studies on the distribution of buried seeds and the nature of tundra plant perennating organs are needed for better prediction of recovery following fire.

3.3.4 Transportation Corridors

3.3.4.1 Bulldozed Tundra and Related Disturbances

During early oil exploration on the North Slope from the 1940s to the 1960s, most of the vehicles and road construction methods were borrowed from temperate regions and adapted to the cold climate. Long trails, such as the Oumalik trail from Barrow to Umiat and the Hickel Highway from the Yukon River to the North Slope (Fig. 3.3b), were commonly bulldozed across the tundra during both summer and winter (Reed 1958). These trails eventually formed deep ruts as ground ice melted and water formed ponds in the depressions. Bulldozing was also done locally for excavations, foundations, drill pads, aircraft runways, sewage disposal, and clearing snow (Lawson et al. 1978; Lawson 1982, 1986). Bulldozers were also used in building peat roads, which were made by removing peat from both sides of the road and piling the material in the center to form a raised surface (Fig. 3.3c). This construction method is used extensively in temperate areas, but was abandoned on the North Slope, because the scraped margins of the roads quickly subsided due to thermokarst along ice wedges. Extensive thermal erosion can be triggered by off-road vehicle movement on hill slopes (Mackay 1970; French 1974; Lawson and Brown 1979; Berg 1980; Lawson 1982). On the North Slope successional processes involved with recovery of bulldozed trails and thermokarst caused by oil development were studied at Fish Creek, Cape Thompson, and Oumalik. (Johnson et al. 1978; Komárková and Webber 1978; Lawson et al. 1978; Lawson 1982; Ebersole 1985; Everett et al. 1985).

The severity of bulldozer damage ranges from compression of microtopographic irregularities to removal of the entire vegetation mat and near-surface sediments. Bulldozing in ice-rich areas inevitably causes the permafrost to melt, the ground surface to subside, and depressions to fill with water, with virtually no possibility of returning the site to its original state. The amount of subsidence is a function of the ice volume, local topographic relief, and parent material (Lawson 1982). At Fish Creek and Cape Simpson, where there are relatively small amounts of massive ground ice and sandy materials, subsidence after 30 years was 0.4-2 m. At East Oumalik, where there are large amounts of ground ice and mainly silty sediments, the amount of subsidence after 30 years was 3-5 m. Hydraulic erosion commonly accompanies subsidence on slopes. At Fish Creek degradational processes probably diminished within 5-10 years after bulldozing had ceased. At East Oumalik, however, thermokarst is still expanding laterally in some areas because of hydraulic and thermal erosion and continued thawing of sediments. After 30 years, the disturbance covered at least twice the area of the initial disturbance (Lawson 1982). Similar expansion has also occurred along many bladed trails throughout the foothills (Fig. 3.3b).

Recovery on bulldozed sites depends largely on local climate and moisture regimes. At Cape Simpson, located within the cold littoral zone near the coast (mean July temperature <7 °C), recovery has been slow in the 30 years following disturbance, whereas sites inland have a lush growth of willows, grasses, and forbs (Ebersole 1985). Thirty years after wet meadows were bulldozed at Fish Creek, Oumalik, and many other old drill sites inland across NPR-A, subsidence has generally stopped, and vascular plant cover is nearly complete (Ebersole and Webber 1983; Komárková 1983). *Carex aquatilis* and *Eriophorum angustifolium* provide essentially all the cover where the soils are still very wet by the end of the summer, except at the arctic coast, where *Dupontia fisheri* is important. In moist tussock-sedge and mixed-shrub tundra, bulldozing often removes the tussocks and the underlying organic mat, which causes water to accumulate and converts the vegetation to wet sedge tundra composed of *Carex aquatilis* and *E. angustifolium*; there is little tendency toward a return to the original community (Johnson et al. 1978). In less severely disturbed areas where the bases of the tussocks and intertussock plants are intact, however, the tussocks slowly regrow. Seedlings establish slowly where the moist vegetation has been scraped away, leaving a thick organic mat. Seedling establishment is especially slow near the coast (Racine 1977; Johnson et al. 1978; Ebersole and Webber 1983; Ebersole 1989). At inland sites, willow seedlings are common. For example, willows are an important component of the vegetation cover on 30-year-old bulldozed areas at Oumalik (Ebersole 1985).

In many areas another consequence of bulldozing was the creation of mounds of vegetation and soil adjacent to the bulldozed sites, with habitats very different from the adjacent tundra. Mounds of bulldozed material up to 1.5 m high form well-drained microsites with higher soil temperatures than the surrounding tundra and thick active layers. Decomposition rates and nutrient availability are consequently higher (Lawson et al. 1978). For example, at Oumalik temperatures at a depth of 10 cm during 1 day averaged 9.4 °C in bulldozed material compared with 2.4 °C in adjacent tussock-sedge, dwarf-shrub tundra and 5.6 °C in wet sedge tundra (Ebersole and Webber 1983). Where the material forming the berms was mixed with mineral soil, plant cover is high, often composed mainly of nitrophilous grasses (*Arctagrostis latifolia, Poa arctica, Festuca brachyphylla*; Ebersole and Webber 1983). After 20–30 years, grasses and erect willows, particularly *A. latifolia, Salix alaxensis, S. glauca, S. lanata,* and *S. pulchra,* form a complete cover and grow more vigorously than in the surrounding undisturbed tundra (Ebersole 1985).

Animal dens and digging sites of bears are possible analogues of some bulldozed tundra sites (Wiggins and Thomas 1962; Peterson and Billings 1978; Gersper et al. 1980; Walker 1985). Many species occurring on these sites are nitrophilous (e.g., *Arctagrostis arundinacea, Festuca baffininsis, Phippsia algida, Poa glauca, Leymus villosissimus, Trisetum spicatum, Draba hirta, Artemisia tilesii, Arnica griscomii, Androsace septentrionalis, Polemonium acutifolium, Valariana capitata, Chrysosplenium tetrandrum, Saxifraga cernua, S. caespitosa, Rumex arctica,* and others) and are not common zonal tundra communities (Matveyeva 1979). The vegetation associated with such disturbances is distinguished by great vegetative vigor, dominance of herbaceous plants, lack of dwarf shrubs, and increased diversity of vascular plant species (Yurtsev and Korobkov 1979).

3.3.4.2 Off-Road Vehicle Trails

Although recent oil exploration and development was accompanied by a trend toward more permanent transportation corridors, all-season off-road transportation is still needed (Radforth and Burwash 1977; Abele et al. 1984). Industrial off-road vehicle (ORV) designs have become increasingly sophisticated to minimize the damage to tundra landscapes, but permanent damage is still possible unless sound guidelines are followed. Another concern is the cumulative effects of recreational one-passenger snow machines and motorized three- and four-wheel off-road vehicles. An estimated 15000 of these vehicles are currently in use in central and northern Alaska (Slaughter et al. 1990).

3.3.4.2.1 Summer Travel

The ORVs are involved in a wide range of summer activities, including seismic operations, drilling, mining, scientific research, reindeer herding, subsistence activities, and recreation. Vehicles vary in type and size and include tracked "weasels" and bulldozers, four-wheel-drive trucks, and vehicles with low-pressured tires, such as "Rolligons" and lightweight three-wheeled all-terrain cycles. Air-cushioned vehicles showed great promise for low-impact transportation on flat tundra (Abele et al. 1972; Sterrett 1976), but they have not been widely used on land, because they are expensive to operate and cannot traverse hilly areas.

Vehicle impact and recovery were monitored for up to 10 years following tests at Barrow, Lonely, and Prudhoe Bay (Abele et al. 1972, 1978a,b, 1984;

Walker et al. 1977). At Cape Thompson, vehicle trails, which were used for research between 1959 and 1961 and then abandoned, were examined in 1981 (Everett et al. 1985). In addition, a variety of 25- to 30-year old off-road-vehicle trails were studied at Fish Creek, Oumalik, and elsewhere on the North Slope (Hok 1969; Lawson et al. 1978; Lawson 1982; Ebersole 1985).

Abele et al. (1984) analyzed 10 years of observations involving six vehicle types (three low-tire-pressure Rolligon vehicles, two lightweight tracked vehicles, and one air-cushioned vehicle) at Barrow and Lonely, Alaska. The air-cushioned vehicle generally produced the least impact. All three Rolligons produced a longer-lasting impact than lightweight-tracked vehicles, mainly because the ground pressure for the Rolligons ($0.25-0.35\,kg\,cm^{-2}$) was higher than for small-tracked vehicles ($0.07-0.1\,kg\,cm^{-2}$). Recent studies of lightweight all-terrain vehicles (ATVs; 0.1 to $0.7\,kg\,cm^{-2}$) at Anaktuvuk Pass show that terrain damage in the mountains is less than on the coastal plain because of better draining soils with less massive ground ice.

Site moisture is particularly important to predict initial impact and recovery. At Prudhoe Bay and Lonely, single-pass Rolligon tracks through wet tundra were initially very visible, but recovered fully in 7 years (Walker et al. 1977; Abele et al. 1978b). In contrast, single-pass trails of overloaded Rolligons in wet tundra are still apparent after 10 years. After 20–30 years, deeply rutted tracks in wet tundra at Cape Thompson, Oumalik, Fish Creek, and Barrow seem to have reached thermal equilibrium and support typical wet tundra plant species (Abele et al. 1972; Lawson et al. 1978; Ebersole 1985; Everett et al. 1985). In some areas, this wet tundra differs in composition from the initial vegetation. In moist tundra at Prudhoe Bay, single-pass tracks of Rolligons compressed hummocks and ice-wedge polygon rims, causing damage that persisted for 5 years (Walker et al. 1985b). Moist tundra generally resists disturbance better than wet tundra, but it is less resilient once disturbed. Nutrient fluxes tend to increase, particularly in wet vehicle trails (Challinor and Gersper 1975; Chapin and Shaver 1981; Shaver and Chapin 1984).

The depth of summer thaw often increases after impact, but tends to rebound in later years. At Barrow the initial depression of a wet meadow surface was 15 cm in a 50-pass weasel track test. Seasonal thaw increased an average of 10 cm after 2 years, but returned to predisturbance levels within 10 years (Abele 1976). In wet meadows at Cape Thompson no significant differences in thaw depths were found in severely disturbed trails and adjacent controls after 20 years (Everett et al. 1985). The depressed surface of tracks in wet sedge tundra at Barrow and Cape Thompson have rebounded, presumably because of subsurface ice buildup (Abele 1976; Everett et al. 1985). Nearly complete vegetation recovery from vehicle traffic is likely within 10 years on flat tundra if the vegetation mat is not broken. This likelihood holds even for apparently serious damage to wet tundra vegetation as long as root systems are not damaged. However, if the mat is broken, or if runoff water is channeled into the track, recovery is likely to be slow. Where off-road vehicle tracks break and churn the organic mat over ice wedges, thaw and melting of the massive ice lead to subsidence and thermokarst formation (Abele 1976; Lawson et al. 1978; Lawson 1982; Abele et al. 1984).

Recent research on 20-year-old vehicle trails in the High Arctic has shown important differences from Low Arctic recovery patterns because of the High Arctic's shorter growing season and lower summer temperatures. Differences included limited reinvasion among rhizomatous graminoids in all but the wettest sites, and minimal seedling establishment. Moreover, strips wider than about 1 m remain mostly unvegetated, except for occasional clumps of mosses and small tufts of invading grasses (Forbes 1992).

3.3.4.2.2 Winter Travel

Snow and Ice Roads. In winter temporary roads are prepared by smoothing or compacting the snow surface to form a snow road or by spraying water on the surface to build up an ice layer that forms the roadway (Gas Arctic-Northwest Project Study Group 1973; Adam and Hernandez 1977; Johnson and Collins 1980; Johnson 1981). Such roads have been used in a variety of situations in northern Alaska. Explorations of NPR-A, for example, took place during the winter over unprepared snow trails, and prepared snow roads and ice roads. In another situation, a snow pad was built adjacent to the TAPS (trans-Atlantic Pipeline System) haul road during construction of a 230-km gas line during the winters of 1975–1976 and 1976–1977 (Brown and Berg 1980; Johnson 1981). Aircraft landing strips have also been prepared using packed snow or ice. Disturbances and recovery resulting from winter trails and snow and ice roads have been studied at Prudhoe Bay, Inigok in NPR-A, Seward Peninsula, along the TAPS haul road, and in the Arctic National Wildlife Refuge (ANWR; Buttrick 1973; Racine 1977; Johnson and Collins 1980; Johnson 1981). If left in place for only 1 year, ice roads are a good technique for minimizing damage, especially considering the alternatives.

A heavily traveled 40-km ice road was used to construct the drill pad at Inigok in spring 1978. When viewed from the air in summer 1978, the impact of the road seemed slight. In part, this impression was because the road traversed mostly tussock-sedge, dwarf-shrub tundra, where the compression of standing dead vegetation is less obvious than in wet sedge tundra. Ground observations 3 years after the disturbance, however, revealed a variety of vegetation impacts ranging from abraded and crushed tussocks to willows with broken or abraded terminal stems. Especially apparent was the alteration or destruction of the intertussock plant community, particularly the mosses (*Sphagnum* and *Dicranum*) and lichens. These species are apparently very susceptible to compression and breakage when frozen (K. R. Everett, unpubl. data). An ice road in the Arctic National Wildlife Refuge (ANWR) was kept in place for two winters, and one summer killed almost 100% of the vegetation (M. T. Jorgenson, pers. comm.).

Where snowpads are used for construction, considerable debris is deposited on their surfaces. For example, along the snowpads used for excavating the

TAPS fuel-gas pipeline and for constructing a section of TAPS as much as 90% of the pad along the fuel-gas line was covered with debris, and 25% of the TAPS snowpad was covered (Johnson and Collins 1980; Johnson 1981). Thaw depths under the pads increased for at least 3 years, reaching maximums of 22 cm deeper than controls along TAPS and 28 cm deeper along the gas line. The total vascular plant cover was reduced under the area of the snowpad during the first growing season after use, with erect shrubs (*Salix* spp. and *Betula nana*) and mosses showing extensive damage (Johnson 1981). Under the TAPS snowpad, many *Eriophorum vaginatum* tussocks were compressed or sheared off. Most of the vascular plants recovered after 3 years, but the amounts of mosses, lichens, and evergreen shrubs were still reduced compared with control areas (Johnson 1981).

Seismic Trails. All seismic activities on the North Slope are now conducted in winter, and most of the recent studies involving vehicle impact have been related to the effects of winter seismic operations (e.g., Reynolds 1982; Densmore 1985; Felix and Raynolds 1989a,b; Raynolds and Felix 1989; Felix et al. 1992). An area where winter vehicle trails are of particular concern is the 60 000 ha coastal plain portion of the ANWR, which was opened to winter seismic surveys in the winters of 1984–1985 and 1985–1986. More than 2000 km of trails were traversed within the refuge (Fig. 3.3d). The U. S. Fish and Wildlife Service (USFWS) studies of ANWR trails showed that 29% of the trails sustained medium-to-high levels of impact (Raynolds and Felix 1989). The most noticeably affected areas were river terraces dominated by *Dryas integrifolia*, terrain with considerable microtopographic relief caused by mounds, tussocks, hummocks, or high-centered polygons (Raynolds and Felix 1989). In places where there was little winter snow cover, as much as 87% of the vegetation cover was destroyed in swaths 10 m wide and up to 50 m long. Riparian shrub tundra was also heavily affected, mainly by breakage of the shrub canopy. Wet or partially vegetated areas were the least affected. Surface depressions to 15 cm deep occurred in wet areas (Felix and Raynolds 1989a). Evergreen shrubs (*Vaccinium vitis-idaea*, *Ledum palustre* ssp. *decumbens*) were among the most sensitive species, showing little recovery potential. In dry sites *Dryas integrifolia*, *Oxytropis* spp., *Astragalus* spp., *Equisetum variegatum*, and numerous mosses were strongly affected (Felix and Raynolds 1989a). Little resilience was seen in any of the vegetation types 4–5 years after the disturbance. The active layer remained deeper on plots in all nonriparian vegetation types, and most areas still had visible trails (Felix et al. 1992).

Disturbance and snow depth are generally inversely related; the amount of snow required to buffer disturbance also varies with vegetation type and impact intensity (e.g., seismic line vs camp move; Felix and Raynolds 1989b). Areas where tracks crossed local snow accumulations, such as the base of low terraces, show none of the damage noted in less-protected sites. During the seismic operations in ANWR, government regulations required an average snow depth of 15 cm before vehicles were permitted on the tundra, but more than 25 cm of snow is needed to minimally protect tussock tundra vegetation (Felix and Raynolds 1989b).

Follow-up studies in ANWR indicate that, despite efforts to minimize damage during the 1984–1985 seismic operations, long-term damage to the tundra did occur and is likely to remain for many years (Felix et al. 1992). Approximately 25% of all the trails showed little recovery between 1985 and 1988, and thaw settlement had not stabilized after 5 years. Once a track is present, a positive feedback is established such that the tracks become wetter, causing more heat gain, melting more ground ice, and further deepening the track. Avoidance of areas with low snow cover, use of lightweight vehicles, dispersed traffic patterns, and minimizing sharp turns and steep grades are the main recommendations to minimize damage (Felix et al. 1992). Revised government regulations regarding minimum snow depths for seismic operations would be effective. Aesthetics is a major consideration in highly protected areas such as national parks, refuges, and wilderness areas. The visible impact of traffic is the most prominent, long-lasting, and difficult-to-measure consequence of current seismic operations (see Fig. 3.3d).

3.3.4.3 Permanent Roads and Pads

The present approach to building roads and construction sites in northern Alaska is to create a thick – often more than 2 m thick – gravel pad to prevent thawing of the underlying permafrost. Although this design maintains the integrity of the permafrost, it causes other environmental impacts including (1) creation of dry elevated sites that are difficult to revegetate once the road or pad is abandoned, (2) creation of gravel-mine sites that also need to be revegetated, (3) blockage of natural drainage patterns, and (4) alteration of snow drift patterns. Other impacts associated with roads include gravel and dust spray, toxic spills, debris, and other disturbances. Sometimes the pads are constructed with rigid polyethylene insulation underneath to reduce the amount of gravel needed to insulate the original ground surface. If the pad is temporary, only a thin layer of gravel or sand is commonly used.

Elevated pads constructed on the open tundra are likely to remain unaltered for many hundreds of years and, for all practical purposes, represent a permanent change to the environment unless intensive site preparation provides adequate soils. Some of the older pads that have been abandoned at Prudhoe Bay with no revegetation efforts have developed very sparse vegetation communities with many of the same taxa that grow on gravel river bars and coastal beaches, e.g., *Arctagrostis latifolia*, *Artemisia alaskana*, *A. borealis*, *A. tilesii*, *Astragalus alpinus*, *Cochlearia officinalis*, *Deschampsia caespitosa*, *Epilobium latifolium*, *Equisetum arvense*, *Festuca rubra*, *Oxytropis campestris*, *Papaver lapponicum*, *Poa glauca*, *Sagina intermedia*, *Salix ovalifolia*, *Saxifraga cernua*, *S. oppositifolia*, and *Trisetum spicatum* (Walker et al. 1987a,b; Jorgenson 1988).

Revegetation of gravel pads at Prudhoe Bay is limited by cold summers, low rainfall during summer, and thick gravel fill that limits plant access to groundwater and nutrients (Johnson 1981; Jorgenson et al. 1991b). Revegetation with seeded grasses has been the most common form of gravel pad rehabilitation. Revegetation on gravel pads seeded with commercially available grasses was generally more successful in the southern portion of the coastal plain and in the foothills (Johnson 1981). Fertilization increases plant colonization rates. At test sites in NPR-A, which were drilled and abandoned in 1976, drill pads, runways, and roads have been revegetated with a mix of exotic and native grasses (*Festuca rubra, Arctagrostis latifolia, Poa glauca, Poa pratense*), and most sites were fertilized (McKendrick 1987). Considerable vegetation cover was observed between 1977 and 1984 at sites that were reseeded and fertilized repeatedly. At Prudhoe Bay, recolonization rates increased from 0.05% cover per year without fertilization to 3.7% after fertilization (Jorgenson 1988). The best development of artificial vegetation cover occurred in depressions on the pads where moisture is more available.

Plant cover is inversely related to fill thickness; it is very low on sites with more than 1 m of gravel, because groundwater availability is low (Jorgenson 1988). Recent rehabilitation of abandoned gravel pads at Prudhoe Bay has incorporated a variety of hydrological manipulations to improve the arid conditions on raised gravel surfaces including (1) creation of 1-m high gravel berms to capture drifting snow to increase water input to the soil, (2) adding organic matter to improve water-holding capacity, and (3) applying mulches to reduce evaporation (Jorgenson et al. 1991b, 1992a). Sewage sludge from wastewater treatment plants is also being considered (Jorgenson et al. 1991b).

Other techniques for rehabilitating thick gravel pads being tested at Prudhoe Bay include fertilization and colonization by local native forbs, native grass cultivars, indigenous wetland sedges, and transplanted tundra plugs combined with gravel removal to reestablish the original hydrological conditions (Jorgenson and Kidd 1991). Early results from these studies indicate that land rehabilitation in the Low Arctic is feasible and can be relatively rapid, even on extreme sites, when adequate soil and hydrological conditions are provided (Jorgenson and Cater 1992). Nevertheless, we do not have enough information presently to develop rehabilitation prescriptions for all sites (Wyant and Knapp 1992).

3.3.4.4 Gravel Mines

Gravel used for pads must sometimes be hauled long distances from rivers or open-pit mines. During the construction of TAPS, more than 78 500 ha of land were disturbed; 31 700 ha, or 40%, was for gravel borrow sites (Pamplin 1979). River bars have historically been the primary gravel source for most of the roads and gravel pads in northern Alaska (Fig. 3.3e). Because they are high-energy environments with a predominance of early successional communities, they can recover relatively quickly from disturbance if proper guidelines are followed (Woodward and Clyde Consultants 1980). Four factors relating to changed hydrology constitute impediments to reestablishing vegetation in mined riparian areas: (1) permanent or annual flooding, (2) increased frequency and duration of temporary flooding, (3) long-term channel changes (increased braiding and channel width and decreased channel stability), and (4) new or increased formation of riving icings (Joyce 1980). Factors that promote recovery of the vegetation and faunal communities include (1) siting the mines in large and medium-width channels flowing in a braided pattern, (2) shallow scraping of surface layers over broad areas, and (3) using organic-rich soils to enhance recovery either by broadcasting it over the ground or forming piles that can become temporary small-mammal and passerine habitat (Joyce 1980).

Gravel borrow pits have many of the same site characteristics as roads. Both nutrients and moisture are frequently limiting, but unlike roads, borrow pits are rarely elevated above the general tundra surface, so revegetation is somewhat easier. If fine-grained material is placed over the gravel, the areas can usually be revegetated (Johnson 1981). Temporary pads and borrow pits in areas with frequent natural disturbance, such as floodplains, are more quickly revegetated, because local seed sources are able to colonize river gravels (Johnson 1981).

Johnson (1987) emphasizes the importance of planning gravel sites carefully, including: (1) selecting the site to minimize visual impacts and promote reinvasion by native species, (2) limiting the area of impact, (3) reducing the need for gravel through innovative engineering designs, and (4) reclamation plans that both ensure erosion control and reestablish as natural a vegetation cover as possible. Important aspects of the reclamation effort include reconstructing soil by spreading stockpiled organic matter and fine-grained mineral soil over the site and seeding or transplanting adapted, preferably native, species.

3.3.4.5 Native Species in Revegetation of Gravel Pads and Mines

Use of native species is increasing for reclamation of gravel pads, mined areas, and disturbed roadsides (Chapin and Chapin 1980; Johnson 1981, 1984, 1987; Shaver et al. 1983; Densmore and Holmes 1987; Densmore et al. 1987; Jorgenson 1988; Jorgenson and Kidd 1991). Early revegetation efforts in the Arctic attempted to identify commercially available grass species that were best adapted to the cold climate to help control erosion (Younkin 1972, 1976; Hernandez 1973; McKendrick and Mitchell 1978). Later consideration focused on native species (Mitchell 1979; Johnson 1981; Jorgenson and Kidd 1991).

Most commercial seed mixes are designed to create a fast-growing cover of grasses, but establishment of native species can be restricted, at least in the short term, by competition from commercial grasses (Johnson 1981; Densmore et al. 1987; Densmore 1992). Recent studies of the long-term effects of seeded grasses in floodplain mines suggest that the grass treatment alters the pattern of succession (Densmore 1992). A gravel mine site on a floodplain of the Atigun River that was scarified, fertilized, and seeded with a commercially available seed mix had only about 5% cover of *Festuca rubra* after 4 years and 1.5% cover after 11 years. Invasion by native species was greatly reduced on seeded areas. Many species, particularly nonlegume forbs such as *Epilobium latifolium*, failed to establish on the seeded areas, and shrubs such as *Salix glauca* and *S. alaxensis* were severely inhibited. Leguminous plants, such as *Astragalus alpinus*, were apparently unaffected by the treatment. Factors contributing to the differences were thought to be (a) initial competition by the grasses for soil moisture and nutrients, and (b) the high moss cover on the fertilized areas, which dried the surface, raised soil temperature, and prevented seedling establishment. Densmore (1992) suggests that seeding be used only on tundra sites where immediate stabilization is needed to prevent erosion. Revegetation is generally best achieved on untreated floodplain sites that are allowed to recolonize by natural riparian successional processes (Bliss and Cantlon 1957; Viereck 1966; Peterson and Billings 1978, 1980; Moore 1983).

Recent studies at Prudhoe Bay have used a variety of native plants including plugs from wet sedge meadows, seed from hydrophytic sedges and grasses, native legumes, and cuttings of *Salix ovalifolia* and *S. arctica* (Jorgenson and Cater 1991; Jorgenson and Kidd 1991). Willows (*Salix alaxensis*) have also been used to restore moose habitat lost to pipeline construction along the Sagavanirktok River (Densmore et al. 1987). Some attempts were made using willows on abandoned TAPS gravel pads, but these attempts were mostly unsuccessful (Johnson 1981).

Surveys of abandoned gravel pads and disturbed sites along the Dalton Highway, Dempster Highway, Canol Pipeline, and Denali National Park have identified several other native species that conceivably could be useful for revegetation efforts on gravel sites. Particularly good candidates include *Arctagrostis latifolia, Artemisia arctica, A. borealis, A. tilesii, A. glomerata, Astragalus alpinus, A. aboriginum, A. eucosmus, Braya purpurascens* (in alkaline coastal areas), *Calamagrostis canadensis, Deschampsia caespitosa, Epilobium latifolium, Festuca altaica, Festuca brachyphylla, Hedysarum alpinum, Oxytropis campestris, O. nigrescens, Potentilla fruticosa, Salix alaxensis, S. ovalifolia,* and *Trisetum spicatum* (Everett et al. 1985; Walker 1985; Densmore and Holmes 1987; Kershaw and Kershaw 1987; McKendrick 1987). There is considerable local variation in the pool of species colonizing disturbed riparian sites and gravel pads. Summer temperatures are particularly important. In a survey of 12 revegetated pads in NPR-A, an average of only three native species was found on pads at the cold coastal sites. An average of 10 native species was found on inland coastal-plain pads, and an average of 24 native species occurred on relatively warm foothill sites

(McKendrick 1987). Few willow species occur near the coast, and slow growth of all woody species limits the height and recovery potential of shrubby communities at the coast (Walker 1987). Detailed studies of the dynamics of succession in relation to the coastal temperature gradient and alkaline vs acidic substrates would be useful.

The buried seed bank is a potentially important source of native seeds. Large numbers of native plant seedlings appear on recently disturbed sites before any external seeds arrive (Chester and Shaver 1982; Gartner et al. 1983). These seedlings are concentrated on organic soils and probably come from buried seeds held in the organic layer of undisturbed tundra (McGraw 1980; Gartner et al. 1983; Ebersole 1989). Replacement of organic soil matter after a disturbance is particularly important to native plant recovery, because the organic layer contains the principal native plant seed source and because native plant growth rates are higher in organic soils than in mineral soils.

The germination strategies of many arctic plants appear to be useful for seedling establishment on disturbances (Densmore 1979; Gartner 1983). The highest germination rates are at high temperatures (25–35 °C) and in sunlight. This observation suggests that many of these species are preadapted for colonizing disturbances.

Studies at cold coastal sites have not found the buried seed bank to be a useful means for revegetating disturbed sites, possibly because of the reduced viability of the seeds (M. T. Jorgenson, pers. comm.). Weeds are not important in the buried seed bank, but they were observed at a number of seeded sites along the trans-Alaska pipeline, where they were introduced in both straw mulch and in the seed mix. On the North Slope weeds generally do not persist (Kubanis 1980; Johnson 1981). An experiment in which grids of toothpicks were implanted in various substrates showed that the mineral substrate was much less stable than the organic substrate (Gartner et al. 1983). Thus, adding organic matter to an artificial seedbed may increase the success of seedlings by reducing needle ice or other unfavorable physical factors. Fertilizing adjacent strips of undisturbed tundra may also promote seedling establishment, but a dependable strategy for encouraging seed production is currently not available.

3.3.4.6 Road Dust

Road dust on arctic tundra is a relatively recent phenomenon associated with high-speed gravel highways (Walker and Everett 1987; Auerbach 1992). Road dust, like the regional loess, is alkaline; calcium and magnesium are the most abundant cations (Chap. 15, this Vol.). Everett (1980) measured dustfall and wind along the Dalton Highway and Prudhoe Bay Spine Road and found that dust loads 1000 m from the road were several times higher in the Prudhoe Bay region than at similar distances from the road at other sites along the Dalton Highway. This difference is an effect of the dense, heavily traveled road network at Prudhoe Bay with road dust coming from many sources. Over the 5 years of Everett's study, the normally high buffering capacity of this tundra, which is due to large amounts of exchangeable hydrogen and aluminum, was neutralized in some areas of heavy dustfall. In the upper 2–5 cm, soil pH has shifted from acid to alkaline. A less dramatic and somewhat erratic buildup of other cations, such as potassium, was also documented. Soil chemical changes induced by road dust were measurable only after 3 or more years of impact in the high dustfall zone (0–30 m). Changes in areas of lower dustfall will take much longer to become recognizable.

In some areas along the most heavily traveled roads at Prudhoe Bay, all vegetation has been totally eliminated within 5 m of the road. Mosses were eliminated to a distance of about 20 m. A few dune and coastal species, such as *Elymus arenarius*, *Braya purpurascens*, *Puccinellia andersonii*, and *Armeria maritima*, are colonizing these areas. Beyond about 100 m in alkaline tundra, the vegetation, although heavily dusted, appears to have survived with little compositional change (Walker and Everett 1987). The loss of vegetation near the road is partially responsible for the extensive thermokarst features and high-centered polygons that have developed along older roads (Fig. 3.3f). Thermokarst features occur mostly within 25 m of the road at Prudhoe Bay, but in some areas, thermokarst is actively expanding into the tundra within 100 m from the road.

The changes at Prudhoe Bay are occurring in a tundra adapted to high influxes of natural loess (Walker and Everett 1991). In acidic tundra, road dust significantly reduces and often eliminates *Sphagnum* moss, especially in the 0–10 m adjacent to the road (Spatt 1978; Werbe 1980; Spatt and Miller 1981; Auerbach 1992). This effect is due to the toxic effects of calcium in the dust (Clymo 1973) and reduced photosynthetic rates of the moss in heavily dusted areas (Spatt and Miller 1981). Ten years after road construction, *Ceratodon purpureus*, *Bryum pseudotriquetrum*, and *Polytrichum juniperinum* had replaced *Sphagnum* as the most common mosses near the road (Walker et al. 1985a). After 15 years, *Sphagnum* spp. has been replaced by *Aulacomnium turgidum* as much as 100 m from the road near Toolik Lake, Alaska (Auerbach 1992). Other plant species also react negatively to road dust (e.g., *Cassiope tetragona* and *Cladonia* spp.). Some plants, on the other hand, especially several minerotrophic moss species (e.g., *Drepanocladus*, *Scorpidium*, and *Catascopium*), respond positively to the increased nutrients.

3.3.4.7 Roadside Impoundments

Within the Prudhoe Bay region, most instances of road-related flooding occur where the road crosses low-lying, vegetated, drained thaw-lake basins. Drainage patterns on the flat tundra are complex, with many unconnected drainage systems. In areas with an intersecting web of roads, flooded areas are more common and often extremely difficult to drain. For example, a 7.0-km road was constructed in 1980–1981 through a flat coastal portion of the Prudhoe Bay oil field, flooding 134 ha (18 ha km^{-1} of road; Fig. 3.3f; Klinger et al. 1983).

Culverts blocked by snow and ice prolong the flooding period during the spring. Detailed hydrological maps and observations during the melt are helpful in deciding where to place culverts.

Another form of flooding occurs in narrow strips along the margins of most Prudhoe Bay roads. This type of flooding is accentuated if the ground subsides, either when the road settles or thermokarst develops. Numerous roadside areas along the Spine Road are covered by thermokarst features flooded 30–100 cm deep above thawed ice wedges (see Fig. 3.3f). The thermal disturbance results from the flooding and the loss of vegetation buried by road dust.

Most flooding is confined to wet and aquatic tundra vegetation. The most notable effect is a greening of the vegetation (Klinger et al. 1983). Many flooded areas drain by midsummer so major damage may not occur, although the plant communities do change in a manner that corresponds to communities along moisture gradients in thaw-lake basins. Mesic species (e.g., *Eriophorum triste*, *Dryas integrifolia*, *Tomenthypnum nitens*, *Thamnolia subuliformis*) are often replaced by hydric taxa (e.g., *Carex aquatilis* and *Eriophorum angustifolium*). Vegetation is killed in areas of prolonged and deep flooding. In other areas moist microsites, such as polygon rims, hummocks, and strangmoor, are eliminated by flooding. Such features are particularly important nesting sites for several species of birds.

An understanding of vegetation succession in thaw lakes, both during their formation and after drainage, is important for properly managing coastal wetlands in areas of development. Ongoing observations of the dominant species in drained thaw lakes at Barrow (*Dupontia fisheri*, *Carex aquatilis*, *Eriophorum angustifolium*, and *Arctophila fulva*) are helping to establish the time scales and mechanisms involved with succession (Billings and Petereson 1980). At Prudhoe Bay techniques are being tested to establish common aquatic grass, *Arctophila fulva*, along disturbed creeks, lake shores, and artificial ponds (McKendrick 1990; Moore 1990). The patterns of vegetation succession in drained thaw lakes are affected by the substrate, local climate, and manner of drainage. For example, *Dupontia fisheri* is an important colonizer at the coast. Where the climate is warmer, a wide variety of forbs, willows, birch, alders, aquatic taxa, and *Sphagnum* play major roles in succession. Substrate also varies across the coastal plain. Areas that have sandy lake bottoms have very different successional processes than the marine clays and peaty substrates in the Barrow region.

Recent efforts by the oil industry to create wetland habitat in impounded gravel mine sites have applied a suite of rehabilitation techniques to (1) restore the sites as fish habitat, (2) create littoral zones for juvenile fish and shorebirds, (3) create wetlands for waterbird habitat, and (4) revegetate overburdened piles with native grasses and forbs for forage and nesting (Jorgenson et al. 1992b). Much of this work relies on detailed knowledge of life history and habitat characteristics of species involved in wetland succession (McKendrick 1990). More work is needed to document variations in successional schemes, not only of the vegetation, but also of other components, such as invertebrates, fish, and wildlife populations, which contribute to total wetland function.

3.3.5 Cumulative Impacts

Perhaps the most severe long-term effects in northern Alaska are the incremental disturbances to vast areas of undisturbed tundra and the tendency for development to focus on sites that also have high wildlife and vegetation diversity. For example, pingos are high points on the flat coastal plain that are often heavily disturbed because they are frequently used for survey bench marks. They deserve full protection from oil-field activities because they are unique landforms containing concentrations of wildlife habitat, rare soils, and forb meadows (Walker 1990). Severe cumulative impacts also commonly occur in biologically rich riparian areas.

Cumulative effects of North Slope oil fields have only recently been explicitly addressed (U. S. Department of Interior 1983; Meehan 1986; Shideler 1986; Walker et al. 1986, 1987b; Speer and Libenson 1988). Cumulative impacts differ from simple additive effects in five major ways: (1) the disturbances are so close in time that the system cannot recover (time-crowded perturbations); (2) the disturbances so are closely spaced that the effects are not fully dissipated between them (space-crowded perturbations); (3) combinations of disturbances produce effects that are different from those of any single disturbance (synergistic effects); (4) the effects are delayed in time or space from the original disturbance (indirect effects); and (5) large regions are affected by many seemingly insignificant perturbations (nibbling; Beanlands et al. 1986). Cumulative impacts go beyond ecology because socioeconomic, legal, jurisdictional, and policy issues are affected by and control ecosystem impacts (Cline et al. 1983; Lee and Gosselink 1988; Stakhiv 1988). Landscape and regional perspectives are needed to expand the area of concern beyond the immediate site of impact (Preston and Bedford 1988; Wyant and Knapp 1992).

On the North Slope, cumulative impact research has focused on enumerating and mapping the history of development in the region (Walker et al. 1986, 1987b), cumulative effects of development on shorebirds (Meehan 1986), caribou (Shideler 1986), and discharge of reserve-pit fluids into tundra wetlands (West and Snyder-Conn 1987). The problems relating to the development of large oil fields include direct and indirect impacts of water pollution, air pollution, hazardous and other wastes, wildlife impacts, restoration problems, regulatory problems, and lack of aggressive enforcement of existing environmental regulations (Speer and Libenson 1988).

Expanding networks of roads and pipelines (Fig. 3.3h) have many cumulative effects (e.g., Berger 1977; Pamplin 1979; Brown and Berg 1980; Johnson 1981). The Prudhoe Bay Oil Field, Kuparuk Oil Field, and several nearby smaller oil fields affect approximately 1000 km², and contain the most extensive road network in the North American Arctic. As of 1991 approximately 12 000 ha had been directly affected by gravel filling or gravel

mines (Wyant and Knapp 1992). The physical disturbances related to roads
include elimination of habitat beneath the roadbed, dustall, roadside trash,
and deposition of plowed material and gravel on the adjacent tundra. In the
Prudhoe Bay region, indirect impacts, such as flooding, road dust, and roadside thermokarst, affect an area equal to nearly 60% of the directly affected
tundra, and in some parts of the region, exceed the area of planned impacts
especially in flat sections of the coastal plain where impoundments are a major
problem (Walker et al. 1986, 1987b). Many extreme examples of cumulative
impacts to tundra systems have been described for the oil and gas fields in the
northern Tyumen Oblast near the Ob River embayment in northwestern Siberia (Vilchek and Bykova 1992). There are currently no accepted scientific or
regulatory means for evaluating or predicting cumulative impacts (Wyant and
Knapp 1992).

3.4 Conclusions

Many of the scientific issues surrounding direct physical disruptions to tundra
ecosystems have been addressed during the past 3 decades (Alexander
and VanCleve 1983). The interconnectedness of physical and biological
components of the ecosystem has been recognized along with the necessity
of first stabilizing the physical system (permafrost and soils) before the
biological components can stabilize. Even seemingly minor physical
perturbations can alter surface hydrology and energy budgets, resulting in
thermokarst or thermal erosion. Nearly all natural disturbances in the Arctic
are either directly or indirectly driven by climate and mediated by hydrological
processes. Snowpack, soil moisture, and runoff patterns control the vast
majority of vegetation patterns, primary production, and the patterns of
animals, yet the links between arctic vegetation, permafrost, energy, and water
are still poorly understood. Because the physical and biological components
are so closely linked in arctic systems, and because both components could
be strongly affected by climate change, it is imperative that we develop
the means to accurately model the linkages between the vegetation and the
physical system.

The advent of environmental laws from the 1960s to the 1990s substantially changed the character of development on the North Slope, substantially
reducing many of the impacts common in the early phase of oil exploration.
On the other hand, some predictions in the 1970s by environmentalists may in
the long-term prove correct. George West, in an early review of environmental
problems associated with Arctic Alaskan oil development, wrote:

Perhaps the greatest impact on arctic ecosystems is simply the increased inter-
vention of the human population. Where native people were previously only
sparsely settled or nomadic in the tundra, and on coasts where they tended to

congregate, now the economic need for resources has resulted in increased pressure overall which will result in decreasing habitats for wildlife, destruction of wilderness areas, and increased access to humans for further exploration and recreation. (West 1976: 223)

While we have seen a heightening of environmental concern, the development of engineering and planning techniques to minimize small-scale direct impacts to the tundra, and the adoption of rehabilitation techniques to return disturbed tundra areas to their original condition, we have simultaneously seen a many-fold increase in the number of impacts and size of the areas affected. A focus on issues relating to cumulative impacts is needed. Part of this will require identifying and protecting areas of high biodiversity and important corridors of migration, as well as large areas representative of the diversity of arctic landscapes.

Much still needs to be done to fine-tune our knowledge of tundra ecosystem processes and to develop the means to rehabilitate tundra regions disturbed by energy development. Just as we cannot apply methods of revegetation developed in the Low Arctic to polar deserts of the High Arctic (Bliss and Peterson 1973; Bliss and Grulke 1988), we need to critically evaluate the variation that exists within Low Arctic tundra regions in order to develop workable strategies for revegetation and restoration. Revegetation of disturbed sites requires a fuller appreciation of the rates and dynamics of succession on a variety of substrates and climates. Perhaps the greatest need is a better understanding of how arctic landscapes function and how this function can best be maintained in the presence of large-scale development.

Basic research is also needed to understand the natural processes that have created existing arctic landscapes. Most anthropogenic perturbations have good natural analogues, although the scale of the analogues may be very different. Arctic landscapes at all scales are in states of natural succession (Churchill and Hanson 1958), and the North Slope is an excellent place to study these processes because large landscapes still exist that are virtually unaltered by human influence. These landscapes are extraordinarily valuable because they are some of the only easily accessible natural scientific laboratories in the world where natural disturbance processes still dominate.

Acknowledgments. This work was inspired by the 7th Conference of the Comité Arctique International, entitled *Restoration and Vegetation Succession in Circumpolar Lands*, held in Reykjavik, Iceland, 7–13 September 1986 (Webber and Fridriksson 1987), which noted a lack of an up-to-date synthesis of arctic vegetation disturbance and recovery research. Jerry Brown, former Chief, Earth Sciences Branch at the US Army Cold Regions Research and Engineering Library (CRREL), organized an earlier review (Walker et al. 1987b), and this chapter updates that work. Without the efforts of many individuals participating in the CRREL-sponsored review, this chapter would not have been possible. Individuals who contributed their time and energies to either the CRREL report or this document include Gunars Abele (deceased), Dianne Andrews, Max Brewer, Dave Cate, Terry Chapin, Fred Crory, Jim Ebersole, Kaye Everett (deceased), Barbara Gartner, Dick Haugen, Al Johnson, Larry Johnson, Torre Jorgenson, Mike Joyce, Vera Komárková, Dan Lawson, Leanne Lestak, Barbara Murray, David Murray, Chuck Racine, Gus

Shaver, John Schindler, Marilyn Walker, and Patrick Webber. The Department of Energy R4D Project (grant no. DE-FG02-84ER60242) sponsored this synthesis.

References

Abele G (1976) Effects of Hovercraft, wheeled and tracked vehicle traffic on tundra. In: Nat Res Counc Can Assoc Comm Geotech Res, Tech Memo 116: 186–215

Abele G, Parrott W, Atwood D (1972) Effects of SK-5 Air Cushion Vehicle operations on organic terrains. US Army Cold Reg Res Eng Lab, Intern Rep 269

Abele G, Johnson LA, Collins CM, Taylor RA (1978a) Effects of winter military operations on cold regions terrains. US Army Cold Reg Res Eng Lab, Spec Rep 78-17

Abele G, Walker DA, Brown J, Brewer MC, Atwood DM (1978b) Effects of low-ground pressure vehicle traffic on tundra at Lonely, Alaska. US Army Cold Reg Res Eng Lab, Spec Rep 78-16

Abele G, Brown J, Brewer MC (1984) Long-term effects of off-road vehicle traffic on tundra terrain. J Terrain Mechanics 21(3): 283–294

Adam KM, Hernandez H (1977) Snow and ice roads: ability to support traffic and effects on vegetation. Arctic 30(1): 13–27

Alexander V, Van Cleve K (1983) The Alaska pipeline: a success story. Annu Rev Ecol Syst 14: 443–463

Atlas RM (1985) Effects of hydrocarbons on microorganims and petroleum biodegradation in arctic ecosystems. In: Engelhardt FR (ed) Petroleum effects in the arctic environment. Elsevier, New York, pp 63–99

Auclair AND (1983) The role of fire in lichen-dominated tundra and forest-tundra. In: Wein WR, MacLean DA (eds) The role of fire in northern circumpolar ecosystems. Wiley, New York

Auerbach NA (1992) Effects of road and dust disturbance in minerotrophic and acidic tundra ecosystems, northern Alaska. MBS Thesis, Univ Colorado, Boulder

Barker M (1985) Two-year study of the effects of winter brine spill on tussock tundra. Final Rep, ARCO Alaska Inc, Anchorage

Beanlands GE, Erckmann WJ, Orians GH, O'Riordan J, Policansky D, Sadar MH, Sadler B (eds) (1986) Cumulative environmental effects: a binational perspective. Can Environ Assess Res Counc, US Natl Res Counc, Ottawa, Ontario and Washington DC

Berg RL (1980) Road performance and associated investigations. US Army Cold Reg Res Eng Lab, Rep 80-19: 53–100

Berger TR (1977) Northern frontier, northern homeland. Report of the Mackenzie Valley pipeline inquiry, vol 2. Terms and conditions. Minister of Supply and Services, Ottawa, Canada

Billings WD, Peterson KM (1980) Vegetational change and ice-wedge polygons through the thaw-lake cycle in arctic Alaska. Arct Alp Res 12: 413–432

Bliss LC, Cantlon JE (1957) Succession on river alluvium in northern Alaska. Am Midl Nat 58: 452–469

Bliss LC, Grulke NE (1988) Revegetation in the High Arctic: its role in reclamation of surface disturbance. In: Kershaw P (ed) Northern environmental disturbances. Boreal Inst North Stud, Edmonton, Canada, pp 43–55

Bliss LC, Matveyeva NV (1992) Circumpolar arctic vegetation. In: Chapin FS, Jefferies RL, Reynolds JF, Shaver GR, Svoboda J (eds) Arctic ecosystems in a changing climate: an ecophysiological perspective. Academic Press, San Diego, pp 59–89

Bliss LC, Peterson EB (1973) The ecological impact of northern petroleum development. In: Proc 5th Int Congr on Arctic oil and gas: problems and possibilities, Le Havre, May 1973

Brendel JE, Eschenbach TG (1985) Check Valve 23 revegetation study. In: Lewis ML (ed) Proc 8th Annu Meet International Society of Petroleum Industry Biologists, Northern hydrocarbon development environmental problem solving. Banff, Alberta

Brown J, Berg RL (eds) (1980) Environmental engineering and ecological baseline investigations along the Yukon River–Prudhoe Bay Haul Road. US Army Cold Reg Res Eng Lab, Rep 80-19

Buttrick SC (1973) The ecological effects of vehicular traffic on frozen tundra. MS Thesis, Ohio State Univ, Columbus

Challinor JL, Gersper PL (1975) Vehicle perturbation effects upon a tundra soil-plant system II. Effects on the chemical regime. Soil Sci Soc Am 39: 689–695

Chapin FS, Chapin MC (1980) Revegetation of an arctic disturbed site by native tundra species. J Appl Ecol 17: 449–456

Chapin FS, Shaver GR (1981) Changes in soil properties and vegetation following disturbance of Alaskan arctic tundra. J Appl Ecol 18: 605–617

Chapin FS, Jefferies RL, Reynolds JF, Shaver GR, Svoboda J (eds) (1992) Arctic ecosystems in a changing climate: an ecophysiological perspective. Academic Press, San Diego, CA

Chester AL, Shaver GR (1982) Seedling dynamics of some cottongrass tussock tundra species during the natural revegetation of small disturbed areas. Holarct Ecol 5: 207–211

Churchill ED, Hanson HC (1958) The concept of climax in arctic and alpine vegetation. Bot Rev 24: 127–191

Cline EW, Vlachos EC, Horak GC (1983) State-of-the-art and theoretical basis of assessing cumulative impacts on fish and wildlife. Rep prepared for the Eastern Energy and Land Use Team, Office of Biological Services, Fish and Wildlife Service. Dynamac Corporation, Ft. Collins, CO

Clymo RS (1973) The growth of *Sphagnum*: some effects of environment. J Ecol 61: 849–869

Deneke FJ, McCown BH, Coyne PI, Rickard W, Brown J (1975) Biological aspects of terrestrial oil spills – US Army CRREL oil research in Alaska, 1970–1974. US Army Cold Reg Res Eng Lab, Res Rep 346

Densmore RV (1979) Aspects of the seed ecology of woody plants of the Alaskan taiga and tundra. PhD Thesis, Duke Univ, Durham

Densmore RV (1985) Effects of dynamite and vibrator seismic exploration on visual quality, soils, and vegetation of the Alaskan north slope. Final Rep Geophys Serv Inc, Envirosphere Co, Anchorage

Densmore RV (1992) Succession on an Alaskan tundra disturbance with and without assisted revegetation with grass. Arct Alp Res 24: 238–243

Densmore RV, Holmes KW (1987) Assisted revegetation in Denali National Park, Alaska USA. Arct Alp Res 19: 544–548

Densmore RV, Nieland BJ, Zasada JC, Masters MA (1987) Planting willow for moose habitat restoration on the North Slope of Alaska, USA. Arct Alp Res 19: 537–543

Ebersole JJ (1985) Medium- and short-term recovery of Low Arctic vegetation, Oumalik, Alaska. PhD Thesis, Univ Colorado, Boulder

Ebersole JJ (1987) Short-term vegetation recovery at an Alaskan arctic coastal plain site. Arct Alp Res 19: 442–450

Ebersole JJ (1989) Role of the seed bank in providing colonizers on a tundra disturbance in Alaska. Can J Bot 67: 466–471

Ebersole JJ, Webber PJ (1983) Biological decomposition and plant succession following disturbance on the Arctic Coastal Plain, Alaska. In: Permafrost: Proc 4th Int Conf. National Academy Press, Washington, DC, pp 266–271

Everett KR (1978) Some effects of oil on the physical and chemical characteristics of wet tundra soils. Arctic 31: 260–276

Everett KR (1980) Distribution and properties of road dust along the northern portion of the Haul Road. US Army Cold Region Res Eng Lab, Rep 80-19: 101–128

Everett KR, Murray BM, Murray DF, Johnson AW, Linkins AE, Webber PJ (1985) Reconnaissance observations of long-term natural vegetation recovery in the Cape Thompson region, Alaska, and additions to the checklist of flora. US Army Cold Reg Res Eng Lab, Rep 85-11

Felix NA, Raynolds MK (1989a) The effects of winter seismic trails on tundra vegetation in northeastern Alaska, USA. Arct Alp Res 21: 188–202

Felix NA, Raynolds MK (1989b) The role of snow cover in limiting surface disturbance caused by winter seismic exploration. Arctic 42: 62–68

Felix NA, Raynolds MK, Jorgenson JC, DuBois KE (1992) Resistance and resilience of tundra plant communities to disturbance by winter seismic vehicles. Arct Alp Res 24: 69–77

Fetcher N, Beatty TF, Mullinax B, Winkler DS (1984) Changes in arctic tussock tundra thirteen years after fire. Ecology 65: 1332–1333

Forbes BC (1992) Tundra disturbance studies, I: long-term effects of vehicles on species richness and biomass. Environ Conser 19(1): 48–58

Freedman W, Hutchinson TC (1976) Physical and biological effects of experimental crude oil spills on Low Arctic tundra in the vicinity of Tuktoyaktuk, NWT, Canada. Can J Bot 54: 2219–2230

French HM (1974) Active thermokarst processes, eastern Banks Island, western Canadian arctic. Can J Earth Sci 11: 785–794

French HM (1985) Surface disposal of waste drilling fluids, Ellef Ringnes Island, NWT: short-term observations. Arctic 38: 292–302

Gabriel HW, Tande GF (1983) A regional approach to fire history in Alaska. US Dep Inter Bur Land Manage, Tech Rep 9

Gartner BL (1983) Germination characteristics of arctic plants. In: Permafrost: Proc 4th Int Conf National Academy Press, Washington, DC, pp 334–338

Gartner BL, Chapin FS III, Shaver GR (1983) Demographic patterns of seedling establishment and growth of native graminoids in an Alaskan tundra disturbance. J Appl Ecol 20: 965–980

Gas Arctic-Northwest Project Study Group (1973) Engineering and environmental factors related to the design, construction and operation of a natural gas pipeline on the arctic region based on Prudhoe Bay, Alaska, research facility. Final Rep, vol 4. Environ Ecol Stud, Battelle Columbus Lab, Columbus

Gersper PL, Alexander V, Barkley SA, Barsdate RJ, Flint PS (1980) The soils and their nutrients. In: Brown J, Miller PC, Tieszen LL, Bunnell FL (eds) An Arctic ecosystem: the coastal tundra on Barrow, Alaska. Dowden, Hutchinson and Ross, Stroudsburg, pp 219–254

Gryc G (1985) The National Petroleum Reserve in Alaska: earth-science considerations. US Geol Surv Prof Pap 1240-C

Hall DK, Brown J, Johnson L (1978) The 1977 tundra fire in the Kokolik River area of Alaska. Arctic 31: 54–58

Hernandez H (1973) Natural plant recolonization of surficial disturbances, Tuktoyaktuk Peninsula region, Northwest Territories. Can J Bot 51(11): 2177–2196

Hinzman L, Kane DL, Gieck RE, Everett KR (1991) Hydrologic and thermal properties of the active layer in the Alaskan arctic. Cold Reg Sci Tech 19(2): 95–110

Hok JR (1969) A reconnaissance of tractor trails and related phenomena on the North Slope of Alaska. US Dep Inter Bur Land Manage Publ

Johnson AW, Murray BM, Murray DF (1978) Floristics of the disturbances and neighboring locales. US Army Cold Reg Res Eng Lab, Rep 78-28, pp 30–40

Johnson L (1981) Revegetation and selected terrain disturbances along the trans-Alaska pipeline, 1975–1978. US Army Cold Reg Res Eng Lab, Spec Rep 81-12

Johnson L (1984) Revegetation along pipeline rights-of-way in Alaska. In: Proc 3rd Int Symp Environ Concerns in Rights-of-way Management, 15–18 February 1982, San Diego, California. Mississippi State Univ, State College, pp 254–264

Johnson L (1987) Management of northern gravel sites for successful reclamation: a review. Arct Alp Res 19: 530–536

Johnson L, Viereck L (1983) Recovery and active layer changes following a tundra fire in northwestern Alaska. In: Permafrost: Proc 4th Int Conf, 17–22 July 1983. National Academy Press, Washington, DC, pp 543–547

Johnson L, Sparrow E, Collins C, Jenkins T, Davenport C, McFadden TM (1980) The fate and effects of crude oil spilled on subarctic permafrost terrain in interior Alaska. US Army Cold Reg Res Eng Lab, Rep 80-29

Johnson PR, Collins CM (1980) Snow pads used in pipeline construction in Alaska, 1976: construction, use and breakup. US Army Cold Reg Res Eng Lab, Rep 80-17

Jorgenson MT (1988) Rehabilitation studies in the Kuparuk Oilfield, Alaska, 1987. Rep prepared for ARCO Alaska, Inc, and Kuparuk River Unit, Alaska Biol Res, Inc, Fairbanks

Jorgenson MT, Cater TC (1991) Land rehabilitation studies in the Kuparuk Oilfield, Alaska, 1990. Final Rep prepared for ARCO Alaska, Inc and Kuparuk River Unit, Alaska Biol Res, Inc, Fairbanks

Jorgenson MT, Cater TC (1992) Land rehabilitation studies in the Kuparuk Oilfield, Alaska, 1991. Rep prepared for ARCO Alaska, Inc and Kuparuk River Unit, Anchorage, AK, Alaska Biol Res, Inc, Fairbanks

Jorgenson MT, Kidd JG (1991) Land rehabilitation studies in the Kuparuk Oilfield, Alaska, 1990. Rep prepared for ARCO Alaska, Inc and Kuparuk River Unit, Anchorage, AK, Alaska Biol Res, Inc, Fairbanks

Jorgenson MT, Robus MA, Zachel CO, Lawhead BE (1987) Effects of a brine spill on tundra vegetation and soil in the Kuparuk Oilfield, Alaska. Rep prepared for ARCO Alaska, Inc and Kuparuk River Unit, Anchorage, AK, Alaska Biol Res, Inc, Fairbanks

Jorgenson MT, Kielland K, Schepart BS, Hyzy JB (1991a) Bioremediation and tundra restoration after a crude oil spill near drill site 2U, Kuparuk Oilfield, Alaska. Rep prepared for ARCO Alaska, Inc and Kuparuk River Unit, Anchorage, Alaska Biol Res, Inc, Fairbanks, AK, and Waste Stream Technol, Inc, Buffalo

Jorgenson MT, Kidd JG, Cater TC (1991b) Rehabilitation of a thick gravel pad using snow capture and topsoil addition, drill site 13, Prudhoe Bay Oilfield, Alaska 1990. Rep prepared for ARCO Alaska, Inc and Kuparuk River Unit, Anchorage, Alaska Biol Res, Inc, Fairbanks

Jorgenson MT, Hefferman TT, Lance BK (1991c) Land rehabilitation studies in the Kuparuk Oilfield, Alaska, 1989. Annu Rep, Alaska Biol Res, Inc, Fairbanks

Jorgenson MT, Kidd JG, Cater TC (1992a) Rehabilitation of a thick gravel pad using snow capture and topsoil addition, drill site 13, Prudhoe Bay Oilfield, Alaska 1990. Report prepared for ARCO Alaska, Inc, Alaska Biol Res, Inc, Fairbanks

Jorgenson MT, Cater TC, Jacobs LL (1992b) Wetland creation and revegetation on an overburden stockpile at mine site D, Kuparuk Oilfield, Alaska, 1991. Rep prepared for ARCO Alaska, Inc and Kuparuk River Unit, Anchorage, AK, Alaska Biol Res, Inc, Fairbanks

Joyce MR (1980) Effects of gravel removal on terrestrial biota. In: Woodward-Clycle Consultants (eds), Gravel removal studies in arctic and subarctic floodplains in Alaska. US Fish and Wildlife Services, Anchorage Tech Rep FWSIOBS-80/08

Kershaw GP, Kershaw LJ (1986) Ecological characteristics of 35-year-old crude-oil spills in tundra plant communities of the Mackenzie Mountains, NWT. Can J Bot 64: 2935-2947

Kershaw GP, Kershaw LJ (1987) Successful plant colonizers on disturbances in tundra areas of northwestern Canada. Arct Alp Res 19: 451-460

Klinger LF, Walker DA, Webber PJ (1983) The effects of gravel roads on Alaskan Arctic Coastal Plain tundra. In: Permafrost: Proc 4th Int Conf, 17-22 July 1983. National Academy Press, Washington, DC, pp 628-633

Komárková V (1983) Recovery of plant communities and summer thaw at the 1949 Fish Creek Test Well 1, arctic Alaska. In: Permafrost: Proc 4th Int Conf, 17-22 July 1983. National Academy Press, Washington, DC, pp 645-650

Komárková V (1985) Plant community recovery 30 years after disturbance at the 1949 Fish Creek Test Well 1, arctic Alaska. US Cold Reg Res Eng Lab, Final Rep

Komárková V, Webber PJ (1978) Geobotanical mapping, vegetation disturbance and recovery. In: Tundra disturbances and recovery following the 1949 exploratory drilling, Fish Creek, northern Alaska. US Cold Reg Res Eng Lab, Report 78-28

Kubanis SA (1980) Recolonization by native plant species and persistence of introduced species along the Yukon River–Prudhoe Bay Haul Road, 1977–1981. Thesis, San Diego State Univ, San Diego

Landers DH, Ford J, Gubala CP, Allen S, Curtis L, Urquhart NS, Omernik JM (1992) US Environ Prot Agency, Arc Contam Res Prog, EPA/600/R-92/210

Lawson DE (1982) Long-term modifications of perennially frozen sediment and terrain at East Oumalik, northern Alaska. US Cold Reg Res Eng Lab, Rep 82-36

Lawson DE (1986) Response of permafrost terrain to disturbance: a synthesis of observations from northern Alaska. Arct Alp Res 18: 1–17

Lawson DE, Brown J (1979) Human-induced thermokarst at old drill sites in northern Alaska. North Eng 10: 16–23

Lawson DE, Brown J, Everett KR, Johnson AW, Komárková V, Murray BM, Murray DF, Webber PJ (eds) (1978) Tundra disturbances and recovery following the 1949 exploratory drilling, Fish Creek, northern Alaska. US Cold Reg Res Eng Lab, Rep 78-28

Lee LC, Gosselink JG (1988) Cumulative impacts on wetlands: linking scientific assessments and regulatory alternatives. Environ Manage 12(5): 591–602

Linkins AE, Fetcher N (1983) Effect of surface-applied Prudhoe Bay crude oil on vegetation and soil processes in tussock tundra. In: Permafrost: Proc 4th Int Conf, 17–22 July 1983. National Academy Press, Washington, DC, pp 723–728

Linkins AE, Johnson LA, Everett KR, Atlas RM (1984) Oil spills: damage and recovery in tundra and taiga. In: Cairns J, Buikema AL (eds) Restoration of habitats impacted by oil spills. Butterworth, Boston, pp 135–155

Mackay D, Mohtadi M (1975) Area affected by oil spills on land. Can J Chem Eng 53: 140–143

Mackay JR (1970) Disturbances to the tundra and forest tundra environment of the western arctic. Can Geotech J 7: 420–432

Matveyeva NV (1979) An annotated list of plants inhabiting sites of natural and anthropogenic disturbances of tundra cover in western Taimyr: the settlement of Kresty. In: West GC (ed) The effect of disturbance on plant communities in tundra regions of the Soviet Union. Biol Pap Univ AK 20

McCown BH, Deneke FJ, Rickard W, Tieszen LL (1973a) The response of Alaskan terrestrial plant communities to the presence of petroleum. In: Proc Symp Impact of Oil Resource Dev on Northern Plant Communities. Univ Alaska, Fairbanks, pp 34–43

McCown BH, Brown J, Barsdate BJ (1973b) Natural oil seeps at Cape Simpson, Alaska: localized influences on terrestrial habitat. In: Proc Symp Impact of Oil Resource Dev on Northern Plant Communities. Univ Alaska, Fairbanks, pp 86–90

McGraw JB (1980) Seed bank size and distribution of seeds in cottongrass tussock tundra, Eagle Creek, Alaska. Can J Bot 58: 1607–1611

McKendrick JD (1987) Plant succession on disturbed sites, North Slope, Alaska, USA. In: Salzberg KA, Fridriksson S, Webber PJ (eds) Proc 7th Conf Comité Actique Int, Sept 1986. Acrt Alp Res 19(4): 544–565

McKendrick J (1990) *Arctophila* feasibility studies – phase I. Abstr of paper presented at North Slope Terrestrial Studies Worksh. BP Exploration (Alaska) Inc, 900 East Benson Blvd, Anchorage, AK, 15–16 Nov 1990

McKendrick JD, Mitchell WM (1978) Fertilizing and seeding oil-damaged arctic tundra to effect vegetation recovery, Prudhoe Bay, Alaska. Arctic 31(3): 296–304

Meehan RH (1986) Impact of oilfield development on shorebirds, Prudhoe Bay, Alaska. PhD Thesis, Univ Colorado, Boulder

Mitchell WW (1979) Three varieties of native Alaskan grasses for revegetation purposes. Univ Alaska Agric Exp Stn Circ 32: 1–9

Moore NJ (1983) Pioneer *Salix alaxensis* communities along the Sagavanirktok River and adjacent drainages. MS Thesis, Univ Alaska, Fairbanks

Moore NJ (1990) Studies on the techniques for revegetation with *Arctophila fulva*. Draft Rep prepared for ARCO Alaska, Inc, Anchorage, AK, Plant Materials Center, Alaska Dep Nat Resources, Palmer

Myers KC, Barker MH (1984) Examination of drilling reserve pit fluids and effects of tundra disposal at Prudhoe Bay, Alaska, 1982–1983. Rep prepared for ARCO Alaska, Inc, Anchorage

Oechel WC (1989) Nutrient and water flux in a small arctic watershed: an overview. Holarct Ecol 12: 229–237

Pamplin WL (1979) Construction-related impacts of the trans-Alaska pipeline system on terrestrial wildlife habitats. Joint State/Federal Fish and Wildlife Advisory Team, Anchorage, Spec Rep No 24

Peterson KM, Billings WD (1978) Geomorphic processes and vegetational change along the Meade River sand bluffs in northern Alaska. Arctic 31: 7–23

Peterson KM, Billings WD (1980) Tundra vegetational patterns and succession in relation to microtopography near Atkasook, Alaska. Arct Alp Res 12: 473–482

Pickett STA, Kolasa J, Armetso JJ, Collins SL (1989) The ecological concept of disturbance and its expression at various hierarchical levels. Oikos 54: 129–136

Pope PR, Hillman SO, Safford L (1982) Arctic Coastal Plains tundra restoration: a new application. In: Proc 5th Arctic marine oilspill program technical seminar. Edmonton, Alberta, pp 93–111

Preston EM, Bedford BL (1988) Evaluating cumulative effects on wetland functions: a conceptual overview and generic framework. Environ Mange 12(5): 565–583

Racine CH (1977) Tundra disturbance resulting from a 1974 drilling operation in the Cape Espenberg area, Seward Peninsula, Alaska. Rep for US Dep Interior, National Park Service, Ord #PX9100-6-140

Racine CH (1981) Tundra fire effects on soils and three plant communities along a hill-slope gradient in the Seward Peninsula, Alaska. Arctic 34: 71–84

Racine CH, Patterson WA, Dennis JG (1983) Permafrost thaw associated with tundra fires in northwest Alaska. In: Permafrost Proc 4th Int Conf, 17–22 July 1983. National Academy Press, Washington, DC, pp 1024–1029

Racine CH, Patterson WA, Dennis JG (1985) Tundra fire regimes in the Noatak River watershed, Alaska: 1956–1983. Arctic 38(3): 194–200

Racine CH, Johnson LA, Viereck LA (1987) Patterns of vegetation recovery after tundra fires in northwestern Alaska, USA. Arct Alp Res 19: 461–469

Radforth JR, Burwash AL (1977) Transportation. In: Radforth NW, Brawner CO (eds) Muskeg and the northern environment in Canada. Univ Toronto Press, Toronto, pp 249–263

Raynolds MK, Felix NA (1989) Airphoto analysis of winter seismic disturbance in northeastern Alaska. Arctic 42(4): 362–367

Reed JC (1958) Exploration of Naval Petroleum Reserve No 4 and adjacent areas, northern Alaska. US Geol Surv Prof Pap 301

Reynolds PC (1982) Some effects of oil and gas exploration activities on tundra vegetation in northern Alaska. In: Rand PJ (ed) Land and water issues related to energy development. Ann Arbor Sci, Ann Arbor, pp 403–417

Shaver GR, Chapin FS (1984) Limiting factors for plant growth in northern ecosystems. In: Moore TR (ed) Future directions for research in Nouveau-Quebec. McGill Univ Press, Montreal, Subarct Res Pap 39: 49–60

Shaver GR, Gartner GL, Chapin FS, Linkins AE (1983) Revegetation of arctic disturbed sites by native tundra plants. In: Permafrost Proc 4th Int Conf 17–22 July 1983. National Academy Press, Washington, DC, pp 1133–1138

Shideler RT (1986) Impacts of human developments and land use on caribou: a literature review. Alaska Dep Fish Game, Tech Rep 86-3

Simmons CL, Everett KR, Walker DA, Linkins AE, Webber PJ (1983) Sensitivity of plant communities and soil flora to seawater spills, Prudhoe Bay, Alaska. US Cold Reg Res Eng Lab, Rep 83-24

Slaughter CW, Racine CH, Walker DA, Johnson LA, Abele G (1990) Use of off-road vehicles and mitigation of effects in Alaska permafrost environments: a review. Environ Manage 14: 63–72

Spatt PD (1978) Seasonal variation of growth conditions on a natural and dust impacted *Sphagnum* (Sphagnacaeae) community in northern Alaska. MS Thesis, Univ Cincinnati, Cincinnati

Spatt PD, Miller MC (1981) Growth conditions and vitality of *Sphagnum* in a tundra community along the Alaska Pipeline Haul Road. Arctic 34(1): 48-54

Speer L, Libenson S (1988) Oil in the Arctic: the environmental record of oil development on Alaska's North Slope. Rep prepared for Trustees for Alaska, Natural Resources Defense Council, and National Wildlife Federation, Natural Resources Defense Council, Washington, DC

Stakhiv EZ (1988) An evaluation paradigm for cumulative impact analysis. Environ Manage 12(5): 725-748

Sterrett KF (1976) The arctic environment and the arctic surface effect vehicle. US Cold Reg Res Eng Lab, Rep 76-1, pp 1-28

US Department of the Interior (1983) Proposed oil and gas exploration within the coastal plain of the Arctic National Wildlife Refuge, Alaska. Final environmental impact statement and preliminary final regulations. US Fish Wild Ser, Washington, DC

Viereck LA (1966) Plant succession and soil development on gravel outwash of the Muldrow Glacier, Alaska. Ecol Monogr 36(3): 181-199

Vilchek GE, Bykova OY (1992) The origin of regional ecological problems within the northern Tyumen Oblast, Russia. Arct Alp Res 24(2): 99-107

Walker DA (1985) Vegetation and environmental gradients for the Prudhoe Bay region, Alaska. US Cold Reg Res Eng Lab, Rep 85-14

Walker DA (1987) Height and growth-ring response of *Salix lanata* ssp. *richardsonii* along the coastal temperature gradient of northern Alaska. Can J Bot 65: 988-993

Walker DA, Everett KR (1987) Road dust and its environmental impact on Alaskan taiga and tundra. Arct Alp Res 19: 479-489

Walker DA, Everett KR (1991) Loess ecosystems of northern Alaska: regional gradient and toposequence at Prudhoe Bay. Ecol Monogr 61: 437-464

Walker DA, Walker MD (1991) History and pattern of disturbance in Alaskan arctic terrestrial ecosystems: a hierarchical approach to analysing landscape change. J Appl Ecol 28: 244-276

Walker DA, Webber PJ, Everett KR, Brown J (1977) The effects of low-pressure wheeled vehicles on plant communities and soils at Prudhoe Bay, Alaska. US Cold Reg Res Eng Lab, Spec Rep 77-17

Walker DA, Webber PJ, Everett KR, Brown J (1978) The effects of crude oil and diesel spills on plant communities at Prudhoe Bay, Alaska and the derivation of oil spill sensitivity maps. Arctic 31: 242-259

Walker DA, Walker MD, Lederer ND (1985a) 1983 observations at the Prudhoe Bay and Franklin Bluffs hydrocarbon spill sites. In: Webber PJ, Walker DA, Komárková V, Ebersole JJ (eds) Baseline monitoring methods and sensitivity analysis of Alaskan Arctic tundra. US Cold Reg Res Eng Lab, Final Rep

Walker DA, Lederer ND, Walker MD (1985b) 1983 observations at the Prudhoe Bay Rolligon Site seven years following the test. In: Webber PJ, Walker DA, Komárková V, Ebersole JJ (eds) Baseline monitoring methods and sensitivity analysis of Alaskan Arctic tundra. US Cold Reg Res Eng Lab, Final Rep

Walker DA, Webber PJ, Walker MD, Lederer ND, Meehan RH, Nordstrand EA (1986) Use of geobotanical maps and automated mapping techniques to examine cumulative impacts in the Prudhoe Bay oilfield, Alaska. Environ Conserv 13: 149-160

Walker DA, Webber PJ, Binnian EF, Everett KR, Lederer ND, Nordstrand EA, Walker MD (1987a) Cumulative impacts of oil fields on northern Alaskan landscapes. Science 238: 757-761

Walker DA, Cate D, Brown J, Racine C (1987b) Disturbance and recovery of arctic Alaskan tundra terrain: a review of recent investigations. US Cold Reg Res Eng Lab, Rep 87-11

Walker MD (1990) Vegetation and floristics of pingos, Central Arctic Coastal Plain, Alaska. Dissertationes Botanicae, 149. Cramer, Stuttgart

Webber PJ, Fridriksson S (1987) Restoration and vegetation succession in circumpolar lands: the conference. Arct Alp Res 19: 343–344

Webber PJ, Ives JD (1978) Recommendations concerning the damage and recovery of tundra vegetation. Environ Conserv 5: 171–182

Wein RW (1975) Vegetation recovery in arctic tundra and forest tundra after fire. Northern Economic Development Branch, Department of Indian Affairs and Northern Development, ALUR Report 74-75-62, Ottawa, Canada

Wein RW, Bliss LC (1973a) Changes in arctic *Eriophorum vaginatum* tussock communities following fire. Ecology 54: 845–852

Wein RW, Bliss LC (1973b) Experimental crude oil spills on arctic plant communities. J Appl Ecol 10: 671–682

Werbe E (1980) Disturbance effects of a gravel highway upon Alaskan tundra vegetation. MA Thesis, Univ Colorado, Boulder

West GC (1976) Environmental problems associated with arctic development especially in Alaska. Environ Conserv 3: 218–224

West RL, Snyder-Conn E (1987) Effects of Prudhoe Bay reserve pit fluids on water quality and macroinvertebrates of arctic tundra ponds in Alaska. US Fish Wild Ser, Biol Rep 87(7)

Westlake DW, Jobson AM, Cook FD (1978) In situ degradation of oil in a soil of the boreal region of the Northwest Territories. Can J Microbiol 24: 254–260

White PS, Pickett STA (1985) Natural disturbance and patch dynamics: an introduction. In: Pickett STA, White PS (eds) The ecology of natural disturbance and patch dynamics. Academic Press, Orlando, pp 3–13

Wiggins IL, Thomas JH (1962) A flora of the Alaskan arctic slope. Toronto Univ Press, Toronto

Woodward-Clyde Consultants I (1980) Gravel removal studies in arctic and subarctic flood plains in Alaska. US Fish Wild Serv, Tech Rep FWS/OBS-80/08

Wyant JG, Knapp CM (1972) Alaska North Slope oil-field restoration research strategy (ANSORRS). US Environ Protect Agency, Res Dev Rep EPA/600/R-92/022

Younkin WE (1972) Revegetation studies of disturbances in the Mackenzie Delta region. In: Bliss LC, Wein RW (eds) Botanical studies of natural and man modified habitats in the eastern Mackenzie Delta region and the Arctic Islands. Dept Indian Affairs and Northern Dev, ALUR Rep 72-73-14, Ottawa

Younkin WE (ed) (1976) Revegetation studies in the northern Mackenzie Valley Region. Arct Gas Biol Rep Ser 38

Yurtsev BA, Korobkov AA (1979) An annotated list of plants inhabiting sites of natural and anthropogenic disturbances of tundra cover; southeasternmost Chukchi Peninsula. Univ Als Biol Pap 20: 1–17

4 Terrain and Vegetation of the Imnavait Creek Watershed

D. A. WALKER and M. D. WALKER

4.1 Introduction

The Imnavait Creek watershed is within a large region of tussock-tundra vegetation that covers much of northern Alaska, northwestern Canada, and northeastern Russia (Bliss and Matveyeva 1992). Tussock tundra has been described in numerous areas of northern Alaska (e.g., Hanson 1951, 1953; Churchill 1955; Bliss 1956, 1962; Spetzman 1959; Douglas and Tedrow 1960; Koranda 1960; Johnson et al. 1966; Lambert 1968; Hettinger and Janz 1974; Holowaychuk and Smeck 1974; Racine 1981; Walker et al. 1982). Some of these studies have touched on the variation that occurs within tussock tundra with regard to topography, hydrology, and soils, but there remains a general impression that tussock tundra is a fairly uniform vegetation type that varies little over vast areas of the Low Arctic. This impression is often reinforced from the air as one flies over the seemingly endless expanse of tussock tundra.

On the ground, however, tussock tundra often defies description because of subtle changes in dominant growth forms and species. Although techniques such as geostatistics (Chap. 1, this Vol.) and gradient analysis (Walker et al. 1994) are useful to study these changes, it is often desirable to define discrete vegetation types and portray that variation in a map format. This is particularly valuable for exploring the spatial patterns of vegetation in relation to terrain as well as hydrological and geological variables using geographic information systems (GIS; e.g., Ostendorf 1994), and a thorough understanding of vegetation variation along natural gradients is an important starting point for developing predictive models of terrain sensitivity to disturbance (Chap. 18, this Vol.). This chapter uses a geoecological approach to describe and map the terrain and vegetation along a hillslope gradient at Imnavait Creek. It includes descriptions of all the major terrain and vegetation types within the Imnavait Creek watershed, and focuses on the changes in vegetation along a west-facing hillslope gradient.

4.2 Terrain

Our studies focused on an area within the Imnavait Creek UTM grid, a 1-km^2 section of watershed with 100 m between grid points, which serves as a

J.F. Reynolds and J.D. Tenhunen (Eds.)
Ecological Studies, Vol. 120
© Springer-Verlag Berlin Heidelberg 1996

reference for much of the data presented in this book. Hereafter, we refer to the portion of the grid that we sampled as the Intensive Research Site (IRS; Fig. 4.1). To place these data into a regional context, we also present comparative data for the Imnaviat Creek watershed (2.2 km²) and the R4D region (22 km²; see Fig. 1.1 in Chap. 1 this Vol.).

The hills near Imnavait Creek rise less than 100 m from the valley bottoms to the crests, and are elongated in SSE to NNW trending ridges. The west-facing aspects of these ridges are much gentler and longer than the east-facing aspects; in fact, over 60% of the terrain in the R4D region has NW/SW-facing aspects.

4.2.1 Glacial Deposits

Hills in this region are covered with smoothly eroded mid-Pleistocene-age glacial deposits, fine colluvium, and tussock-tundra vegetation. Shallow peat deposits are found in the basins between the ridges. Small bedrock outcrops,

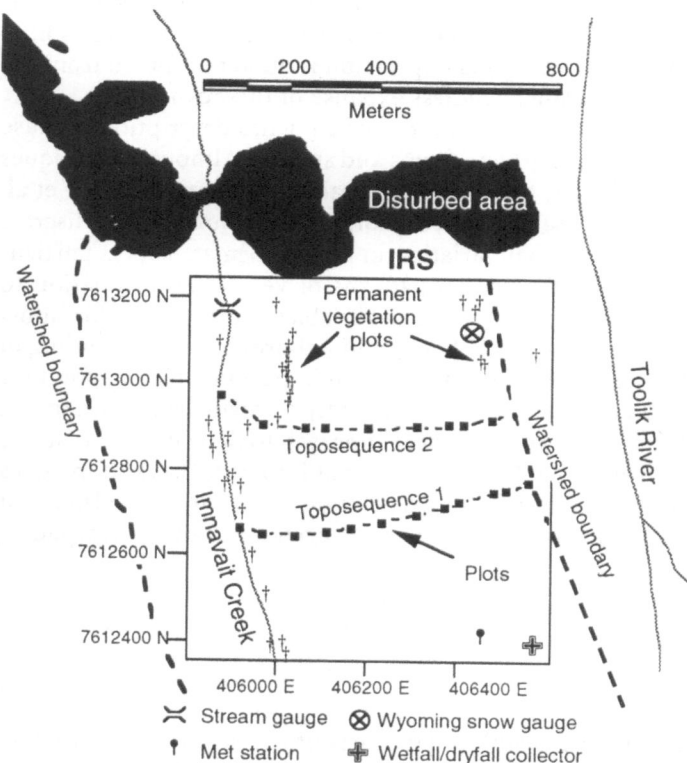

Fig. 4.1. Map of the intensive research site, showing location of the permanent vegetation plots, toposequences, and meteorological sites. UTM grid as in Fig. 1.1, Chap. 1 (this Vol.)

glacial till of late-Pleistocene age, and glaciofluvial outwash surround the Imnavait Creek watershed (Fig. 4.2). The glacial till in the Imnavait Creek watershed was deposited during the Sagavanirktok Glaciation (deglaciated about 125 000 years ago; Hamilton 1986). Glacial till is exposed primarily on ridgecrests and scattered sites on the hillslopes. About 9% of the R4D region, 4% of the watershed, and 5% of the IRS contains exposed till deposits (Fig. 4.2). These till surfaces are generally rocky and gently undulating with small blockfields (areas with greater than 50% cover of moderate-to-large-sized angular rocks; Washburn 1980) and nonsorted circles (patterned ground features composed of regularly spaced circular patches of bare ground with diameters 0.5–3 m that are formed by frost heaving of the soil; Washburn 1980). The till on most slopes is covered by fine-grained colluvium that has been retransported from upslope.

4.2.2 Retransported Hillslope Deposits

About 70% of the R4D region, 76% of the watershed, and 83% of the IRS consist of retransported hillslope deposits (Fig. 4.2). The till on the hillslopes of the watershed is covered by a clay loam that has been moved from upslope and topped with about 20–40 cm of peat. This unit is frozen and commonly contains massive ice (Kreig and Reger 1982). Scattered glacial erratics protrude up to 1 m above the tundra surface.

A variety of surficial geomorphological features are superimposed on the hillslope deposits as shown in a detailed map of the IRS (Fig. 4.3). Nonsorted stripes occur on the upper slopes and shoulders of most of the hillslopes in this region. From the air these features have a pattern oriented toward the downslope (Fig. 4.4d). The more barren stripes are relatively dry with nonsorted circles. The vegetation on these stripes, which are probably caused by combinations of cryoturbation, erosion, and gelifluction (Washburn 1980), is dominated by dwarf-shrubs and fruticose lichens (see Chap. 5, this Vol.). The interstripe areas are poorly drained, peaty, and covered with tussock-tundra vegetation. Ten percent of the R4D region, 15% of the watershed, and 4% of the IRS are covered by nonsorted stripes.

Hillslope water track, shallow drainage channels spaced tens of meters apart, are common features on the mid-to-lower portions of most hills giving them a ribbed appearance (Fig. 4.4). We distinguish hillslope water tracks (also termed horsetail drainages; Cantlon 1961) from the lowland water tracks described in mires of Minnesota, Labrador, and northern Europe where the term *water track* applies to minerotrophic drainage channels in flat peatlands (Heinselman 1970; Glaser et al. 1981; Glaser 1983). Lowland water tracks also occur in the colluvial basins of our study area. We distinguish well-defined hillslope water tracks, which have distinct channels, from weakly defined tracks, which are more subtle and do not have incised channels (Fig. 4.4b,c). Weakly defined tracks are discernible because of a greater abundance of dwarf

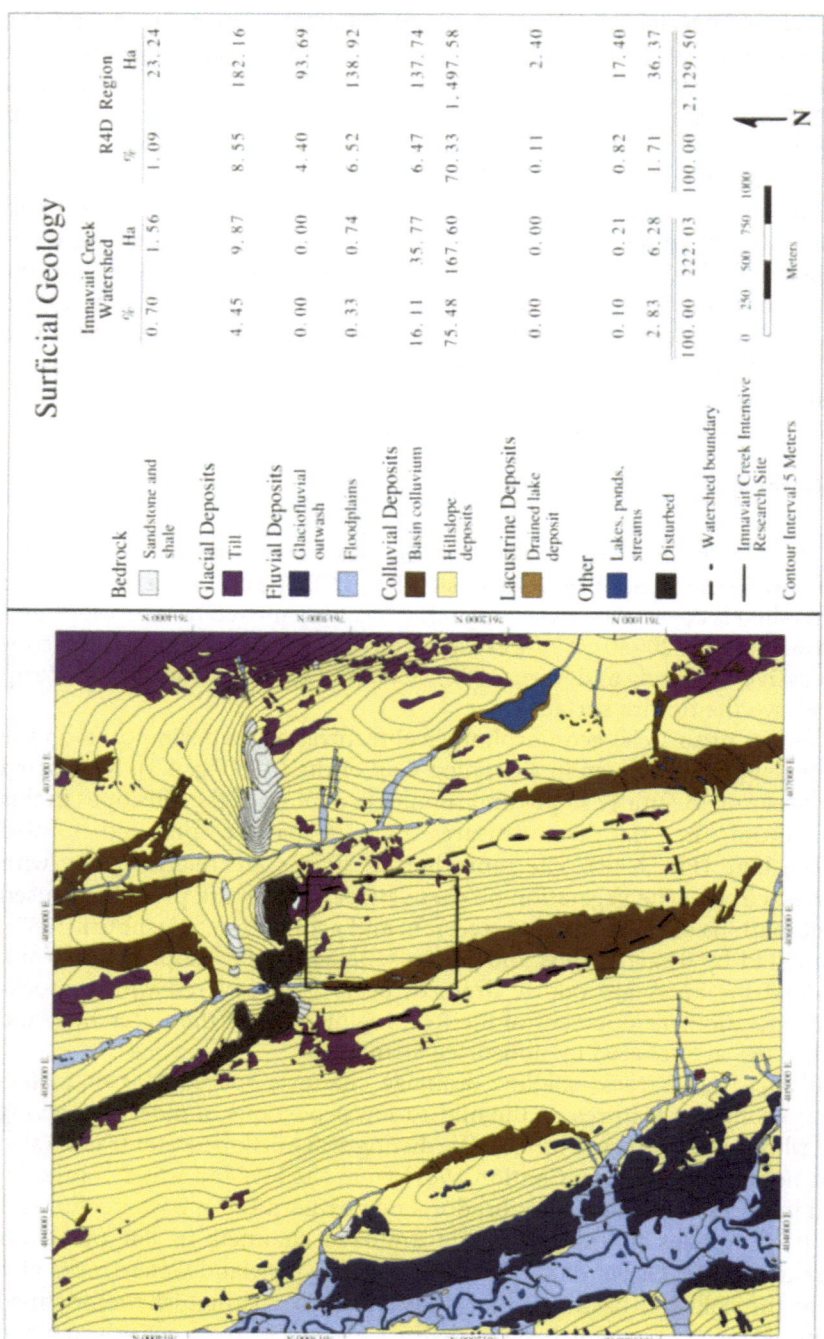

Surficial Geology

	Immavait Creek Watershed		R4D Region	
	%	Ha	%	Ha
Bedrock				
Sandstone and shale	0.70	1.56	1.09	23.24
Glacial Deposits				
Till	4.45	9.87	8.55	182.16
Fluvial Deposits				
Glaciofluvial outwash	0.00	0.00	4.40	93.69
Floodplains	0.33	0.74	6.52	138.92
Colluvial Deposits				
Basin colluvium	16.11	35.77	6.47	137.74
Hillslope deposits	75.48	167.60	70.33	1,497.58
Lacustrine Deposits				
Drained lake deposit	0.00	0.00	0.11	2.40
Other				
Lakes, ponds, streams	0.10	0.21	0.82	17.40
Disturbed	2.83	6.28	1.71	36.37
	100.00	222.03	100.00	2,129.50

Watershed boundary

Immavait Creek Intensive Research Site

Contour Interval 5 Meters

0   250   500   750   1000

Meters

N

Fig. 4.2. Surficial geology of the R4D region. *Dashed line* indicates Immavait Creek watershed boundary. *Solid line* indicates intensive research site of Fig. 4.1

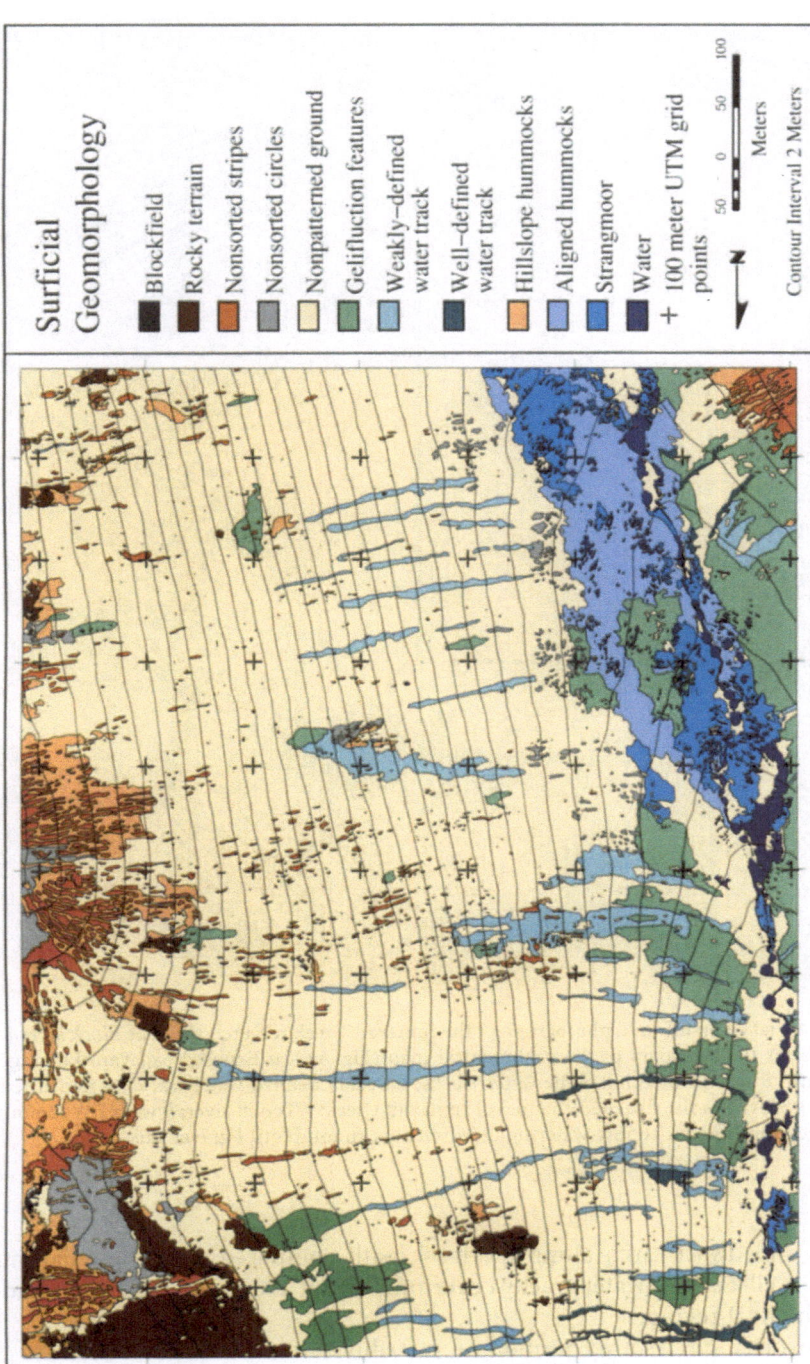

Fig. 4.3. Surficial geomorphology of the R4D intensive research site shown in Fig. 4.1

Fig. 4.4. Aerial color-infrared photograph of the Imnavait Creek research site. The *black rectangle* delineates the boundary of the geobotanical map (Fig. 4.6). *a* Exposed Sagavanirktok-age glacial drift; *b* well-defined hill-slope water tracks; *c* weakly defined water tracks; *d* nonsorted stripes; *e* colluvial basin; *f* beaded channel of Imnavait Creek; *g* Toolik river; *h* weir; *i* Wyoming snow gage; *j* pipeline for tundra watering experiment; *k* runoff plots; *l* gravel pad

willows and dwarf birch along the track path compared with the adjacent tundra. Well-defined water tracks cover only about 1% of the watershed and 6% of the R4D region.

Nonsorted circles are common on interfluve areas between water tracks, and are more common on upper hillslopes. Even many areas we describe as "nonpatterned" have small nonsorted circles between tussocks, i.e., they are concealed by vegetation and are therefore not detectable on aerial photographs (Walker et al. 1989a).

Gelifluction features are earth lobes, terraces, and benches formed by the slow downslope flow saturating soil in permafrost regions (Washburn 1980). Gelifluction features are common on wetter hillslopes at the IRS and are predominantly small features, less than 20 cm high, consisting of terraces or aligned hummocks oriented perpendicular to the slope.

4.2.3 Colluvial Basin Deposits

Most of the hillslope water tracks drain into a gently sloping basin that forms the headwater of Imnavait Creek (Fig. 4.4). Similar basins occur in the bottom of valleys between smoothly rounded slopes throughout the arctic foothills. They consist of fine-grained, organic-rich deposits that appear to have moved into smaller basins from the surrounding slopes by solifluction, creep, and/or slope wash (Kreig and Reger 1982). The basins have a complex micro-topography consisting of string bogs (peatlands characterized by low ridges of peat and vegetation interspersed with depressions that often contain shallow ponds; Washburn 1980), palsas (small ice-cored mounds), high-centered ice-wedge polygons, and wet areas with lowland water-track patterns. Basin colluvium covers 6% of the R4D region, 16% of the watershed, and 11% of the intensive research site.

4.2.4 Floodplain Deposits

Imnavait creek is a beaded stream (Chap. 13, this Vol.), composed of a series of small pools connected by short water courses (Fig. 4.4). The pools result from the thawing of ice masses that occur at ice-wedge polygon intersections, and the connecting drainage is commonly along the thawing ice wedges (Washburn 1980). The creek has a narrow floodplain, and only about 0.3% of the watershed has floodplain deposits. Within the surrounding R4D region extensive alluvial and glaciofluvial deposits occur along the Kuparuk River.

4.3 Vegetation

4.3.1 Flora

Imnavait Creek is in the hypoarctic zone, which lies between the boreal forest to the south and the arctic tundra zone to the north (Yurtsev 1994). The hypoarctic zone is south of the 7 °C mean July isotherm (the mean July temperature at Imnavait Creek is 10.9 °C). This zone is characterized by a closed vegetation cover compared with the generally open and interrupted vegetation cover of the arctic tundra zone. Low and dwarf shrubs are a major component of the plant canopy over much of the landscape, whereas they are much less common in the arctic tundra zone (e.g., *Betula nana, Empetrum hermaphroditum, Ledum palustre* ssp. *decumbens, Vaccinium uliginosum, V. vitisidaea, Andromeda polifolia*). Other important boreal species include *Eriophorum vaginatum, Calamagrostis canadensis, Comarum paluster, Linnaea borealis,* and *Sparganium hyperboreum.* Willow thickets (e.g., *Salix planifolia* ssp. *pulchra*) occur in protected riparian areas. Tussock tundra, dominated by the cottongrass, *Eriophorum vaginatum,* or *Carex bigelowii* and the boreal shrub species mentioned above, cover the uplands and interfluves between streams.

Along an east–west floristic gradient, Imnavait Creek lies within the North Alaska subprovince of the Alaska province of the Beringian sector, which includes arctic Chukotka, east of the Indigarika River, and arctic North America, west of the Mackenzie River delta (Yurtsev 1994). During the Pleistocene glacial interval, the eastern and western parts of Beringia were linked by an unglaciated land bridge that was a corridor for many Asiatic species moving into North America (Hopkins 1982). It is one of the floristically richest areas in the Arctic. Many of the Beringian species that occur in the Imnavait Creek flora are found on dry, exposed, south-facing sandstone outcrops outside the watershed (e.g., *Astragalus umbellatus, Douglasia ochotensis, Eritrichium aretioides, Festuca altaica, Oxytropis bryophila, Smelowskia calycina, Anemone drummondii, Bupleurum triradiatum, Phlox sibirica, Spiraea stevenii*). Although these outcrops cover only about 1% of the region, they account for 32 species (20% of the known flora) including the endemic-to-Alaska and threatened plant *Erigeron muirii* (Murray and Lipkin 1987). The documented flora of the R4D region consists of 174 vascular plants (Appendix A). The combined flora of the Toolik Lake and Imnavait Creek regions is about 300 species (Walker, unpubl. data), of which only about 120 occur within the Imnavait Creek watershed.

4.3.2 Vegetation Types

The diversity of the vegetation physiognomy in the R4D region is shown in Fig. 4.5. This vegetation classification is based on 72 study plots at Imnavait Creek and 81 plots at the Toolik Lake site (Walker et al. 1994). Community types and subtypes were defined using the Braun-Blanquet tabular analysis approach (Westhoff and van der Maarel 1978).

A total of 22 vegetation map units occur within the IRS (Fig. 4.6). Community types are characterized by groups of species that are normally found in one type of community and are uncommon or absent in other community types. The names of the community types include two species associated with them; the second is usually the dominant species. Many types are divided into subtypes and facies. The subtypes have no characteristic taxa that separate

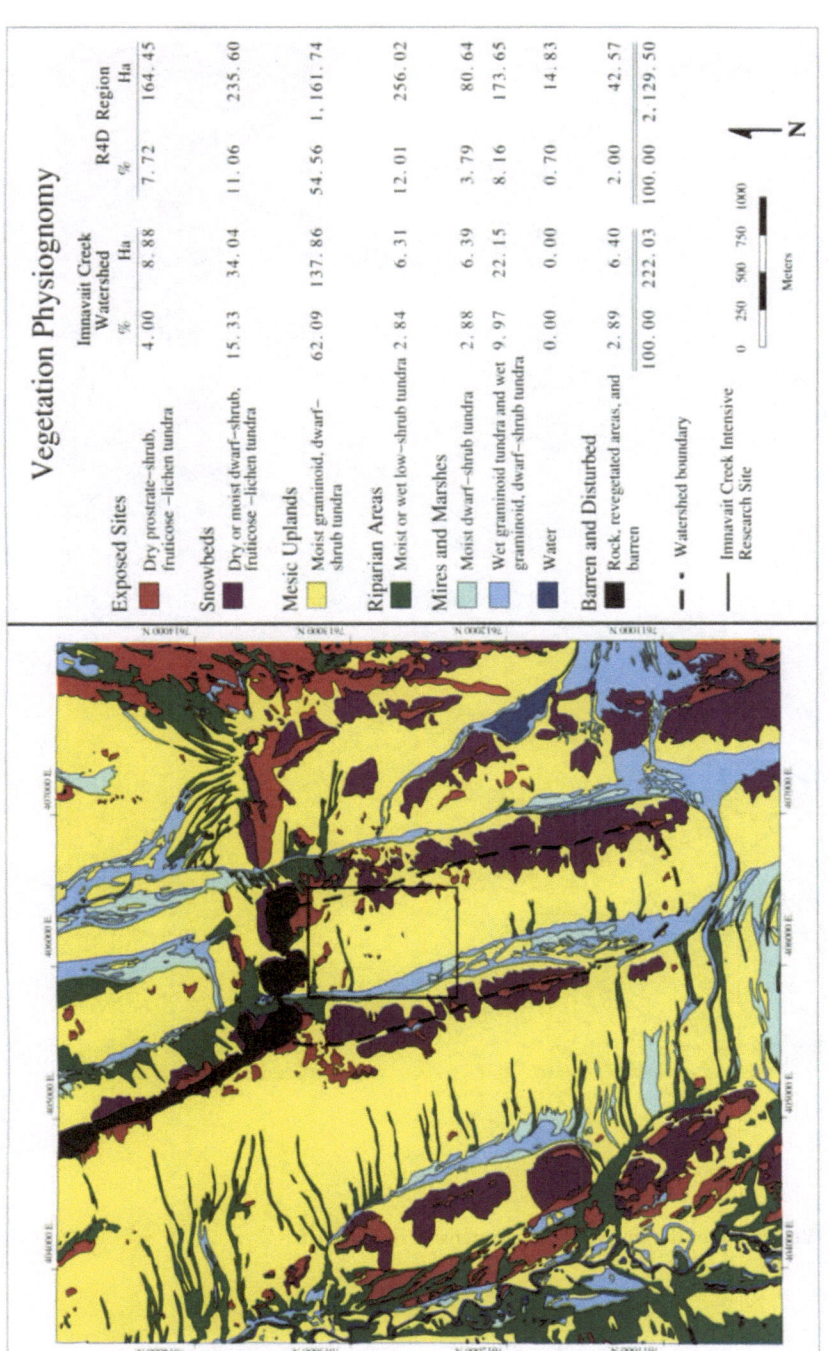

Vegetation Physiognomy

	Imnavait Creek Watershed		R4D Region	
	%	Ha	%	Ha
Exposed Sites				
Dry prostrate–shrub, fruticose–lichen tundra	4.00	8.88	7.72	164.45
Snowbeds				
Dry or moist dwarf–shrub, fruticose–lichen tundra	15.33	34.04	11.06	235.60
Mesic Uplands				
Moist graminoid, dwarf–shrub tundra	62.09	137.86	54.56	1,161.74
Riparian Areas				
Moist or wet low–shrub tundra	2.84	6.31	12.01	256.02
Mires and Marshes				
Moist dwarf–shrub tundra	2.88	6.39	3.79	80.64
Wet graminoid tundra and wet graminoid, dwarf–shrub tundra	9.97	22.15	8.16	173.65
Water	0.00	0.00	0.70	14.83
Barren and Disturbed				
Rock, revegetated areas, and barren	2.89	6.40	2.00	42.57
	100.00	222.03	100.00	2,129.50

- - Watershed boundary
— Imnavait Creek Intensive Research Site

Fig. 4.5. Vegetation of the R4D region. *Dashed* and *solid lines* indicate watershed boundary and intensive research site, respectively

Fig. 4.6. Vegetation of the Imnavait Creek intensive research site

Vegetation, Imnavait Creek Intensive Study Area

Community Type (GIS codes)	Ha	% of Map
Cetraria nigricans-Rhizocarpon geographicum (07)	0. 15	0. 2 2
Hierochloe alpina-Arctous alpina, subtype *Arctous alpina* (01)	0. 89	1. 2 9
Hierochloe alpina-Arctous alpina, subtype *Salix phlebophylla* (02)	0. 06	0.09
Hierochloe alpina-Arctous alpina, subtype *Vaccinium vitis-idaea* (03, 04)	0. 82	1. 1 8
Hierochloe alpina-Betula nana (06)	1. 52	2. 2 1
Diapensia lapponica-Cassiope tetragona, subtype *Calamagrostis inexpansa* (05)	3. 34	4. 8 5
Diapensia lapponica-Cassiope tetragona, subtype *Nephroma arcticum* (09)	0. 17	0. 2 5
Sphagnum rubellum-Eriophorum vaginatum, subtype *Eriophorum vaginatum* (10)	20. 50	2 9. 7 4
Sphagnum rubellum-Eriophorum vaginatum, subtype *Carex bigelowii* (11)	1. 04	1. 5 1
Sphagnum rubellum-Eriophorum vaginatum, subtype *Cassiope tetragona* (14)	3. 09	.4. 4 8
Sphagnum rubellum-Eriophorum vaginatum, subtype *Betula nana* (tussock tundra facies) (12)	15. 17	2 2. 0 1
Sphagnum rubellum-Eriophorum vaginatum, subtype *Salix planifolia* ssp. *pulchra* (tussock tundra facies) (13)	6. 16	8. 9 4
Sphagnum rubellum-Eriophorum vaginatum, subtype *Betula nana* (shrubland facies) (15)	6. 21	9. 0 1
Sphagnum rubellum-Eriophorum vaginatum, subtype *Salix planifolia* ssp. *pulchra* (shrubland facies) (16, 17)	4. 96	7. 2 0
Eriophorum angustifolium-Salix planifolia ssp. *pulchra* (26)	0. 33	0. 4 8
Salix chamissonis-Carex aquatilis (25, 19)	0. 23	0. 3 4
Eriophorum angustifolium-Carex aquatilis (21, 22,)	0. 08	0. 1 2
Sphagnum lenense-Salix fuscescens (23, 24)	3. 04	4. 4 1
Sphagnum orientale-Eriophorum scheuchzeri (20, 28)	0. 85	1. 2 2
Hippuris vulgaris-Sparganium hyperboreum (29)	0. 05	0. 0 8
Barren and miscellaneous vegetation types (08, 18, 30, 33, 34)	0. 23	0. 3 4
Water (31)	0. 03	0. 0 3
Totals	68. 92	100. 00

Table 4.1. Environmental and soil information for community types at the Imnavait Creek intensive research site (values are mean ± standard error.)

	Cetnig-Rhigeo[a] (n = 2)	Hiealp-Arcalp (n = 12)	Hiealp-Betnan (n = 4)	Dialap-Castet (n = 13)	Sphrub-Erivag (n = 33)	Eriang-Salpla (n = 6)	Sphlen-Salfus (n = 6)	Sphori-Erisch (n = 3)	Eriang-Caraqu (n = 5)	Hipvul-Spahyp (n = 3)
Environment										
Slope (°)	1±1	5±2	2±1	17±4	4±0	10±4	2±1	1±0	1±0	0±0
Aspect (°)	90	210±39	90±90	161±38	199±22	173±54	316±14	325±25	355±5	n.d.
Exposure (scalar)[b]	2.0±0.0	3.0±0.2	2.4±0.4	2.1±0.2	2.0±0.1	1.7±0.1	1.8±0.1	1.8±0.0	1.7±0.1	1.0±0.0
Site moisture (scalar)[c]	1.5±0.0	3.8±0.1	4.3±0.3	4.6±0.2	6.8±0.1	8.3±0.5	7.7±0.3	8.3±0.3	8.7±0.4	10.0±0.0
Snow duration (scalar)[d]	3.8±0.3	3.2±0.4	3.9±0.5	5.0±0.4	4.3±0.1	4.2±0.6	4.8±0.4	4.0±0.0	4.4±0.2	4.0±0.0
Stability (scalar)[e]	1.0±0.0	1.3±0.1	1.1±0.1	2.2±0.2	2.1±0.2	3.7±0.2	2.1±0.4	1.5±0.3	3.0±0.8	3.7±0.9
Thaw depth (cm)	n.d.	82±18	45±15	71±7	36±3	50±4	30±5	40±2	51±8	43±13
Bare soil (%)	0.0±0.0	0.6±0.3	0.0±0.0	0.1±0.1	0.4±0.3	7.0±4.8	0.3±0.3	9.3±3.5	11.0±6.0	20.0±5.8
Rock cover (%)	95.0±0.0	6.1±2.1	0.0±0.0	1.4±0.7	0.0±0.0	0.0±0.0	0.0±0.0	0.0±0.0	0.0±0.0	0.0±0.0
Soil characteristics at 10 cm depth										
Organic matter (%)	n.d	9.10±2.76	39.15±13.46	29.62±5.03	53.81±5.46	29.15±8.86	76.27±4.12	79.40±8.36	55.91±11.99	70.83
Sand (%)	n.d.	54.1±3.0	58.0	39.0±6.3	15.9±4.5	11.8±5.5	2.1±0.5	4.1±2.3	30.9±21.4	n.d.
Silt (%)	n.d.	30.4±2.2	30.0	38.5±3.4	40.6±4.9	37.1±4.3	51.9±6.6	52.6±6.8	43.4±14.9	n.d.
Clay (%)	n.d.	15.5±3.1	12.0	22.5±4.5	42.4±5.1	51.0±8.1	46.1±6.3	43.3±9.1	25.7±6.5	n.d.
Soil moisture (%)[f]	n.d.	19.3±4.1	111.2±48.5	84.0±25.1	318.8±48.2	302.7±109.0	507.4±35.2	631.5±14.5	508.2±100.5	407.0
pH[g]	n.d.	4.0±0.1	3.8±0.2	4.4±0.2	4.6±0.1	4.0±0.2	4.4±0.2	4.3±0.2	5.1±0.4	4.1

n.d., no data.

[a] Community types: abbreviations use the first three letters of the genus name plus the first three letters of the species name, e.g. Cetnig-Rhigeo is community type *Cetraria nigricans – Rhizocarpon geographicum*.
[b] 1 (protected) to 4 (very exposed).
[c] 1 (extremely xeric) to 10 (hydric).
[d] 1 (snow-free all year) to 10 (deep snow all year).
[e] 1 (stable) to 5 (disturbed more than once annually).
[f] Gravimetric soil moisture.
[g] Saturated paste method.

them from other subtypes; they are defined mainly on the basis of differences in species dominance. Facies are finer divisions of subtypes that relate to the general physiognomy of the plant canopy.

The dominant vegetation types that generally follow a moisture gradient from dry-to-wet sites along an idealized hillslope gradient within the IRS (Fig. 4.7) are described in Sections 4.3.2.1–4.3.2.6. A summary of key environmental and soil variables is given in Table 4.1.

4.3.2.1 Lichen-Covered Rocks

Crutose and foliose lichens are abundant on most rocks and dominate vegetation communities on blockfields, sorted stone polygons, and isolated glacial erratics. Most of the rocks within the IRS are acidic conglomerate sandstone derived from the Kanayut formation. We sampled only two of these rocky sites, and defined a single community type, *Cetraria nigricans-Rhizocarpon geographicum*. This unit comprises only about 0.2% of the IRS. A detailed analysis of lichen communities and non-rock substrates, including species composition and aboveground biomass, is given in Chap. 5 (this Vol).

4.3.2.2 Dry Heath

Dry heaths occur on windblown ridge tops and on early-melting snowbed areas. Prostrate shrubs and fruticose lichens are the primary growth forms. Winter snow cover strongly affects the relative dominance of species within the heath vegetation types. Very exposed windblown sites with *Selaginella sibirica-Dryas octopetala* communities are common on sandstone outcrops within the R4D region (Chap. 5, this Vol.), but do not occur within the IRS. Somewhat less exposed sites with shallow snow cover occupy about 4.8% of the IRS; the main plant communities are *Hierochloë alpina-Arctous alpina*, and *Hierochloë alpina-Betula nana*. Areas with deeper snow (50–150 cm) have the *Diapensia lapponica-Cassiope tetragona* community type.

Fig. 4.7a–f. Common vegetation types in the intensive research site. **a** Dry acidic glacial till with community type *Hierochloë alpina-Arctous alpina*; subtype *Arctous alpina*; **b** moderately deep snowbeds with community type *Nephroma arcticum-Cassiope tetragona*; **c** moist uplands with shrub-rich tussock tundra, community type *Sphagnum rubellum-Eriophorum vaginatum* subtype *Betula nana*; **d** well-developed water track with community type *Eriophorum angustifolium-Salix planifolia* ssp. *pulchra*; **e** Elevated microsites in colluvial basins have community type *Sphagnum lenense-Salix fuscescens*, and wet microsites between hummocks have community type *Sphagnum orientale-Eriophorum scheuchzeri*; **f** Beaded pond with *Eriophorum angustifolium-Carex aquatilis* around margin of pond (vegetation in the pond is *Sparganium hyperboreum-Hippuris vulgaris*)

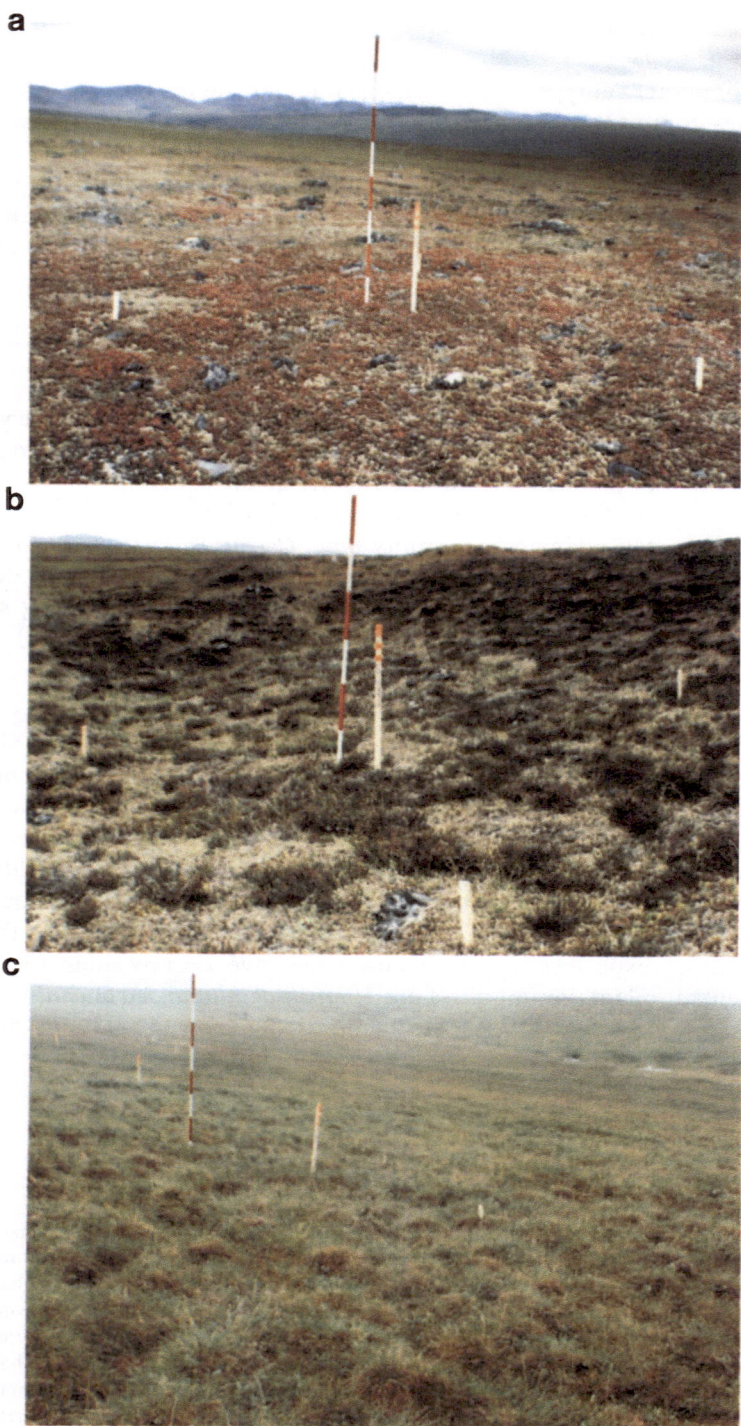

Fig. 4.7

Fig. 4.7 (*continued*).

4.3.2.2.1 Exposed Sites

The *Hierochloë alpina-Arctous alpina* community (Fig. 4.7a) occurs on dry, rocky, windblown, acidic glacial deposits that cover about 2.6% of the IRS (Fig. 4.6); common species include *Arctous alpina, Hierochloë alpina, Salix phlebophylla, Vaccinium vitis-idaea, V. uliginosum* ssp. *microphyllum, Ledum palustre* spp. *decumbens*, the mosses *Dicranum* cf. *elongatum, Polytrichum strictum, P. piliferum*, and the lichens *Alectoria ochroleuca, Asahinea chrysantha, Bryocaulon divergens, Cetraria cucullata, Cetraria nivalis, Cetraria islandica, Cladonia amaurocraea, C. coccifera, C. gracilis, Dactylina arctica, Masonhalea richardsonii, Sphaerophorus globosus*, and *Thamnolia subuliformis*. The soils are rocky, deeply thawed, acidic, mineral soils with shallow organic-rich surface horizons. The depth of thaw at the end of summer is 82 ± 18 cm; mean soil pH is 4.0 ± 0.1 (Table 4.1). Three physiognomically similar subtypes are recognized on the basis of the dominant species: *Arctous alpina* (Fig. 4.7a), *Salix phlebophylla*, and *Vaccinium vitis-idaea*.

The *Hierochloë alpina-Betula nana* community is dominated by dwarf birch. This community type is found in dry, somewhat protected sites such as depressions in the till or in the lee of larger rocks on till deposits. It covers about 2.2% of the IRS. It has a suite of moist species that thrive in the protection of the birch overstory including *Pedicularis labradorica, Aulacomnium turgidum, Dicranum angustum*, and *Hylocomium spendens*. The soils are moister and have thicker organic surface horizons (Table 4.1) than those of the *Hierochloë alpina-Arctous alpina* community, due to the presence of moss carpets.

4.3.2.2.2 Snowbeds

The *Diapensia lapponica-Cassiope tetragona* community occurs in acidic sites with shallow to moderately deep snowbeds (0.3–4.0 m of snow). It is characterized by a high cover of dwarf and prostrate shrubs (e.g., *Cassiope tetragona, Diappensia lapponica* ssp. *obovata, Ledum palustre* ssp. *decumbens, Loiseleuria procumbens, Salix phlebophylla, Vaccinium vitis-idaea, V. uliginosum*) and fruticose lichens (e.g., *Cladina rangiferina, C. arbuscula, Cetraria cucullata, C. nivalis, C. islandica, Cladonia gracilis, C. amaurocraea, Dactylina arctica, Peltigera aphthosa, Stereocaulon tomentosum*, and *Sphaerophorus globosus*). Common forbs include *Pedicularis capitata, P. langsdorffii* ssp. *arctica*, and *Polygonum bistorta. Cassiope tetragona* is typically found in moderately deep snowbeds in much of the Arctic (Nordhagen 1943; Churchill 1955; Rönning 1965; Alexandrova 1980; Cooper 1986; Evans et al. 1989; Walker 1990). Although it occurs outside of snowbeds as well, it reaches its greatest cover and abundance on moderately deep snowbeds. It is relatively insensitive to substrate pH (Böcher 1954). Two subtypes were recognized: *Calamagrostis inexpansa* and *Nephroma arcticum*.

The *Calamagrostis inexpansa* subtype occurs primarily on the dry mineral element of nonsorted stripes, which often also holds shallow snow drifts. Highly variable, this type is intermediate between the dry heath and the true snowbed communities. It covers about 4.9% of the IRS (Fig. 4.6). This subtype has a high cover of dry species and species occurring on mineral soils (e.g., *Cladonia coccifera*, *Dicranum* cf. *elongatum*, *Petasites frigidus*, *Salix phlebophylla*, and *Sphaerophorus globosus*), and a relatively low cover of mesic species (e.g., *Cassiope tetragona*, *Hylocomium splendens*, and *Aulacomnium turgidum*). The soils are deeply thawed (>100 cm).

Subtype *Nephroma arcticum* occurs in moderately deep snowbeds, often on north-facing slopes where the snow cover may persist through late June to early July. A few poorly developed versions of this type occur on east-facing depressions within the study area, where snow depths are about 75 cm in early May (Fig. 4.7b). This unit covers only 0.3% of the IRS. The moss carpet is better developed than in the *Calamagrostis inexpansa* subtype, and therefore harbors a relatively high cover of mesic taxa such as *Hylocomium splendens*, and a relatively low cover of dry taxa such as *Salix phlebophylla*. Other common taxa include *Carex microchaeta. C. tetragona*, *Dicranum scoparium*, *Dactylina ramulosa*, *Huperzia selago*, *Nephroma arcticum*, *Novosieversia glacialis*, *Parrya nudicaulis*, *Pogonatum urnigerum*, *Ptilium crista-castrensis*, *Pyrola grandiflora*, *Rhytidium rugosum*, *Saxifraga nelsoniana*, *Senecio atropurpureus*, and *Abietinella abietina*.

4.3.2.3 Tussock Tundra

Tussock tundra is the zonal vegetation on mesic slopes throughout the foothills. Considerable variations in species and growth-form dominance within this broad physiognomic unit correspond to variations in substrate pH, site stability, snow, and soil moisture (Walker et al. 1994). *Sphagnum rubellum-Eriophorum vaginatum* is the most common community type on acidic uplands (Fig. 4.7c), and covers about 58% of the IRS (Fig. 4.6). Important vascular species associated with this community include *Eriophorum vaginatum*, *Betula nana*, *Salix planifolia* ssp. *pulchra*, *Vaccinium uliginosum*, *V. vitis-idaea*, *Rubus chamaemorus*, *Carex bigelowii*, *Cassiope tetragona*, *Ledum palustre* ssp. *decumbens*, *Petasites frigidus*, *Bistorta vivipara*, and *Pedicularis lapponica*. Common mosses include *Sphagnum angustifolium*, *S. balticum*, *S. lenense*, *S. rubellum*, *S. teres*, *Dicranum* spp., *Polytrichum strictum*, *Hylocomium splendens*, *Aulacomnium turgidum*, and *A. palustre*. Common lichens are *Cetraria cucullata*, *C. islandica*, *Cladina rangiferina*, *C. arbuscula*, *Cladonia amaurocraea*, *C. gracilis*, *Dactylina arctica*, and *Peltigera aphthosa*.

The soils are wet with shallow organic horizons generally 10–25 cm thick. The mineral portion of the soils are generally high in clay (42 ± 5%) and are gleyed (neutral gray color due to waterlogging and lack of oxygen) with abundant orange mottles due to intermittent saturation. The average depth of the active layer is 36 ± 3 cm (Table 4.1).

Nonacidic uplands (community type *Tomentypnum nitens-Carex bigelowii*) are rare within the Imnavait Creek region, but they are found in few sites outside of the watershed where the groundwater pH is relatively high, such as at the base of some steep slopes. Although the physiognomies of acidic and nonacidic moist tundra are similar, few species occur in both types (e.g., *Eriophorum vaginatum, Carex bigelowii, Aulacomnium turgidum*, and several lichen species). Minerotrophic species dominate the nonacidic sites (e.g., *Dryas integrifolia, Salix reticulata, Salix arctica, Eriophorum triste, Lagotis glauca, Tofieldia pusilla, Rhododendron lapponicum, Pedicularis oederi, Equisetum variegatum, Saussurea angustifolia, Tomentypnum nitens, Orthothecium chryseum*, and *Ditrichum flexicaule*).

Five acidic tussock tundra subtypes can be consistently recognized on the basis of species and growth-form dominance: subtypes *Eriophorum vaginatum, Betula nana, Carex bigelowii, Salix planifolia* ssp. *pulchra*, and *Cassiope tetragona*. These subtypes show distinct distribution patterns with regard to micro- and mesotopographic slope positions, moisture, snow, and site stability. The *Cassiope tetragona* subtype occurs mainly on gentle (<4 %) slope shoulders in interstripe elements of nonsorted stripe complexes and in association with shallow snow drifts. Subtype *Eriophorum vaginatum* characterizes gentle upper slopes and footslopes, reaching its peak abundance on slopes of less than 9%, whereas subtypes *Betula nana* and *Salix planifolia* ssp. *pulchra* are more common near water tracks and on steeper slopes (>12%). Similar trends have been noted for the distribution of *Eriophorum vaginatum, Carex bigelowii*, and dwarf shrubs on a southwest-facing toposequence on the Seward Peninsula (Racine 1981).

4.3.2.4 Riparian Areas

Most of the IRS consists of a mosaic of poorly-defined water track communities and tussock tundra – only a few well-defined water tracks are present. Vegetation in the water tracks varies with the degree of channel development. Weakly developed water tracks have communities that are scarcely distinguishable from tussock tundra, whereas well-developed tracks contain distinctive willow and dwarf-birch communities. The common sequence of communities associated with well-defined water tracks is (1) *Sphagnum rubellum-Eriophorum vaginatum* subtype *Eriophorum vaginatum* on the interfluves, grading into (2) subtype *Betula nana* (tussock-tundra faces), grading into (3) subtype *Betula nana* (shrubland facies), grading into (d) subtype *Salix planifolia* ssp. *pulchra* (shrubland facies), grading into (5) community-type *Eriophorum angustifolium-Salix planifolia* ssp. *pulchra* in the channel of the water track (Fig. 4.7d). In addition, tracks where the water flows all summer long harbor a sixth community in the track channel dominated by almost pure stands of *Eriophorum angustifolium*.

Sphagnum rubellum-Eriophorum vaginatum, subtype *Betula nana* (shrubland facies), forms distinctive margins along many well-developed water tracks. It occupies about 9% of the IRS. It is easily recognized by the dominance of *Betula nana* and *Rubus chamaemorus*, and thick *Sphagnum* moss carpet. It is particularly noticeable in the fall when the *Betula nana* turns red and contrasts with the yellow willows in the center of the tracks. Although this community type is characterized by the high constancy and abundance of *B. nana* and *R. chamaemorus*, and the sporadic occurrence and low cover of *Eriophorum vaginatum* and *Carex bigelowii*, there are no characteristic taxa that distinguish it from other tussock tundra subtypes. Soil pH is low, averaging 4.4. The soils have deep organic horizons that cause reduced summer thaw relative to the adjacent tussock tundra areas (e.g., thaw in the birch subtype averaged 35 ± 1 cm and that in the *Eriophorum vaginatum* tussock tundra averaged 48 ± 4 cm). This may tend to hydrologically isolate the water track from the surrounding tussock tundra.

Subtype *Salix planifolia* ssp. *pulchra* (shrubland facies) covers 7.2% of the IRS. It occurs in upland water tracks and on lower slopes, areas with a somewhat higher seasonal water flux than areas with subtypes *Eriophorum vaginatum* and *Betula nana*. These stands are characterized by high constancy and a cover of *Aulacomnium palustre*, *Petasites frigidus*, *Rubus chamaemorus*, *S. planifolia* ssp. *pulchra*, and *Sphagnum rubellum*, as well as the differential taxa *Saxifraga nelsoniana*, *Bistorta vivipara*, *Poa arctica*, *Sanionia uncinatus*, *Scapania paludicola*, and *Sphagnum warnstorfii*. Soil characteristics differ between the *Betula* and *Salix* subtypes, reflecting their different positions in the landscape. Greater water flux in the *S. planifolia* ssp. *pulchra* subtype results from a deeper summer thaw (56 ± 2 cm) and soils that are relatively minerotrophic and aerobic. Soil pH is relatively high (5.1 ± 0.9 compared with 4.2 ± 0.2 for the *Betula nana* subtype).

The distinctive community type *Eriophorum angustifolium-Salix planifolia* ssp. *pulchra* (Fig. 4.7d) occurs in the channels of well-defined water tracks, and covers about 0.5% of the IRS. These stands have medium-height (50–100 cm) *S. planifolia* ssp. *pulchra* with *Eriophorum angustifolium* understories. Species diversity is relatively low, and the stands are floristically distinct from tussock tundra areas, with several differential taxa, including *Calliergon stramineum*, *C. giganteum*, *Eriophorum angustifolium* var. *subarcticum*, *Valeriana capitata*, and *Polemonium acutiflorum*. These sites accumulate snow because of depressions associated with the water tracks (mean snow depth in early May is 66 ± 3 cm).

Riparian areas along upper Imnavait Creek are representative of headwaters of many high-elevation tundra streams. They lack well-developed willow shrublands that occur along the lower reaches of these streams. The riparian vegetation grades into the mire vegetation types found in the colluvial basin at the head of Imnavait Creek (see Sect. 4.3.2.5). The streamside community *Salix chamissonis-Carex aquatilis* covers about 0.3% of the IRS and is physiognomically similar to the mire communities, but it has species that are rare elsewhere including *Aconitum delphinifolium* ssp. *paradoxum, Anemone richardsonii, Aster sibiricus, Calamagrostis canadensis, Climacium dendroides, Dodecatheon frigicum, Gentianella propinqua* ssp. *arctophila, Festuca altaica, Lycopodium annotinum, Paludella squarrosa, Philonotis fontana* var. *pumila, Plagiomnium ellipticum, Polemonium acutiflorum, Rubus arcticus, Senecio lugens, Salix chamissonis, Solidago multiradiata* var. *multiradiata, Valeriana capitata, Wilhelmsia physodes,* and *Zygadenus elegans.* These species become more common downstream from the IRS. Better-developed riparian willow shrublands (*Salix lanata* spp. *richardsonii, S. planifolia* ssp. pulchra, and *S. alaxensis*) occur along the meandering Kuparuk River.

4.3.2.5 Mires

The Imnavait Creek colluvial basin exhibits complex patterns of plant communities that are related to microtopography and the relative position of microsites with regard to the water table. The basin is relatively acidic compared with, for example, colluvial basins near Toolik Lake. Most wetlands in the Imnavait Creek watershed have soil pH ranging from 4.1 to 4.8; at Toolik Lake, soil pH ranges from 4.6 to 6.6. The acidity of the soil and water is presumably derived from the cation exchange system of *Sphagnum* (Clymo 1963; Clymo and Hayward 1982) and the concentration of organic acids from decomposition (Gorham et al. 1985).

The colluvial basin at the head of Imnavait Creek is a mosaic of poor fens and bogs. The pH of the water in Imnavait Creek averages 5.9 ± 2.0, and Ca ion concentration averages 55.3 ± 11.2 μeq l^{-1} (Chap. 9, this Vol.). If these values can also be used for the surface waters in the nearby colluvial basin, then Ca ion concentrations are sufficiently low to satisfy the criteria for a bog, but the pH values are those of an intermediate fen (Heinselman 1970). Some low microsites even have a few rich-fen plant species such as *Sphagnum warnstorfii, Scorpidium scorpioides,* and *Aneura pinguis.* Permafrost hydrologically isolates some palsas and raised microsites above the ground-water table due to permafrost, and pH of these sites is very low (<4).

We mapped three broad mire community types, one that occurs in slightly minerotrophic sites, and two in acidic mires. *Eriophorum angustifolium-Carex aquatilis* occurs in somewhat minerotrophic water tracks and the channel of Imnavait Creek between beaded ponds. This type consists of almost pure stands of *C. aquatilis* and *E. angustifolium* ssp. *subarcticum.* These stands are floristically depauperate. There are no unique faithful taxa nor any differential taxa. Similar communities inhabit the shallow water of the mires in the colluvial basin. These communities occupy only about 0.1% of the IRS.

Acidic mires predominate in most of the colluvial basin. Community-type *Sphagnum lenense-Salix fuscescens* (Fig. 4.7e) occurs on raised microsites, such as strangs and hummocks, and covers about 4.4% of the IRS. It includes the characteristic acidic-wetland taxa: *Carex rariflora, C. rotundata,*

Eriophorum scheuchzeri, Polytrichastrum alpinum, Sphagnum fimbriatum, S. imbricatum, and *S. rubellum,* plus the differential species found in slightly less-wet microsites, *Andromeda polifolia, Salix fuscescens, Sphagnum aongstroemii,* and *S. lenense.*

Community-type *Sphagnum orientale-Eriophorum scheuchzeri* (Fig. 4.7e) occurs in low microsites of wet meadows, pond margins, and lowland water tracks. It has a single faithful taxon, *Sphagnum orientale;* it covers 1.2% of the IRS. *Pedicularis albolabiata* and *Carex rotundata* reach their peak abundance in this type.

4.3.2.6 Beaded Ponds

Several beaded ponds in Imnavait Creek have aquatic communities. The *Hippuris vulgaris-Sparganium hyperboreum* community occurs in water up to 2 m deep, and occupies less than 0.1% of the IRS. The shallower pond margins have a group of characteristic aquatic taxa including *Caltha palustris, Comarum palustre, Hippuris vulgaris, Sparganium hyperboreum, Sphagnum squarrosum,* and *S. lindbergii* (Fig. 4.7f). The soils are generally thick mats of undecomposed peat moss.

4.4 West-Facing Toposequence

The trends in surface forms, vegetation structure, species composition, and soil characteristics along west-facing hillslope gradients are summarized in the idealized toposequence (Fig. 4.8; see Fig. 4.1 for location of the transects). Generally, the crests of the hills have rocky glacial till with frost scars, acidic mineral soils with organic horizons surface horizons (Pergelic Cryumbrepts in the US soil taxonomy; Soil Survey Staff 1975), and dry, dwarf-shrub, fruticose-lichen tundra (dry heath vegetation). The hillslope shoulders have nonsorted stripes with frost-scar complexes, Pergelic Cryumbrept soils, and dry, dwarf-shrub, fruticose-lichen tundra on the stripes. Between the stripes are wet nonacidic soils with thin organic horizons (Pergelic Cryaquepts) and moist, sedge, dwarf-shrub tundra. The upper backslopes have nonpatterned surfaces, acidic Pergelic Cryaquept soils, and tussock tundra. The interfluves between water tracks show an abundance of gelifluction features and thicker surficial organic layers downslope, so that soils near the footslope are classified as acidic Histic Pergelic Cryaquepts (organic horizons >25 cm thick). The vegetation becomes increasingly dominated by dwarf shrubs (e.g., *Betula nana* and *Rubus chamaemorus*) and *Sphagnum* moss. The gradual hillslope interfluve transition often culminates on the footslope with a moist or wet dwarf-shrub moss tundra. The transition from the hillslope to the colluvial basin or stream floodplain is generally abrupt.

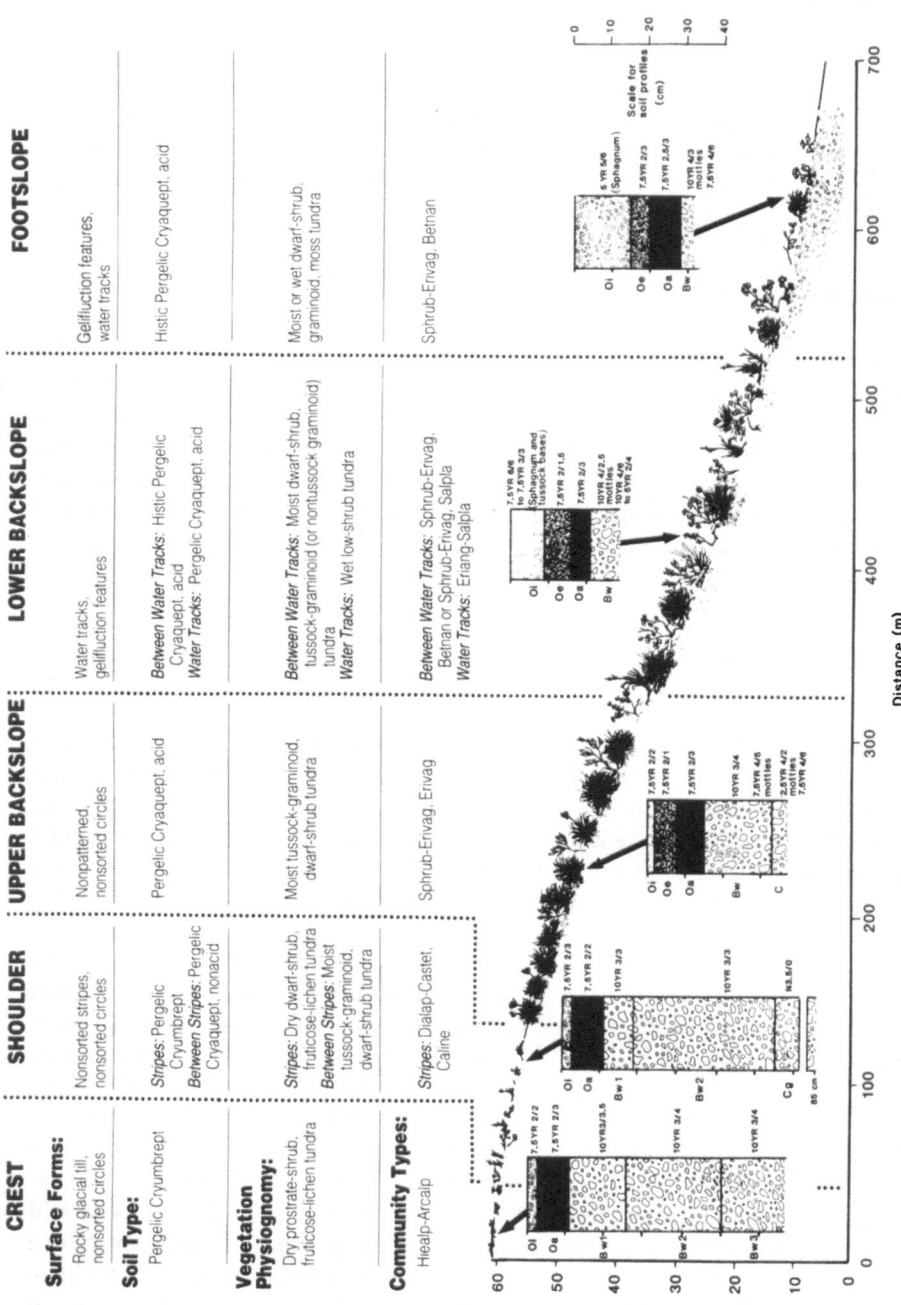

Fig. 4.8. Idealized toposequence for the west-facing slope at the intensive research site. (Modified from Walker et al. 1989b)

There are a number of downslope trends in vegetation along the toposequence. These are illustrated in Fig. 4.9a–e: (a) an overall decrease in species richness, due mainly to (b) a bimodal distribution for dwarf shrubs; (c) a maximum of graminoid species (mostly sedges) on the upper backslope; (d) the abundance of lichens on the hillcrest and shoulder; and (e) a general increase in mosses, primarily *Sphagnum*. On the ridges prostrate ericaceous species (e.g., *Arctous alpina*, *Vaccinium vitis-idaea*, *Diapensia lapponica*) are abundant, whereas deciduous dwarf shrubs (e.g., *Betula nana* and *Salix planifolia* ssp. *pulchra*) dominate the footslope.

The abundant mosses on the lower slopes strongly affects the hydrological regimes, soil nutrient regimes, and permafrost conditions. The soils are more organic downslope; active layers are thinner, and pH and nutrients decline (Fig. 4.9f–j). *Sphagnum* moss is thought to be a keystone species controlling the patterns of several soil and vegetation parameters along the slope gradient. The moss is able to establish on hillslopes with sufficient soil moisture and low soil pH. Older landscapes have more leached nutrient-poor and organic-rich soils, ideal conditions for the establishment of *Sphagnum*. Once it becomes established, it tends to further promote the conditions for *Sphagnum* growth, because it can hold 10–25 times its dry weight in water (Vitt et al. 1975) and acidify the soil (e.g., Clymo 1963). This process by which peatlands develop on previously well-drained sites is termed paludification, and has been described from boreal forested ecosystems worldwide (e.g., Auer 1928; Lawrence 1958; Heinselman 1970; Ugolini and Mann 1979; Noble et al. 1984; Klinger 1990). Viereck (1966) has described a similar process leading to *Sphagnum*-rich tussock-tundra vegetation on the outwash gravels of the Muldrow Glacier.

An important consequence of the thicker organic mats and shallower active layers downslope is that thawed mineral material decreases to an almost negligible thickness on the footslope (Fig. 4.8). Runoff passes through the organic horizons, which contributes to the decline in pH and soil nutrients downslope. On the Imnavait Creek hillslope, the pH on the hillcrest at the top of the mineral horizon is 4.7 rising to 5.3 on the mineral-rich slope shoulder, and then decreases steadily to pH 3.8 on the footslope. There is also a trend of lower pH in the lower soil horizons and higher pH for the interstitial waters (Fig. 4.9h). The soil water pH values on the footslope are comparable to values reported from Imnavait Creek waters (Oswood et al. 1989). Soil nutrients (e.g., Ca and NO_3; Figs. 4.9i,j) measured at the top of the Bw horizon generally peak in the shoulder area and then decrease downslope with very low values on the footslope. For example, nitrate-nitrogen declined from 2.6 $\mu g\,g^{-1}$ on the shoulder to negligible values on the lower backslope and footslope (Fig. 4.9j). The formation of *Sphagnum*-rich peat on the hill slopes is, thus, a prerequisite to the development of poor-fen vegetation in the lowlands, because the water entering the colluvial basins is acidified by passing through the *Sphagnum* peat on the hill slopes.

The toposequence at Imnavait Creek has many of the same characteristics that Hamilton (1986) described for Sagavanirktok-age (mid-Pleistocene)

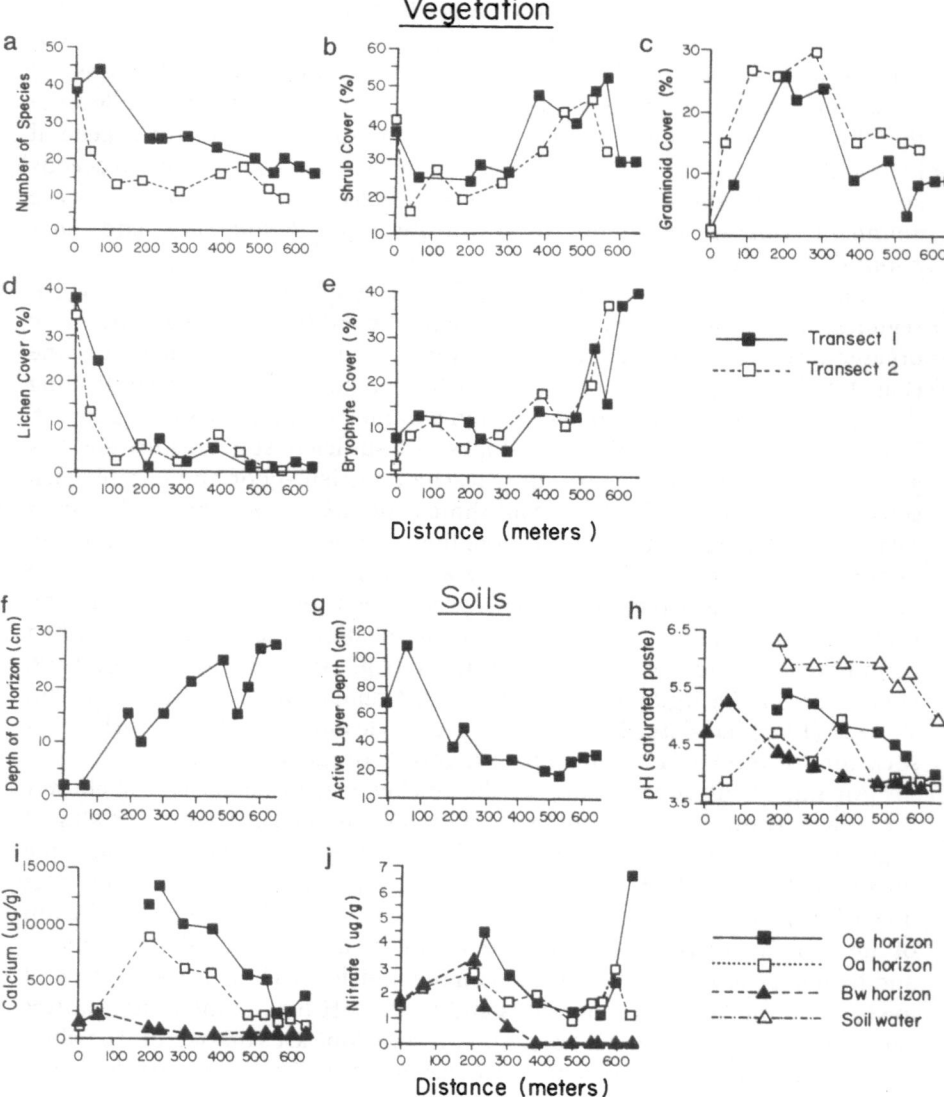

Fig. 4.9a–j. Trends in key vegetation and soil characteristics downslope along the west-facing toposequence. See text for explanation. (Modified from Walker et al. 1989b)

moraines on the North Slope. Terrain of this age is dominated by acidic tussock tundra, thaw depressions, bogs, and water-track complexes (Jorgenson 1984; Hamilton 1986). The hill crests are generally sites of dry heath vegetation and are usually free of tussock tundra. Older landscapes to the north of Imnavait Creek that were glaciated during Anaktuvuk Glaciation (early Pleistocene, deglaciated about 250 000 years ago) have broader, more rounded moraine crests with tussock tundra overtopping the hill crests, few glacial erratics protruding above the tundra surface, better integrated stream networks, and few lakes.

The nearby Sagavanirktok River valley and the Toolik Lake region have Itkillik glacial till deposits (late-Pleistocene, deglaciated 11 500 to 60 000 years ago). The geomorphological contrast between the Itkillik and Sagavanirktok drifts is strong. The Sagavanirktok surfaces are generally more highly eroded with smooth slopes, broad hill crests, few glacial erratics, mature drainage systems, and a more extensive tussock tundra. In contrast, the younger Itkillik surfaces have a more irregular topography, stony surfaces, steeper slopes, deranged drainage systems, and more heterogeneous vegetation cover (Hamilton 1986). These areas are dominated by dry heath vegetation, nonacidic tundra on moist sites, nonsorted stripe complexes, and less tussock tundra (Jorgenson 1984).

Based on our experience while mapping the region's vegetation, the vegetation trends described for this toposequence are representative of other long west-facing slopes in the region. Slopes of other aspects are less common and show distinctively different zonal vegetation with regard to slope position. For example, east facing slopes tend to be shorter and steeper, and accumulate snow because of the westerly and southwesterly winter storm winds and less solar heating (Chap. 6, this Vol.). Consequently, east-facing slopes tend to be wetter with more snowbed plant communities and more abundant shrubs. North-facing slopes also accumulate snow because of the prevailing southerly winds. More than 4 m of snow can accumulate on the steeper north-facing slopes, and snowbed plant communities are common. Gentle north-facing slopes downslope of large sonwdrifts are generally covered by well-developed water tracks formed by meltwater from the drifts. In contrast, the few south-facing slopes occurring on bedrock outcrops are steep, dry, blown free of winter snow, and offer mineral substrates for communities dominated by *Dryas octopetala* (Chap. 5, this Vol.).

4.5 Terrain Sensitivity to Disturbance

Thermokarst is the melting of massive ground ice to form a topography of depressions, hummocks, and small ponds (Washburn 1980), often an undesirable consequence from construction in arctic regions (see Chap. 3, this Vol.). The distribution of ground ice along hillslope gradients is related to landscape age and slope position. Older glacial deposits on the North Slope generally have greater amounts of massive ground ice because older till is generally finer grained and has had more time to form segregated ice (Brown and Kreig 1983). According to Jorgenson's (1984) hypothesis of landscape evolution for the glaciated foothills, the accumulation of organic matter on well-drained sites reduces heat flux and the depth of summer thaw, and leads to the accumulation of aggradation ice at the permafrost/active-layer boundary.

On the Imnavait Creek toposequence the organic layer increases from about 2 cm on the hill crest to about 30 cm on the footslope. The thawed zone (active layer) is over 100 cm thick on the hill shoulder and declines to about 25 cm on the footslope (Figs. 4.8 and 4.9f–j). Soil pits along the toposequence were dug to penetrate the top of the permafrost/active-layer boundary. Although no quantitative assessment of ice was made, there was an obvious trend toward greater amounts of clear ice encountered at the permafrost boundary on the lower slope positions. Thickening of the organic mat and reduced heat flux are known to contribute to an aggrading permafrost table, which, combined with large amounts of water flowing from upslope, favor the accretion of ice layers near the top of the permafrost table (Washburn 1980). The combination of thin active layers and large amounts of ice near the soil surface make the lower slopes particularly sensitive to disturbance. This sensitivity is enhanced by the thick moss mats that are easily compressed or removed by physical disturbance. This alters the local microtopography and hydrology, and influences the heat budget and leads to thermokarst.

4.6 Conclusions

The Imnavait Creek IRS is a high-altitude, acidic tussock-tundra site that is representative of large regions of foothills that were glaciated during the mid-Pleistocene. The site contains good examples of typical features of the Arctic Foothills physiographic province including broad smooth hill slopes, glacial till surfaces, water tracks, nonsorted stripes, colluvial basins, and beaded streams.

The known flora for the region is 174 vascular plants, 81 mosses, 20 liverworts, and 95 lichens. Imnavait Creek is in the hypoarctic zone, and the North Alaska subprovince of the Alaska province of the Beringian sector (Yurtsev 1994). The relative depauperate flora of the watershed is due primarily to the acidic substrates and low habitat diversity.

A detailed vegetation map of the research site contains 22 map units, 7 of which are tussock-tundra subtypes and facies. Imnavait Creek is dominated by acidic tussock tundra that has few species in common with nonacidic tussock tundra areas. Tussock tundra is much more variable than has been generally assumed; there is considerable variation in species and growth-form dominance related to local site factors including soil pH, snow regime, site stability, and hydrological regime.

An idealized toposequence on the west-facing slope of the study area is useful for examining geoecological interactions between terrain, soil, vegetation, and ground-ice conditions. Paludification is an important process controlling vegetation distribution on hillslopes in the Low Arctic, and strongly influences the occurrence of massive ground ice. The older foothill landscapes glaciated during the early- and mid-Pleistocene generally have greater amounts of ground ice and are more susceptible to disturbance (Kreig and Reger 1982). Within these old landscapes, upper slopes, particularly hill shoulder areas with nonsorted stripes, are less susceptible to disturbance because of course-grained, deeply thawed soils and less ice in the soils, whereas the footslopes and areas with deep *Sphagnum* moss mats are more susceptible because of thin active layers and the presence of considerable pure ice near the soil surface.

Acknowledgments. This work was supported by the Department of Energy R4D Program (grant no. DE–84ER60242) and the Arctic LTER project (BSR 8705816). Many people have contributed to this work. Special thanks go to Alan Batten, Barbara Murray, and Dave Murray for plant determinations, editing of species lists, and data from the Northern Plant Documentation Center. We are also very grateful for the help and support of the following individuals: William Acevedo, Nancy Auerbach, Liz Arnold, Carl Benson, John Dietz, Barry Evans, Kaye Everett (deceased), Mike Figgs, Kim Furry, Allan Hope, Torre Jorgenson, Tom Hamilton, David Kallenbach, Doug Kane, Rolf Kihl, Lee Klinger, Vera Komárková, Nancy Lederer, Leanne Lestak, Brad Lewis, Glenn Liston, Walt Oechel, Gary Petersen, Scott Randolph, Jim Reynolds, Kristi Rose, Doug Stow, John Tenhunen, Dale Vitt, Curt Wessburg, and Pat Webber.

Appendix A. List of Plants for Imnavait Creek, Alaska[1]

Vascular Plants (174)

Acomastylis rossii (R. Br.) Greene
(=*Geum rossii*)
Aconitum delphinifolium DC. ssp.
delphinifolium
Aconitum delphinifolium DC. ssp.
paradoxum (Reichb.) Hult.[2]
Alopecurus pratensis L.
Andromeda polifolia L.
Anemone drummondii S. Wats.
Anemone parviflora Michx.
Anemone richardsonii Hook[2]
Antennaria alpina (L.) Gaertner
var. *media* (Greene) Jepson (=*A. friesiana*)

Arctagrostis latifolia (R. Br.) Griseb.
var. *latifolia*
Arctophila fulva (Trin.) Anderss.
Arctous alpina (L.) Niedenzu
(=*Arctostaphylos alpina*)
Arnica angustifolia M. Vahl ssp.
angustfolia (=*A. alpina* ssp.
angustifolia)
Arnica griscomii Fern. ssp. *frigida*
(C. A. Mey.) S. J. Wolf (= *A. Frigida*)
Arnica lessingii Greene
Artemisia arctica Less. ssp. *arctica*
Artemisia tilesii Ledeb. ssp. *tilesii*
Aster sibiricus L.

---

[1] Nomenclature is according to species list provided by the University of Alaska Herbarium, Fairbanks, Alaska.
[2] Species recorded in sample plots, but not documented with collection at the University of Alaska Herbarium.

Astragalus alpinus L. ssp. *alpinus*
Astragalus umbellatus Bunge
Betula glandulosa Michx.
Betula nana L. ssp. *exilis*
  (Sukatsch.) Hult.
Bistorta plumosa (Small) Greene
  (=*Polygonum bistorta* ssp.
  plumosum)
Bistorta vivipara (L.) S. F. Gray
  (=*Ploygonum viviparum*)
Bupleurum triradiatum Adams ssp.
  arcticum (Regel) Hult.
Calamagrostis canadensis (Michx.)
  Beauv.
Calamagrostis inexpansa Gray
Calamagrostis purpurascens R. Br.
Caltha palustris L. ssp. *arctica*
  (R. Br.) Hult.[2]
Campanula lasiocarpa Cham.
Cardamine bellidifolia L.
Cardamine digitata Richards.
  (=*C. hyperborea*)
Cardamine pratensis L.
Carex aquatilis Wahlenb.
Carex atrofusca Schkuhr.
Carex bigelowii Torr.
Carex chordorrhiza Ehrh.
Carex lachenalii Schkuhr.
Carex michrochaeta Holm
Carex misandra R. Br.
Carex obtusata Lilj.
Carex podocarpa R. Br.
Carex rariflora (Wahlenb.) J. E. Smith
Carex rotundata Wahlenb.
Carex rupestris All.
Carex saxatilis L. ssp. *laxa* (Trautv.)
  Kalela
Carex scirpoidea Michx.
Carex vaginata Tausch
Cassiope tetragona (L.) D. Don ssp.
  tetragona
Chrysosplenium tetrandrum (Lund)
  T. Fries
Comarum palustre L. (=*Potentilla*
  palustris)

Crepis nana Richards.
Diapensia lapponica L. ssp. *obovata*
  (F. W. Schmidt) Hult.
Dodecatheon frigidum Cham. &
  Schlecht.
Douglasia ochotensis (Willd.) Hult.
Dryas integrifolia M. Vahl
Dryas octopetala L. var. *octopetala*
Empetrum hermaphroditum (Lange)
  Hagerup (=*E. nigrum* ssp.
  hermaphroditum)
Epilobium angustifolium L.
Epilobium davuricum Fisch.
Epilobium latifolium L.
Equisetum arvense L.
Equisetum palustre L.
Equisetum scirpoides Michx.
Equisetum variegatum Schleich.
Erigeron muirii Gray (=*E.*
  grandiflorus ssp. *muirii*)
Eriophorum angustifolium Honck.
  ssp. *subarcticum* (Vassiljev) Hult.
Eriophorum callitrix Cham.
Eriophorum russeolum E. Fries
Eriophorum scheuchzeri Hoppe var.
  scheuchzeri
Eriophorum triste (Th. Fries) Hadac
  & Love (=*E. angustifolium* ssp.
  triste)
Eriophorum vaginatum L.
Eritrichium aretoides (Cham.) DC.
Eutrema edwardsii R. Br.
Festuca altaica Trin.
Festuca brachyphylla Schult.
Festuca rubra L.
Gastrolychnis macrosperma (Pors.)
  Tolm. & Kozh. (=*Melandrium*
  macrospermum)
Gentiana glauca Pall.
Gentianella propinqua (Richards.)
  Gillett ssp. *arctophila* (Griseb.)
  Hult. (=*Gentiana propinqua* ssp.
  arctophila)[2]
Hierochloë alpina (Sw.) Roem. &
  Schult.

Hippuris vulgaris L.
Huperzia selago (L.) C. Martius ssp.
  appressa (Desv.) D. Löve
  (=*Lycopodium selago* ssp.
  appressum)
Juncus biglumis L.
Kobresia myosuroides (Vill.) Fjori &
  Paol.
Koeleria asiatica Domin
Lagotis glauca Gaertner
Ledum palustre L. ssp. *decumbens*
  (Ait.) Hult.
Linnaea borealis L.
Loiseleuria procumbens (L.) Desv.
Luzula arctica Blytt
Luzula confusa Lindeb.
Luzula kjellmaniana Miyabe &
  Kudo (=*L. tundricola*)
Luzula multiflora (Retz.) Lej.
Luzula wahlenbergii Rupr.
Lycopodium annotinum L. ssp.
  pugens (La Pyl.) Hult.[2]
Minuartia arctica (Stev.) Aschers. &
  Graebn.
Minuartia macrocarpa (Pursh)
  Ostenf.
Minuartia obtusiloba (Rydb.) House
Minuartia rubella (Wahlenb.)
  Graebn.
Novosieversia glacialis (Adams)
  Bolle (=*Geum glaciale*)
Orthilia secunda (L.) House ssp.
  obtusata (Turcz.) Böcher
  (=*Pyrola secunda* ssp. *obtusata*)
Oxytropis bryophila (Greene)
  Yurtsev (=*O. nigrescens* ssp.
  bryophila)
Papaver macounii Greene
Parrya nudicaulis (L.) Regel
Pedicularis albolabiata (Hult.) Kozh.
  (=*P. sudetica* ssp. *albolabiata*)
Pedicularis capitata Adams
Pedicularis labradorica Wirsing
Pedicularis lanata Cham. & Schlecht
  (=*P. kanei* ssp. *kanei*)

Pedicularis langsdorffii Fisch.
Pedicularis lapponica L.
Pedicularis oederi M. Vahl[2]
Pentaphylloides floribunda (Pursh)
  Löve (=*Pontentilla fruticosa*)[2]
Petasites frigidus (L.) Franch.
Phlox sibirica L.
Poa alpigena (E. Fries) Lindm.
Poa alpina L.
Poa arctica R. Br.
Poa glauca M. Vahl
Poa lanata Scribn. & Merr.
Poa paucispicula Scribn. & Merr.
Poa pseudoabbreviata Roshev.
  (=*P. branchyanthera*)
Polemonium acutiflorum Willd.
Potentilla uniflora Ledeb.
Pyrola grandiflora Radius
Ranunculus eschscholtzii Schlecht.
Rhododendron lapponicum (L.)
  Wahlenb.
Rubus arcticus L. ssp. *acaulis*
  (Michx.) Focke[2]
Rubus chamaemorus L.
Salix alaxensis (Anderss.) Cov.
Salix arbusculoides Anderss.
Salix arctica Pall.
Salix brachycarpa Nutt. ssp.
  niphoclada (Rydb.) Argus
  (=*S. niphoclada*)
Salix chamissonis Anderss.
Salix fuscescens Anderss.
Salix lanata L. ssp. *richardsonii*
  (Hook) A. Skvortsov[2]
Salix phlebophylla Andress.
Salix planifolia Pursh ssp. *pulchra*
  (Cham.) Argus (=*S. pulchra*)
Salix reticulata L. ssp. *reticulata*
Salix rotundifolia Trautv. ssp.
  rotundifolia
Saussurea angustifolia (Willd.)
  DC.
Saxifraga bronchialis L. ssp.
  funstonii (Small) Hult.
Saxifraga cernua L.

Saxifraga flagellaris Willd. ssp.
   setigera (Pursh) Tolm.
Saxifraga foliolosa R. Br. var.
   foliolosa
Saxifraga hieracifolia Waldst. & Kit.
Saxifraga hirculus L.
Saxifraga nelsoniana D. Don ssp.
   nelsoniana (=*S. punctata* ssp.
   nelsoniana)
Saxifraga nivalis L.
Saxifraga reflexa Hook.
Saxifraga rivularis L.
Saxifraga tricuspidata Rottb.
Selaginella sibirica (Milde) Hieron.
Senecio atropurpureus (Ledeb.)
   Fedtsch. ssp. *frigidus* (Richards.)
   Hult.
Senecio kjellmanii Pors. (=*S.
   atropurpureus* ssp. *tomentosus*)
Senecio lugens Richards.[2]
Silene acaulis L.
Smelowskia calycina (Stephen) C. A.
   Mey.
Solidago multiradiata Ait. var.
   multiradiata[2]
Sparganium hyperboreum Laest.
Spiraea stevenii (Schneid.) Rydb.
   (=*S. beauverdiana*)
Stellaria crassifolia Ehrh.
Stellaria edwardsii R. Br.
Stellaria laeta Richards.
Stellaria longipes Goldie
Tofieldia pusilla (Michx.) Pers.
Trisetum spicatum (L.) Richter
Vaccinium uliginosum L. ssp.
   alpinum (Bigel.) Hult.
Vaccinium uliginosum L. ssp.
   microphyllum Lange
Vaccinium vitis-idaea L. ssp. *minus*
   (Lodd.) Hult.
Valeriana capitata Pall.
Viola epipsila Ledeb.
Wilhelmsia physodes (Fisch.)
   McNeill
Zygadenus elegans Pursh[2]

Mosses [81]

Abietinella abietina (Hedw.)
   Fleisch. (=*Thuidium abietinum*)
Aulacomnium palustre (Hedw.)
   Schwaegr.
Aulacomnium turgidum (Wahlenb.)
   Schwaegr.
Brachythecium groenlandicum
   (C. Jens.) Schljak.
Brachythecium turgidum (Hartm.)
   Kindb.
Bryum algovicum Sendtn.
Bryum psuedotriquetrum (Hedw.)
   Gaertner et al.
Calliergon giganteum (Schimp.)
   Kindb.
Calliergon stramineum (Brid.)
   Kindb.
Campylium stellatum (Hedw.) C.
   Jens.
Catoscopium nigritum (Hedw.)
   Brid.
Ceratodon purpureus (Hedw.) Brid.
Cirriphyllum cirrosum (Schwaegr.)
   Grout
Dicranella varia (Hedw.) Schimp.
   (=*Anisothecium varium*)
Climacium dendroides (Hedw.)
   Web. & Mohr[2]
Dicranum acutifolium (Lindb. &
   Arnell) C. Jens.
Dicranum angustum Lindb. (=*D.
   laevidens*)
Dicranum elongatum Schleich.
Dicranum groenlandicum Brid.
Dicranum muehlenbeckii Bruch &
   Schimp.
Dicranum scoparium Hedw.
Dicranum spadiceum Zett. (=*D.
   neglectum*)
Drepanocladus brevifolius (Lindb.)
   Warnst. (=*D. lycopodioides* var.
   brevifolius)

Encalypta brevicolla Bruch &
Schimp.
Encalypta rhaptocarpa Schwaegr.
(=*E. vulgaris* var *rhaptocarpa*)
Hylocomium splendens (Hedw.)
Schimp. (=*H. splendens* ssp
obtusifolium)
Hypnum bambergeri Schimp.
Hypnum procerrimum Mol.
(=*Pseudostereodon procerrimum*)
Limprichtia revolvens (Sw.) Loeske
(=*Drepanocladus revolvens*)
Loeskypnum badium (Hartm.) Paul
(=*Drepanocladus badius*)
Meesia uliginosa Hedw.
Orthothecuim chryseum (Schwaegr.)
Schimp.[2]
Paludella squarrosa (Hedw.) Brid.
Philonotis fontana (Hedw.) Brid.
var. *pumila* (Turn.) Brid. (=*P.
tomentella*)[2]
Plagiomnium ellipticum (Brid.) T.
Kop. (=*Plagiomnium rugicum*)[2]
Plagiomnium medium (Bruch &
Schimp.) T. Kop. (=*Mnium
medium*)
Pleurozium schreberi (Brid.) Mitt.
Pogonatum urnigerum (Hedw.)
Beauv.
Pohlia andrewsii Shaw
Pohlia crudoides (Sull. & Lesq.)
Broth.
Pohlia elongata Hedw. var. *greenii*
(Brid) Shaw
Pohlia nyuans (Hedw.) Lindb. (=*P.
schimperi*)
Polytrichastrum alpinum (Hedw.)
G. L. Sm. var. *alpinum*
(=*Pogonatum alpinum*)
Polytrichum commune Hedw.
Polytrichum hyperboreum R. Br.
Polytrichum juniperinum Hedw.
Polytrichum longisetum Brid
(=*Polytrichastrum longisetum*)
Polytrichum piliferum Hedw.

Polytrichum sexangulare Brid.
Polytrichum strictum Brid
(=*P. juniperinum* var. *gracilius*)
Polytrichum swartzii Hartm.
(=*P. algidum*)
Pseudobryum cinclidioides (Hüb.)
T. Kop.
Ptilium crista-castrensis (Hedw.) De
Not.
Racomitrium lanuginosum (Hedw.)
Brid.
Rhizomnium andrewsianum
(Steere) T. Kop.
Rhytidium rugosum (Hedw.) Kindb.
Sanionia uncinata (Hedw.) Loeske
(=*Drepanocladus uncinatus*)
Sarmenthypnum sarmentosum
(Wahlenb.) Tuom & T. Kop.
(=*Calliergon sarmentosum*)
Scorpidium scorpioides (Hedw.)
Limpr.[2]
Sphagnum angustifolium C. Jens.
(=*S. recurvum* var. *tenue*)
Sphagnum aongstroemii Hartm.
Sphagnum balticum (Russ.) C.
Jens.
Sphagnum capillifolium (Ehrh)
Hedw. (=*S. nemoreum*)
Sphagnum compactum DC.
Sphagnum fimbriatum Wils. var.
fimbriatum
Sphagnum girgensohnii Russ.
Sphagnum imbricatum Hornsch.
(=*S. steerei*)
Sphagnum lenese H. Lindb.
Sphagnum lindbergii Schimp.
Sphagnum magellanicum Brid.
Sphagnum obtusum Warnst.
Sphagnum orientale Sav.-Ljub. (=*S.
perfoliatum*)
Sphagnum rubellum Wils. (=*S.
Capillifolium* var *tenellum*)
Sphagnum squarrosum Crome
Sphagnum subsecundum Nees var.
subscundum

Sphagnum teres (Schimp.) Ångstr.
Sphagnum warnstorfii Russ.
Splachnum sphaericum Hedw. (=*S. ovatum*)
Tetraplodon pallidus Hag.
Tomentypnum nitens (Hedw.) Lowske
Tortula ruralis (Hedw.) Gaertner et al.

Liverworts [20]

Anastrophyllum minutum (Schreb.) Schust.
Aneura pinguis (L.) Dum.
Barbilophozia binsteadii (Kaal.) Loeske (=*Lophozia binsteadii*)
Barbilophozia quadriloba (Lindb.) Loeske (=*Lophozia quadriloba*)
Blepharostoma trichophyllum (L.) Dum. ssp. *brevirete* (Bryhn & Kaal.) Schust. (=*B. trichophyllum* var. *brevirete*)
Chandonanthus setiformis (Ehrh.) Lindb. (=*Tetralophozia setiformis*)
Diplophyllum albicans (L.) dum.
Diplophyllum plicatum Lindb. (=*Microdiplophyllum plicatum*)
Gymnomitrion concinnatum (Lightf.) Corda
Lophozia guttulata (Lindb. & Arnell) A. Evans (=*L. porphyroleuca*)
Lophozia opacifolia Culm.
Lophozia ventricosa (Dicks.) Dum.
Mylia anomala (Hook.) S. F. Gray
Plagiochila arctica Bryhn & Kaal.
Pseudolepicolea fryei (Perss.) Grolle & Ando
Ptilidium ciliare (L.) Hampe
Radula prolifera S. Arnell
Scapania paludicola Loeske & K. Müll.
Scapania simmonsii Bryhn & Kaal.
Tritomaria quinquedentata (Huds.) Buch

Lichens [95]

Alectoria nigricans (Ach.) Nyl.
Alectoria ochroleuca (Hoffm.) Massal.
Arctoparmelia centrifuga (L.) Hale (=*Xanthoparmelia centrifuga*)
Arctoparmelia separata (L.) Hale (=*Xanthoparmelia separata*)
Asahinea chrysantha (Tuck.) Culb. & C. Culb.
Asahinea scholanderi (Llano) Culb. & C. Culb.
Bryocaulon divergens (Ach.) Kärnef (=*Cornicularia divergens*)
Caloplaca jungermanniae (Vahl) Th. Fr.
Catapyrenium lachneum (Ach.) R. Sant. (=*Dermatocarpon lachneum*)
Cetraria andrejevii Oxner
Cetraria commixta (Nyl.) Th. Fr.
Cetraria cucullata (Bellardi) Ach.
Cetraria delisei (Bory) Nyl. (=*C. hiascens*)
Cetraria fastigiata (Del.) Kärnef.
Cetraria hepatizon (Ach.) Vainio
Cetraria inermis (Nyl.) Krog
Cetraria islandica (L.) Ach.
Cetraria kamczatica Savicz.
Cetraria laevigata Rass.
Cetraria nigricans Nyl.
Cetraria nivalis (L.) Ach.
Cetraria tilesii Ach.
Cladina arbuscula (Wallr.) Hale & Culb. (=*Cladonia arbuscula*)
Cladina mitis (Sandst.) Hustich (=*Cladonia mitis*)
Cladina rangiferina (L.) Nyl. (=*Cladonia rangiferina*)
Cladina stellaris (Opiz) Brodo (=*Cladonia alpestris*)
Cladonia alaskana A. Evans
Cladonia amaurocraea (Flörke) Schaerer

Cladonia carneola (Fr.) Fr.
Cladonia cenotea (Ach.) Schaerer
Cladonia chlorophaea (Flörke)
    Sprengel
Cladonia coccifera (L.) Willd.
Cladonia cornuta (L.) Hoffm.
Cladonia deformis (L.) Hoffm.
Cladonia ecmocyna Leight.
Cladonia fimbriata (L.) Fr.
Cladonia gracilis (L.) Willd.
Cladonia macrophylla (Schaerer)
    Stenh.
Cladonia pleurota (Flörke)
    Schaerer
Cladonia pocillum (Ach.) O. Rich
Cladonia pyxidata (L.) Hoffm.
Cladonia subulata (L.) Weber
Cladonia sulphurina (Michx.) Fr.
Cladonia uncialis (L.) Weber
Coleocaulon aculeatum (Schreb.)
    Link (=Cornicularia aculeata)
Dactylina arctica (Richards.) Nyl.
Dactylina beringica Bird & Thomson
Dactylina madreporiformis (Ach.)
    Tuck.
Dactylina ramulosa (Hook.) Tuck.
Haematomma lapponicum Räsänen
Hypogymnia subobscura (Vainio)
    Poelt (=Parmelia subobscura)
Lecanora epibryon (Ach.) Ach.
Lecidoma demissum (Rutstr.) G.
    Schneider & Hertel (=Lecidea
    demissa)
Lobaria linita (Ach.) Rabenh.
    (=Sticta linita)
Masonhalea richardsonii (Hook.)
    Kärnef (=Cetraria richardsonii)
Melanelia septentrionalis (Lynge)
    Essl. (=Parmelia septentrionalis)
Melanelia stygia (L.) Essl.
    (=Parmelia stygia)
Mycoblastus sanguinarius (L.)
    Norman
Nephroma arcticum (L.) Torss
Nephroma expallidum (Nyl.) Nyl.

Ochrolechia frigida (Swartz) Lynge
    (=O. gonatodes)
Ochrolechia upsaliensis (L.) Massal.
Parmelia omphoalodes (L.) Ach.
Parmelia sulcata Taylor
Peltigera aphthosa (L.) Willd.
Peltigera canina (L.) Willd.
Peltigera horizontalis (Huds.)
    Baumg.
Peltigera leucophlebia (Nyl.) Gyelnik
Peltigera malacea (Ach.) Funck.
Peltigera polydactyla (Necker)
    Hoffm.
Peltigera scabrosa Th. Fr.
Pertusaria bryontha (Ach.) Nyl.
Pertusaria dactylina (Ach.) Nyl.
Pertusaria panyrga (Ach.) Massal.
Physconia muscigena (Ach.) Poelt
Porpidia flavocaerulescens
    (Hornem.) Hertel & Schwab.
    (=Lecidea flavocaerulescens)
Polyblastia gelatinosa (Ach.) Th. Fr.
Pseudephebe pubescens (L.) M.
    Choisy
Psoroma hypnorum (Vahl) S.F.
    Gray
Rhizpcarpon geographicum (L.) DC.
Rinodina turfacea (Wahlenb.)
    Körber
Solorina bispora Nyl.
Solorina crocea (L.) Ach.
Solorina saccata (L.) Ach.
Sphaerophorus fragilis (L.) Pers.
Sphaerophorus globosus (Huds.)
    Vainio
Stereocaulon alpinum Laur.
Stereocaulon paschale (L.) Hoffm.
Stereocaulon tomentosum Fr.
Thamnolia subuliformis (Ehrh.)
    Culb.
Tuckermannopsis pinastri (Scop.)
    Hale (=Cetraria pinastri)
Tuckermannopsis sepincola (Ehrh.)
    Hale (=Cetraria sepincola)
Umbilicaria caroliniana Tuck.

Umbilicaria hyperborea (Ach.)   *Umbilicaria proboscidea* (L.)
Hoffm.   Schrader

References

Alexandrova VD (1980) The Arctic and Antarctic: their division into geobotanical areas. Cambridge Univ Press, Cambridge
Auer V (1928) Some future problems of peat bog investigations in Canada. Commentat For 1: 1–31
Bliss LC (1956) A comparison of plant development in microenvironments of arctic and alpine tundras. Ecol Monogr 26: 303–337
Bliss LC (1962) Adaptation of arctic and alpine plants. Biol Rev 43: 481–529
Bliss LC, Matveyeva NV (1992) Circumpolar arctic vegetation. In: Chapin FS III, Jefferies RL, Reynolds JF, Shaver GR, Svobada J, Chu EW (eds) Arctic ecosystems in a changing climate: an ecophysiological perspective. Academic Press, San Diego, pp 59–89
Böcher TW (1954) Oceanic and continental vegetational complexes in southwest Greenland. Medd Grönl 148: 1–336
Brown J, Kreig RA (1983) Guidebook to permafrost and related features along the Elliot and Dalton highways, Fox to Prudhoe Bay, Alaska. In: Brown J, Kreig RA (eds) 4th Int Conf Permafrost, Univ Alaska, Fairbanks (Guidebook 4)
Cantlon JE (1961) Plant cover in relation to macro-, meso- and micro-relief. Final Rep, Office of Naval Research, Washington, DC, grants ONR-208 and 216
Churchill ED (1955) Phytosociological and environmental characteristics of some plant communities in the Umiat region of Alaska. Ecology 36: 606–627
Clymo RS (1963) An experimental approach to part of the calcicole problem. J Ecol 50: 707–731
Clymo RS, Hayward PM (1982) The ecology of *Sphagnum*. In: Smith AJF (ed) Bryophyte ecology. Chapman Hall, New York, pp 229–289
Cooper DJ (1986) Arctic-alpine tundra vegetation of the Arrigetch Creek Valley, Brooks Range, Alaska. Phytocoenologia 14: 467–555
Douglas LA, Tedrow JCF (1960) Tundra soils of arctic Alaska. In: Proc 7th Int Congr of Soil Scientists. Int Congr Soil Sci 4, Soc Soil Sci Am, Madison, pp 291–304
Evans BM, Walker DA, Benson CS, Nordstrand EA, Petersen GW (1989) Spatial interrelationships between terrain, snow distribution and vegetation patterns at an arctic foothills site in Alaska. Holarct Ecol 12: 270–278
Glaser PH (1983) Vegetation patterns in the North Black River peatland, northern Minnesota. Can J Bot 61: 2085–2104
Glaser PH, Wheeler GA, Gorham E, Wright HE Jr (1981) The patterned mires of the Red Lake peatland, northern Minnesota: vegetation, water chemistry, and landforms. J Ecol 69: 575–599
Gorham E, Eisenreich SJ, Ford J, Santelmann MV (1985) The chemistry of bog waters. In: Stumm W (ed) Chemical processes in lakes. Wiley, New York, pp 339–363
Hamilton TD (1986) Late Cenozoic glaciation of the Central Brooks Range. In: Hamilton TD, Reed KM, Thorson RM (eds) Glaciation in Alaska: the geologic record. Alaska Geol Soc, pp 9–49
Hanson HC (1951) Characteristics of some grassland, marsh, and other plant communities in western Alaska. Ecol Monogr 21: 317–378
Hanson HC (1953) Vegetation types in northwestern Alaska and comparisons with communities in other arctic regions. Ecology 34: 111–148
Heinselman ML (1970) Landscape evolution, peatland types, and the environment in the Lake Agassiz Peatlands Natural Area, Minnesota. Ecol Monogr 40: 235–261

Hettinger LR, Janz AJ (1974) Vegetation and soils of northeastern Alaska. Arct Gas Biol Rep Ser 21

Holowaychuk N, Smeck NE (1974) Soils of the Chukchi-Imuruk Area, Seward Peninsula, Alaska. Final Rep, Cooperative Park Studies Unit, Univ Alaska, Fairbanks

Hopkins DM (1982) Aspects of the paleogeography of Beringia during the late Pleistocene. In: Hopkins DM, Matthews JV, Schweger CE, Young SB (eds) Paleoecology of Beringia. Academic Press, New York, pp 3–28

Johnson AW, Viereck LA, Johnson RE, Melchior H (1966) Vegetation and Flora. In: Wilimovsky NJ, Wolfe JN (eds) Environment of the Cape Thompson Region, Alaska. US Atomic Energy Comm, Washington, DC, pp 277–354

Jorgenson MT (1984) The response of vegetation to landscape evolution on glacial till near Toolik Lake, Alaska. In: Laban VJ and Kerr CL (eds) Inventorying forest and other vegetation of the high latitude and high altitude regions, Proc of an international symposium. Soc Am Foresters Regional Tech Conf, Fairbanks, pp 134–141

Klinger LF (1990) Global patterns in community succession. 1. Bryophytes and Forest Decline. Mems Torrey Bot Club 24: 1–50

Koranda JJ (1960) The plant ecology of the Franklin Bluffs area, Alaska. Dissertation, Univ Tennessee, Memphis

Kreig RA, Reger RD (1982) Air-photo analysis and summary of landform and soil properties along the route of the trans-Alaska pipeline system. Division of Geological and Geophysical Surveys, State of Alaska, Geol Rep 66

Lambert JDH (1968) The ecology and successional trends in the Low Arctic subalpine zone of the Richardson and British Mountains of the Canadian western arctic. Thesis, Univ British Columbia, Vancouver, Canada

Lawrence DB (1985) Glaciers and vegetation in southeastern Alaska. Am Sci 46: 89–122

Murray DF, Lipkin R (1987) Candidate threatened and endangered plants of Alaska with comments on other rare plants. Univ Alaska Museum, Fairbanks, 76 PP

Noble MG, Lawrence DB, Streveler GP (1984) *Sphagnum* invasion beneath an evergreen forest canopy in southeastern Alaska. Bryologist 87: 119–127

Nordhagen R (1943) Sikilsdalen og Norges Fjellbeiter: en plantesosiologisk monografi. Bergens Museums Skrifter, vol 22

Ostendorf B (1994) Spatial explicit modeling of arctic tundra landscapes. Thesis, Univ California, Davis, and San Diego State Univ, San Diego

Oswood MW, Everett KR, Schell DM (1989) Some physical and chemical characteristics of an arctic beaded stream. Holarct Ecol 12: 290–295

Racine CH (1981) Tundra fire effects on soils and three plant communities along a hill-slope gradient in the Seward Peninsula, Alaska. Arctic 34: 71–84

Rönning OI (1965) Studies on Dryadion on Svalbard. Nor Polarinst Skr 134

Spetzman LA (1959) Vegetation of the Arctic Slope of Alaska. US Geol Surv Prof Pap 302-B

Ugolini FC, Mann DH (1979) Biopedological origin of peatlands in southeast Alaska. Nature 281: 366–368

US Soil Taxonomy Soil Survey Staff (1975) Soil taxonomy, a basic system of soil classification for making and interpreting soil surveys. Dep Agric Handb 436

Viereck LA (1966) Plant succession and soil development on gravel outwash of the Muldrow Glacier, Alaska. Ecol Monogr 36(3): 181–199

Vitt DH, Crum H, Snider J (1975) The vertical distribution of *Sphagnum* species in hummock-hollow complexes in northern Michigan. Mich Bot 14: 190–200

Walker DA, Everett RR, Acevedo W, Gaydos L, Browm J, Webber PJ (1982) Landsat-assisted environmental mapping in the Arctic National Wildlife Refuge, Alaska. US Army Cold Regions Res, CRREL Rep 82-27

Walker DA, Binnian E, Evans BM, Lederer ND, Nordstrand E, Webber PJ (1989a) Terrain, vegetation and landscape evolution of the R4D research site, Brooks Range Foothills, Alaska. Holarct Ecol 12: 238–261

Walker MD, Walker DA, Everett KR (1989b) Wetland soils and vegetation, Arctic Foothills, Alaska. US Fish Wildl Serv Biol Rep 89(7)

Walker MD (1990) Vegetation and floristics of pingos, Central Arctic Coastal Plain, Alaska. Dissertationes Botanicae 149, Cramer, Stuttgart

Walker MD, Walker DA, Auerbach NA (1994) Plant communities of a tussock tundra landscape in the Brooks Range foothills, Alaska. J Veg Sci 5 (6): 843–866

Washburn AL (1980) Geocryology: a survey of periglacial processes and environments. Halsted Press, Wiley, New York

Westhoff V, van der Maarel E (1978) The Braun-Blanquet approach. In: Whittaker RH (ed) Classification of plant communities. W Junk, Den Haag, pp 287–399

Yurtsev BA (1994) The floristic division of the Arctic. J Veg Sci 5(6): 765–776

5 Vegetation Structure and Aboveground Carbon and Nutrient Pools in the Imnavait Creek Watershed

S. C. HAHN, S. F. OBERBAUER, R. GEBAUER, N. E. GRULKE, O. L. LANGE, and J. D. TENHUNEN

5.1 Introduction

The North Slope of the Brooks Range in Alaska spans three physiographic provinces: the Coastal Plain, the Brooks Range Foothills, and the Brooks Range proper (Hultén 1968). The vegetation of the foothills province, which includes the Imnavait Creek watershed, is composed of a complex mosaic of communities that is tightly coupled to environmental gradients. This mosaic encompasses exposed upland ridges and mountain slopes, which have communities similar to those found in large areas of the High Arctic (Bliss and Matveya 1992), and, at the other extreme, the communities in valley basins resemble the sedge meadows of the Coastal Plain (Webber 1978). The distribution of these tundra communities is influenced by glacial history and topography, soil development, cryoturbation effects, and local erosion (Walker and Everett 1991; Chap. 4 this Vol.), as well as by summer thaw depth and soil water mobility (Giblin et al. 1992).

As part of the R4D study (Chap. 1, this Vol.), we undertook a detailed inventory of vegetation structure and aboveground carbon and nutrient pools of the representative communities found in the Imnavait Creek watershed. Our goal was to eluciate landscape patterns in plant types, cover, standing biomass, tissue nutrient content, and total aboveground carbon and nutrient pools. These data also provided crucial input for the development of landscape models (see Chaps. 11, 17, and 18, this Vol.). We previously described vegetation structure and aboveground nutrient pools for tussock tundra and water-track communities (Hastings et al. 1989), and for wet sedge and riparian communities (Oberbauer and Tenhunen, unpubl. data). In this chapter we present data on the largely ignored, but species-rich, heath communities of bedrock outcrops, glacial till and blockfields, and ridge summits in the Imnavait Creek watershed. Using these data as well as those from our previous work, we summarize landscape patterns of vegetation structure and nutrient stores along a topographic moisture gradient from the dry ridgetop to the wet riparian zone.

5.2 Description of Vegetation

We adopted Walker's (Walker et al. 1989; Chap. 4, this Vol.) hierarchical classification of plant communities for the Imnavait Creek watershed, al-

J.F. Reynolds and J.D. Tenhunen (Eds.)
Ecological Studies, Vol. 120
© Springer-Verlag Berlin Heidelberg 1996

Table 5.1. Dominant species of major vegetation types in the Imnavait Creek watershed identified in this study and equivalence with Walker's classification schemes

Vegetation type	Dominant species[a]	Equivalence	
		Walker et al. (1989)	Vegetation (GIS) codes[b]
Dryas heath	*Selaginella sibirica, Dryas octopetala*, and *Vaccinium vitis-idaea*	D1a, D1b	Occurs outside of IRS
Lichen heath	*Hierochloë alpina, Arctous alpina*; and *Vaccinium uliginosum, Arctous alpina*	D1c, D1d	01, 03, 04, 06
Dry *Cassiope*-dwarf shrub heath	*Diapensia lapponica, Cassiope tetragona, Calamagrostis inexpansa*	D1e	05, 09 (Minor)
Moist *Cassiope*-dwarf shrub heath	*Sphagnum rubellum, Eriophorum vaginatum, Cassiope tetragona, Dryas integrifolia*, and *Equisetum scirpoides; Cassiope tetragona* and *Carex microchaeta*	D1f, M1a	14, 10 (Minor)
Tussock tundra (upslope)	*Sphagnum rubellum, Eriophorum vaginatum*	M2, M4a, M4b	10, 12, 13, 11 (Minor)
Tussock tundra (downslope)	*Sphagnum rubellum, Eriophorum vaginatum*	M2, M4a, M4b	10, 12, 13
Water-track vegetation	*Sphagnum rubellum, Eriophorum vaginatum, Salix planifolia* ssp. *pulchra*	W1, W3	16, 17, 26
Well-drained riparian	*Sphagnum rubellum, Eriophorum vaginatum, Salix planifolia* ssp. *pulchra, Betula nana*	M4c, M4d	13, 15, 16, 17
Wet sedge meadows	*Eriophorum angustifolium, Carex aquatilis*	W1	25, 19, 21, 22

IRS, R4D intensive research site.
[a] Grouped by vegetation codes in Chap. 4, Fig. 4.6.
[b] From Chap. 4, Fig. 4.6 IRS, GIS = Geographic Information System.

though we grouped several recognizable variants of a particular community. We considered the drier upland tundra – Walker's "dry dwarf shrub-fruticose lichen-tundra" – to consist of four major heath communities: (1) *Dryas* heath, (2) Lichen heath, (3) Dry *Cassiope*-dwarf shrub (CDS) heath, and (4) Moist *Cassiope*-dwarf shrub heath. The dominant species of each community type are given in Table 5.1 along with reference to Walker's classification schemes. At these sites vascular plant growth is strongly restricted by mechanical action of ice and snow, temperature extremes due to lack of complete snow cover in winter, physical abrasion from windblown particles, frost drought, solifluction that may dislodge seedlings, and soil drying in summer (Sakai and Larcher

1987). Lichens tend to dominate where vascular species maintain much of their biomass below the surface, succumb to stress, or are unable to establish themselves (Bliss 1971; Kappen 1988).

Dryas heath covers small areas just outside of the Imnavait Creek watershed, occurring on siltstone outcrops, primarily on south-facing slopes with exposed clay shale and siltstone substrates as well as glacial debris. However, this community type has a high species diversity. The dominant species, *Dryas octopetala*, *Selaginella sibirica*, and *Vaccinium vitis-idaea*, occur on thin soil and are scattered sparsely together with *Salix phlebophylla*, dicotyledonous rosettes and cushions, and tuft-forming or prostrate graminoids (Grulke and Bliss 1985); they also occur with desiccation-resistant moss species such as *Polytrichum piliferum*, and with the lichens *Cetraria* spp., *Thamnolia vermicularis*, *Cornicularia* spp., *Alectoria* spp., *Asahinea chrysantha*, *Sphaerophorus globosus*, and *Stereocaulon* spp.

Lichen heath occurs on glacial till that is covered with a thin layer of snow during winter and where snowbeds form in larger depressions. Microrelief appears to influence the presence or absence of individual species. The major character species are *Arctous alpina*, *Hierochloë alpina*, *Vaccinium uliginosum*, *Vaccinium vitis-idaea*, and *Salix phlebophylla*; the moss species *Dicranum elongatum* and *Polytrichum strictum*; and the lichens *Cetraria* spp., *Thamnolia vermicularis*, *Cornicularia divergens*, *Alectoria ochroleuca*, *Sphaerophorus globosus*, *Stereocaulon* spp., *Masonhalea richardsonii*, *Dactylina arctica*, *Peltigera malacea*, and *Cladonia* spp.

Both dry and moist CDS heath occur as part of nonsorted, stone-stripe complexes on ridge summits (Washburn 1956) with relatively dry, well-vegetated stony (1.5–2 m wide) surfaces alternating with moist interstripe areas (Walker et al. 1989; Chap. 4, this Vol.). The complex develops in those areas where lateral drainage appears slow. The dry CDS heath areas are raised, thus accumulating less snow during winter; moist heath often has a high standing water table. In dry CDS heath, the main species, *Cassiope tetragona* and *Vaccinium uliginosum*, are accompanied by *Vaccinium vitis-idaea*, *Calamagrostis inexpansa*, and *Carex bigelowii*; the herbaceous species *Pedicularis lanata* and *Petasites frigidus*; moss species typical of dry warm habitats, such as *Rhacomitrium lanuginosum*; and a lichen community with the same elements as found in lichen heath. Moist CDS heath is distinguished by greater species richness than dry heath, including the dwarf shrubs *Ledum palustre* and *Empetrum nigrum*; the graminoids *Carex bigelowii* and *Carex microchaeta*; the herbs *Polygonum bistorta* and *Polygonum vivipara*; *Equisetum scirpoides*; and moss species such as *Hylocomium splendens*, *Aulacomnium turgidum*, *Rhytidium rugosum*, and *Ptilidium ciliare*. Lichen diversity is low and includes mainly *Cetraria* spp., *Thamnolia vermicularis*, *Stereocaulon* spp., *Masonhalea richardsonii*, *Dactylina arctica*, *Peltigera aphthosa*, *Cladonia* spp., and *Nephroma arcticum*.

5.3 Sampling Methods

5.3.1 Cover

We estimated cover for all vascular plant species in 30 20- × 50-cm plots in
Dryas heath, 45 20- × 50-cm plots in lichen heath, and 40 20- × 20-cm plots each
in moist and dry CDS heath. In *Dryas* heath and lichen heath we recorded
percent cover for mosses, lichens, rocks, bare soil, and litter in a 20- × 20-cm
subplot within the larger plot. In the CDS heath communities we estimated
cover for deciduous shrubs, evergreen shrubs, graminoid species, herbaceous
dicots, and rosette-forming species; mosses, lichens, and surface and standing
dead litter. Samples harvested for biomass determination (see Sect. 5.3.2) were
obtained from the 20- × 20-cm plots.

Species were identified according to Hultén (1968), Thomson (1979, 1984),
and Vitt et al. (1988), as well as from a field herbarium for the site established
by D. Murray at the Museum of the University of Alaska, Fairbanks. We
assigned a growth form to the major species as shown in Table 5.2, treating
Dryas octopetala as a deciduous species based on copious *Dryas* leaf litter at
this site. Although Miller (1982) considered *Dryas* as evergreen, perhaps based
on observations of Svoboda (1972) in the high arctic, Berg et al. (1975) found
relatively large seasonal changes in the leaf area index of *Dryas* at
Hardangervidda, Norway.

5.3.2 Biomass and Nutrient Pools

For the biomass harvests, all material above the transition between the green
living and brown dead moss layers was considered as aboveground biomass
(Bliss et al. 1977). All samples of vascular plants were sorted by species and
separated into leaf and stem material. The samples were dried and weighed
separately for each plot, but plot samples were pooled and subsampled for
nutrient analyses. Samples were subjected to a Kjeldahl digest; total P and N

Table 5.2. Growth forms of major shrub species in upland tundra heath

Species	Growth form
Vaccinium uliginosum	Deciduous
Vaccinium vitis-idaea	Evergreen
Empetrum nigrum	Evergreen
Cassiope tetragona	Evergreen
Betula nana	Deciduous
Dryas octopetala[a]	Deciduous
Arctous alpina	Deciduous

[a] See discussion in text.

were determined colorimetrically with a Technicon autoanalyzer (Technicon Industrial Systems, Tarrytown, NY).

Samples from the *Dryas* heath and lichen heath were taken between 20 July and 3 August 1987. Three 10- × 10-m plots were delimited in each vegetation type. Because of impacts such as foot trails, the plots were nonrandom. Within the large plots, 20- × 20-cm plots for harvesting were designated by selecting 20 sets of random coordinates. Lichens and mosses were separated from vascular plants at the field laboratory, air dried, frozen, and later sorted under a dissecting microscope according to species. Our work focused on macrolichens; smaller crustose species and those occurring on stones contributed little biomass and were not included. If species identification could not be made, material was allocated to morphological groups, e.g., "yellow *Cetraria*" or "brown *Cetraria*." Unidentifiable material, especially small lichen parts, was combined as "unknown." In total, 50 possible categories for cryptogams were considered during sorting. Subsamples of the material in all major categories were analyzed for carbon, nitrogen, and phosphorus content according to Schramel (1988).

Samples from CDS heath were obtained between 28 July and 1 August 1989. Because of the large mosaic patterning of dry and moist CDS heath vegetation, 40 20- × 20-cm plots were randomly selected from a 100- × 100-m area. We harvested paired samples (20 pairs) in the center of both dry and moist heath, closest to the randomly determined coordinate point. Processing of samples was the same as described above for *Dryas* heath and lichen heath.

5.4 Cover

Total cover ranged from 86% in *Dryas* heath to 150% in moist CDS heath (Table 5.3). Cover values in the CDS heath exceeded 100% because of the multiple layers in the canopy of dwarf shrubs that grow to ca. 20 cm in height. (We did not consider multiple layers in mosses.) In addition to dwarf shrubs and several graminoid species, herbaceous species, such as *Polygonum bistorta*, *Petasites frigidus*, and *Pyrola grandiflora*, represented substantial cover in CDS heath. In *Dryas* heath and lichen heath, the vegetation was between 5 and 10 cm in height.

Dead organic litter occupied a relatively large portion of the ground cover in all four communities (Table 5.3). It was possible to identify the major contributors of litter in each community: In *Dryas* heath most litter derived from leaves of *Dryas octopetala* and *Salix phlebophylla*; in lichen heath *Arctous alpina* was the major source; in dry CDS heath standing dead material consisted mainly of leaves from *Vaccinium uliginosum* and *Salix reticulata*; and in moist CDS heath, from *Betula nana* and *Salix pulchra* and a variety of graminoid species.

Mosses were a significant component of the CDS heath communities, with a mean cover of ca. 21% in dry and 54% in moist (Table 5.3). *Dicranum* spp.,

Table 5.3. Mean cover (%) and mean aboveground biomass (%) (living only and total) for different plant life forms and organic litter in heath communities. Rocks and bare soil were 14% of total cover in *Dryas* heath and 5% in lichen heath, but were not important in other types

	Dryas heath	Lichen heath	Dry CDS heath	Moist CDS heath
Cover (%)				
Vascular plants	30	46	44	70
Mosses	8	10	21	54
Lichens	31	26	62	8
Litter	17	13	9	18
Total	86	95	136	150
Living aboveground biomass (%)				
Vascular plants	35	39	19	39
Mosses	17	17	39	50
Lichens	48	44	41	11
Total	100	100	100	100
Total aboveground biomass (%)				
Vascular plants	15	20	13	26
Mosses	8	9	26	33
Lichens	21	22	28	7
Litter	55	49	32	34
Total	100	100	100	100

CDS, *Cassiope*-dwarf shrub.

Hylocomium splendens, *Rhythidium rugosum*, and *Rhacomitrium lanuginosum* prevailed in drier locations, while *Hylocomium splendens*, *Ptilium crista-castrensis*, *Drepanocladus* spp., *Aulacomnium turgidum*, and *Aulacomnium palustre* grew in moist locations. In *Dryas* heath mosses (mainly *Dicranum* spp., *Pohlia nutans*, and *Encalypta rhaptocarpa*) made up only 8% of the surface cover and in lichen heath, only 10% (mainly *Polytrichum strictum*, *P. piliferum*, and *P. commune*).

Lichens covered ca. 30% of the ground surface in both *Dryas* heath and lichen heath. However, the lichen mat was 2–3 cm thick in *Dryas* heath vs 5–6 cm in lichen heath. Lichen surface cover in dry CDS heath was over 60%; in moist CDS heath, where they are easily overgrown by mosses, it was only 8%.

5.5 Aboveground Biomass

5.5.1 Live Biomass

Total live aboveground biomass averaged $490\,g\,m^{-2}$ in the four upland heath communities, ranging from $321\,g\,m^{-2}$ in *Dryas* heath to $622\,g\,m^{-2}$ in dry CDS

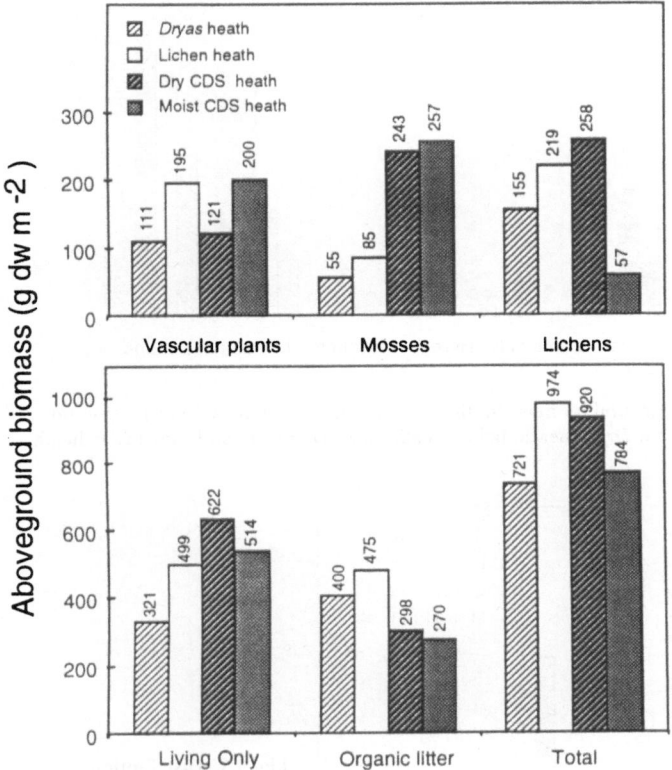

Fig. 5.1. Mean live aboveground (above transition between green and brown moss layers; see methods) biomass in different plant life forms and organic litter for *Dryas* heath, lichen heath, dry *Cassiope* dwarf shrub (CDS) heath, and moist CDS heath

heath (Fig. 5.1). Lichen heath had ca. 40% higher biomass than *Dryas* heath, but the relative contribution of the different growth forms to total live biomass was almost identical (Table 5.3). Vascular plants and lichens represented ca. 83% of the total biomass in these two communities, and mosses 17%. Mosses comprised nearly 40–50% of the live biomass in CDS heath. In dry CDS, vascular plants (primarily deciduous and evergreen shrubs) comprised only 20% of the biomass compared with 40% in moist CDS. Lichen biomass, on the other hand, was much greater in dry CDS heath (Table 5.3), and, when combined with dwarf shrubs and dense moss cushions, resulted in the largest amount of living biomass per square meter found within the watershed. Moist CDS heath, with a 70% cover of vascular plants, would seem to have greater biomass than the dry variant. However, the vascular canopy shades the ground surface; which stimulates production by moss (cf. Murray et al. 1989; Tenhunen et al. 1992). Greater water availability stimulates production by moss (cf. Murray et al. 1989; Tenhunen et al. 1992) and permits moss types having a less dense structure to replace lichens.

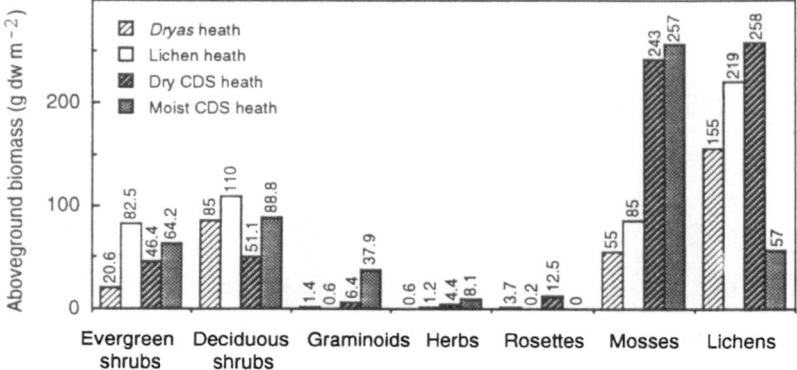

Fig. 5.2. Mean live aboveground biomass in different vascular plant growth forms as compared with mosses and lichens for *Dryas* heath, lichen heath, dry CDS heath, and moist CDS heath

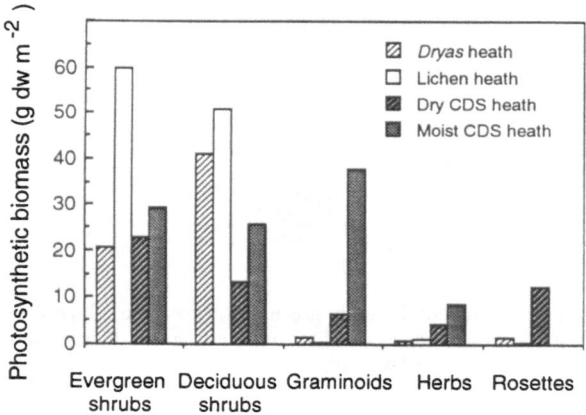

Fig. 5.3. Distribution of photosynthetic biomass among vascular plant growth forms for *Dryas* heath, lichen heath, dry CDS heath, and moist CDS heath

Deciduous and evergreen shrubs were important elements of these heath communities (Fig. 5.2). Herbs and graminoid species were significant components only in moist CDS heath. A considerable number of rosette species occur in *Dryas* heath, but were significant contributors to biomass only in dry CDS heath.

5.5.2 Photosynthetic Biomass

In terms of photosynthetic biomass contributed by evergreen shrubs per se, *Dryas* heath and dry and moist CDS heath were similar (Fig. 5.3). Although less obvious than in the CDS heath due to acaulescence, deciduous shrubs in *Dryas* heath and lichen heath have high photosynthetic biomass. Despite relatively low total biomass, graminoid species were important contributors to carbon fixation in moist CDS heath (see also Chaps. 11 and 17, this Vol.).

Table 5.4. Biomass (mean ± SE; g m^{-2}) of principal lichen species

Species	Dryas heath	Lichen heath	Dry CDS heath	Moist CDS heath
Cetraria cucullata[a]	13.2 ± 1.9	36.5 ± 3.5	10.2 ± 2.5	1.5 ± 0.9
Cetraria, dark brown	9.0 ± 1.6	14.8 ± 1.8	2.4 ± 1.0	–
Cetraria islandica	0.1 ± 0.0	1.9 ± 0.9	11.1 ± 2.0	13.4 ± 6.4
Cladonia arbuscula	11.0 ± 3.8	22.1 ± 5.3	101.4 ± 12.7	9.9 ± 3.6
Cladonia gracilis	12.3 ± 3.1	34.5 ± 6.2	76.4 ± 8.2	22.2 ± 7.8
Cladonia uncialis	5.1 ± 1.5	17.8 ± 3.8	18.9 ± 3.9	0.0 ± 0.0
Dactylina arctica	0.7 ± 0.2	4.4 ± 0.9	7.3 ± 1.2	2.1 ± 0.7
Masonhalea richardsonii	1.6 ± 0.6	4.2 ± 2.0	5.3 ± 2.8	0.8 ± 0.4
Peltigera spp.	9.9 ± 3.4	6.6 ± 2.4	8.8 ± 3.0	7.9 ± 4.2
Sphaerophorus globosus	9.2 ± 2.1	15.8 ± 3.6	7.4 ± 6.2	0.1 ± 0.1
Stereocaulon alpinum	3.8 ± 1.1	10.6 ± 7.3	1.2 ± 0.8	0.1 ± 0.1
Thamnolia vermicularis	5.6 ± 0.6	4.7 ± 0.8	1.6 ± 0.5	0.7 ± 0.3
Other lichens[b]	73.0 ± 16.1	45.1 ± 10	5.7 ± 3.3	0.1 ± 0.1
Total	154.5 ± 19.6	218.9 ± 20.8	257.6 ± 24	57.0 ± 13.6

CDS, Cassiope-dwarf shrub.

[a] Data for Cetraria cucullata and Cetraria nivalis were combined.

[b] A minor crustose and foliose lichen component including Icmadophila ericetorum, Lecidea atrata, Pertusaria spp., Psoroma hypnorum, Micarea incrassata; folious species, e.g., Asahinea chrysantha, Hypogymnia sub-obscura; dark Parmelia species, e.g., Parmelia omphalodes; light-yellow Parmelia species, e.g., Xanthoparmelia separata section; filamentous species, e.g., Alectoria nigricans, Alectoria ochroleuca, Bryoria nitidula, Cornicularia divergens, and Pseudephebe minuscula.

5.5.3 Lichen Biomass

Biomass estimates of the principal lichen species are given in Table 5.4. The largest component of lichen biomass was contributed by *Cladonia* species, with species composition depending on heath type. *Stereocaulon alpinum* and *Thamnolia vermicularis* occurred in greater quantities in both *Dryas* heath and lichen heath. Dark brown *Cetraria* species, e.g., *C. allankinii, C. andrejevii, C. kamczatica,* and *C. nigricans* (also accompanied by *Cornicularia aculeata*), occurred in the driest exposed sites, whereas *Cetraria islandica* was most common in more protected locations.

5.5.4 Organic Litter

A large amount of organic litter was found in all vegetation types (Fig. 5.1). Litter represented from 30 to 55% of the total aboveground biomass in the heath communites (Table 5.3). Although this organic material is undoubtedly subjected to leaching, it represents significant carbon and nutrient stores (see Sect. 5.6). Shaver and Chapin (1991) established turnover times for aboveground vegetation in lichen heath (ca. 3.4 years) and tussock tundra (ca.

2.6 years). Hence, the standing dead litter remains in place unchanged for considerable periods of time.

5.5.5 Watershed Patterns

We combined the live aboveground biomass data obtained for the heath types with our previous estimates for other community types in the Imnavait Creek watershed to examine landscape patterns. A comparison of nine communities is given in Table 5.5. Whereas the heath communities had both the lowest (*Dryas* heath) and the highest (dry CDS heath) total live biomass in the watershed, there is a striking similarity for the different communities. However, there were significant shifts in community composition along this gradient (see also Tenhunen et al. 1992). Evergreen shrub biomass was lower in moist situations such as water tracks and riparian areas, but in tussock tundra it remained equal to that found in CDS heath (Hastings et al. 1989). Deciduous shrub biomass was relatively constant along the moisture gradient, even occurring in similar amounts in riparian areas. Deciduous shrub biomass may double in water track communities where flowing water may reduce soil aeration, but also provides nutrients (Hastings et al. 1989). Living aboveground graminoid biomass was low in tussock tundra, but was dominant in riparian areas, where a mixture of deciduous shrubs and graminoid elements occurred in better drained microsites and dense monospecific sedge communities occurred in waterlogged soils (see also Oberbauer and Dawson 1992). Once water availability is relatively high, as found in moist CDS heath, moss biomass remains high (greater than $200\,g\,m^{-2}$) as communities change along the slope. Lichen biomass was not determined in communities other than the heaths. Changes in photosynthetically active biomass along this slope gradient have been described elsewhere (Tenhunen et al. 1992).

5.6 Nutrient Pools

5.6.1 N and P in Heath Cryptogams

Little variation was found in carbon content among lichen and moss species, and among heath vegetation communities for the cryptogams (overall average was 43.3% dry weight). Organic litter had a similar carbon content. In contrast, large differences in N content of cryptogams occurred. The differences were not apparent among vegetation communities, but seem related to species physiology. Mean percent total N, mean percent total P, and the P/N ratio of tissue obtained by pooling all replicates for a particular species are shown in Fig. 5.4. Nitrogen content of lichens with green algal photobionts varied between 0.17 and 0.44% of dry weight. *Polytrichum*, *Hylocomium* and *Aulacomnium* had nitrogen contents between 0.56 and 0.6%. *Stereocaulon*

Table 5.5. Aboveground biomass (g m^{-2}, mean ± SE) in different plant types and organic litter for plant communities occurring within the Imnavait Creek watershed

Toposequence locations[c]:	Dryas heath	Lichen heath	Dry CDS heath	Moist CDS heath	Tussock tundra (upslope)[a]	Tussock tundra (downslope)[a]	Water track[a]	Drained riparian[b]	Wet sedge[b]
	–[d]	Crest	Shoulder	Shoulder	Upper backslope	Lower backslope	Lower backslope	Footslope	–[d]
Vascular plants									
Evergreen shrubs	20.6 ± 6.1	82.5	46.4 ± 11	64.2 ± 26	62.1	60.0	44.5	14.0	–
Leaves	20.6 ± 6.1	59.8	22.7 ± 5	29.4 ± 10	40.4	37.6	29.0	14.0	–
Axes	–	22.7	23.7 ± 7.8	34.8 ± 15	21.7	22.4	15.5	–	–
Deciduous shrubs	85.0	110	51.1 ± 7.8	88.8 ± 11	56.6	43.1	149	69.5	–
Leaves	41 ± 5.8	50.9	13.6 ± 1.8	25.5 ± 2.6	11.0	9.5	31.3	37.0	–
Axes	44.0	59.2	37.5 ± 6.3	63.3 ± 8.6	45.6	33.6	118	32.4	–
Graminoids	1.3 ± 0.7	0.6 ± 0.3	6.4 ± 1.5	37.9 ± 7.4	14.0	15.0	12.9	91.1	441
Herbs	0.6 ± 0.3	1.2 ± 0.8	4.4 ± 1.4	8.1 ± 1.5	–	–	–	6.0	0.3
Rosettes	3.7 ± 1.6	0.2 ± 0.1	12.5 ± 4.1	–	–	–	–	0.2	–
Total	111	195	121 ± 12	200 ± 27	133	118	207	181	441
Cryptograms									
Mosses	55 ± 10	85 ± 30	243 ± 25	257 ± 24	414	443	288	195	5.0
Lichens	155 ± 20	219 ± 21	258 ± 23	57 ± 12	–	–	–	–	–
Total	210 ± 23	303 ± 44	504 ± 40	316 ± 28	414	443	288	195	5.0
Total live	321	498	628 ± 42	516 ± 36	547	561	495	376	446
Organic litter	400	475	298 ± 21	270 ± 26	–	–	–	–	–
Grand total (live + litter)	721	973	933 ± 50	786 ± 40	–	–	–	–	–

CDS, *Cassiope*-dwarf shrub.
[a] From Hastings et al. (1989).
[b] From Oberbauer and Tenhunen (unpubl. data).
[c] From Fig. 4.8 in Chap. 4, this Vol.
[d] Not part of toposequence.

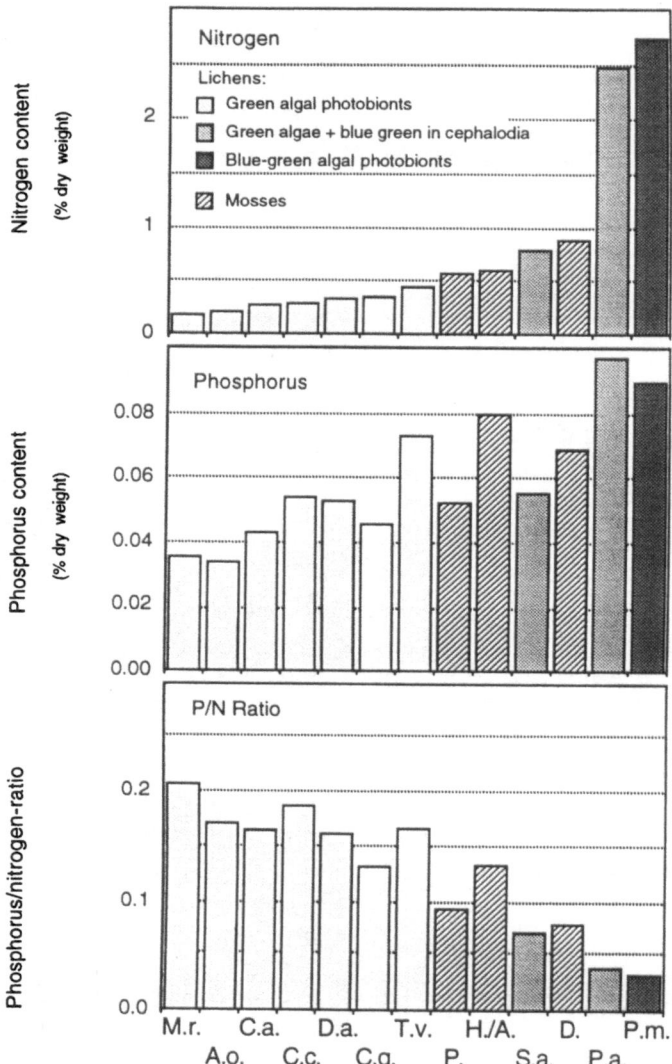

Fig. 5.4. Mean nitrogen and phosphorus composition and tissue ratio of P/N of principal lichen and moss species found in the heath communities within the Imnavait Creek watershed. *A.o.*, *Alectoria ochroleuca*; *C.c.*, *Cetraria cucullata*; *C.a.*, *Cladonia arbuscula*; *C.g.*, *Cladonia gracilis*; *D.a.*, *Dactylina arctica*; *D.*, *Dicranum* spp.; *H./A. Hylocomium/Aulacomnium* spp.; *M.r.*, *Masonhalea richardsonii*; *P.a.*, *Peltigera aphthosa*; *P.m.*, *Peltigera malacea*; *P.*, *Polytrichum* spp.; *S.a.*, *Stereocaulon alpinum*; *T.v.*, *Thamnolia vermicularis*

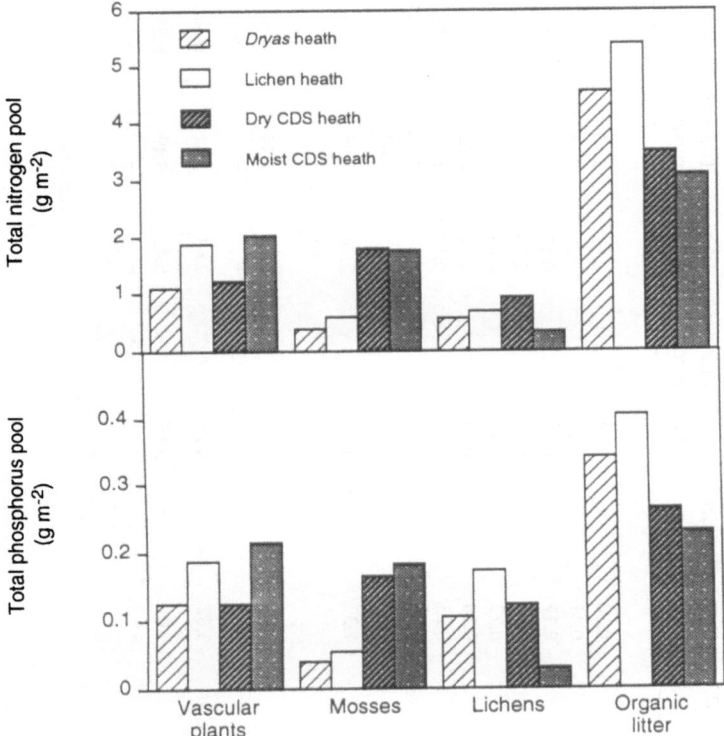

Fig. 5.5. Mean total aboveground nitrogen and phosphorus pools per square meter in different plant types and organic litter for *Dryas* heath, lichen heath, dry CDS heath, and moist CDS heath

alpinum, which has N-fixing blue–green algal photobionts in cephalodia in addition to the green-algal photobiont, exhibited an elevated N content of ca. 0.77%; the moss *Dicranum* spp. also had an N content in this range. The effective N fixation in cephalodia of *Peltigera aphthosa* results in very high tissue N content, established here at approximately 2.47%. Similarly, *Nephroma* and *Lobaria* spp. with cephalodia had tissue N contents above 2% of dry weight (not included in graph because they were not sampled from all communities). The highest N content of approximately 2.74% was found in the *Peltigera malacea*, a species with a blue – green algal photobiont. The species with high N content are those nonfruticose elements that remain a part of the lichen community in moist CDS heath.

5.6.2 N and P in Communities

Mean total aboveground N and P pools in the different plant types and in organic litter of heath vegetation are shown in Fig. 5.5 and Table 5.6. The total

Table 5.6. Mean aboveground total nitrogen and phosphorus pools (g m⁻²) in different plant types and organic litter for plant communities occurring within the Imnavait Creek watershed (based on mean values from Tables 5.4 and 5.5; toposequence locations given in Table 5.5)

Nitrogen	Dryas heath	Lichen heath	Dry CDS heath	Moist CDS heath	Tussock tundra (upslope)	Tussock tundra (downslope)	Water track	Drained riparian	Wet sedge
Vascular plants									
Evergreen shrubs	0.18	0.71	0.4	0.55	0.58	0.54	0.42	0.15	–
Leaves	0.18	0.53	0.22	0.28	0.41	0.36	0.3	0.15	–
Axes		0.18	0.18	0.27	0.17	0.18	0.12	–	–
Deciduous shrubs	0.87	1.14	0.56	0.92	0.58	0.46	2.15	1.01	–
Leaves	0.52	0.66	0.27	0.46	0.22	0.19	0.63	0.75	–
Axes	0.35	0.47	0.29	0.46	0.36	0.26	1.51	0.25	–
Graminoids	0.015	0.006	0.072	0.42	0.23	0.24	0.18	1.68	8.81
Herbs	0.011	0.02	0.067	0.14	–	–	–	0.11	0.005
Rosettes	0.013	0.002	0.12	–	–	–	–	0.002	–
Total	1.09	1.88	1.22	2.04	1.39	1.24	2.74	2.96	8.82
Cryptograms									
Mosses	0.37	0.58	1.79	1.76	3.62	4.24	2.68	2.54	0.04
Lichens	0.55	0.68	0.92	0.33	–	–	–	–	–
Total	0.92	1.26	2.71	2.09	3.62	4.24	2.68	2.54	0.04
Total live	2.01	3.14	3.93	4.13	5.01	5.48	5.42	5.49	8.85
Organic litter	4.54	5.39	3.5	3.07	–	–	–	–	–
Grand total (live + litter)	6.55	8.53	7.43	7.2	–	–	–	–	–

Phosphorus	Dryas heath	Lichen heath	Dry CDS heath	Moist CDS heath	Tussock tundra (upslope)	Tussock tundra (downslope)	Water track	Drained riparian	Wet sedge
Vascular plants									
Evergreen shrubs	0.018	0.075	0.039	0.057	0.058	0.053	0.041	0.017	–
Leaves	0.018	0.055	0.020	0.027	0.040	0.035	0.029	0.017	–
Axes	–	0.020	0.019	0.029	0.017	0.018	0.012	–	–
Deciduous shrubs	0.101	0.108	0.055	0.089	0.054	0.042	0.207	0.081	–
Leaves	0.062	0.056	0.032	0.036	0.016	0.014	0.046	0.054	–
Axes	0.039	0.053	0.032	0.053	0.038	0.028	0.161	0.027	–
Graminoids	0.002	0.001	0.009	0.053	0.024	0.025	0.020	0.176	0.899
Herbs	0.001	0.002	0.009	0.016	–	–	–	0.010	0.001
Rosettes	0.001	–	0.012	–	–	–	–	–	–
Total	0.123	0.186	0.124	0.215	0.135	0.120	0.269	0.284	0.899
Cryptograms									
Mosses	0.038	0.055	0.165	0.181	0.290	0.310	0.202	0.137	0.004
Lichens	0.104	0.172	0.121	0.030	–	–	–	–	–
Total	0.142	0.227	0.286	0.211	0.290	0.310	0.202	0.137	0.004
Total live	0.265	0.413	0.410	0.426	0.425	0.430	0.470	0.420	0.903
Organic litter	0.340	0.404	0.262	0.230	–	–	–	–	–
Grand total (live + litter)	0.605	0.817	0.672	0.656	0.425	0.430	0.470	0.420	0.903

CDS, *Cassiope*-dwarf shrub.

pools reflect the biomass distribution shown in Fig. 5.1, although lichen contribution to the N pool is less than to the carbon pool. This results from vascular plant and moss N concentration greater than 1% of dry weight, whereas the most prevalent lichen types have an N concentration less than 0.5% (Fig. 5.4). When lichen species were considered in the order indicated for increasing N content (Fig. 5.4), P also generally increased. Nevertheless, tissue N content changes were so large in comparison that the ratio of P/N decreased by a factor of ca. 10. Nevertheless, because lichen species with low N content dominate in all heath types except moist CDS heath, lichen biomass may be considered to be relatively rich in P as compared with N. Except in moist CDS heath, therefore, contribution to the P pool (Fig. 5.5) is similar to the contribution to the carbon pool (Fig. 5.1). Although not statistically significant, there was a trend within lichen species toward higher tissue P content in lichen heath, which might be related to close contact with the mineral soil.

Along the topographic gradient from heath to riparian vegetation within the Imnavait Creek watershed, the importance of moisture availability and water movement through the vegetation becomes apparent in the standing aboveground pool of N (Table 5.6). The N pool increases continuously as one moves along the moisture gradient from *Dryas* heath to moist CDS heath, along with development of the deciduous shrub and graminoid plant canopy. The N pool increases further as one moves downslope, increasing from 2–4 $g m^{-2}$ in upslope heath communities to 5.5 $g m^{-2}$ in drained riparian areas at the base of the slope. In contrast, the aboveground pool for P remains surprisingly constant, except at the extremes of the gradient. Although in a somewhat special situation, absolute maxima are found in the N pool in waterlogged sedge communities, where aboveground N and P may increase to as much as two times that in water tracks or drained riparian areas. Strong gradients in standing N and P pools probably occur along short distances and with variation in tiller density in the sedge communities bordering Imnavait Creek. Vegetation dynamics in these moist community types obviously play a special role in movement of nutrients between the tundra terrestrial and aquatic ecosystems.

5.7 Discussion and Conclusions

In these upland heath communities, habitat conditions, such as temperature and water availability, are strongly influenced by surface microrelief, a factor that promotes high lichen diversity. The number of macrolichen species exceeds 50 in *Dryas* heath and decreases to ca. 40, 20, and 15 in lichen heath, dry CDS, and moist CDS heath, respectively. The importance of the morphological groups (e.g., fruticose, foliose, or crustose lichens) changes in the different heath types. Microenvironmental conditions in the wind-exposed *Dryas* heath selects for crustose, small foliose, or umbilicate forms, rather than fruticose growth forms. In less extreme situations, lichen cover is greater and fruticose lichens form dense tufts and carpets. As water availability increases in the CDS heath types, vascular plants become dominant and mosses begin to replace the lichen elements. Sheltered by the dwarf shrubs, fruticose lichens continue to increase until in moist CDS heath the nonfruticose element is represented only by the genus *Peltigera* (rarely accompanied by *Nephroma* or *Lobaria*). Possibly only those lichen species with nitrogen-fixing ability have high enough growth rates to compete with the mosses and vascular plant species.

Deciduous and evergreen shrubs are approximately equal in importance, except in *Dryas* heath, where evergreen shrubs have lower biomass. Evergreen shrubs exhibit relatively shallow rooting, which may be a disadvantage in *Dryas* heath, where the bare soil is exposed and surface soil may become very dry. Another possibility, however, is that evergreen shrubs cannot withstand extreme winter conditions in *Dryas* heath, where exposure to abrasive winds is greater and the temperatures are lower, due to lack of snowcover as compared with other heath communities.

Different growth forms represent different energy investments (Shaver and Chapin 1991), which influence productivity. Adjusting for photosynthetic capacity differences on a dry-weight basis (cf. Tenhunen et al. 1992 – poikilohydric plant capacity ca. 20–30% of vascular plant capacity), equal carbon flux from the poikilohydric and vascular components of these communities would be expected during favorable (moist) conditions in *Dryas* heath, lichen heath, and moist CDS heath. Changes in structure suggest that approximately twice as much carbon fixation might occur during favorable periods via nonvascular plants compared with vascular plants in dry CDS heath. On the basis of photosynthetic biomass, the highest carbon-fixation potential would be expected in dry CDS heath, and the lowest, in *Dryas* heath.

Although aboveground biomass and nutrient pool comparisons among tundra communities have been made previously, this study provides an example of change along a landscape continuum with gradual increases in water availability. Shaver and Chapin (1991) have compiled the most extensive data on production, biomass, and nutrient cycling in contrasting tundra vegetation types. Of the four sites studied, their lichen heath was very similar to our lichen heath and dry CDS heath, although standing biomass was ca. 60% of that described here. Vegetation at their graminoid-dominated "wet" site was similar to that found in the drained riparian at Imnavait Creek, although total biomass was again much less. This may have resulted from vegetation development on a floodplain, rather than in a basin where the total seasonal amount of nutrients transported in flowing water may have been higher. "Wet" tundra communities apparently very similar to that described by Shaver and Chapin (1991) were studied at Barrow during the International Biological Program (IBP) (Dennis et al. 1978; Chap. 2, this Vol.). The "shrub" tundra described by

Shaver and Chapin is not represented at Imnavait Creek, but occurs on gravel bars and in eroded drainages, rather than along beaded streams associated with deep organic soils. We would have expected the tussock tundra described by Shaver and Chapin to be similar to that at Imnavait Creek; however, they found a tenfold greater graminoid biomass at their site, much less moss, and 50% greater total biomass.

Sonesson et al. (1975) examined two heath communities ("lichen heath" and "dry meadow") located near Stigstuv, Norway, and an alpine heath near Kevo, Finland. The first of these sites was similar in aspect, vegetation composition, and biomass to the lichen heath types described from Imnavait Creek watershed. The latter two had similar vascular plant biomass, but low moss and lichen biomass. Miller (1982) examined vegetation zones of high elevation tundra in a snow accumulation area near Eagle Creek, Alaska. The "fellfield" area examined was in some ways similar to *Dryas* heath, but with lower lichen biomass. The "lichen heath" described by Miller was similar to lichen heath at Imnavait Creek, but lichen biomass was extremely high, approximately twice as high as found in the heath types at Imnavait Creek. The *"Cassiope"* vegetation described by Miller was not at all similar to the CDS heath, but was apparently a dense *Cassiope* stand as often seen on slopes covered by snowbeds. Finally, the downslope "sedge moss" described by Miller was similar to vegetation at lower slope sites of the Imnavait Creek watershed, although moss biomass was high.

At the landscape scale total live biomass was similar in all vegetation units within the Imnavait Creek watershed. The compact and dense tissues of heath species are replaced in lower slope communities by multilayered canopies. There was generally a gradual transition in composition according to plant types from the heath sites at the ridge tops to the riparian vegetation in the valley bottom, which led to greater aboveground concentrations of nutrients, but presumably greater and longer physiological activity, greater turnover of biomass, and higher production as well. The studies of community structure and composition described here support the conclusions of spatial remote sensing and modeling studies in the watershed (Ostendorf and Reynolds 1993; Chaps. 14 and 17, this Vol.), that understanding water transport and availability along topographic gradients provides a key to understanding ecosystem function in tundra regions. The degree and potential of biological response to altered resource availability is likely to remain within the same range along other environmental axes imposed on the tundra system. Studies of this type along existing gradients may contribute strongly to our ability to examine questions related to future climate change in tundra regions.

Acknowledgments. This study was partially supported by funds from the U.S. Department of Energy R4D Program, a project of the Deutsche Forschungsgemeinschaft (Sonderforschungsbereich 251), the Leibniz Program (O.L. Lange), and the German Federal Ministry for Science and Technology (grant no. BEO51-0339476A).

References

Berg A, Kjelvik S, Wielgolaski FE (1975) Measurement of leaf areas and leaf angles of plants at Hardangervidda, Norway. In: Wielgolaski FE (ed), Fennoscandian tundra ecosystems. Part 1. Plants and microorganisms. Ecological Studies 16. Springer, Berlin Heidelberg New York, pp 101–110

Bliss LC (1971) Arctic and alpine life cycles. Annu Rev Ecol Syst 2: 405–438

Bliss LC, Matveya NV (1992) Circumpolar arctic vegetation. In: Chapin FS III, Jeffries RL, Reynolds JF, Shaver GR, Svoboda J (eds) Arctic ecosystems in a changing climate. Academic Press, San Diego, pp 59–89

Bliss LC, Kerik J, Peterson W (1977) Primary production of dwarf shrub heath communities, Truelove Lowland. In: Bliss LC (ed) Truelove Lowland, Devon Island, Canada: a high arctic ecosystem. Univ Alberta Press, Edmonton, Alberta, Canada, pp 217–224

Dennis JG, Tieszen LL, Vetter MA (1978) Seasonal dynamics of above- and belowground production of vascular plants at Barrow, Alaska. In: Tieszen LL (ed) Vegetation and production ecology of an Alaskan Arctic Tundra. Ecological Studies 2. Springer, Berlin Heidelberg New York, pp 113–140

Giblin AE, Nadelhoffer KJ, Shaver GR, Laundre JA, McKerrow AJ (1992) Biochemical diversity along a riverside toposequence in arctic Alaska. Ecol Monogr 61: 415–435

Grulke NE, Bliss LC (1985) Environmental control of the prostrate growth form in two high arctic grasses. Holarct Ecol 8: 204–210

Hastings SJ, Luchessa SA, Oechel WC, Tenhunen JT (1989) Standing biomass and production in water drainages of the foothills of the Philip Smith Mountains, Alaska. Holarct Ecol 12: 304–311

Hultén E (1968) Flora of Alaska and neighboring territories. Stanford Univ Press, Stanford, 1008 pp

Kappen L (1988) Ecophysiological relationships in different climatic regions. In: Galun M (ed) CRC handbook of lichenology, vol 2. CRC Press, Boca Raton, pp 37–100

Miller PC (1982) Environmental and vegetational variation across a snow accumulation area in montane tundra in central Alaska. Holarct Ecol 5: 85–98

Murray KJ, Tenhunen JD, Kummerow J (1989) Limitations on *Sphagnum* growth and net primary production in the foothills of the Philip Smith Mountains, Alaska. Oecologia 80: 256–262

Oberbauer SF, Dawson TE (1992) Water relations of Arctic vascular plants. In: Chapin FS III, Jeffries RL, Reynolds JF, Shaver GR, Svoboda J (eds) Arctic ecosystems in a changing climate. Academic Press, San Diego, pp 259–279

Ostendorf B, Reynolds JF (1993) Relationships between a terrain-based hydrologic model and patch-scale vegetation pattern in an arctic tundra landscape. Landscape Ecol 8: 229–237

Sakai A, Larcher W (1987) Frost survival of plants. Springer, Berlin Heidelberg New York, 321 pp

Schramel P (1988) ICP and DCP spectrometry for trace element analysis in biomedical and environmental samples. Spectrochim Acta 43B: 881–896

Shaver GR, Chapin FS III (1991) Production: biomass relationships and element cycling in contrasting arctic vegetation types. Ecol Monogr 61: 1–31

Sonesson M, Wielgolaski FE, Kallio P (1975) Description of fennoscandian tundra ecosystems. In: Wielgolaski FE (ed) Fennoscandian tundra ecosystems. Part 1: Plants and microorganisms. Ecological Studies 16. Springer, Berlin Heidelberg New York, pp 3–28

Svoboda J (1972) Vascular plant productivity studies of raised beach ridges (semi-polar desert) in the Truelove Lowland. In: Bliss LC (ed) Devon Island IBP Project. High Arctic Ecosystem, Project Report 1970 and 1971. Univ Alberta Press, Edmonton, Alberta, Canada, pp 146–184

Tenhunen JD, Lange OL, Hahn SC, Siegwolf R, Oberbauer SF (1992) The ecosystem role of poikilohydric tundra plants. In: Chapin FS III, Jeffries RL, Reynolds JF, Shaver GR, Svoboda J (eds) Arctic ecosystems in a changing climate. Academic Press, San Diego, pp 193–237

Thomson JW (1979) Lichens of the Alaskan Arctic Slope. Univ Toronto Press, Toronto, Canada, 314 pp

Thomson JW (1984) American arctic lichens. I. The macrolichens. Columbia Univ Press, New York, 504 pp

Vitt DH, Marsh JE, Bovey RB (1988) A photographic field guide to the mosses, lichens and ferns of northwest North America. Lone Pine Publishing, Edmonton, Alberta, 296 pp

Walker DA, Everett KR (1991) Loess ecosystems of northern Alaska: regional gradient and toposequence at Prudhoe Bay. Ecol Monogr 61: 437–464

Walker DA, Binnian E, Evans BM, Lederer ND, Nordstrand E, Webber PJ (1989) Terrain, vegetation, and landscape evolution of the R4D research site, Brooks Range Foothills, Alaska. Holarct Ecol 12: 238–261

Washburn AL (1956) Classification of patterned ground and review of suggested origins. Geol Soc Am Bull 67: 823–866

Webber PJ (1978) Spatial and temporal variation of the vegetation and its production, Barrow, Alaska. In: Tieszen LL (ed) Vegetation and production ecology of an Alaskan Arctic Tundra. Ecological Studies 2. Springer, Berlin Heidelberg New York, pp 37–112

II Physical Environment, Hydrology, and Transport

6 Energy Balance and Hydrological Processes in an Arctic Watershed

L. D. Hinzman, D. L. Kane, C. S. Benson, and K. R. Everett

6.1 Introduction

Major efforts in recent years to understand global energy and water balances have focused attention on thermal and hydrological processes in high latitudes (Kane et al. 1992). One of our objectives in the R4D program (Chap. 1, this Vol.) was to develop a quantitative understanding of hydrological processes in the Imnavait Creek watershed and the energy flows that drive them. In this chapter we present monitoring data on energy balance, evapotranspiration, precipitation, snow distribution, snowmelt, runoff, and snow damming of runoff during the spring melt. We use these data first to develop budgets and elucidate seasonal and annual patterns; subsequently, we present physical process models that further quantify the dynamics and interactions between thermal and hydrological regimes and provide additional insight with regard to water and energy budgets in tundra ecosystems.

6.2 Radiation and Thermal Regimes

The Arctic receives much less solar radiation than lower latitudes and also experiences higher annual variation, both of which affect all aspects of arctic hydrological and thermal regimes. At the Imnavait Creek watershed we measured inputs and outputs of radiation over tussock tundra, the dominant type of vegetation (see Fig. 4.7c in Chap. 4, this Vol.). These data along with air and soil temperatures have been collected between early spring (April) and late fall (October) since 1985. Data collection during the winter is more limited, due to difficulties in maintaining and servicing the station.

From October to May and before the initiation of spring snowmelt – normally a few weeks before summer solstice – the tundra surface is characterized by a homogeneous high albedo near 0.8 (Weller and Holmgren 1974). Due to the uneven distribution of snow, the surface albedo varies greatly as the melt progresses (Liston 1986). Between the period of spring snowmelt and fall snow accumulation, the tundra surface has its lowest albedo of ca. 0.2, which results in maximum energy exchange (Ohmura 1981). Short-term increases in albedo may occur during midsummer, due to snowfall, which can occur on any day of the year. Initial snow accumulation in the autumn is usually near the equinox,

J.F. Reynolds and J.D. Tenhunen (Eds.)
Ecological Studies, Vol. 120
© Springer-Verlag Berlin Heidelberg 1996

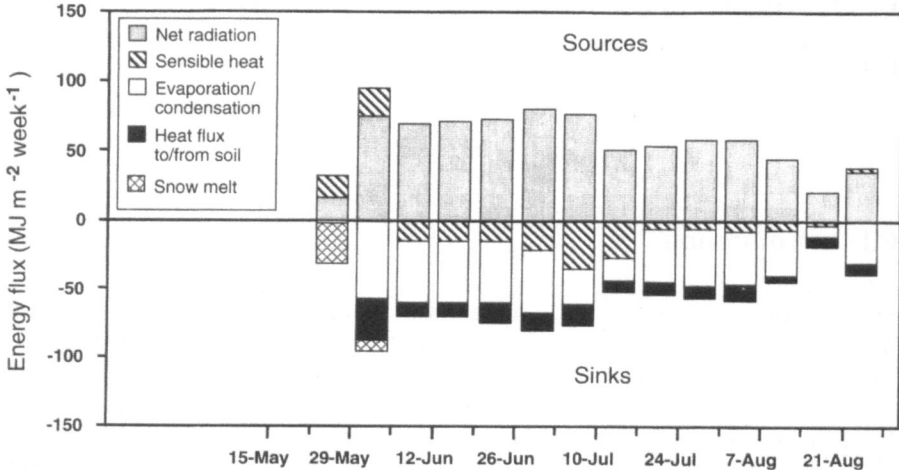

Fig. 6.1. Surface energy balance calculated in weekly increments at a point on a west-facing slope of the Imnavait Creek watershed during summer 1989

and because solar radiation is considerably less at this time, arrival of new snow cover does not produce the dramatic changes in surface energy and water fluxes that occur during spring-snow ablation (Fig. 6.1; Kane et al. 1990).

The Imnavait Creek watershed receives no direct solar radiation between 5 December and 8 January, and although several hours of diffuse radiation are incident on each day throughout the winter, the energy input is small. About 77% of the annual sunlit hours in the watershed occur between 21 March and 21 September. Incident shortwave radiation is governed by sun angle, but is greatly reduced on cloudy days. Even on clear summer days the low solar angle (maximum at summer solstice is only 45°) means that incoming solar radiation is highly attenuated by the atmosphere. The amount of shortwave radiation reflected depends on the presence of snowcover, soil moisture levels, sun angle, and radiation intensity. Atmospheric longwave radiation is primarily influenced by sky conditions and temperature. Snow ablation in spring and snow accumulation are significant factors determining the longwave radiation emitted from the surface. Net radiation becomes positive during daylight in March, increases substantially after snowmelt, as surface albedo and reflected radiation sharply decrease, and varies around a value of approximately $10\,MJ\,m^{-2}\,day^{-1}$ (see also Kane et al. 1990) during midsummer (Fig. 6.1), as has been described at a number of other tundra sites (Barry 1981).

During the winter, arctic tundra climate is affected primarily by radiative heat loss and atmospheric circulation (Weller and Holmgren 1974; Ohmura 1981). Low incoming radiation and high albedo determine that little energy is input to the active layer (the shallow layer of soil above the permafrost that thaws – and then freezes – seasonally as a function of the net energy balance).

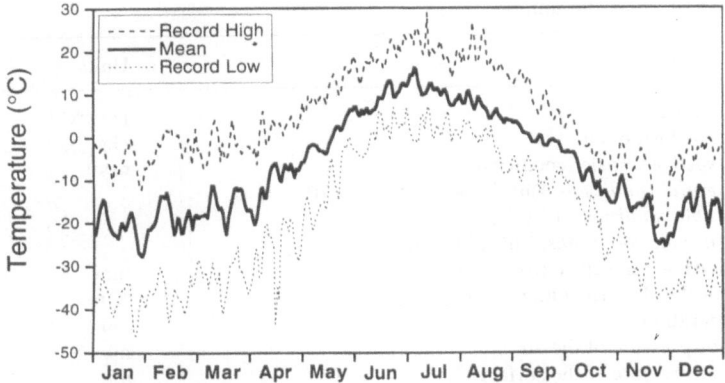

Fig. 6.2. Seven-day running mean of daily air temperature and record high and low average hourly air temperatures from January 1985 through May 1993

Mean daily air temperature and record high and low temperatures (average hourly temperatures from January 1985 to May 1993 are used in the comparison) in the Imnavait Creek watershed are presented in Fig. 6.2. The average annual air temperature during this period was −7.4 °C. Mean monthly air temperatures exhibit similar trends to data collected at Barrow (ca. 400 km NW; Dingman et al. 1980), i.e., interannual variability (expressed as standard deviation of mean monthly temperature) for the winter months is usually >3 °C, and for the summer months, usually <2 °C. This difference has been attributed to northward shifts in the arctic frontal zone during the summer (Dingman et al. 1980).

6.2.1 Surface Energy Balance

The surface energy balance is given by (terms defined in Table 6.1):

$$Q_n + Q_h + Q_c + Q_e + Q_m = 0. \tag{1}$$

Energy transferred by rain or running water was not included in this analysis, because precipitation and soil water temperature are usually close to soil temperature and, consequently, over short periods of time add little energy. This energy source may be significant when rain falls on existing snow, but this was rare during the period of our study.

Convection (Q_h), the sensible heat transferred between air and soil or snow surface, was calculated as a function of temperature gradient, wind speed, and surface conditions:

$$Q_h = C_p \cdot p_a \cdot D_h \cdot (T_a - T_{sur}), \tag{2}$$

where it is assumed that the momentum exchange coefficient between the atmosphere and ground equals the heat exchange coefficient (D_h; Szeicz et al.

Table 6.1. Definition of terms and units

Variable	Definition	Units
C	Soil heat capacity	$J\,m^{-3}\,°C^{-1}$
C_p	Specific heat of air	$J\,kg^{-1}\,°C^{-1}$
D_h	Convective transfer coefficient	$m\,s^{-1}$
	Neutral conditions, $Dh(n) = k^2 u_{z2}/[\ln(z_2/z_o)]^2$	
	Stable conditions, $Dh(n)/[1 + (\sigma * Ri)]$	
	Unstable conditions, $Dh(n)/[1 - (\sigma * Ri)]$	
dS	Active layer moisture storage and surface storage	mm
dT/dx	Vertical soil thermal gradient over distance x	$°C\,m^{-1}$
E	Evaporation	mm
e_a	Vapor pressure of the air	mb
e_{sur}	Vapor pressure of the surface	mb
F	Melt factor	$mm\,°C^{-1}\,h^{-1}$
F_t	Temperature melt factor	$mm\,°C^{-1}\,h^{-1}$
F_r	Radiation melt factor	$mm\,m^2\,W^{-1}\,h^{-1}$
g	Gravitational constant	$m\,s^{-2}$
K	Thermal conductivity	$W\,m^{-1}\,°C^{-1}$
k	von Karman's constant	0.41
L_v	Latent heat of vaporization	
M	Snowmelt rate	$mm\,h^{-1}$
P	Air pressure	mb
p	Precipitation or snowmelt	mm
Q_c	Heat conduction into soil	$W\,m^{-2}$
Q_e	Energy utilized for evapotranspiration or condensation	$W\,m^{-2}$
Q_h	Convective heat transfer	$W\,m^{-2}$
Q_m	Energy utilized for snowmelt	$W\,m^{-2}$
Q_n	Net radiation	$W\,m^{-2}$
R	Runoff	mm
R_i	Richardson number, $[g * z_2 * (Ta - T_{sur})/[u_{z2} * (Ta + 273.15)]$	
SW_r	Reflected shortwave radiation	$W\,m^{-2}$
SW_s	Incident shortwave radiation	$W\,m^{-2}$
t	Time	s
T_a	Temperature of air at height z	°C
T_{sur}	Temperature of the surface	°C
u_{z1}	Wind speed at height 1	$m\,s^{-1}$
u_{z2}	Wind speed at height 2	$m\,s^{-1}$
z	Height of the wind speed and temperature measurement	m
z_0	Roughness length, $\exp[\{(u_{z2} * \ln(z_1)) - (u_{z_i} * \ln(z_2))\}/\{u_{z2} - u_{z1}\}]$	m
ρa	Density of air	$kg\,m^{-3}$
σ	Constant	10
θ	Temperature	°C

1969), a relationship that holds only for neutral (isothermal) conditions. Daily heat exchange coefficients were adjusted based on the air temperature profile between the surface and 10 m using different values of D_h for neutral, stable, and unstable conditions (see Table 6.1; Braun 1985). The average surface roughness length, z_0, was determined from wind-speed profiles between 1.5 and 10 m. For tussock tundra we estimated z_0 to be 0.02 m by averaging calculations made from several hundred wind profile measurements during near-neutral conditions. This value falls well within the reported values for short grasses and tundra (Szeicz et al. 1969; Weller and Holmgren 1974). The roughness length over snow can vary a great deal as the windswept surface changes from very smooth to very rough. During the spring melt period we estimated z_0 as 0.0013 m. Effective surface temperatures were obtained from emitted terrestrial longwave radiation.

Conductive heat transfer (Q_c) is the energy passed from particle to particle within the soil matrix. Because the energy budget is calculated in a plane at the surface, it is not necessary to consider latent heat associated with the freezing or thawing of the active layer. Conductive heat transfer is calculated as:

$$Q_c = -K \cdot dT/dx. \tag{3}$$

The thermal conductivity (K) of these soils was determined for several soil types under a variety of moisture and temperature conditions (Hinzman et al. 1991b). Values of K were selected depending on whether the surface soil was wet or dry, thawed or frozen. Heat conduction into the soil was measured with heat flux plates located near the surface and at the organic-mineral soil interface. Heat flux at the soil surface largely mirrors net radiation except that diurnal variation in surface heat flux is greatly reduced during the winter because of the insulative properties of the snowpack and low variation in energy input from radiation. Daily amplitude and instantaneous magnitude of heat flux decrease with soil depth. Each spring and autumn, heat flux approaches zero as the surrounding soil progresses through a phase change. As the soil water freezes in September, energy flows from the warm soil to the colder air, and the near-surface soil temperature remains isothermal at 0 °C, because energy released as latent heat offsets surface heat loss.

Evapotranspiration can be estimated as the remainder term in a water or energy balance equation (Eq. (1)) or calculated using techniques based on vapor and wind gradients:

$$Q_e = L_v \cdot p_a \cdot D_h \cdot (0.622/P) \cdot (e_a - e_{sur}). \tag{4}$$

Because it is difficult to determine surface vapor pressure when soils are less than saturated, Eq. (4) works best during snowmelt or immediately after rainfall events. Our calculated values of Q_e compare well with values estimated from water-balance studies at Imnavait Creek (Hinzman et al. 1991a).

The importance of each of the terms in Eq. (1) varies considerably throughout the year. The rates of all energy fluxes are reduced during winter. Transpiration is zero. Conductive heat transfer is reduced by the insulating effect of the snow layer or reversed because of the temperature gradient. Evaporation is normally considered to be low during the winter because of energy constraints (Ohmura 1982). Nevertheless, the loss of snow by sublimation can be high during large wind events (Tabler 1975). Convective heat transfer is affected least by the season, but can be lower because of the

Table 6.2. Summer and annual water balances (cm)

	Precipitation			Runoff			Annual evapotranspiration	
Year	Maximum snowpack water equivalent	Summer[a]	Annual total	Snowmelt	Summer[a]	Annual total	Actual	Potential[b]
1985	10.2	25.1	35.3	6.6	c	c	c	c
1986	10.9	16.3	27.2	5.7	6.2	11.9	15.3	31.0
1987	10.8	27.2	38.0	7.1	17.9	25.0	13.0	32.0
1988	7.8	25.2	33.0	3.9	7.2	11.1	21.9	33.2
1989	15.5	25.7	41.2	9.4	7.8	17.2	24.0	42.0
1990	10.6	16.3	26.9	6.4	2.8	9.2	13.3	39.4
1991	8.2	24.9	33.1	5.6	3.3	8.9	24.2	37.7
1992	18.1	24.1	42.2	14.4	6.3	20.7	21.5	32.8
1993	12.5	20.8	33.3	7.9	14.6	22.5	10.8	32.1

[a] Late June–August.
[b] Estimated from pan evaporation.
[c] No data.

smoother winter snow surface and smaller temperature gradients between snow and air.

Weekly magnitudes and variations in the various components of the surface energy budget during spring and summer 1989 are illustrated in Fig. 6.1. The spring snowpack in 1989 was the second greatest of the 9 years measured (1985–1993), but the energy balance was similar in most regards to those calculated for 1986–1992. Solar radiation is the primary source of energy during the summer months, especially after spring snowmelt. During spring melt the effect of convective heat transfer is an especially important complement to radiative heat transfer. The contribution of energy advected over the nearby Brooks Range and transferred via longwave radiation or heat convection is of primary importance in determining the initiation and rate of snowmelt. The fundamental sinks of energy are warming of the snow and soil, snowmelt, evaporation, and thawing of the active layer. In the summers of 1987, 1988, 1989, and 1990 about 39, 46, 65, and 38%, respectively, of net radiation was used in evapotranspiration (Table 6.2; Kane et al. 1990; see also Chap. 17, this Vol.). During midsummer, the surface is normally warmer than the air, due to radiative heating, and sensible heat is lost by longwave radiation and convection from the warm surface to the cooler air. Comparing the energy balance during spring melt at Imnavait watershed with other sites on the North Slope, convective heat transfer exerts more influence near the Brooks Range, and decreases in importance north onto the Coastal Plain, where net radiation plays a greater role (Weller and Holmgren 1974; Dingman et al. 1980; Hinzman et al. 1991a).

6.2.2 Snow Cover and Soil Thermal Regime

Snow distribution and snowpack volumes in the Imnavait watershed are extremely variable both in time (year to year) and space (within the watershed; Benson 1982; Liston 1986), and are largely a function of wind and topography. At the end of the accumulation season, snow depths can range from a few centimeters on windswept ridgetops to more than 1 m in the bottom of the valley. The average depth over the whole watershed is about 50 cm. This region has primarily north-flowing katabatic winds that result from downslope drainage of denser air from the Brooks Range to the south. However, large wind events can originate from any direction causing extensive drifts and wind slabs throughout the watershed. Wind slabs are characterized by very hard, fine-grained layers of snow with densities between 0.35 and 0.54 g cm^{-3}. In Imnavait watershed the direction and force of the larger wind events vary from year to year, changing the orientation of the drift sequence. Nevertheless, the consistency of the predominantly southeast wind yields similar snow distribution each year, i.e., deposition in valley bottoms and on the lee side of slopes (Fig. 6.3). The orientation of wind slabs depends on the direction of strong winds, whereas the density of the drift depends on the magnitude of the wind events.

Snow depths, measured along selected traverses, were combined with pit studies to measure snow density, temperature, and hardness profiles. In addition to providing the water equivalent of the snowpack (Fig. 6.3), pit studies allowed us to measure extreme snow types such as wind-slab and depth-hoar layers. Photographs were taken from control points on the ground at selected time intervals during each melt season. These data permitted us to extrapolate detailed point measurements over broad areas, producing maps of the maximum end-of-winter snow cover.

Whereas wind direction did not vary, year-to-year wind strengths differed significantly. In 1985 there was maximum wind slab formation and the highest snow density with maximum values of 0.54 g cm^{-3}. The sensitivity of snow distribution to topography is pronounced (Fig. 6.4). Accumulation on lee slopes was about 65% more than on windward slopes, although slope angles differed by only 2–3 degrees.

Snowpack distribution and density affect runoff processes in several ways. Because of snowdrifts, snowpack water content can vary by a factor of 2–3 over a distance of a few meters. The result is a fast melt where snowpack is thin, and the development of bare patches, with considerable edge effect around the drifts during melting. On the valley floor, where snowpack is thick and dense, it functions as a dam, holding until the force of the water overcomes the bonding strength of the snow (Kane et al. 1989, 1991a). In addition, snow accumulation near stream channels and water tracks yields a higher proportion of runoff and less evaporation than if the snow were uniformly distributed throughout the watershed. Deeper snowpacks also greatly influence the thermal regime of the underlying soil by increasing thermal resistance to heat flow.

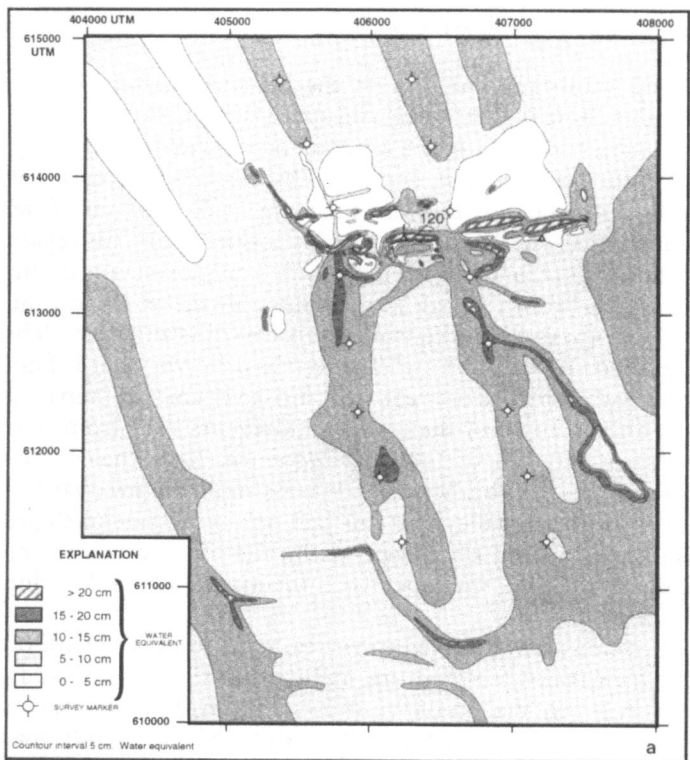

Fig. 6.3a,b. Areal snow distribution over R4D region in Spring 1988 (**a** one of the lowest snowpacks) and Spring 1989 (**b** one of the highest snowpacks). UTM grid location is illustrated in Chap. 1 (this Vol.)

Soil stratigraphy also affects the thermal regime through the insulative properties of the organic soil at the surface. The thermal conductivity of this layer is one-half to one-sixth that of the underlying mineral soil (Hinzman et al. 1991b). The albedo is also higher than mineral soil and, thus, if this surface layer is removed, heat flux increases in the exposed mineral soil.

Although air temperatures normally reach their annual minimum in January or February (Fig. 6.2), the annual minimum soil temperature normally occurs in late March or April (Fig. 6.5). Surface soil warms rapidly by 6–7 °C within a few days in late May or June when solar radiation and soil heat fluxes are near annual maximum. The primary reason for the very rapid spring warming of the surficial soil layer is infiltration and freezing of snow meltwater in the still-cold soils and the release of substantial amounts of latent heat. The daily and hourly temperature variability of each layer is greatest in the summer. This variation decreases with depth. The thermal gradient reverses during autumn freeze-up and spring melt, with the soil at 40 cm warmer in winter and cooler in summer than the surface soil.

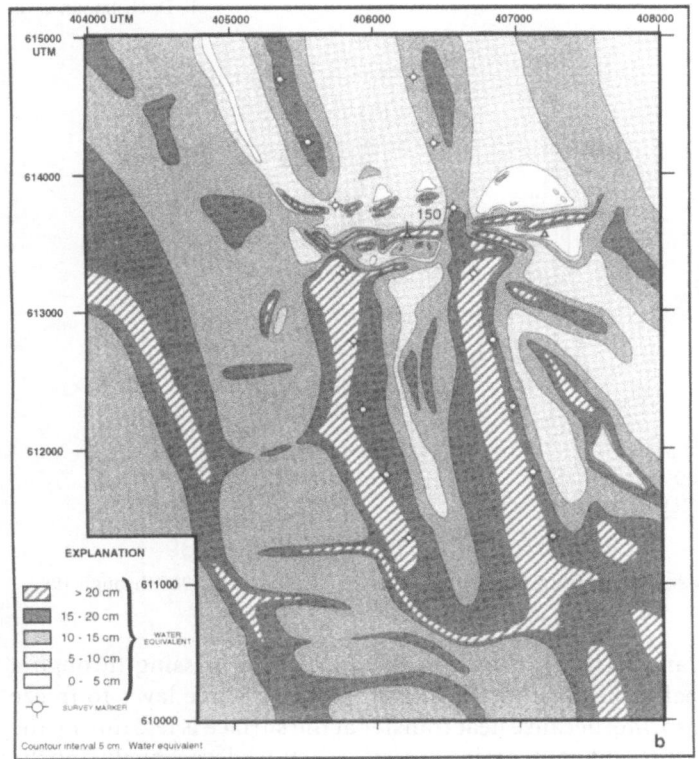

Fig. 6.3. (*continued*)

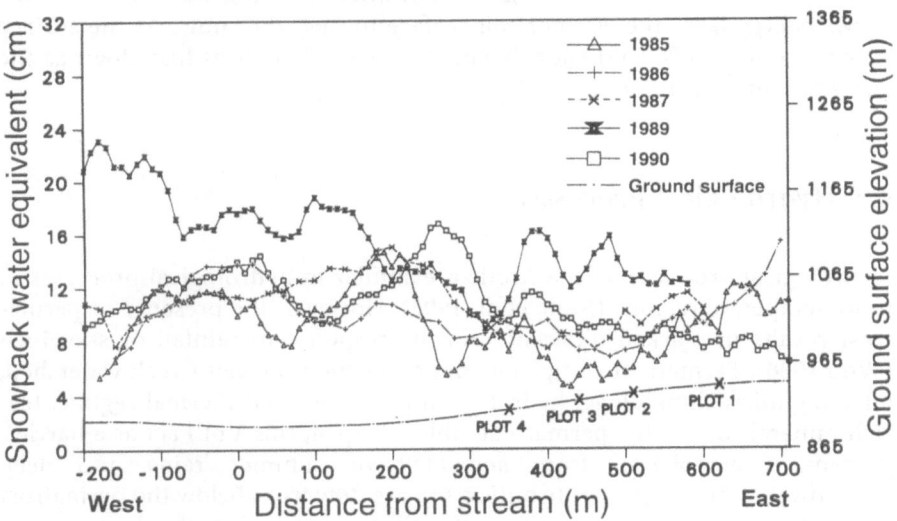

Fig. 6.4. Snowpack distribution across Imnavait Creek watershed. The four runoff plots constructed along the east slope of the watershed are depicted at their respective elevation on the ground surface line

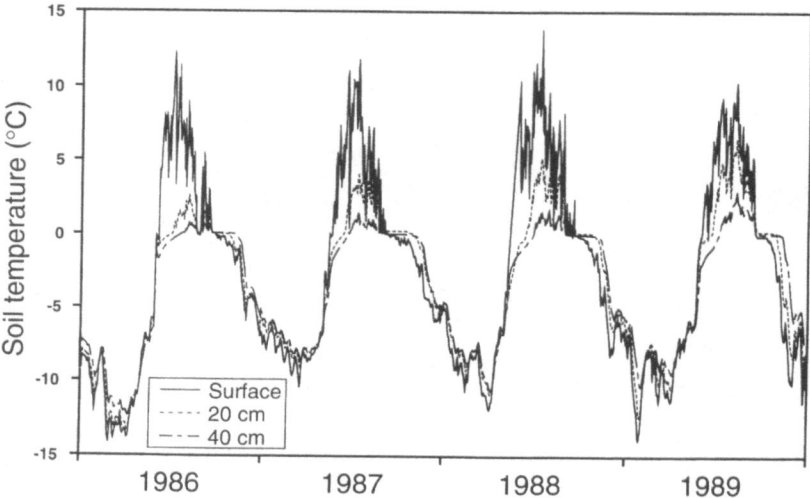

Fig. 6.5. Average daily soil temperatures measured at several depths from 1986 through 1989

In September and October, the soil begins to freeze, passing through a period of isothermal conditions. It takes longer for the active layer to freeze than to thaw in the spring, because heat transfer at the surface is less during the fall. In October, the amount of incoming radiation is much less than during the spring thaw, but the net radiation balance is still positive. The primary reason that the heat transfer rate is lower during the autumn is because of snowfall. Early-season snow will usually melt soon after touching the surface, which draws energy from the warmer soil surface to melt the snow. As the surface quickly cools to 0 °C and snow begins to accumulate, heat loss slows as the snow provides insulation.

6.3 Hydrological Processes

Hydrological processes in the Arctic are similar to hydrological processes in more temperate regions (Kane et al. 1992). However, the presence of permafrost results in marked differences in the response to rainfall or snowmelt (Woo 1986). Permafrost completely underlies the Imnavait Creek watershed, affecting microclimatology, hydrological processes, and thermal regime. Ice-rich mineral soils at the permafrost table (Chap. 4, this Vol.) act as a barrier, preventing percolation from snowmelt or summer rains into deep groundwater; hence, the contribution to base flow from below the permafrost table at Imnavait Creek is zero, effectively simplifying the hydrological dynamics. Although deep springs provide water for base flow throughout the year in some places on the North Slope of Alaska (such as the adjacent

Kuparuk River system), the Imnavait Creek watershed is isolated from this subpermafrost water source. Because water is not lost to groundwater recharge, all water leaves the basin either through near-surface runoff or evapotranspiration.

The depth of the active layer ranges from 25 to 100 cm. The highly stratified soils – classified as pergelic cryaquepts or histic pergelic cryaquepts (Rieger et al. 1979; Chap. 4) – strongly influence both hydrological and thermal regimes (Hinzman et al. 1991b; Kane et al. 1992). There is a thick, porous layer of organic matter on the surface consisting of partially decomposed mosses, sedges, and other plants, which saturates and drains quickly. The underlying mineral soil is usually saturated with water. The hydraulic conductivity of the organic layer is 10–1000 times greater than that of mineral soil (Hinzman et al. 1991b). Consequently, water from snowmelt or summer rainfall moves rapidly downslope above the organic–mineral interface. An analysis of streamflow hydrographs reveals that as summer progresses, the recession curves of stream discharge following a rain event increase slightly. This observation indicates that the percentage of hillslope runoff flowing through the mineral soils increases slightly as the active layer increases in thickness, although most of the runoff still flows through the organic mat (Hinzman et al. 1993).

6.3.1 Snowmelt

The start of spring snowmelt in the Imnavait Creek watershed varies greatly depending on the depth of snowpack and meteorological conditions. From 1985 to 1993 the date of initial melt ranged from 8 May to 1 June with an average of only 10 days required for complete ablation (Fig. 6.6). An analysis of snowmelt onset as a function of initial snowpack water equivalent showed that 67% of the variation in the initiation of snowmelt could be explained by the amount of snow (dashed line in Fig. 6.6). During snowmelt, net radiation along with convective heat transfer dominate the surface energy balance (Fig. 6.1; Hinzman et al. 1991a). Although incoming shortwave radiation is very high, no melt will usually occur until convective heat transfer becomes positive.

Within a few days of a sustained melt the entire watershed becomes a patchwork of snow-covered and bare tundra. Liston (1986) found temperatures averaging 15 °C in 1985 and 24 °C in 1986 with an extreme temperature of 42 °C on bare tundra surfaces surrounded by snow, while the bordering snow was isothermal at 0 °C. Complex patterns of stable and unstable air result from strong thermal gradients. Longwave radiation emitted from snow-free ground warms the overlying air, and turbulent transfer in these locations accelerates melting of the surrounding snow. The west-facing slope melts off sooner than the rest of the watershed, because it retains less snow (see Figs. 6.3 and 6.4). The west-facing slope also receives more direct solar radiation in the afternoon when air temperatures are highest. The result is a complementary summation of radiant and convective heat transfer that results in a greater positive surface

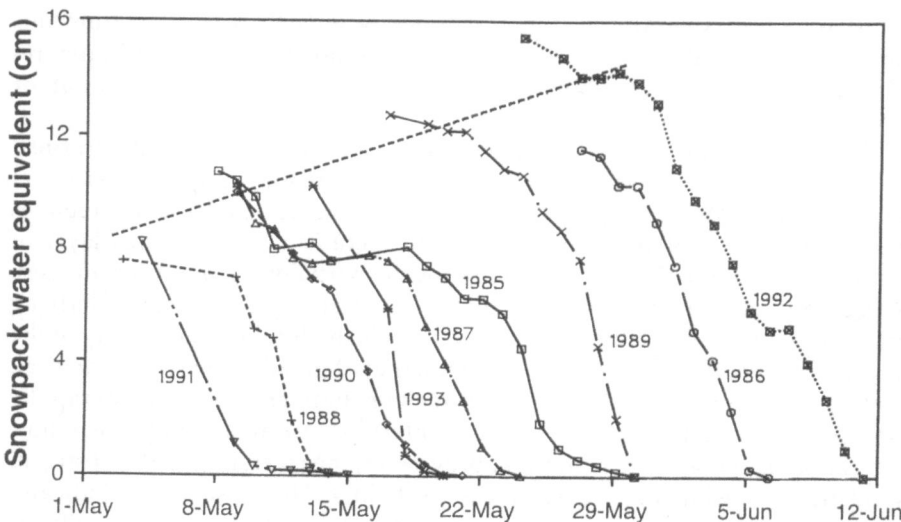

Fig. 6.6. Rates of snowpack ablation characterized by amount of snow still present as a function of time. Regression of dependence of snowmelt initiation (y) on initial snowpack water equivalent (x) is $y = 0.225\,x + 7.80$, $r^2 = 0.67$. Date of initial melt determined by the first date to show decrease in snowpack water equivalent

heat flux. The east-facing slope, in contrast, receives its maximum irradiance in the morning while convective heat transfer is smaller.

Problems in modeling snowmelt for the Imnavait watershed from surface energy balance relate to the large spatial variability in snow depth, which affects albedo, longwave emittance, and sensible heat flux. The energy balance presented in Fig. 6.1 was calculated based on instrumentation located at a single point on the west-facing slope. The measured and calculated snowpack ablation (Fig. 6.6) also represents processes occurring only on the west-facing slope (78% of the basin). A comprehensive formulation of watershed snowmelt runoff would require measurement of the surface energy balance at many points throughout the watershed. Figure 6.1 probably overestimates the magnitude of net radiation on the east-facing slope. The difficulty in calculating heat transfer and a surface energy balance in such situations has stimulated the development of a number of simplified methods for estimating rates of snowmelt (Bergström 1986; Sand 1990).

6.3.2 Plot and Basin Water Balance

Rainfall each autumn ensures that the active layer will be saturated at the beginning of each winter. Because the moisture stored in the soil is nearly the same at the end of each thaw season, it is straightforward to estimate annual water balances. Four 89-m² plots were constructed along a diagonal on the west-facing slope in tussock tundra to directly measure surface runoff and subsurface water flow above the permafrost (see Fig. 9.2 in Chap. 9 for location of plots). Each plot was bounded with heavy plastic, which isolated the area from upslope water. A gutter system at the base of each plot routed the water into a holding tank instrumented with a water-level recorder that provided data on runoff rates and volumes. Although these plots provided valuable information on the components of the water balance at these sites (Hinzman et al. 1993), they cannot be considered replicates, due to the large spatial heterogeneity within the watershed. The proportion of a snowpack or soil moisture that is lost either to runoff or evaporation is affected by slope and aspect; consequently, these plots may not reflect the magnitude of processes occurring elsewhere in the watershed.

The snowpack water equivalent of each plot and the basin average were determined before snowmelt began and daily during spring melt. Runoff was measured directly. The proportion of moisture going into storage in the desiccated organic layer during snowmelt was estimated in laboratory and field studies to average 1.5 cm over the entire basin (Kane et al. 1989). Evaporation was calculated as:

$$E = p - R - dS. \tag{5}$$

During this study the initial snowpack on each plot ranged between 7.2 and 14.7 cm, and basin averages ranged between 7.8 and 18.1 cm. Evaporation varied greatly depending on the initial water content of the snowpack, proportionally increasing as the snowpack decreased. On the basin scale the amount of runoff depended primarily on the amount of snow, but distribution of the snow was important in determining the fraction going into runoff or evaporation (Table 6.2). When more snow was deposited in the bottom of the watershed, near the stream, runoff was greater and evaporation was lower. The processes associated with spring melt and runoff are discussed in more detail in Kane and Hinzman (1988) and Kane et al. (1989, 1991a).

6.3.3 Runoff and Basin Discharge

Frequently, there were large differences in the depth of snowpack along the hillsides (Figs. 6.3 and 6.4), which strongly affected basin runoff. Compared with plots having an average snowpack water equivalent, runoff on plots with a thin snowpack began earlier, probably because of a lower albedo, greater shortwave absorbance by the underlying soil, and greater longwave emittance from shrubs protruding through the snow surface. Plots with thinner snowpacks also lost a greater proportion of water to evaporation. Plots with the deepest snowpacks had the greatest runoff volumes and the greatest peak flows.

During spring melt most downslope water movement occurs in the top 10 cm of the organic layer (Hinzman et al. 1991b). Moisture content at 5 cm

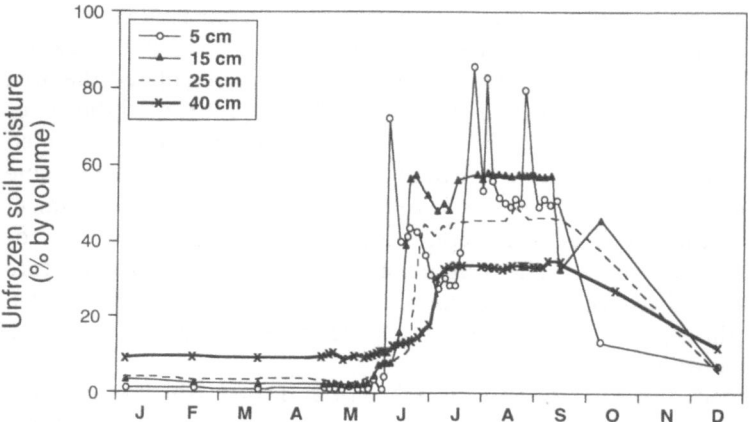

Fig. 6.7. Unfrozen soil moisture content during 1986 measured at several depths with time-domain reflectometry (Hinzman et al. 1991b)

increases abruptly as meltwater begins to infiltrate (Fig. 6.7), but the moisture content below this layer did not substantially increase until snowmelt was complete and the depth of thaw reached below 10 cm. As the surficial organic layer becomes saturated, an intricate flow occurs in discontinuous rills between sedge tussocks. Over time the ice-rich soil thaws, releasing water and maintaining the mineral soil near saturation throughout the year. However, the moisture content of the organic soil varies greatly throughout the year. The moisture content of surface organic soils varies by as much as 60% by volume throughout the summer (Fig. 6.7), saturating and drying quickly in response to summer rains or snowmelt. Subsurface mineral soils, in contrast, remain near saturation (Hinzman et al. 1991b).

Flow alternates between the surface and subsurface depending on soil conditions, topography, and depth of thaw. The highly porous moss layer and well-developed intertussock channels are conducive to rapid downslope movement. The hillside is drained by small water tracks that are nearly parallel to each other (ca. 50 m apart; see Fig. 4.4 in Chap. 4, this Vol.). Following a precipitation event, these water tracks fill with water and quickly drain the hillslope (Kane et al. 1991a). As the melt season progresses and the depth of thaw increases, the amount of moisture storage increases. Thus, runoff and soil moisture dynamics are a function of seasonal time, an attribute that has significant implications for modeling the hydrological regime especially considering the effects of disturbance (Hinzman and Kane 1991).

The drainage is a beaded ephemeral stream that flows from spring melt until freeze-up in September. Spring runoff is usually the dominant hydrological event of the year (Kane and Hinzman 1988), producing the annual peak flow and about 50% of the total annual runoff volume. Streamflow almost ceases after extended periods of low precipitation, which may occur in late

Table 6.3. Average daily evapotranspiration calculated from
an hourly energy balance in the Imnavait Creek watershed
(mm day^{-1})

	1987	1988	1989	1990	Average
May	0.6	1.0	0.5	1.7	1.0
June	2.7	1.5	3.0	1.1	2.1
July	0.8	2.0	2.0	0.8	1.4
August	0.6	2.1	1.9	1.5	1.5

May or June, whereas intense summer rainfall events produces substantial
stream flow (see also Chaps. 9 and 17, this Vol.).

6.3.4 Precipitation, Evaporation, and Evapotranspiration

Based on the 8-year record from 1985 until 1992, the average annual precipitation is ca. 35 cm, two-thirds of which falls in late June, July, and August (Table
6.2). The period in the Arctic with the least sea ice falls in late July and August
(Labelle et al. 1983) corresponding to the period of greatest precipitation. The
total number of rain events recorded for the summers of 1985–1988 was 256. A
substantial number of these reflect discontinuous periods of precipitation
(many of them too light to be recorded by a standard rain gauge) within
protracted periods of rainy weather. Most rainfall is light (95% at rates
<10 mm h^{-1} and 82% < 1.0 mm h^{-1}), appears evenly distributed over the catch-
ment, and is associated with the dissipating phase of convective storms generated over the Brooks Range or with air masses moving from the North Pacific
Ocean. Maximum rainfall intensities generally occur in the first 4 or 5 hours of
the event. High-intensity (>20 mm h^{-1}), short-duration rainfall is associated
with convective storms which generally occur earlier in summer.

Actual and potential evapotranspiration (Table 6.2) and average daily
evapotranspiration per month (Table 6.3) indicate that both evapotranspiration and pan evaporation are generally greatest after snowmelt and decrease throughout the summer. On hillslopes with good drainage the rate of
evaporation is possibly limited by the degree of hydration of mosses. In
marshy areas of the valley bottom, the free water surface frequently lies above
the soil surface so that the rate of evaporation is limited by the amount of
available energy. Rates of evapotranspiration were calculated according to the
energy balance (Sect. 6.2.1) and with the Priestley-Taylor method (Priestley
and Taylor 1972; Kane et al. 1990). The predictions for summer 1987 are
compared with precipitation and pan evaporation in Fig. 6.8. During the early
summer, evapotranspiration rates are greater than precipitation indicating
watershed drying. This drying of the active layer occurred regularly following
spring snowmelt and significantly modifies decomposition (Chap. 16), nutrient availability (Chap. 10), and carbon balance (Chaps. 11 and 17).

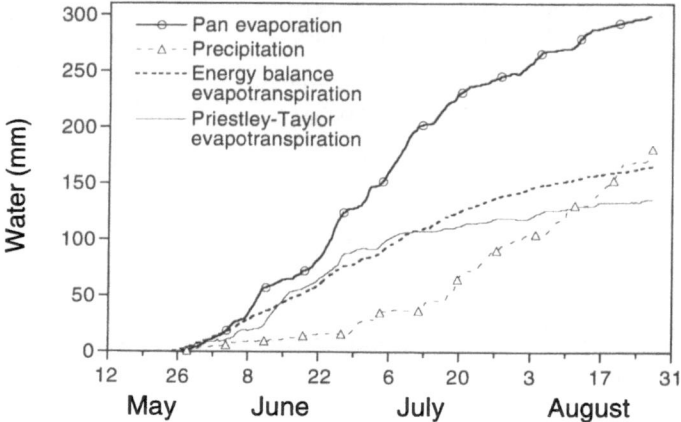

Fig. 6.8. Comparison of cumulative precipitation, pan evaporation and calculations of evapotranspiration during the summer of 1987. (from Kane et al. 1990)

6.4 Energy Balance and Hydrology Models

6.4.1 Simulation of the Thermal Regime

The thermal regime of the top 60 m of soil in tussock-tundra plots was simulated using the Two-Dimensional Heat Conduction (TDHC) model, a finite element computer simulator that allows non-steady-state analysis (Goering and Zarling 1985). This is a physical model based on heat conduction and phase change. The TDHC model utilizes the following formulation of the heat-diffusion equation.

$$\frac{\partial}{\partial x}\left[K\frac{\partial\theta}{\partial x}\right] + \frac{\partial}{\partial y}\left[K\frac{\partial\theta}{\partial y}\right] = C\frac{\partial\theta}{\partial t}, \tag{6}$$

where x and y are the coordinates (m) in the horizontal and vertical directions, respectively, θ is temperature, K is the thermal conductivity, C is heat capacity, and t is time (see Table 6.1). A two-dimensional grid of linear triangular elements of variable size was created vertically in the soil profile where the top of the grid corresponded to the soil surface. The elements were smallest near the surface where soil thermal properties and soil temperatures changed quickly and increased in size with depth. The grid contained 255 nodes and 336 elements; the simulation time step was 1 day. The lower boundary of the grid was selected as 61 m to ensure only negligible change in the temperature of the lower boundary in response to surface perturbations.

Equation (6) is solved using a finite element technique based on the Galerkin weighted-residual process. The simulations are driven by the surface temperature and heat flux at the bottom of the grid. Information must also be

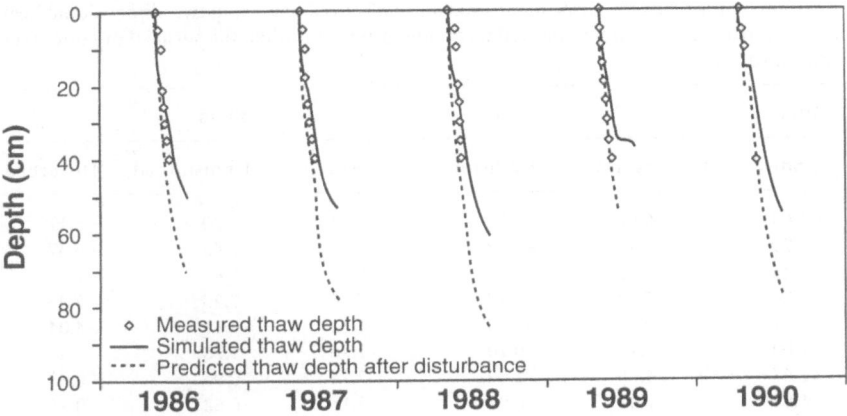

Fig. 6.9. Comparison of simulated thaw depths with and without surface disturbance (removal of organic soil layer)

provided on the frozen and thawed thermal conductivity, frozen and thawed heat capacity, and the amount of latent heat required for phase change in each soil element. The amount of heat transferred during phase change in each element is calculated using a Dirac delta function (O'Neill 1983). Heat flux at the lower boundary was estimated at $0.050\,W\,m^{-2}$ based on an analysis of soil thermal properties and the geothermal gradient of a deep well 15 km from the Imnavait watershed (Osterkamp et al. 1987). The model accuracy was verified by testing it against closed form exact solutions (Goering and Zarling 1985; Zarling et al. 1989).

The model was calibrated and used to simulate rates of thaw of the active layer in the Imnavait Creek watershed during the years 1985–1990 (Fig. 6.9; Kane et al. 1991b). Simulated temperatures at several soil depths were compared with continuous measurements of actual subsurface temperatures. In the active layer, predicted temperatures were within 0.25 °C of observed temperatures 75% of the time, within 0.50 °C of observed temperatures 92% of the time, and within 1 °C of observed temperatures 96% of the time. Thus, reasonable agreement was also obtained between observed and simulated thaw in the active layer (Fig. 6.9). The TDHC model simulated the observed temperature gradients extremely well during periods when the soil was completely frozen and heat flow was entirely via conduction.

To examine sensitivity of the thaw processes to surface modification as described by TDHC, we simulated an extreme case of disturbance in which the surficial organic layer is stripped off (Fig. 6.9), a situation that occurred during the early years of development in Alaska (Chap. 4, this Vol.; Parker 1929). Due to the greater heat conductivity of the mineral soil, thaw proceeds rapidly and reaches greater depth by the end of the season (ca. 20–30 cm). The subsurface temperatures near the surface of the disturbed soil profile are colder in the coldest part of winter and warmer in the warmest part of summer, in effect

Table 6.4. Soil temperatures at three depths during 1987 simulated using two-dimensional heat conduction (TDHC) in cases where the soil was undisturbed and when the surficial organic layer was removed (disturbed)

	10 cm		20 cm		30 cm	
	Undisturbed	Disturbed	Undisturbed	Disturbed	Undisturbed	Disturbed
Jan	−6.63	−6.62	−6.24	−6.21	−5.30	−5.17
Feb	−7.79	−7.82	−7.52	−7.55	−6.42	−6.37
Mar	−8.46	−8.52	−8.41	−8.45	−7.47	−7.45
Apr	−7.88	−7.89	−7.77	−7.78	−7.25	−7.23
May	−0.31	−0.20	−1.50	−1.38	−4.11	−4.04
June	1.00	3.15	−0.10	1.76	−1.78	−1.48
July	2.67	4.06	1.00	2.74	−0.95	−0.63
Aug	1.89	2.84	0.97	2.13	−0.62	−0.25
Sept	−0.02	0.01	0.02	0.01	−0.49	−0.19
Oct	−0.22	−0.19	0.00	0.03	−0.45	−0.18
Nov	−2.11	−2.01	−0.99	−0.61	−0.45	−0.19
Dec	−5.61	−5.07	−5.06	−3.52	−3.59	−0.11

changing the amplitude and the lag time of the annual temperature cycle (Table 6.4). Although thaw depth changes are small, such changes can lead to the melting of ground ice, the formation of depressions in which surficial runoff and infiltration water may accumulate, and thermal erosion. On hillsides, the fine-grained mineral soils offer little resistance to the erosion process (Mageau and Rooney 1984).

6.4.2 Simulation of Snowmelt

The physically correct method to quantify snowmelt is by determining a complete energy balance for the snowpack (Sand 1990). However, the processes of heat transfer within and to the snowpack are highly variable, both spatially and temporally, and are extremely difficult to define (Male and Granger 1981). The complexity of this approach has led to development of many simplified snowmelt equations (Bengtsson 1976; Bengtsson 1982).

We computed rates of snowmelt within the Imnavait Creek catchment using three different approaches (Fig. 6.10): an energy balance model (Price and Dunne 1976), a "degree day method", and a modified "degree day method" that includes radiation influences. The second approach used the following empirical equation:

$$
M = \begin{cases} F\,(T_a - T_o) & \text{when } T_a > T_o \\ 0 & \text{when } T_a \le T_o, \end{cases} \tag{7}
$$

Where M is the snowmelt rate, F is the melt factor, and T_a and T_o are air and threshold temperatures, respectively (Table 6.1). When $T_a < T_o$, melt will not

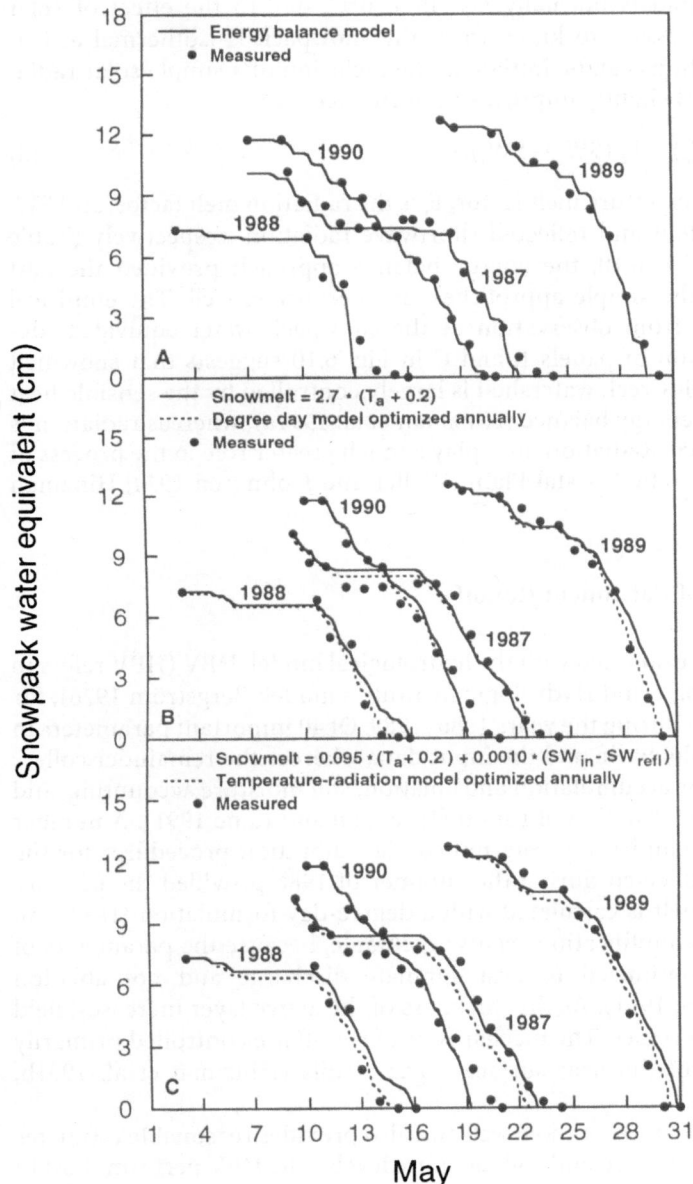

Fig. 6.10A–C. Simulated and observed snowpack ablation in Imnavait Creek watershed during spring melt 1987–1990 using **A** an energy balance model; **B** a degree-day model [Eq. (7)]; and **C** a temperature/radiation index model [Eq. (8)]

occur. The threshold is normally less than 0 °C, due to the effect of solar radiation. It is necessary to know when the snowpack is isothermal at 0 °C before the calculations can be initiated. The inclusion of a simple solar radiation term in Eq. (7) slightly improves performance, i.e.:

$$M = F_t (T_a - T_o) + F_r (SW_s - SW_r),\tag{8}$$

Where F_t is the temperature melt factor, F_r is the radiation melt factor, and SW_s and SW_r are incident and reflected shortwave radiation, respectively (Table 6.1). As seen in Fig. 6.10, the energy balance approach provided the best results; however, the simple approaches also performed well. The empirical methods deviated from observation as the snowpack water equivalent decreased. Comparison of panels B and C in Fig. 6.10 suggests that snowmelt within the Imnavait Creek watershed is largely controlled by the sensible heat component of the energy balance (Hinzman et al. 1991a), whereas radiation is of lesser importance. Radiation does play a much greater role in the process of snowmelt on the Arctic Coastal Plain (Weller and Holmgren 1974; Hinzman et al. 1991a).

6.4.3 Simulation of Catchment Runoff

Water balance was examined with the hydrological model, HBV (HBV refers to Swedish Meteorologic and Hydrologic Institute's model; Bergström 1976), for the summer season during the years 1986–1989. Of 40 important parameters in HBV, 26 describe the basin and the input data, whereas the remainder collectively describe snow accumulation and ablation, soil moisture accounting, and generation or transformation of runoff (Hinzman and Kane 1991). A number of parameters are empirically determined via calibration procedures for the watershed. Data collected during the summer of 1986 provided the primary calibration. Snowmelt is calculated with a degree-day formulation (Hinzman et al. 1991a). This simplification seems acceptable, because the parameters of the model were optimized to best simulate discharge and not ablation (Hinzman and Kane 1991). As the thickness of the active layer increases, field storage capacity increases. The mechanisms of runoff are controlled primarily by the properties of the near-surface organic soils (Hinzman et al. 1991b, 1993).

The HBV predicts stream flow well and also provides reasonable estimates of daily evapotranspiration and soil moisture levels. The HBV performs best in reconstructing simple discharge events such as the snowmelt or rainfall events of 1986 (see hydrograph in Chap. 17, this Vol.). The three major summer precipitation events were very distinct and not complicated by overlapping recession curves. The first two events after the spring melt were caused by rainfall, and the third event was a summer snowmelt event. The HBV considers the state of summer precipitation, i.e. snow or rain, and determines runoff appropriately. It proved difficult to adequately describe many of the 1987

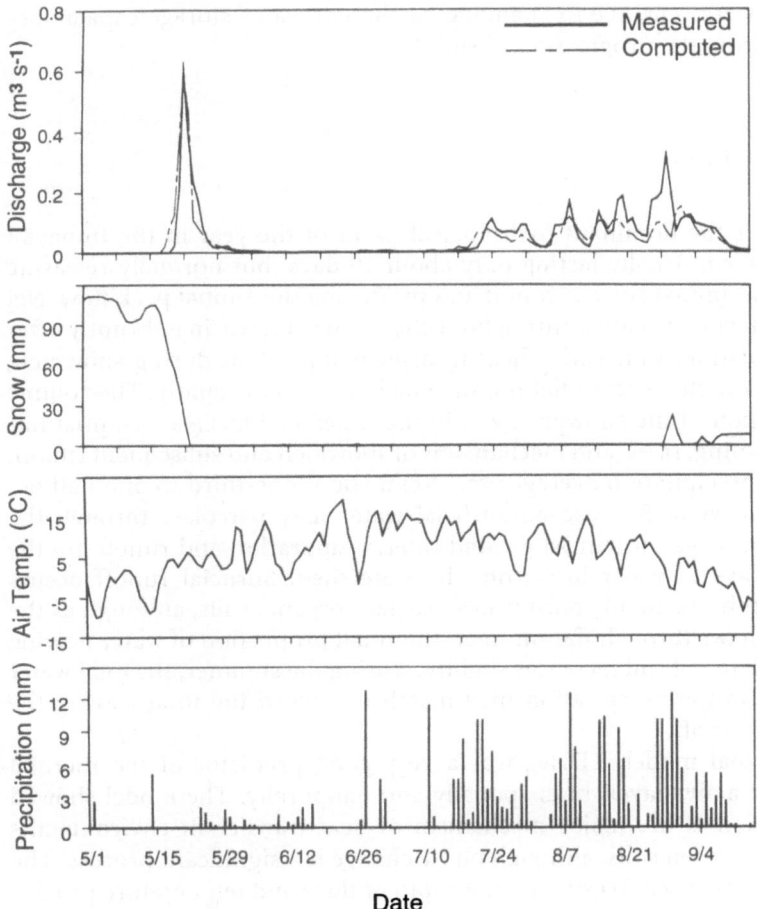

Fig. 6.11. Stream flow observed and simulated by HBV model, predicted snowpack ablation, measured air temperature and precipitation in 1987

summer events (Fig. 6.11), because precipitation fell as mixed snow and rain, and a series of storms resulted in overlapping recession curves.

An important objective in this analysis was to identify basin characteristics that appeared to be sensitively related to runoff discharge as described by model parameters. To predict discharge four key parameters were varied during the summer: (1) the threshold temperature for onset of snowmelt; (2) maximum soil moisture storage in the model; (3) maximum soil moisture storage in the upper reservoir; and (4) a parameter in the soil moisture routine that is influenced by the thickness of the active layer (Hinzman and Kane 1992). The threshold temperature must change after spring melt, because energy from the surficial soils contribute to the melting of summer snow. The simulations were sensitive to changing catchment water-storage capacitance that results from the progression of soil thawing.

6.5 Conclusions

Spring melt is the dominant hydrological event of the year in the Imnavait Creek watershed, usually lasting only about 10 days, but normally releasing about half the annual surface runoff and producing the annual peak flow. Net radiation is low or negative throughout the winter, increasing abruptly after snowpack ablation. Convective heat transfer is important during snowmelt, and when it complements radiation, snowmelt may occur rapidly. The volume and distribution of the snowpack within the watershed has a substantial impact on the timing, rates, and mechanisms of snowmelt and subsequent runoff. Total yearly precipitation averages ca. 35 cm, about one-third to one-half occurring as snowfall. Because no surficial water may percolate through the permafrost into subpermafrost groundwater, evaporation and runoff are the major pathways of water loss from the watershed. Surficial runoff occurs primarily within the highly porous near-surface organic soils, although as the active layer thaws through the summer, the small proportion of water flowing through the mineral soil increases slightly. During the summer, the total water loss through evapotranspiration may match or exceed the total leaving the watershed as runoff.

The thermal model, TDHC, was a very good predictor of the thermal regime of the active layer, both spatially and temporally. The model showed that conduction is the major mechanism of heat transfer in environments dominated by frozen soils, although phase change is a significant process. The TDHC model was used to estimate the depth of thaw and temperature profiles in soils following surface disturbance. As expected the disturbed soils experience greater thaw depths and greater year-to-year variability. The hydrological model, HBV, satisfactorily described the significant hydrological processes of this arctic watershed including evapotranspiration, infiltration, soil moisture fluctuations, runoff, snowmelt, and snow damming. In this model, a simple temperature-index method was used to predict snowmelt.

Acknowledgments. This research was sponsored by the US Department of Energy, Office of Health and Environmental Research, Ecological Research Division. Elizabeth K. Lilly and Robert E. Gieck assisted in every aspect of this research.

References

Barry RG (1981) Tundra climates. In: Bliss LC, Heal OW, Moore JJ (eds) Tundra ecosystems: a comparative analysis. Cambridge Univ Press, Cambridge, pp 241-256

Bengtsson L (1976) Snowmelt estimated from energy budget studies. Nordic Hydrol 7: 3-18

Bengtsson L (1982) The importance of refreezing on the diurnal snowmelt cycle with application to a northern Swedish catchment. Nordic Hydrol 13: 1-12

Benson, CS (1982) Reassessment of winter precipitation on Alaska's arctic slope and measurements on the flux of wind-blown snow. Research Report UAG R-288. Geophysical Institute, University of Alaska, Fairbanks. 26p

Bergström S (1976) Development and application of a conceptual runoff model for Scandinavian catchments. Swed Meteorol Hydrol Inst, Norrköping, Sweden, Rep RHO7

Bergström S (1986) Recent developments in snowmelt-runoff simulation. In: Kane D (ed) Proc Symp: Cold regions hydrology. Am Water Resources Assoc, Fairbanks, Alaska, pp 461-468

Braun LN (1985) Simulation of snowmelt-runoff in lowland and lower alpine regions of Switzerland. Züricher Geogr Schr 21, Geogr Inst, Eidgenössische Technische Hochschule, Zürich, Switzerland

Dingman SL, Barry RG, Weller G, Benson C, LeDrew EF, Goodwin CW (1980) Climate, snowcover, microclimate and hydrology. In: Brown J, Miller PC, Tieszen LL, Bunnell FL (eds) An arctic ecosystem. The Coastal Tundra at Barrow, Alaska. Dowden, Hutchinson and Ross, Stroudsburg

Goering DJ, Zarling JP (1985) Geotechnical thermal analysis with a microcomputer. Civil Engineering in the Arctic Offshore. Am Soc Civil Engineers, New York, pp 604-616

Hastings SJ, Luchessa SA, Oechel WC, Tenhunen JD (1989) Standing biomass and production in water drainages of the foothills of the Philip Smith Mountains, Alaska. Hol Ecol 12: 304-311

Hinzman LD, Kane DL (1991) Snow hydrology of a headwater arctic basin 2. Conceptual analysis and computer modeling. Water Resour Res 27: 1111-1121

Hinzman LD, Kane DL (1992) Potential response of an arctic watershed during a period of global warming. J Geophys Res - Atmospheres 97: 2811-2820

Hinzman LD, Kane DL, Gieck RE (1991a) Regional snow ablation in the Alaskan Arctic. In: Prowse T, Ommanney C (eds) Northern hydrology: selected perspectives. NHRI Symp 6, Natl Hydrol Res Inst, Saskatoon, Saskatchewan, pp 121-140

Hinzman LD, Kane DL, Gieck RE, Everett KR (1991b) Hydrologic and thermal properties of the active layer in the Alaskan Arctic. Cold Reg Sci Technol 19: 95-110

Hinzman LD, Kane DL, Everett KR (1993) Hillslope hydrology in an arctic setting. Proc Permafrost 6th Conf. Beijing, China, South China Univ Technol Press, pp 267-271

Kane DL, Hinzman LD (1988) Permafrost hydrology of a small arctic watershed. In: Senneset K (ed) Proc 5th Int Conf Permafrost, Tapir, Trondheim, Norway, pp 590-595

Kane DL, Hinzman LD, Benson CS, Everett KR (1989) Hydrology of Imnavait Creek, an arctic watershed. Holarct Ecol 12: 262-269

Kane DL. Gieck RE, Hinzman LD (1990) Evapotranspiration from a small Alaskan arctic watershed. Nordic Hydrology 21: 253-272

Kane DL, Hinzman LD, Benson CS, Liston GE (1991a) Snow hydrology of a headwater arctic basin. 1. Physical measurements and process studies. Water Resour Res 27: 1099-1109

Kane DL, Hinzman LD, Zarling JP (1991b) Thermal response of the active layer in a permafrost environment to climatic warming. Cold Reg Sci Technol 19: 111-122

Kane DL, Hinzman LD, Woo M, Everett KR (1992) Arctic hydrology and climate change. In: Chapin FS III, Jefferies RL, Reynolds JF, Shaver GR, Svoboda J (eds) Arctic ecosystems in a changing climate. Academic Press, New York, pp 35-57

Labelle JC, Wise JL, Voelker RP, Schulze RH, Wohl GM (1983) Alaska marine ice atlas. Arctic Environ Inf Data Center, Univ Alaska, Anchorage

Liston GE (1986) Seasonal snowcover of the foothills region of Alaska's arctic slope: a survey of properties and processes. MS Thesis, Univ Alaska, Fairbanks

Mageau DW, Rooney JW (1984) Thermal erosion of cut slopes in ice-rich soil. State of Alaska, Dept Trans Public Facilities, Rep FHWA-AK-RD-85-02

Male DH, Granger RJ (1981) Snow surface energy exchange. Water Resour Res 17: 609–627

Ohmura A (1981) Climate and energy balance on arctic tundra, Canadian Arctic Archipelago, spring and summer 1969, 1970 and 1972. Geogr Ins, Eidgenössische Technische Hochschule Zürich, Heft 3

Ohmura A (1982) Evaporation from the surface of the arctic tundra on Axel Heiberg Island. Water Resour Res 18: 291–300

O'Neill K (1983) Fixed mesh finite element solution for cartesian two dimensional phase change. J Energy Resour Technol 105: 436–441

Osterkamp TE, Gosink JP, Kawasaki K (1987) Measurements of permafrost temperatures to evaluate the consequences of recent climate warming. State of Alaska, Dept Trans Public Facilities, Rep AK-RD-88-05

Parker GA (1929) The evolution of placer mining methods in Alaska. BS Thesis, Alaska Agric College, Fairbanks

Price AG, Dunne T (1976) Energy balance computations of snowmelt in a subarctic area. Water Resour Res 12: 686–694

Priestley CHB, Taylor RJ (1972) On the assessment of surface heat flux and evaporation using large-scale parameters. Monthly Weather Rev 100: 81–92

Rieger S, Schoephorster DB, Furbush CE (1979) Exploratory soil survey of Alaska. USDA Soil Conserv Serv, Washington DC

Sand K (1990) Modeling snowmelt runoff processes in temperate and arctic environments. PhD Thesis, Norwegian Inst Technol, Univ Trondheim

Szeicz G, Endrdi G, Tajchman S (1969) Aerodynamic and surface factors in evaporation. Water Resour Res 5: 380–394

Tabler RD (1975) Estimating the transport and evaporation of blowing snow. In: Snow Management on the Great Plains Symposium. Univ Nebraska Agric Exp Station, Lincoln, Nebraska Great Plains Agric Council Publ 73, pp 85–104

Weller G, Holmgren B (1974) The microclimates of the arctic tundra. J Appl Meteorol 13: 854–862

Woo MK (1986) Permafrost hydrology in North America. Atmos-Ocean 24: 201–234

Zarling JP, Braley WA, Pelz C (1989) The modified Berggren method: a review. Proc 5th Int Conf Cold Regions Engineering, Am Soc Civil Engineers, New York, pp 263–273

7 Shortwave Reflectance Properties of Arctic Tundra Landscapes

A. S. Hope and D. A. Stow

7.1 Introduction

Studies of shortwave reflectance properties fall into two broad categories: (1) those dealing with the total broad band reflectance (albedo) and (2) those concerned with spectral reflectance characteristics of terrestrial surfaces. Despite the importance of shortwave radiation as the primary energy source for most physical and biological processes, information regarding the shortwave reflective properties of Arctic landscapes is scarce. Albedo is a key variable affecting the surface energy balance (Chap. 6, this Vol.), whereas spectral reflectance may be used to infer biophysical quantities such as biomass or vegetation composition (Chap. 18, this Vol.). Spectral radiances also constitute the basic data used in classical remote sensing studies (e.g., land cover classification) (Stow et al. 1989; Walsh and Davis 1994).

Given that shortwave reflectance is the fraction of incident solar radiation reflected from a surface, both incident and exitant radiances must be measured. Albedo measurements are commonly made at the surface using a set of two pyranometers, one facing up and the other down. In contrast, most remotely sensed data consist of digital imagery collected by down-looking sensors on satellite or aircraft platforms and represent spectral radiances, rather than reflectances. Using nonimaging spectral radiometers – hand-held or mounted on an aircraft platform – is often the first step in multilevel remote sensing studies. By collecting concurrent spectral radiance data over a calibration panel, one can determine incident spectral radiances and use these values to calculate spectral reflectances.

Both albedo and spectral reflectance data can be used as indicators of ecosystem disturbance. Anthropogenic disturbance may influence the reflectance properties of tundra vegetation either directly, e.g., a physical alteration of vegetative surfaces, or indirectly, e.g., a gradual change in the type of vegetation cover (Lawson 1986; Walker and Acevedo 1987; Chap. 3, this Vol.). In this chapter we describe some practical considerations and problems in conducting shortwave reflectrance studies in the Arctic and present some of our results on the use of albedo and spectral reflectance in the R4D program (Chap. 1, this Vol.).

J.F. Reynolds and J.D. Tenhunen (Eds.)
Ecological Studies, Vol. 120
© Springer-Verlag Berlin Heidelberg 1996

7.2 Shortwave Reflectance Studies in Arctic Environments

7.2.1 Environmental Considerations

Frequent cloud cover and low solar elevations have a major influence on radiometric and remote sensing studies on the North Slope of Alaska. Whereas these are important causal variables in albedo studies, they are an impediment in relating spectral reflectances to biophysical quantities. Spectral reflectances are affected by variations in solar elevation, because natural surfaces have non-Lambertian (magnitude varies with direction) reflectance properties (e.g., Goel and Reynolds 1989). Deering and Middleton (1990) found that the bidirectional reflectance properties of vegetation tend to vary the most at low sun and view angles. This is of significant concern at the Imnavait Creek watershed where the maximum solar elevation is approximately 45°.

7.2.2 Radiometric Data

Non-imaging spectral radiometers, held by hand or mounted on a light aircraft, permit great flexibility to take advantage of the brief periods of clear sky conditions. These instruments are valuable because:

1. The spectral data are collected close to the Earth's surface, which avoids problems associated with the scattering and absorption of radiation by the atmosphere.
2. The target can be identified with great certainty. (Misregistration problems associated with aircraft or satellite imagery leads to uncertainty when one attempts to relate spectral reflectance properties to surface quantities).
3. The area sampled by a hand-held instrument is small enough for measurements to be made at the surface, enabling accurate characterization of the biophysical variable of interest (Curran and Williamson 1985).

7.2.3 Image Data

Although data collected using non-imaging radiometers are valuable for studying basic relationships between biophysical properties and spectral reflectances, it is often necessary to use multispectral imagery for landscape or regional studies. Remotely sensed images can be acquired with photographic, videographic, or imaging radiometer sensors from airborne or satellite platforms. The two-dimensional, quantitative sensing capabilities of imaging radiometers (e.g., multispectral scanners or linear array devices) hold great promise for use in sampling spectral reflectance properties over the vast expanses of the North Slope. In particular, the large-area, synoptic, and repetitive coverage capabilities of imaging radiometers on polar-orbiting satellites enable frequent "wall-to-wall" sampling of shortwave radiance over Arctic lands. For example, shortwave radiance data can be acquired with 20-m spatial resolution from the Systeme Probatoire d'observation de la Terre (SPOT) high-resolution visible (HRV) satellite system for all of the North Slope once every 3 days, and 1–2 km resolution data can be acquired several times daily with the National Oceanic and Atmospheric Administration (NOAA) Advanced Very-High Resolution Radiometer (AVHRR).

In practice such frequent and synoptic sampling can rarely be achieved, and the conversion of satellite spectral radiances to spectral reflectance quantities is difficult and prone to errors. The persistent cloud cover over Arctic skies during the summer growing season affects the frequency at which shortwave radiance data can be acquired, despite the highly repetitive coverage of polar orbiters at high latitudes. Because satellite imaging radiometers only sample upwelling spectral radiance, no direct means for deriving spectral reflectances exists. Spectral reflectances can be indirectly estimated from satellite-derived spectral radiances based on sensor calibration factors, corrections for atmospheric scattering, and modeled or measured solar irradiance at the surface.

7.3 Spectral Reflectance

A major effort in biophysical remote sensing over the past 15 years has been to relate quantities such as leaf area index or biomass to the spectral reflectance properties of vegetation communities (e.g., Drake 1976; Tucker 1979; Tucker et al. 1981; Aase et al. 1986). The prospect of making large-area estimates of biophysical variables using spectral radiances collected from satellites or aircraft is attractive for regional ecosystem studies. Another valuable application of remotely sensed radiometric measurements, which may involve hand-held instruments, is to obtain rapid and nondestructive estimates of plant canopy attributes (Asrar et al. 1989).

7.3.1 Aboveground Biomass

Spectral vegetation indices used to estimate abovegound biomass reduce multispectral data to a single value that is sensitive to changes in vegetation quantities, minimizing the effects of varying view angles and illumination conditions. Relationships between biophysical quantities and spectral vegetation indices do, however, vary with vegetation type, soil background, season, and the amount of dead material in the plant canopy (Sellers 1985). The normalized difference vegetation index (NDVI), widely used in biophysical remote sensing studies, is obtained by dividing the difference between near-infrared and red reflectances by the sum of these reflectances.

Hope et al. (1993) considered the NDVI a potential means for estimating aboveground biomass over large areas of the Arctic. Radiometric and biomass data were collected at the Imnavait Creek watershed in three major communities (dry heath, moist tussock, and water tracks) (see Fig. 4.7 in Chap. 4, this Vol.) during three periods in 1989. All biomass above the surface moss layer was destructively harvested and separated into photosynthetic and nonphotosynthetic categories. Regression analyses revealed that the nonphotosynthetic biomass fraction in each community had no effect on NDVI, whereas up to 50% of the variance in NDVI was associated with the fraction of photosynthetic biomass.

7.3.2 Vegetation Composition

Vegetation composition, rather than biomass, may be a more significant determinant of the spectral reflectance properties of tundra landscapes. Hope et al. (1993) found that vegetation composition accounted for up to 90% of the variance in NDVI and individual spectral reflectances. These relationships were, however, specific to the individual communities. The fraction of lichens within the moist tussock community was a significant variable in predicting the blue and green reflectances of this community. Petzold and Goward (1988) found that lichens of the eastern Canadian Subarctic exhibit spectral reflectance characteristics noticeably different from those of vascular plants. These authors reported that the dominant lichens were strong absorbers of ultraviolet and shortwave blue radiation relative to their absorption in other wavelengths.

7.3.3 Landscape Patterns

It is clear that general vegetation type is the first effect to be considered in biophysical remote sensing studies in tundra landscapes. Thus, the spatial variation of remotely sensed radiances should correspond primarily to variations in vegetation composition across the landscape. This is the basis for classifying and mapping Arctic tundra vegetation from satellite image data, which has been accomplished with varying degrees of success for portions of the North Slope (Walker and Acevedo 1987; Talbot and Markon 1988; Stow et al. 1989). An accurate, detailed classification of tundra vegetation community types can be difficult to achieve, because the dominant communities often comprise different proportions of the same plant species. This means that multispectral radiances processed with multidimensional pattern-recognition routines are more useful for discriminating community types and mapping community patterns than simply thresholding a single vegetation index. The spatial pattern of a single vegetation index, however, can be quantified more readily via spatial statistical measures (Musick and Grover 1991).

Several investigators have inferred tundra landscape patterns within the Imnavait Creek watershed by spatial analyses of NDVI values extracted from imaging and nonimaging remote sensor data. Stow et al. (1993a) found that the

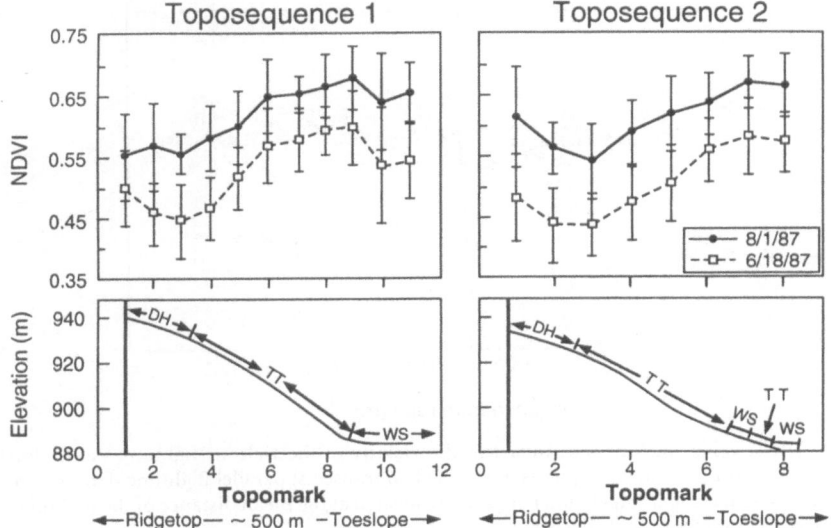

Fig. 7.1. Mean and standard deviation of hand-held normalized difference vegetation index (NDVI) for two toposequences (location given in Fig. 4.1, Chap. 4, this Vol.) and a hillslope profile with corresponding topographic gradient and vegetation types. Plots were spaced approximately even and roughly oriented along toposequences. *DH*, Dry heath; *TT*, moist tussock tundra; *WS*, wet sedge. (From Stow et al. 1993a)

one-dimensional pattern of NDVI values from hand-held radiometer measurements taken along two 500-m toposequences (see Fig. 4.1 in Chap. 4, this Vol.) corresponded to slope position (Fig. 7.1). The major pattern was associated with site moisture status and variations in plant communities along these moisture gradients (see Fig. 4.9 in Chap. 4, this Vol.). This interpretation is supported by Ostendorf and Reynolds (1993), who found that two-dimensional patterns of NDVI from June 1987 SPOT multispectral-image data corresponded to estimates of discharge generated by a digital terrain-based hydrological model.

7.3.4 Effects of Dust Deposition

Data from a spectral radiometer mounted in a light aircraft were acquired by Stow et al. (1993b) along flightlines perpendicular and parallel to the Dalton Highway. The NDVI values were calculated and plotted along these transects to determine whether dust deposition from the highway affected surface reflectance (Fig. 7.2). Stow et al. (1993b) found evidence of dust disturbance in the pattern of NDVI approaching the highway, which corroborates findings from other studies (see Sect. 7.4.2). At a coarser scale the spatial pattern of NDVI appears to be controlled by the various ages of glacial surfaces of the

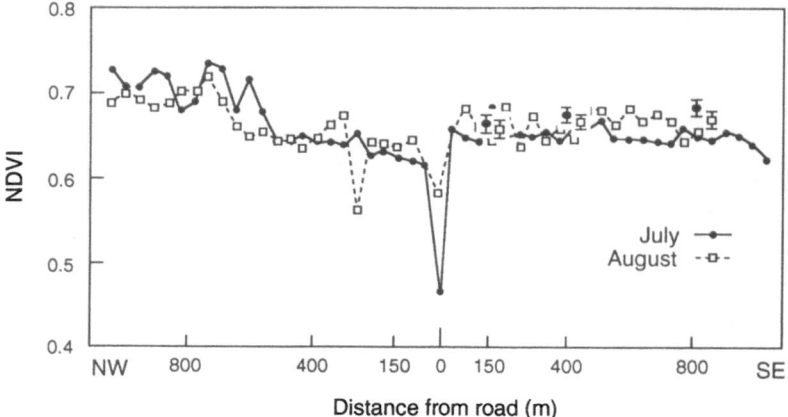

Fig. 7.2. NDVI values in July and August in the vicinity of the Dalton Highway. Values for perpendicular flightlines shown as points along the two transects; parallel flightline data shown as mean values with standard deviation bars and are located at the mean distance of the flightline from the road. (From Stow et al. 1993b)

North Slope foothills with their differing soil acidity, moisture, and organic contents (Auerbach et al. 1992; Chap. 4, this Vol.).

7.4 Albedo

Illumination conditions in the Arctic are unique, characterized by long periods of daylight, a large ratio of diffuse-to-direct solar radiation resulting from frequent cloud cover and low solar elevations (Stoner et al. 1978). Differences in albedo from location to location imply different microclimatic responses in evaporation, soil moisture, air temperature, and soil temperature (Johnston and Fuggle 1988). To model adequately the effects of anthropogenic disturbances on tussock-tundra landscapes, one must establish the natural variability of albedo for undisturbed vegetation and assess possible changes due to disturbances. Small changes in albedo may be significant at high latitudes during the summer months because of long periods of solar irradiance and integrated effects of the change in radiation balance.

7.4.1 Undisturbed Tussock Tundra

Studies by Miller (1981) indicate that the albedo of tundra is generally lower than that of agricultural crops, grass, and other vegetated surfaces of equal stature. Variability in the albedo of undisturbed tussock tundra may be caused by changes in illumination conditions or by spatial variability in the vegetation characteristics. Hope et al. (1991) conducted a study during the summer of 1988 to characterize tussock-tundra albedo at an undisturbed site near Toolik Lake. This study was also designed to determine the extent to which different illumination conditions caused by variations in cloud cover and vegetation composition affect albedo. Measured albedos over the undisturbed tussock-tundra site ranged from 0.142 to 0.183, and the mean albedo and coefficient of variation (percent) were 0.171 and 5.26, respectively. This range in observed values is consistent with the range of values calculated by McFadden and Ragotzkie (1967) using aircraft observations of incident and reflected shortwave radiation over tundra (0.123–0.193).

Hope et al. (1991) used a cloud index (CI) described by Nkemdirim (1972) to quantify variations in illumination conditions caused by clouds. This index is calculated by dividing the global radiation received at the surface by the radiation incident at the top of the atmosphere. The relationship between albedo and CI was significant when observed albedos were regressed on the corresponding CI, although the r^2 value was only 0.42. The fractions of vegetation cover types estimated using a point-quadrat sampling scheme were included in these regression analyses, but were not found to significantly affect the albedo of undisturbed tussock tundra (Hope et al. 1991).

7.4.2 Effects of Dust Deposition

Eller (1977) and Walker and Everett (1987) have suggested that dust deposition may affect the albedo of tussock tundra either directly (dust coating of leaves) or indirectly through induced change in species composition (see Chap. 18, this Vol.). Hope et al. (1991) collected albedo data during the summer of 1988 at a site adjacent to the Dalton Highway where the tussock tundra had been visibly affected by dust deposition. The mean albedo values for each dust transect were plotted against the corresponding CI values along with the points plotted for the undisturbed site (Fig. 7.3). The relationship between albedo and CI was different for the two sites with the disturbed site having generally smaller albedo values at the same illumination level. The slopes of the two regressions were significantly different at the 95% confidence limit.

The reduction in albedo associated with dust deposition from the Dalton Highway may affect a substantial area of Alaskan tundra adjacent to the highway. A reduction in albedo implies an increase in absorbed shortwave radiation at the ground surface, and this increase may affect the substrate's thermodynamic behavior. Possible consequences are changes in the permafrost depth or early thawing in the spring (Chap. 3, this Vol.). Although the observed differences in albedo over the disturbed and undisturbed sites were not large in absolute terms, the cumulative effects of the differences on albedo could be significant.

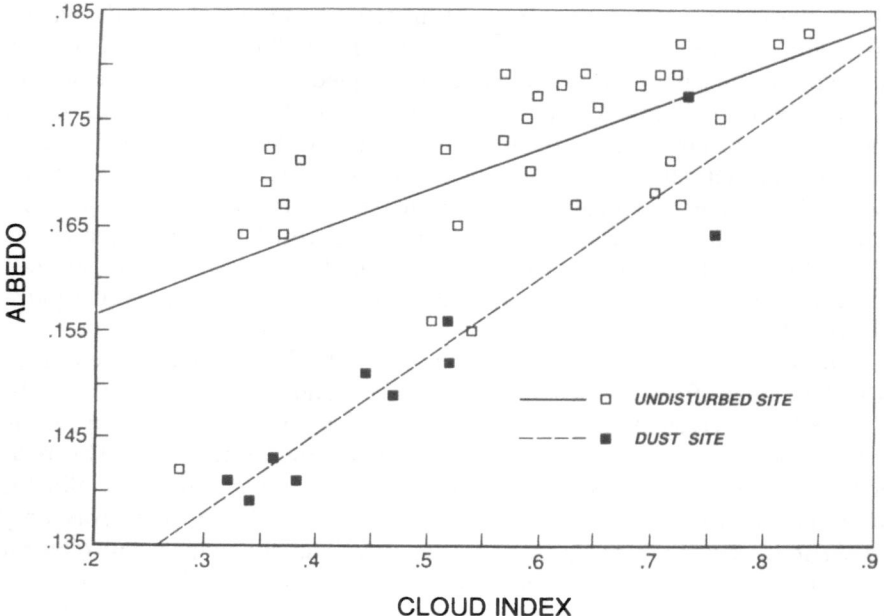

Fig. 7.3. Relationship between albedo and cloud index for tussock-tundra sites that were undisturbed or affected by dust deposition from the Dalton Highway. (From Hope et al. 1991)

7.5 Conclusions

Cloud cover is the primary environmental variable influencing remote sensing and other radiometric studies in Arctic regions. Our observation that spectral reflectance patterns across the landscape were similar – regardless of the system used to collect the data – suggests the possibility of using hand-held or aircraft-mounted radiometers more extensively in the future. It should be possible to extrapolate findings from these systems to satellite observations covering larger areas, although this will require considerable research effort.

In many cases, obtaining radiometric data (i.e., imagery) over entire study areas may not be necessary. Using nonimaging radiometers in a well-designed sampling scheme may be adequate for determining model parameters or input variables, verifying model predictions (Chaps. 14 and 18, this Vol.), or evaluating vegetation condition or disturbance effects (Chap. 3). An aircraft-mounted radiometer system appears to be particularly well-suited to this approach in the Arctic environment. Although such a system offers greater flexibility, extended periods of cloud cover may still preclude collecting spectral reflectance data for long periods, yet changes in vegetation characteristics during these "down times" may be critical to an experiment. Therefore, future optical remote sensing studies in these Arctic locations should consider the possibility of conducting these studies under variable cloud cover.

We found that the spectral reflectance (using NDVI) of tussock-tundra communities was significantly affected by the relative proportions of different plant types, particularly shrubs. These effects were apparent regardless of the sensor system employed. In contrast, the variability in albedo was caused mainly by variations in illumination conditions, not vegetation composition.

The spectral reflectance (aircraft NDVI) and albedo properties of tussock tundra suggest that they are quantities sensitive to disturbance, particularly dust deposition. Other disturbances that affect large areas of the landscape may also alter the shortwave reflection properties of tundra landscapes.

References

Aase JK, Millard JP, Brown BS (1986) Spectral radiance estimates of leaf area and leaf phytomass of small grains and native vegetation. IEEE Trans Geosci Remote Sens GE-24: 685–691

Asrar G, Myneni RB, Kanemasu ET (1989) Estimation of plant-canopy attributes from spectral reflectance measurements. In: Asrar G (ed) Theory and applications of optical remote sensing. Wiley, New York, p 252

Auerbach NA, Walker DA, Walker MD, Hope AS, Stow DA (1992) Landscape-scale vegetation production in glaciated surfaces of different ages in the Arctic Foothills, Alaska. Paper presented at the Ecological Society of America Symp, Honolulu, Hawaii, 9–13 Aug 1992

Curran PJ, Williamson HD (1985) The accuracy of ground data used in remote sensing investigations. Int J Remote Sens 6: 1637–1651

Deering DW, Middleton EM (1990) Spectral bidirectional reflectance and effects on vegetation indices for a prairie grassland. Symp on FIFE, 7–9 Feb, Am Meteorol Soc, Anaheim, California. AMS, Boston

Drake BG (1976) Seasonal changes in reflectance and standing crop biomass in three salt marsh communities. Plant Physiol 58: 696–699

Eller BM (1977) Road dust-induced increase of leaf temperature. Environ Pollut 13: 99–107

Goel SG, Reynolds NE (1989) Bidirectional canopy reflectance and its relationship to vegetation characteristics. Int J Remote Sens 10: 107–132

Hope AS, Fleming JB, Stow DA, Aguado E (1991) Tussock tundra albedos on the north slope of Alaska: effects of illumination, vegetation composition, and dust deposition. J Appl Meteorol 8: 1200–1206

Hope AS, Kimball JS, Stow DA (1993) The relationship between tussock tundra spectral reflectance properties and biomass, and vegetation composition. Int J Remote Sens 14(10): 1861–1874

Johnston PA, Fuggle RF (1988) Variations in albedo in the southwestern Cape Province. South Afr Geogr 15: 37–42

Lawson DE (1986) Response of permafrost terrain to disturbance: a synthesis of observations from northern Alaska, USA. Arct Alp Res 18: 1–17

McFadden JD, Ragotzkie RA (1967) Climatological significance of albedo in central Canada. J Geophys Res 72: 1135–1143

Miller DH (1981) Energy at the surface of the earth. Academic Press, New York

Musick HB, Grover HD (1991) Image textural measures as indices of landscape pattern. In: Turner MG, Gardner RH (eds) Quantitative methods in landscape ecology. Ecological Studies 82. Springer, Berlin Heidelberg New York, pp 77–103

Nkemdirim LC (1972) A note on the albedo of surfaces. J Appl Meteorol 11: 867–874

Ostendorf B, Reynolds JF (1993) Relationships between a terrain-based hydrologic model and patch-scale vegetation pattern in an arctic tundra landscape. Landscape Ecol 8: 229–237

Petzold DE, Goward SN (1988) Reflectance spectra of subarctic lichens. Remote Sens Environ 24: 481–492

Sellers PJ (1985) Canopy reflectance, photosynthesis and transpiration. Int J Remote Sens 8: 1335–1372

Stoner WA, Miller PC, Oechel W (1978) Simulation of the effect of the tundra vascular plant canopy on the production of four moss species. In: Tieszen LL (ed) Vegetation and production ecology of Alaskan Arctic Tundra. Springer, Berlin Heidelberg New York, PP 371–387

Stow DA, Burns B, Hope AS (1989) Mapping arctic tundra vegetation using digital SPOT/HRV-XS data: a preliminary assessment. Int J Remote Sens 10: 1451–1457

Stow DA, Burns B, Hope AS (1993a) Spectral, spatial and temporal characteristics of arctic tundra reflectance. Int J Remote Sens 14(13): 2445–2462

Stow DA, Hope AS, George T (1993b) Reflectance characteristics of arctic tundra vegetation from airborne radiometry. Int J Remote Sens 14: 1239–1244

Talbot ST, Markon CJ (1988) Intermediate-scale vegetation mapping of Innoko National Wildlife Refuge, Alaska, using Landsat MSS digital data. Photogramm Eng Remote Sens 54: 377–383

Tucker CJ (1979) Red and photographic infrared linear combinations for monitoring vegetation. Remote Sens Environ 8: 127–150

Tucker CJ, Holben BN, Elgin JH, McMurtrey J (1981) Remote sensing of total dry matter accumulation in winter wheat. Remote Sens Environ 11: 171–190

Walker DA, Acevedo W (1987) Vegetation and Landsat-derived landcover map of the Beechey Point quadrangle, Arctic Coastal Plain, Alaska. CRREL Rep 87-5, Univ Colorado, Boulder

Walker DA, Everett KR (1987) Road dust and its environmental impact on Alaskan taiga and tundra. Arct Alp Res 19: 479–489

Walsh SJ, Davis FW (1994) Applications of remote sensing and geographic Information systems in vegetation science: introduction. J Veg Sci 5: 610–613

8 Isotopic Tracers for Investigating Hydrological Processes

L. W. Cooper, I. L. Larsen, C. Solis, J. M. Grebmeier, C. R. Olsen, D. K. Solomon, and R. B. Cook

8.1 Introduction

The use of natural and anthropogenic isotopic tracers – both stable and radioactive – is a well-developed biogeochemical technique used in evaluating water-borne chemical fluxes (see Broecker and Peng 1982; Fry and Sherr 1984; Rundel et al. 1988; Griffiths 1991). Watersheds in temperate regions have been studied extensively using stable isotopes (oxygen-18 and deuterium) to follow water flow paths (e.g., Dinçer et al. 1970; Sklash and Farvolden 1979; Bottomley et al. 1986; Hooper and Shoemaker 1986; Kennedy et al. 1986; Obradovic and Sklash 1976; Pearce et al. 1986; Sklash et al. 1986; DeWalle et al. 1988; Swistock et al. 1989; Wels et al. 1991). However, there are few examples of simultaneous work tracing the fate of particle-reactive isotopes in snow, and even fewer isotopic applications in watersheds underlain by permafrost.

In arctic watersheds, such as Imnavait Creek, questions of fate and transport of precipitation-borne materials are particularly important, because material transfer reaches a peak during snowmelt, the dominant yearly hydrological event controlling stream flow (Chap. 6, this Vol.). During snowmelt, particles and solutes that have accumulated over many months of snowfall are released into watersheds over short periods of time. The combination of long accumulation times and short, episodic release makes the fate of chemical constituents stored within the snowpack of arctic watersheds particularly important to understand. Any type of disturbance that changes hydrological processes, e.g., water flow paths and thaw-layer depth (see Chap. 4, this Vol.), could fundamentally alter the fate of atmospherically derived materials (McCauley 1991). Another consideration is the higher cumulative retention of contaminants in high latitude watersheds including chlorinated hydrocarbons (Calamari et al. 1991) and radionuclides released as a result of weapons testing and nuclear accidents (Taylor et al. 1985).

In our work at Imnavait Creek we used a suite of natural and anthropogenic isotopic tracers to study the retention and flux of precipitation-borne chemicals during the annual hydrological cycle. The combination of short-lived (days to weeks) and long-lived nuclides (years) has provided insights into differences in hydrological transport of precipitation-borne materials during snowmelt vs summer rainfall (Cooper et al. 1991b).

J.F. Reynolds and J.D. Tenhunen (Eds.)
Ecological Studies, Vol. 120
© Springer-Verlag Berlin Heidelberg 1996

8.1.1 Units

Stable isotope data for oxygen-18 and deuterium are expressed as $\delta^{18}O$ and δD, respectively, where $\delta^{18}O$ or $\delta D = [(R_{sample}/R_{SMOW})-1] \times 1000$ per mil ($R = {}^{18}O/{}^{16}O$ or D/H; SMOW = Standard Mean Ocean Water, distributed by the International Atomic Energy Agency, Vienna). Radioisotope data are presented as either an inventory of radioactivity present at the time of collection per cm^2, regardless of the vertical distribution of the radioisotope in the snow or tundra, or as a concentration per liter or per gram dry weight of soil or organic matter. Data are provided in the SI unit for radioactivity, the becquerel (Bq), equivalent to one decay per second.

8.1.2 Conservative vs Nonconservative Isotopes

If during hydrological transport the concentration of an isotope does not change, it can be considered "conservative." For example, ^{18}O is normally considered a conservative tracer. Oxygen-18 is present in approximately 2 of every 1000 molecules of water. Its concentration in natural water varies geographically and seasonally because of isotopic fractionation during evaporation and condensation (Dansgaard 1964; Siegenthaler 1979). Chemically, $H_2^{18}O$ is almost identical to $H_2^{16}O$, so variations in time and space in $^{18}O/^{16}O$ ratios in water can be used to study mixing and transport of isotopically distinct reservoirs. Of course, the scientific question being investigated determines the utility of the isotope chosen, whether conservative or nonconservative (Table 8.1).

Table 8.1. Examples of isotope tracers used in the Arctic

Isotope (half-life)	Source	Chemical characteristics	Example of use
Cesium-137 (30.2 years)	Nuclear testing, nuclear accidents (e.g., Chernobyl)	Particle-reactive, binds to minerals, exchanges with potassium	Baskaran et al. (1991) Taylor et al. (1988) Hermanson (1990)
Sulfur-35 (87 days)	Cosmic ray spallation especially of Argon-40	Can be incorporated into organisms	Cooper et al. (1991b)
Lead-210 (22.3 years)	Radon gas derived from uranium-238 decay	Particle-reactive, adheres to soils, vegetation	Cornwell (1985) Hermanson (1990)
Beryllium-7 (53 days)	Cosmic ray spallation especially of nitrogen-14	Particle-reactive, adheres to soils, vegetation	Cooper et al. (1991b)
Oxygen-18 (Stable)	Naturally present, ca. 0.2%, with variations due to evaporation or condensation	Normally considered conservative; chemically identical to oxygen-16	Cooper et al. (1991b)

8.2 Nonconservative Tracers

In contrast to ^{18}O, the cosmogenic radioisotope beryllium-7 (^{7}Be), derived from the effects of atomic shattering (spallation) during cosmic ray interactions with nitrogen and oxygen, is distinctly nonconservative because of its strong particle-reactive chemical behavior. Estimates of the concentration of ^{7}Be in air range from 10 to 25 atoms 1^{-1}, but in precipitation, concentrations are on the order of 6×10^{6} atoms 1^{-1} (Hörnmann 1978). At Imnavait Creek concentrations of ^{7}Be in precipitation or snowpack are as much as two orders of magnitude higher than concentrations in stream flow. Additional evidence from tundra inventories after snowmelt corroborates a very high retention on vegetation of this radioisotope derived exclusively from the atmosphere (Cooper et al. 1991b). Concentrations of ^{7}Be in stream flow are slightly higher during snowmelt (<10% of snow concentrations) than after summer rainfall events (<1% of rainfall concentrations), although inventories contributed by summer rainfall are higher than the amounts present in the premelt snowpack (Cooper et al. 1991b). This difference is likely due to higher water transport during snowmelt, which facilitates movement of beryllium without chemical contact with vegetation or soil.

Some redistribution of ^{7}Be was apparent during snowmelt in both 1989 and 1990 (Cooper et al. 1991b). All decreases in ^{7}Be inventories between premelt snow and postmelt tundra were observed in upslope areas, and all increases in ^{7}Be inventories between premelt snow and postmelt tundra were observed in downslope or riparian zones (see Fig. 4.7 in Chap. 4, this Vol.). Nevertheless, the patterns were not consistent, and in a majority of cases no significant changes in ^{7}Be inventories were observed (Cooper et al. 1991b) These data are consistent with the possibility that particle-reactive materials and contaminants behave differentially during the peak of snowmelt relative to summer rainfall, leading to enhanced transport from upland areas into riparian zones. This effect may be due to greater water flow during snowmelt, and suggests that riparian zones in arctic watersheds can be more vulnerable to contamination from heavy metals, contaminants, and other materials with chemical behavior similar to beryllium.

8.3 Sulfur-35

The application of natural abundance measurements of ^{35}S to hydrological transport was reported for the first time at Imnavait Creek (Cooper et al. 1991b). ^{35}S is formed from cosmic ray spallation of ^{40}Ar (Peters 1959; Lal and Peters 1967). There are no other significant natural sources. Fallout of ^{35}S in precipitation is rapid and on the order of days (Goel et al. 1959; Lal and Peters 1967). Because of its 87-day half-life, at the time of snowmelt almost all ^{35}S present within Imnavait Creek is stored within the snowpack. These circumstances provide a significant advantage over conventional comparisons of the chemical content of precipitation and stream water (e.g., SO_4^{2-} concentration), because these methods cannot readily separate the direct contributions of precipitation from the indirect contributions of leachates released from soils or vegetation. This new application of ^{35}S has provided insights into the transport and fate of sulfur derived from atmospheric deposition.

The ^{35}S data show that the majority of sulfur deposited in snowfall is retained by watershed vegetation. Approximately one-third of the total inventory present in the snowpack was exported from the watershed in the stream flow during snowmelt (Cooper et al. 1991b). Also, the maximum ^{35}S and maximum total sulfate concentrations occurred during the initial snowmelt and decreased to minimum values as flow increased (Cooper et al. 1991b). The concentration of ^{35}S was related to total dissolved sulfate ($r^2 = 0.98$; $p < 0.0001$), although the certainty of this relationship was constrained by the size of the data set ($n = 5$). Considered over the seasonal hydrological cycle in which sulfur is released through snowmelt and stream flow, the ^{35}S data indicate that export of sulfur from the watershed depends on processes that have a time scale longer than the two to three half-life periods (174–261 days) in which ^{35}S deposited in snow can continue to be detected. As a result, fluxes of sulfur out of the watershed are weakly linked to input fluxes, and sulfate cannot be considered a conservative indicator in this watershed. Plant uptake of sulfur deposited in snow tends to lengthen its residence time. Most sulfate released in stream flow during snowmelt is leached from longer-term storage in soils, vegetation, or geological reservoirs within the watershed.

8.4 Oxygen-18

Ratios of [18]O to [16]O and deuterium to hydrogen are now widely used to study stream flow generation as a result of storm events or snowmelt (see Sect. 8.1). Because isotopic compositions are assumed not to change during stream flow generation, this technique has been used to evaluate the proportional contributions to stream flow of "old" water already within the soil to "new" water originating from the snowmelt or storm event. Although the vast majority of stable isotope studies of stream flow generation have been undertaken in humid temperate watersheds without permafrost, we did not expect any complications in applying this technique to an arctic watershed. Our preliminary use of [18]O assays (Cooper et al. 1991b) to follow water flow in this permafrost-underlain watershed did in fact assume that the [18]O content of snowmelt-derived water was conservative and could be used unequivocally to follow snowmelt contributions to stream flow. This work suggested that [18]O or deuterium would serve as a useful hydrological tool for tracing the varying contributions of snowmelt, thawing-soil moisture, and summer precipitation to surface waters within the watershed. For example, significant differences were observed in the $\delta^{18}O$ values of Imnavait Creek after snowmelt in 1989, from −25.0‰ at peak stream discharge on 29 May, to −18.9‰ on 29–30 June (precision of the isotopic determination was ±0.1‰ SD). Snow collected before melt began was similar in isotopic composition (−25.9‰) to the peak discharge, and $\delta^{18}O$ values of thawed-soil moisture at the end of June 1989 were also similar (−19.7‰) in isotopic composition to the stable oxygen isotope composition of the stream at that time (Cooper et al. 1991b).

Samples analyzed since 1989 for $\delta^{18}O$ values have included stream water, snowpack, precipitation, soil waters, permafrost, and waters quantitatively extracted from plants. These detailed isotopic and stream flow measurements demonstrate differing conditions in which the assumption of conservative stable isotope behavior is both true as well as apparently false. As a result this work has implications for direct application of stable isotope methodologies for permafrost hydrology, as well as the general application of these techniques for evaluating snowmelt contributions to stream flow generation in other watersheds with long-term snow cover. A portion of this isotopic data has been interpreted in conjunction with stream flow data collected during the 1990 field season (Cooper et al. 1993). We also discuss interpretations of the isotopic data alone.

8.4.1 Oxygen-18 Content of Snowpack

Whole snow cores collected on 12–13 May 1990, before significant runoff occurred, had a mean $\delta^{18}O$ value of −27.6 ± 1.6‰ SD; $n = 13$. Although the spatial pattern was not completely consistent, four of five snow core samples with more ^{18}O-depleted values than average were collected within 100 m of Imnavait Creek. Less ^{18}O-depleted values were generally observed on the east ridge and hillslope of the watershed at much greater distances from the creek. These observations are consistent with isotopic fractionation associated with evaporation and sublimation (Friedman et al. 1991; Sommerfield et al. 1991). Snow surveys at Imnavait Creek indicate that snow is significantly redistributed during the winter, and this redistribution is greater on the ridgeline than in the more sheltered valley bottom (Kane et al. 1991; see Fig. 6.3 in Chap. 6, this Vol.). As a result snow deposited on the ridgelines of the watershed tends to be more subject to evaporative enrichment of heavy isotopes than snow in the more protected valley bottom.

8.4.2 Oxygen-18 Content of Imnavait Creek

Imnavait Creek $\delta^{18}O$ values changed dramatically during the first 8 days of stream flow in 1990, increasing from −30.3‰ on 14 May, to −22.5‰ on 22 May (Fig. 8.1). On the basis of the oxygen isotopic content (−20.6 ± 0.1‰) of waters collected at depths of 3, 10, and 17 m on 11 May, in Toolik Lake, a 20-m-deep lake 16 km west of Imnavait Creek (Cooper and Kipphut, unpubl.) we estimate

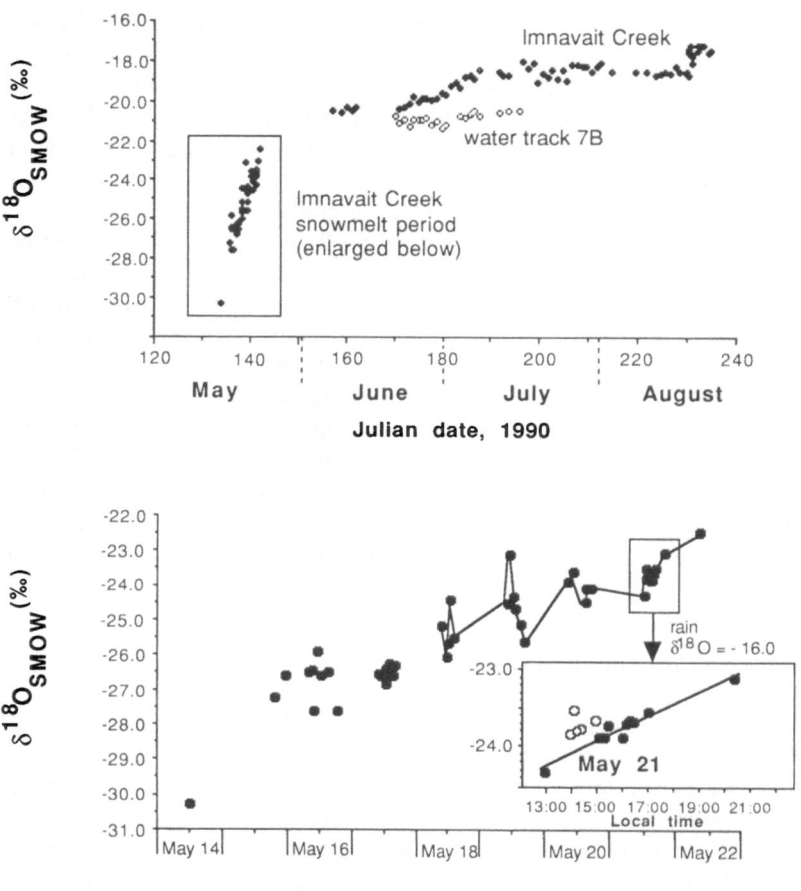

Fig. 8.1. *Above* Oxygen isotope composition of Imnavait Creek, Alaska, during stream flow, May-August, 1990, *Darkened crosses*, Imnavait Creek main channel; *circles*, water track 7B. *Below* Oxygen isotope composition of Imnavait Creek, Alaska, during snowmelt, May 1990. Average bulk snow $\delta^{18}O$ value before snowmelt started, 12–13 May 1990, was −27.6 ± 1.6‰ (SD), n = 13. Estimated "old" water $\delta^{18}O$ value for frozen soil moisture was ca. −20‰ based upon measurements at 3 m, 10 m, and 17 m on 11 May 1990 from Toolik Lake; $\delta^{18}O$ = −20.6 ± 0.1‰ (SD). **Inset:** Oxygen isotope composition of Imnavait Creek during convective rainstorm, 21 May 1990, *Open circles* Stream during 2 mm rain; *closed circles* other stream samples; *line* is least-squares fit for stream samples collected on 21 May, except during rainstorm; $r^2 = 0.95$

that soil moisture frozen in the soil had a $\delta^{18}O$ value of approximately −20‰. The lake has a 1.5-km² surface area, and the water residence time within the lake is approximately 1–2 years (George Kipphut, pers. comm.). The mean isotopic content of Toolik Lake, while still ice-covered and before snowmelt, is probably close to the seasonally averaged isotopic content of frozen water in tundra at Imnavait Creek, and falls within the range of the isotopic content of soil water we observed toward the end of the summer season in 1990 (Cooper et al. 1993).

The large shift in $\delta^{18}O$ values during snowmelt suggests a high degree of mixing of snow meltwater with frozen soil moisture. In the case of Imnavait Creek, however, an alternative explanation is necessary, because measurements of snow redistribution, ablation, snow and soil moisture content, and snowmelt runoff indicate that meltwater cannot mix to a high degree with underlying ice-rich soils (Kane et al. 1989; Hinzman et al. 1991; Kane et al. 1991). The large shift in ^{18}O content we observed over the snowmelt period in 1990 is actually more consistent with isotopic fractionation during melt than with mixing with underlying soil moisture. Vapor pressures and freezing points for the most common heavy isotopic forms of water ($HH^{18}O$ and $DH^{16}O$) are slightly lower than for $HH^{16}O$ (Friedman et al. 1964); thus, in any phase change at equilibrium from solid to liquid, the heavier isotopic forms of water tend to be retained in the solid phase. After the initial release of meltwater is depleted in heavy isotopes relative to the whole snowpack, the remaining snowpack becomes more and more isotopically enriched as melt proceeds. The initial release of isotopically depleted meltwater, followed by progressive increases in $\delta^{18}O$ values, is consistent with our observations of the isotopic shift in stream flow (see Fig. 8.1).

Isotopic fractionation between ice and water has been well studied in both simulated and natural experiments (Arnason 1969; Herrmann et al. 1981; Stichler et al. 1981; Jouzel and Souchez 1982; Souchez and Jouzel 1984). Nevertheless, observation of stream discharge that is enriched by heavy isotopes relative to original premelt snowpack has usually been interpreted as a mixing between groundwaters and snowmelt, with ice-liquid fractionation, if considered, playing a relatively minor role (Dinçer et al. 1970; Martinec et al. 1974; Rodhe 1981; Bottomley et al. 1986; Lawrence 1987; Ingraham and Taylor 1989; Wels et al. 1991). Our data indicate that previous studies in watersheds where soil water mixing was physically possible may have underestimated the importance of this isotopic fractionation. For the Imnavait Creek watershed, and probably many other permafrost-underlain watersheds, the use of ^{18}O as a conservative tracer of snowmelt can only be supported for carefully delimited conditions. For instance, we did observe apparently conservative behavior during a convective rainstorm during snowmelt when stream flow $\delta^{18}O$ temporarily increased as a result of direct contributions of isotopically distinct rain to the stream floodplain (Fig. 8.1; Cooper et al. 1993). Also, under dry antecedent conditions in late summer, rainfall behaved in a conservative manner by forcing isotopically distinct "old" water already within the soil into the stream before the "new" water contributed by rain changed the isotopic composition of the increasing stream flow (Cooper et al. 1993).

The slowly rising $\delta^{18}O$ values in Imnavait Creek throughout the summer season (Fig. 8.1) indicate that evaporative heavy isotope enrichment occurs throughout the summer within the stream channel. The water-track tributary of Imnavait Creek, which flowed through mid-July, was not as markedly affected by evaporation; it exhibited significantly more negative $\delta^{18}O$ values for samples collected on the same dates after mid-June. $\delta^{18}O$ values for the water-track and Imnavait Creek were similar in mid-June, at the same time as the last major stream flow generation event of 1990 (Cooper et al. 1993). This indicates that evaporation was thereafter the major factor affecting isotopic changes in Imnavait Creek. Increasing differences in the $\delta^{18}O$ values of the stream water relative to water track 7B were observed during June 1990 (Fig. 8.1). The steady increase in $\delta^{18}O$ values of the stream during this time suggests that water within the stream underwent heavy isotope enrichment through evaporation, whereas this effect was modest with water in water track 7B. The isotopic composition of Imnavait Creek was relatively unchanged from July until mid-August (Fig. 8.1). This consistency may have been a consequence of decreasing evaporation rates within the watershed. Evaporative loss reaches a maximum soon after snowmelt and gradually decreases throughout the summer (Kane et al. 1989; Chap. 6, this Vol.).

8.4.3 Oxygen-18 Content of Soil Moisture

The ^{18}O content of soil moisture is significantly lower than the content of surface waters of Imnavait Creek sampled on the same dates (Cooper et al. 1991b; Cooper et al. 1993). The likely explanation is that heavy isotopes are enriched in open water in the stream channel because evaporation proceeds there with less resistance than in vegetated portions of the watershed. Pan evaporation experiments indicate that water losses at Imnavait Creek range from 2–4 mm day^{-1} during June, July, and August (Chap. 6, this Vol.). Within individual soil cores the ^{18}O content of moisture varied over about 3‰ in the six cores that were sampled on 16, 22, and 23 August 1990 (Cooper et al. 1993). Permafrost $\delta^{18}O$ values recovered from four cores revealed no obvious differ-ence from thaw-zone moisture (Cooper et al. 1993), although a sample core recovered near the stream channel exhibited the least negative values, as did active layer samples. The pattern of less negative $\delta^{18}O$ values in samples of permafrost and thaw-zone moisture collected near the stream suggests that similar evaporative isotopic enrichment processes were occurring when water in the permafrost originally froze.

8.4.4 Covariance of Oxygen-18 and Deuterium in Watershed Compartments

We also have assayed many water samples for δD values because of the infor-mation stable oxygen and hydrogen isotope measurements together can pro-vide for evaluating the evaporative history of the water sample (Craig 1961; Dansgaard 1964; Siegenthaler 1979). The isotopic composition of meteoric water is controlled by equilibrium effects associated with water-vapor conden-sation. These effects cause most rain and snow to cluster along a meteoric water line described by the equation $\delta D = 8\delta^{18}O + 10‰$ (Craig 1961). Water that has undergone increasing evaporation has a slope increasingly lower than 8 in this relation. This phenomenon can be explained qualitatively through the more-rapid evaporation of $HD^{16}O$ (atomic mass 19) than $H_2^{18}O$ (atomic mass 20) under kinetically dominant conditions, and as a result, the concentration of $H_2^{18}O$ increases faster than $HD^{16}O$ in the residual pool of evaporating water (Gat 1971). Local summer rainfall that we sampled in the vicinity of Imnavait Creek appears to adhere largely to this meteoric water line, being described by the least-square-fit line, $\delta D = 8.28\ \delta^{18}O + 16.2‰$. If the rainfall had resulted from reevaporated continental moisture, we would expect the slope to be lower than 8.

One result of this is that we can use $\delta^{18}O - \delta D$ values to evaluate the importance of evaporation in determining the isotopic content of various water compartments within the watershed. Comparison of $\delta^{18}O - \delta D$ relationships for various waters within the Imnavait Creek watershed (Table 8.2) show that only meteoric precipitation and possibly the runoff in the water-track tributary is relatively unchanged as a result of evaporation. The slope coefficient, m, in the equation $\delta D = m\delta^{18}O + b$ is below 8 (meteoric water) for all other waters including the snowpack before melt, stream flow, permafrost, thaw-zone moisture, and water distilled from living mosses and vascular plants. For example, the regression equation for $\delta^{18}O - \delta D$ values for the snowpack before melt shows that δD values are related to $\delta^{18}O$ values by a line with a slope of 6.77 and a y-intercept of $-25.4‰$ (Table 8.2). Meltwater derived from that snowpack collected in the stream channel within a week after we sampled the unmelted snow, however, had an even lower slope and y-intercept of 6.04 and -42.12, respectively (Table 8.2), indicating that evaporative heavy-isotope enrichment continued throughout snowmelt. It is likely that the generally arid climate, coupled with sublimation of the snowpack throughout the winter, causes almost all watershed waters to exhibit isotopic evidence of significant postprecipitation evaporation.

Table 8.2. Relationships in Imnavait Creek watershed for model: $\delta D = m\delta^{18}O + b$

Sample	m	b	r^2	P	n
Water track	8.31	12.90	46	0.2072	5
Summer rain	8.28	16.23	92	0.0001	10
Snowpack before snowmelt	6.77	−25.37	95	0.0048	5
Summer stream flow	6.70	−23.14	95	0.0051	5
Permafrost	6.12	−31.37	93	0.0001	8
Snowmelt stream flow	6.04	−42.12	93	0.0001	9
Thaw-zone moisture	4.88	−61.37	95	0.0001	11
Polytricum and *Sphagnum* spp.					
Day site, ridge top	3.34	−85.38	64	0.0055	10
Wet site, at flume	2.15	−102.71	54	0.0237	9
Petasites frigidus					
Day site	1.66	−81.42	80	0.0012	9
Wet site	1.20	−87.26	86	0.0004	9

8.4.5 Covariance of Oxygen-18 and Deuterium in Plant Water

Studies of diurnal variations in the stable isotopic composition of plant water composition have linked evapotranspiration on a watershed scale to water fluxes within plants. Because of the extensive vegetation cover in arctic tundra, it is reasonable to assume that isotopic variability in plant waters, as well as meteoric, surface, and soil waters, can provide insights into larger hydrological processes in arctic ecosystems and serve as a mechanism for linking plant canopy transpiration, water loss from bryophytes, and evaporation on a watershed scale.

We determined the stable isotopic composition of water (δD and $\delta^{18}O$) on a diurnal basis in mosses (mainly *Polytricum* and *Sphagnum* spp.) and a vascular plant, *Petasites frigidus* (L.) Fries. Plant and soil-water sampling was done at a dry upslope site at the ridgetop on the northeastern boundary of the watershed study site and also within the wet riparian zone at the flume, on the north–central boundary of the study site, within 10 m of the stream. Moisture extracted from soil cores at both sites showed an evaporative profile with $\delta^{18}O$ values increasing from $-22‰$ near the boundary with underlying permafrost (15 cm), to $-17.5‰$ at 2 cm below the surface, with an additional enrichment to $-13.3‰$ at the tundra surface. Heavy isotope enrichment was slightly greater at the dry site indicating strong evapotranspiration was taking place at both sites. The mosses did not show a typical diurnal heavy isotope pattern, which is similar to other astomatal plants, including marine algae, seagrasses, and specialized epiphytes (Cooper and DeNiro 1989; Cooper et al. 1991a). Least-square-fit line plots of δD vs $\delta^{18}O$ values indicated that the moss waters seem to fall on evaporation lines that are extensions of thawed soil moisture evaporation lines, i.e., evaporation effects on the isotopic content of living mosses undergoing water loss do not appear to differ significantly from evaporation effects on the isotopic content of soil moisture, much of which consists of preserved mosses in various stages of decay. This similarity is no doubt due to the lack of stomatal control of water loss in living mosses. By contrast, the vascular plant *P. frigidus* showed a typical diurnal isotopic fractionation pattern with a maximum heavy-isotope enrichment of at least 20‰ for $\delta^{18}O$ and 75‰ for δD relative to unfractionated soil water. It also exhibited very low slopes, m, (1.2 and 1.6) in the equation $\delta D = m\delta^{18}O + b$, relative to mosses (2.1 and 3.3) and other waters within the watershed (see Table 8.2).

8.5 Long-Lived Radioisotopes: Lead-210 and Cesium-137

Like [7]Be, both ^{210}Pb and ^{137}Cs are particle-reactive in freshwater and terrestrial systems (Olsen et al. 1985; Olsen et al. 1989). Therefore, they provide a measure of multiannual deposition processes, although differential summer and winter deposition cannot be separated as in the case of [7]Be and ^{35}S inventories. The major input of fallout ^{137}Cs occurred during the mid-1960s, with some additional contributions to the North American Arctic resulting from the Chernobyl nuclear accident in the Ukraine (Taylor et al. 1988). ^{210}Pb is derived from the natural ^{238}U radioactive decay series, and that which is "unsupported" results from atmospheric decay of ^{222}Rn daughter products, and is, in addition to ^{238}U decay products, in the soil.

8.5.1 Distributions of ^{137}Cs on Tundra and in Lake Sediments

In contrast to enhanced ^{7}Be activity in riparian tundra, the two longer-lived isotopes, ^{137}Cs and ^{210}Pb, were not concentrated in riparian zones (Cooper et al. 1991b). This apparent differential behavior between a short-lived isotope (^{7}Be) and the longer-lived nuclides may be due to differences in input fluxes, temporal and seasonal differences in deposition, or possibly the higher solubility of ^{7}Be at low pH (<5) during snowmelt. Variability in ^{137}Cs inventories were observed throughout the Imnavait Creek watershed; however, radioisotope activities ranged from 77 to 177 mBq cm^{-2} (Cooper et al. 1991b; Grebmeier et al. 1993; Cooper et al. 1995). A transect of the Dalton Highway corridor north from Imnavait Creek to Prudhoe Bay indicates that total ^{137}Cs inventories decline to ~60 mBq cm^{-2} on the Arctic coastal plain near Prudhoe Bay (Grebmeier et al. 1993; Cooper et al. 1995). Atmospheric deposition of ^{137}Cs decreased with increasing latitude in the Arctic, but the shift in deposition would have been relatively small over this distance (ca. 200 km), implying the recent loss of ^{137}Cs and associated organic material from arctic tundra over the northern portions of the transect between Imnavait Creek and Prudhoe Bay. Corroborating this evidence is the presence of maximum ^{137}Cs accumulations in near-surface layers of the more northern tundra that was sampled (Grebmeier et al. 1993), rather than at depth, as at Imnavait Creek (Cooper et al. 1991b).

In most cases ^{137}Cs strongly adsorbs to soil particles, so losses of ^{137}Cs relative to expected deposition suggest watershed-scale soil erosion (Ritchie and McHenry 1990). Mineral soil is poorly developed at Imnavait Creek, and most ^{137}Cs is associated with peat and other organic materials near the tundra surface. Relatively large masses of organic matter are lost from the watershed of Imnavait Creek, and elsewhere on the North Slope (Everett et al. 1989; Oswood et al. 1989; Oechel et al. 1993). Recently Kling et al. (1991) determined that organic materials exported from arctic tundra are being deposited in lakes, a widespread landscape feature in the Arctic. If ^{137}Cs is bound to these organic materials, we would expect that inventories of ^{137}Cs would be elevated in Arctic lake sediments. Studies of Toolik Lake sediments indicate, however, that ^{137}Cs inventories per unit area are significantly lower than atmospheric deposition and current inventories in upland tundra, although high levels were found near the lake inflow (Grebmeier et al. 1993). This result suggests that the high observed carbon dioxide saturation of North Slope arctic lakes (Kling et al. 1991) derives from dissolved organic materials, rather than particulates.

Very low sedimentation rates in Toolik Lake (Cornwell 1985; Grebmeier et al. 1993; Cooper et al. 1995) indicate few particulate mechanisms for transporting particle-reactive radionuclides to lake sediments. More extensive sampling of arctic coastal plain lake sediments would help clarify if the apparent loss of ^{137}Cs from coastal plain tundra is leading to enhanced deposition in lakes located on the coastal plain, as opposed to Toolik, which is in the Brooks Range foothills. One mechanism for cesium transport, which has not been investigated, is that ion exchange and remobilization of ^{137}Cs by NH^{3+}, Fe^{2+}, or Mn^{+2} can be accentuated under less oxic conditions (Evans et al. 1983) such as those prevailing on the wetter coastal plain.

8.5.2 Cycling of ^{137}Cs in Annual Berries

Cycling of ^{137}Cs by biological activity has also been investigated by assaying radioactivity in annual berries growing within the Imnavait Creek watershed. We observed significant translocation of ^{137}Cs from vegetation into annual fruit of two arctic berries that are important food sources throughout circumboreal latitudes: the lingonberry, or low-bush cranberry, *Vaccinium vitis idaea* L., and the cloudberry, *Rubus chamaemorus* L. Radioassayed samples of a crow berry, *Empetrum nigrum* L. showed comparatively low concentrations of ^{137}Cs. These findings hold significance for understanding remobilization of radiocesium within arctic ecosystems, where slow-growing plants, cold temperatures, and poor soil development may slow the removal of the radionuclide into soil strata where it is less accessible for biological uptake (Hutchison-Benson et al. 1985). In addition, the local importance of these berries as food for the people within the circumpolar north (Cooperative Extension Service 1981) as well as for arctic wildlife suggest that public health implications should be taken into account when considering the potentially greater impacts of radiocesium released into arctic vs temperate ecosystems.

Ripe fruit of *V. vitis-idaea* and *R. chamaemorus*, and other vegetation, from surface to 2 cm depth, were collected in August 1990. We detected no evidence of radiocesium inputs in precipitation collectors at the Imnavait Creek watershed during 1989 or 1990, and expect that almost all ^{137}Cs was deposited before 1989. On a dry-weight basis, vegetation from the surface to a depth of 2 cm shows ^{137}Cs activities of 20.0 to 171 mBq g^{-1} (Fig. 8.2). The annual fruits of *V. vitis-idaea* and *R. chamaemorus* exhibit generally lower, but nevertheless significant, activities. Because no ^{137}Cs was deposited by precipitation in two summers of sampling at this site, radiocesium most likely was translocated into the fruit by the plant vascular system from inventories available in the top 10 cm of soil consisting largely of peat, moss, and other organic materials. One other fruit assayed as part of this sampling effort, that of *E. nigrum*, in contrast, incorporated very low levels of ^{137}Cs. These fruits were collected near Toolik Lake approximately 16 km west of Imnavait Creek, from plants growing on a dry, glacial till with exposure to rocks and gravel, where there are greater

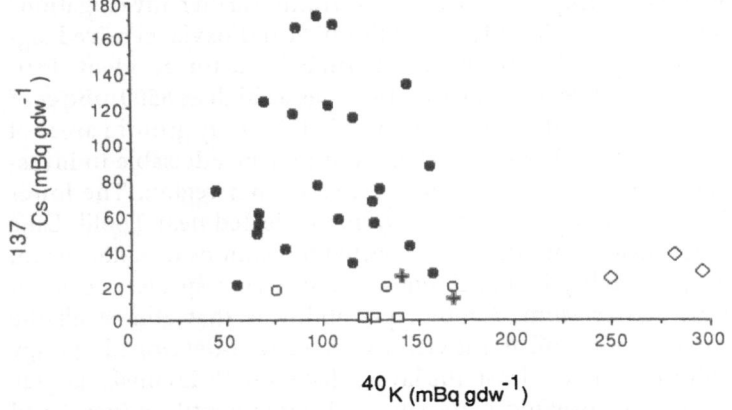

Fig. 8.2. Radioisotopic activities of ¹³⁷Cs and ⁴⁰K for surface vegetation to 2 cm depth (*darkened circles*; $n = 23$), and four species of ripened berries: *Vaccinium vitis-idaea* (*open circles*; $n = 3$), *Empetrum nigrum* (*squares*; $n = 3$), *Rubus chamaemorus* (*diamonds*; $n = 3$) and *Vaccinium uliginosum* (*crosses*; $n = 2$). All samples were collected at Imnavait Creek, Alaska, except for *Empetrum nigrum* berries, which were collected at Toolik Lake, approximately 16 km to the west, and *V. uliginosum*, growing in tussock tundra at Salmon Lake, near Nome, on Alaska's Seward Peninsula. All samples were collected in August 1990, except for *V. uliginosum*, collected in August 1992. Radioactivity is corrected to date of collection. Radioactivity was assayed using low-background, high-resolution, germanium gamma-ray detectors equipped with a Nuclear Data Model 9900 microprocessor system programmed to record gamma spectra in 4096 channels

mineral binding sites for cesium. Since we completed this work, we have also assayed the radioactivity of an additional widely used fruit, the blueberry, *Vaccinium uliginosum* L., growing in tussock tundra at Salmon Lake, near Nome, on Alaska's Seward Peninsula. These data, showing ¹³⁷Cs as of August 1992, resemble those from *V. vitis-idaea* collected at Imnavait Creek in 1990 (Fig. 8.2).

Because the chemical behavior of cesium is similar to potassium, and natural ⁴⁰K is also assayed during gamma-ray analyses, inventories of ⁴⁰K on a dry-weight basis were also determined (Fig. 8.2). The combination of potassium and cesium isotope data indicate that although the activities of ¹³⁷Cs and ⁴⁰K in the tundra are not directly related, radiocesium concentrations are significantly correlated ($P < 0.05$) to the activities of ⁴⁰K in the ripened fruit. This correlation suggests that competition between potassium, an essential nutrient, and cesium during plant uptake may be the mechanism by which ¹³⁷Cs is translocated to the berries.

Radiocesium inventories in these Alaskan fruits are modest and probably constitute no significant health hazard. Taylor et al. (1988) report ¹³⁴Cs and ¹³⁷Cs activities in caribou meat of up to 710 mBq gdw⁻¹, but conclude that consumption of such meat is "safe by international standards." The British National Radiological Protection Board has recommended that a level of radioactivity above $1000\,mBq\,g^{-1}$ foodstuffs "warrants further investigation" (Stockbridge 1986). Northern latiudes, specifically Scandinavia, received significant radiocesium deposition from the Chernobyl reactor accident. Estimated deposition of ^{137}Cs in Sweden in May 1986 was as high as $8500\,mBq\,cm^{-2}$ (Persson et al. 1987). Because the berries assayed in this study grow in areas of Scandinavia affected by the Chernobyl accident, it may be advisable to investigate radioisotope concentrations in wild berries in that region. The lower activities of ^{137}Cs observed in berries of *E. nigrum* collected near Toolik Lake may provide further insight into these unexpected indications of radiocesium translocation. One possibility is that *E. nigrum*, a different species, does not actively translocate radiocesium. Another possibility is that almost all the cesium is bound to the rocky soil at this vicinity. In the well-developed, spongy tundra at Imnavait Creek, in contrast, the layers that store ^{137}Cs consist largely of peat, moss, and other organic materials, and contain little mineral soil to which cesium can bind. Thus, this work, as well as previous studies (Hutchison-Benson et al. 1985; Taylor et al. 1988), indicate that arctic tundra and boreal forests are particularly vulnerable to radiocesium pollution, and that assumptions about cesium uptake and cycling by plants at temperate latitudes do not necessarily apply at higher latitudes.

8.5.3 Distribution of ^{210}Pb in Tundra

[210]Pb inventories in the Imnavait Creek watershed are considerably smaller than inventories measured at temperate latitudes, although our measurements – ranging widely from 5 to $104\,mBq\,cm^{-2}$ (Cooper et al. 1991b) – are consistent with the mean inventory reported for 12 lake sediment cores from Toolik Lake (mean = $75\,mBq\,cm^{-2}$; Cornwell 1985). In comparison, annual fluxes of atmospherically derived ^{210}Pb, average ca. $15\,mBq\,cm^{-2}$ along the eastern coastline of the United States (Krishnaswami et al. 1980). This annual flux will support a steady-state inventory of ca. $480\,mBq\,cm^{-2}$ based on a 32-year mean life of [210]Pb. These low fluxes of ^{210}Pb within the Arctic apparently come from low rates of precipitation and inhibition of [222]Rn exhalation from frozen soils (Robbins 1978).

8.6 Conclusions

Although stable isotopes of oxygen and hydrogen are being widely applied in watershed research, and the use of isotopic tracers such as [7]Be, ^{35}S, ^{137}Cs, and ^{210}Pb is common in geochemistry research, our work demonstrates that additional insights may come from combining elements of both approaches. Other particle-reactive metals and contaminants can be expected to behave in the same manner as [7]Be in this study. Thus, although snowmelt can occur over short periods and generate high stream flow, high water flow itself may have little impact on the export of materials deposited by precipitation. Redistribution of particle-reactive materials to riparian zones or out of coastal plain tundra may be influenced, however, and this indication merits further study. Biologically active elements, such as sulfur, also deposited by precipitation may be taken up in varying degrees by surface vegetation. Most radiosulfur deposited by precipitation at Imnavait Creek is not immediately exported, indicating that the input of atmospheric sulfate at this site is only indirectly coupled to export processes. Radiocesium and possibly other contaminants may lack suitable mineral binding sites in tundra ecosystems, and may be cycled biologically to a greater extent than previously recognized.

Our stable isotope analyses of waters within the Imnavait Creek watershed indicate that one should be careful in applying this technique to hydrological studies of arctic watersheds. Studies of the stable isotope composition of various watershed compartments indicate the importance of evaporation in controlling the isotopic composition of various watershed reservoirs. Apparent isotopic fractionation during snowmelt would rule out a straightforward use of stable isotopes to follow snow contributions to this watershed. Nevertheless, ^{18}O assays appear to be sensitive to small contributions of rain during snowmelt, and we have observed apparently conservative contributions of ^{18}O stream flow generation under dry antecedent conditions.

Whereas our work does not invalidate the large number of isotopic studies of snowmelt runoff that have previously been done, it also suggests that it is possible to overestimate mixing of snowmelt with underlying soil water, because isotopic fractionation during the ice–liquid phase transition can be similar in direction and magnitude to mixing between snowmelt and groundwaters. Future work should incorporate studies of the release of cations, anions, and other water-borne materials in relation to the isotopic fractionation we observed here.

Acknowledgments. We thank Chirk Chu, Kathy Turco, George Kipphut, Peter McRoy, and Dennis Hansell for providing logistical support or collecting precipitation samples. We thank Kaye Everett, Julie Hambrook, and Rob Gieck for sampling snow cores and stream water. Permafrost coring was done with the assistance of Skip Walker and Kaye Everett. Doug Kane and Larry Hinzman have generously shared stream flow and snow ablation data that have helped in interpretation of our results. The Logistical Services Office of the Institute of Arctic Biology, University of Alaska, Fairbanks, also provided invaluable assistance throughout our field work. Ernie Bondietti and Jerry Brantley assisted with ^{35}S analyses in the laboratory. Comments from Phil Jardine, Pat J. Mulholland, Robb Turner, and John Tenhunen helped improve previous drafts of the manuscript. This research was supported by the Environmental Sciences Division, Office of Health and Environmental Research, U. S. Department of Energy, under contract DE-AC05-840R21400 with the Lockheed Martin Energy Systems, Inc., publication no. 4128, Environmental Sciences Division, Oak Ridge National Laboratory.

References

Arnason B (1969) The exchange of hydrogen isotopes between ice and water in temperate glaciers. Earth Planet Sci Lett 6: 423–430

Baskaran M, Kelley JJ, Naidu AS, Holleman DF (1991) Environmental radiocesium in subarctic and arctic Alaska following Chernobyl. Arctic 44: 346–350

Bottomley DJ, Craig D, Johnston LM (1986) Neutralization of acid runoff by groundwater discharge to streams in Canadian Precambrian shield watersheds. J Hydrol 88: 213–234

Broecker WS, Peng T-H (1982) Tracers in the sea. Eldigio Press, Lamont-Doherty Earth Observatory, Palisades, NY

Calamari D, Bacci E, Focardi S, Gaggi C, Morosini M, Vighl M (1991) Role of plant biomass in the global partitioning of chlorinated hydrocarbons. Environ Sci Technol 25: 1489–1495

Cooper LW, DeNiro MJ (1989) Depletion of heavy isotopes of oxygen and hydrogen in tissue water of intertidal plants: implications for water economy. Mar Biol 101: 397–400

Cooper LW, DeNiro MJ, Keeley JE (1991a) The relationship between stable oxygen and hydrogen isotope ratios of water in astomatal plants. In: Taylor HP, O'Neil JR, Kaplan IR (eds) Stable isotope geochemistry: a tribute to Samuel Epstein. The Geochemical Society, San Antonio, pp 247–255

Cooper LW, Olsen CR, Solomon DK, Larsen IL, Cook RB, Grebmeier JM (1991b) Stable isotopes of oxygen and natural and fallout radionuclides used for tracing runoff during snowmelt in an arctic watershed. Water Resour Res 27: 2171–2179

Cooper LW, Solis C, Kane DL, Hinzman LD (1993) Application of oxygen-18 tracer techniques to arctic hydrological processes. Arct Alpine Res 25: 247–255

Cooper LW, Grebmeier JM, Larsen IL, Solis C, Olsen CR (1995) Evidence for redistribution of cesium-137 in Alaskan tundra, lake and marine sediments. Sci Total Environ 160/161: 295–306

Cooperative Extension Service U of Alaska (1981) Wild edible and poisonous plants of Alaska. Univ Alaska Cooperative Extension, Fairbanks

Cornwell JC (1985) Sediment accumulation rates in an Alaskan arctic lake using a modified ^{210}Pb technique. Can J Fish Aquat Sci 42: 809–814

Craig H (1961) Isotopic variations in meteoric waters. Science 133: 1702–1703

Dansgaard W (1964) Stable isotopes in precipitation. Tellus 16: 436–468

DeWalle DR, Swistock BR, Sharpe WE (1988) Three-component tracer model for stormflow on a small Appalachian forested catchment. J Hydrol 104: 301–310

Dinçer T, Payne BR, Florkowski T, Martinec J, Tongiorgi T (1970) Snowmelt runoff from measurements of tritium and oxygen-18. Water Resour Res 6: 110–118

Evans DW, Alberts JL, Clark RA (1983) Reversible ion-exchange fixation of cesium-137 leading to mobilization from reservoir sediments. Geochim Cosmochim Acta 47: 1041–1049

Everett KR, Marion GM, Kane DL (1989) Seasonal geochemistry of an arctic tundra drainage basin. Holoarct Ecol 12: 279–289

Friedman I, Redfield AC, Schoen B, Harris J (1964) The variation of the deuterium content of natural water in the hydrologic cycle. Rev Geophys 2: 177–224

Friedman I, Benson C, Gleason J (1991) Isotopic changes during snow metamorphism. In: Taylor HP, O'Neil JR, Kaplan IR (eds) Stable isotope geochemistry: a tribute to Samuel Epstein. The Geochemical Society, San Antonio, pp 211–221

Fry B, Sherr EB (1984) δ^{13}C measurements as indicators of carbon flow in marine and freshwater ecosystems. Contrib Mar Sci 27: 13–47

Gat J (1971) Comments on the stable isotope method in regional groundwater investigation. Water Resour Res 7: 980–993

Goel PS, Narasappaya N, Prabhakara C, Thor R, Zutshi PK (1959) Study of cosmic ray produced short-lived P^{32}, P^{33}, Be^7, and S^{35} in tropical latitudes. Tellus 11: 91–100

Grebmeier JM, Cooper LW, Larsen IL, Solis C, Olsen CR (1993) Cesium-137 inventories in Alaskan tundra, lake and marine sediments: an indicator of recent organic material transport? In: Applications of isotope techniques in studying past and current environmental changes in the hydrosphere and the atmosphere. International Atomic Energy Agency, Vienna, pp 147–159

Griffiths H (1991) Applications of stable isotope technology in physiological ecology. Funct Ecol 5: 254–269

Hermanson MH (1990) ^{210}Pb and ^{137}Cs chronology of sediments from small, shallow Arctic lakes. Geochim Cosmochim Acta 54: 1443–1451

Herrmann A, Lehrer M, Stichler W (1981) Isotope input into runoff systems from melting snow covers. Nordic Hydrol 12: 309–318

Hinzman LD, Kane DL, Gieck RE, Everett KR (1991) Hydrological and thermal properties of the active layer in the Alaskan Arctic. Cold Reg Sci Technol 19: 95–110

Hooper RP, Shoemaker CA (1986) A comparison of chemical and isotopic hydrograph separation. Water Resour Res 22: 1444–1454

Hörnmann PK (1978) Beryllium. In: Wedepohl KH (ed) Handbook of geochemistry. Springer, Berlin Heidelberg New York

Hutchison-Benson E, Svoboda J, Taylor HW (1985) The latitudinal inventory of ^{137}Cs in vegetation and topsoil in northern Canada, 1980. Can J Bot 63: 784–791

Ingraham NL, Taylor BE (1989) The effect of snowmelt on the hydrogen isotope ratios of creek discharge in Surprise Valley, California. J Hydrol 106: 233–244

Jouzel J, Souchez RA (1982) Melting-freezing at the glacier sole and the isotopic composition of the ice. J Glaciol 28: 35–42

Kane DL, Hinzman LD, Benson CS, Everett KR (1989) Hydrology of Imnavait Creek, an arctic watershed. Holoarct Ecol 12: 262–269

Kane DL, Hinzman LD, Benson CD, Liston GE (1991) Snow hydrology of a headwater arctic basin 1. Physical measurements and process studies. Water Resour Res 27: 1099–1109

Kennedy VC, Kendall C, Zellweger GW, Wyerman TA, Avanzino RJ (1986) Determination of the components of storm-flow using water chemistry and environmental isotopes, Mattole Basin. California. J Hydrol 84: 107–140

Kling GW, Kipphut GW, Miller MC (1991) Arctic lakes and streams as gas conduits to the atmosphere: implications for tundra carbon budgets. Science 251: 298–301

Krishnaswami S, Benninger LK, Aller RC, Van Damme KL (1980) Atmospherically derived radionuclides as tracers of sediment mixing and accumulation in near-shore marine and lake sediments. Earth Planet Sci Lett 47: 307–318

Lal D, Peters B (1967) Cosmic ray produced radioactivity on earth. Handb Physik 46: 551–612

Lawrence JR (1987) Use of contrasting D/H ratios of snows and groundwaters of eastern New York State in watershed evaluation. Water Resour Res 23: 519–521

McCauley LL (1991) The Arctic Research Consortium of the United States: creating a synergy for the Arctic. Arct Res U. S. 5: 17–25

Martinec J, Siegenthaler U, Oeschger H, Tongiorgi E (1974) New insights into the run-off mechanism by environmental isotopes. In: Isotope techniques in groundwater hydrology. Proc Symp, International Atomic Energy Agency, Vienna

Obradovic MM, Sklash MG (1986) An isotopic and geochemical study of snowmelt runoff in a small arctic watershed. Hydrol Processes 1: 15–30

Oechel WC, Hastings SJ, Vourlitis G, Jenkins M, Riechers G, Grulke N (1993) Recent change of Arctic tundra ecosystems from a net carbon dioxide sink to a source. Nature 361: 520–523

Olsen CR, Larsen IL, Lowry PD, Cutshall NH, Todd JF, Wong GTF, Casey WH (1985) Atmospheric fluxes and marsh-soil inventories of ^{7}Be and ^{210}Pb. J Geophys Res 90: 10487–10495

Olsen CR, Thein M, Larsen IL, Lowry PD, Mulholland PJ, Cutshall NH, Byrd JT, Windom HL (1989) Plutonium, lead-210, and carbon isotopes in the Savannah Estuary: riverborne versus marine sources. Environ Sci Technol 23: 1475–1481

Oswood MW, Everett KR, Schell DM (1989) Some physical and chemical characteristics of an arctic beaded stream. Holoarct Ecol 12: 290–303

Pearce AJ, Stewart MK, Sklash MG (1986) Storm runoff generation in humid headwater catchments, I. Where does the water come from? Water Resour Res 22: 1263–1272

Persson C, Rodhe H, De Geer L-E (1987) The Chernobyl accident: a meteorological analysis of how raionuclides reached and were deposited in Sweden. Ambio 16: 20–31

Peters B (1959) Cosmic ray produced radioactive isotopes as tracers for studying large-scale atmospheric circulation. J Atmos Terr Phys 13: 351–370

Ritchie JC, McHenry JR (1990) Application of radioactive fallout Cesium-137 for measuring soil erosion and sediment accumulation rates and patterns: a review. J Environ Qual 19: 215–233

Robbins JA (1978) Geochemical and geophysical applications of radioactive lead. In: Nriagu JO (ed) Biogeochemistry of lead in the environment. Elsevier, New York, pp 285–293

Rodhe A (1981) Spring flood: meltwater or groundwater? Nordic Hydrol 12: 21–30

Rundel PW, Ehleringer JR, Nagy KA (1988) Stable isotopes in ecological research. Springer, Berlin Heidelberg New York

Siegenthaler U (1979) Stable hydrogen and oxygen isotopes in the water cycle. In: Jäger E, Hunziker JC (eds) Lectures in isotope geology. Springer, Berlin Heidelberg New York, pp 264–273

Sklash MG, Farvolden RN (1979) The role of groundwater in storm runoff. J Hydrol 43: 45–65

Sklash MG, Farvolden RN, Fritz P (1976) A conceptual model of watershed response to rainfall developed through the use of oxygen-18 as a natural tracer. Can J Earth Sci 13: 271–283

Sommerfield RA, Judy C, Friedman I (1991) Isotopic changes during the formation of depth hoar in experimental snowpacks. In: Taylor HP, O'Neil JR, Kaplan IR (eds) Stable isotope geochemistry: a tribute to Samuel Epstein. The Geochemical Society, San Antonio, pp 205–211

Souchez RA, Jouzel J (1984) On the isotopic composition in δD and $\delta^{18}O$ of water and ice during freezing. J Glaciol 30: 369–372

Stichler W, Rauert W, Martinec J (1981) Environmental isotope studies of an alpine snowpack. Nordic Hydrol 12: 297–308

Stockbridge FB (1986) Chernobyl: the consequences in Europe. Ambio 15: 332–334

Swistock BR, DeWalle DR, Sharpe WE (1989) Sources of acidic storm flow in an Appalachian headwater stream. Water Resour Res 25: 2139–2147

Taylor HW, Hutchison-Benson E, Svoboda J (1985) Search for latitudinal trends in the effective half-life of fallout 137_{Cs} in vegetation of the Canadian Arctic. Can J Bot 63: 792–796

Taylor HW, Svoboda J, Henry GHR, Wein RW (1988) Post-Chernobyl ^{134}Cs and ^{137}Cs levels at some localities in northern Canada. Arctic 41: 293–296

Wels C, Taylor CH, Cornett RJ, Lazerte BD (1991) Streamflow generation in a headwater basin on the Precambrian shield. Hydrol Processes 5: 185–199

III Nutrient and Carbon Fluxes

9 Surface Water Chemistry and Hydrology of a Small Arctic Drainage Basin

K. R. Everett, D. L. Kane, and L. D. Hinzman

9.1 Introduction

Spatial and temporal variation in the composition of water in ecosystems is governed by many biogeochemical processes that differ in magnitude, direction, and rate. An understanding of water fluxes and water chemistry is critical to predicting ecosystem dynamics in arctic landscapes (cf. Bormann and Likens 1979; Molenaar 1987; Chapin et al. 1988; Swanson et al. 1988; Chaps. 17 and 18, this Vol.). Changes in the chemical composition of water that affect ecosystem function within the Imnavait Creek watershed may be examined with regard to biogeochemical processes and background atmospheric inputs for two distinctly different periods, during snowmelt and the post-snowmelt summer growth season.

Hydrological transport of materials within the Imnavait Creek basin is coupled to exchanges in distinct soil-vegetation units (or established plant communities) as water passes along the topographic gradient from ridge crests to the valley bottom. The plant communities (see Fig. 4.8 in Chap. 4; Tables 5.1 and 5.5 in Chap. 5, this Vol.) include the heath vegetation on the ridgetops, tussock tundra on the upper and lower backslopes, and the riparian and wet sedge meadows adjacent to Imnavait Creek. In addition, the slopes are characterized by numerous water tracks (Everett and Ostendorf 1988; Chap. 4, this Vol.) that act as conduits. Thus, transport – via either overland flow or flow within the active layer above the permafrost – occurs between adjacent communities, between a community and water flowing in a water track, or between a community and Imnavait Creek (Fig. 9.1). Following the conceptual model presented in Fig. 9.1, the chemistry of the water moving as overland flow within the active layer and in the stream at the valley bottom was analyzed. The observation series is one of the longest and most complete records assembled for a watershed in a tundra region. Based on these data records we summarize the element flows occurring in the tundra landscape at Imnavait Creek during the summer growth period.

9.2 Watershed Instrumentation

The catchment of Imnavait Creek comprises ca. 210 ha (see Fig. 1.1 in Chap. 1, this Vol.). The headwaters of the creek are found in a nearly level string bog, or

J.F. Reynolds and J.D. Tenhunen (Eds.)
Ecological Studies, Vol. 120
© Springer-Verlag Berlin Heidelberg 1996

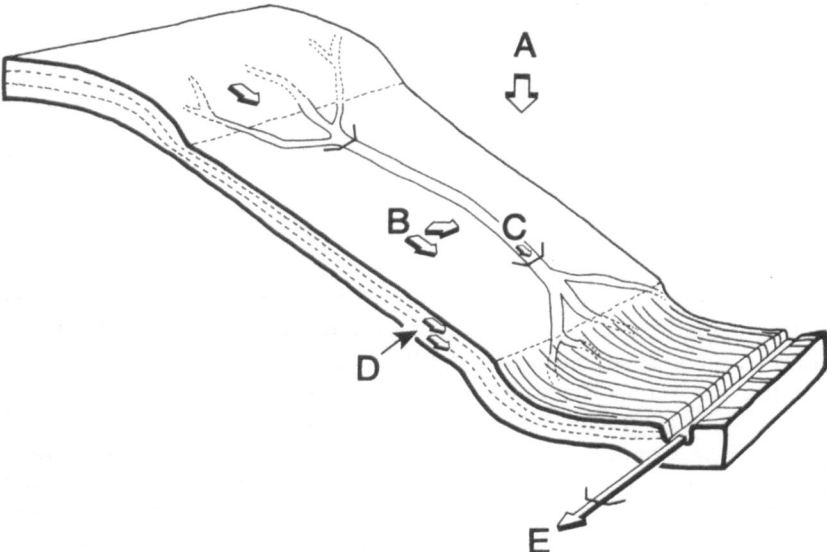

Fig. 9.1. Idealized section of the tundra slope illustrating structure and indicating relationships among the fluxes for which water chemistry was determined at Imnavait Creek. *A*, Atmospheric input; *B*, overland flow; *C*, water-track flow (diagram shows the diffuse nature of the tracks both at the ridge crests and in the riparian meadows; water-track paired weirs installed on channeled portion indicated by symbols); *D*, active-layer flow (*dashed lines* indicate the levels of the organic-mineral soil interface and permafrost); *E*, Imnavait Creek flow (flume indicated)

strangmoor, with many poorly defined and interconnecting waterways (Oswood et al. 1989; Chap. 13, this Vol.). In the lower part of the valley the strangmoor narrows and the creek gradually develops a single, relatively straight channel in a narrow floodplain. The slopes flanking the creek are underlain by frozen acid till (pH ca. 4) and are occupied by a sequence of plant communities that develop above- and belowground structure in correlation with topography, water availability, and degree of water saturation (Chap. 5, this Vol.).

The principal measurement sites, transects, and instrumentation for hydrological and geochemical monitoring within the intensive research site (IRS) of the watershed are shown in Fig. 9.2. The primary monitoring device for the catchment was a 1-m "H" flume with wing dams that channeled the discharge of Imnavait Creek. A stationary water-level gauge and a Stevens water-level recorder were affixed to the flume. There were two meteorological stations, a wetfall–dryfall collector, a Wyoming snow gauge, four overland flow/shallow subsurface flow plots, and a transect for sampling soil solution in the different communities. Several water tracks were monitored with pairs of weirs. One of each pair was placed high on the slope at the microtopographic break, separating the upslope heath tundra from tussock tundra of the upper and lower

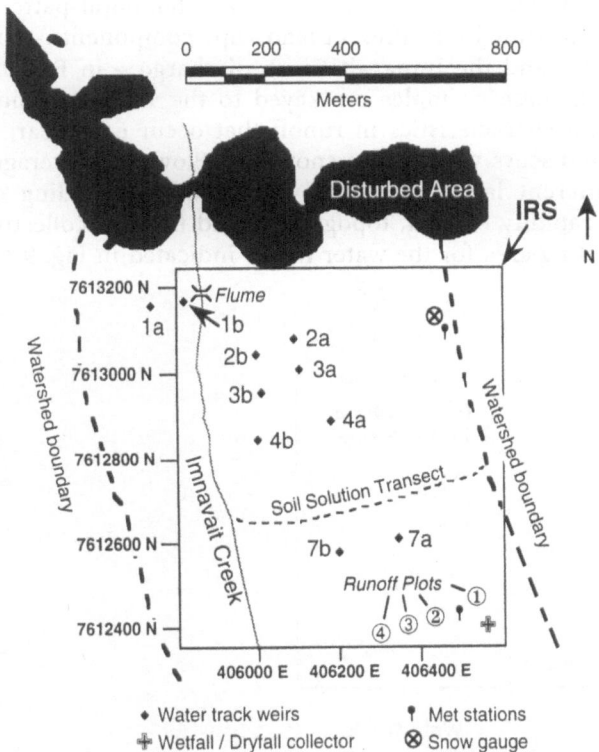

Fig. 9.2. The principal measurement sites, transects, and instrumentation for hydrological and geochemical monitoring within the intensive research site (IRS) of the watershed (see also Fig. 1.1 and figures in Chap. 4, this Vol.)

backslopes (e.g., weirs 1a, 2a, 3a, 4a, and 7a in Fig. 9.2). The lower weirs were placed at the last constriction in the channel before moving from the toe slope into the riparian zone of Imnavait Creek, where water tracks become multichanneled (e.g., weirs 1b, 2b, 3b, 4b, and 7b in Fig. 9.2).

9.3 Snowmelt Period

9.3.1 Snowmelt Hydrology

The snowpack in spring is spatially variable in thickness (Fig. 6.3, this Vol.), density (Liston 1988), and chemical composition (Zukowski 1990), due to redistribution of snow by strong winter winds. Snowmelt runoff in the watershed usually begins with the start of snowpack ablation in early May, and is terminated with the disappearance of all nondrift snow cover in late

May to early June (Liston 1988; Chap. 6, this Vol.). The temporal patterns of snowmelt runoff are illustrated for different landscape components – tussock tundra, water tracks, and the Imnavait Creek discharge – in Fig. 9.3. Although the patterns in these examples are keyed to the 1989 snowmelt, they illustrate the general characteristics in runoff that occur each year; in Chapter 6 (this Vol.) we discuss variation in snowmelt. Flow rates averaged over the surface of different landscape units vary greatly depending on snowpack distribution, rapidity of melt, topography, and upslope collecting area. The exact contributing area for the water tracks indicated in Fig. 9.3 is difficult to define.

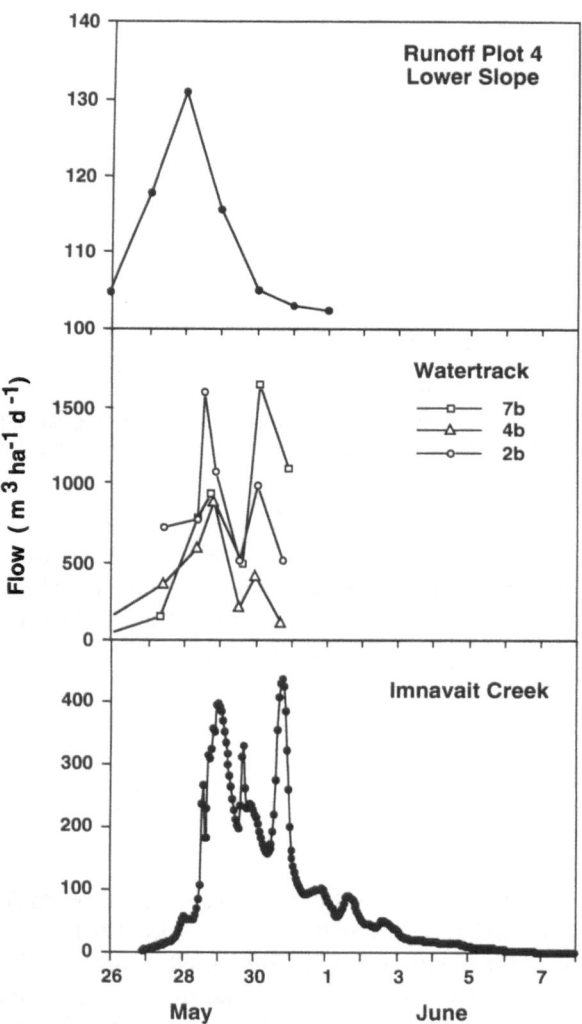

Fig. 9.3. Snowmelt hydrographs in 1989 for overland, shallow active-layer flow in 90 m² runoff plot 4; for water tracks 2b (collecting area estimated as ca. 0.1 ha), 4b (ca. 0.2 ha), and 7b (ca. 0.33 ha); and for Imnavait Creek catchment (ca. 218 ha). Locations of plots, weirs, and flume shown in Fig. 9.2

Three periods of snowmelt can be differentiated. Phase I is defined by the ascending limb of the discharge hydrograph beginning with the start of flow until the first major discharge peak is attained (compare Fig. 9.3: May 28–29 at different measurement locations). Phase II includes all discharge during peak flow. Variation in flow during phase II occurs due to short-term changes in weather conditions and diurnal temperature cycles. Phase III is viewed as the period during final recession of the snowmelt hydrographs. A shift in these phases occurs between runoff plots (data for plot 4, which is low on the slope, is shown in Fig. 9.3), water tracks, and Imnavait Creek, which reflects the difference in timing of melt between the southwest- and northeast-facing slopes of the watershed as well as the melting of snowbanks in the larger water tracks. For example, by noon on 30 May 1989, the catchment of water track 2b was essentially snow-free, and little snow remained in the track; in contrast, the catchment at 7b was approximately 85% snow-free, but the track itself retained considerable snow. Late-melting snow causes a protracted recessional limb (phase III) of the Imnavait Creek hydrograph.

9.3.2 Snowmelt Chemistry

9.3.2.1 Overland Flow

Overland flow commences once the snowpack has become isothermal and the pore spaces in the upper few centimeters of the soil beneath the snowpack are filled. These pores are a result of upward heat flux from the soil that moves moisture out of the near-surface materials, causing recrystallization of the lower 10–15 cm of the snowpack (Woo 1986). Kane et al. (1989) estimated that the pore space in the Imnavait Creek basin represented a storage capacity of ca. 15 mm of water. The chemistry of water reaching and filling these pores differs significantly from that of the snow because of chemical fractionation (Johannessen and Henricksen 1978; Brimblecombe et al. 1985; Tsiouris et al. 1985; Tranter et al. 1987). At Imnavait Creek, Zukowski (1990) found concentrations for Cl^-, NO_3^-, and SO_4^{2-} in initial snowmelt water to be three-to-nine times higher than the concentrations in the snowpack, with most of the enrichment taking place during the first 40% of the ablation period.

Subnival meltwater chemistry (phase I) differed from both snow and later runoff chemistry (runoff plot 4 in Table 9.1). At this time K^+ was elevated, probably due to desorption from organic exchange sites as well as release from lysed cells of leaf litter, where it is in soluble ionic form (Gauch 1972). An enrichment in Mg^{2+} relative to Ca^{2+} was also observed. There was a slight but nonsignificant decrease in both NO_3^- and SO_4^{2-} concentrations in phase I. Adsorption on organic exchange sites or biological uptake may occur, because respiration of microbes and roots has been measured at temperatures as low as $-0.3\,°C$ (Chapin 1974; Flanagan and Bunnel 1980). High concentrations of dissolved organic carbon (DOC) were present during early snowmelt, also

Table 9.1. Composition of runoff at different times and locations (see Fig. 9.2). Phases as described in Section 9.3.1. Snowmelt occurred from June 1–11 and post-snowmelt is June 12–September 7, 1986. Mean ion concentrations in $\mu Eq\,l^{-1}$; other units as noted. Concentrations with asterisks are statistically different ($p < 0.05$) from concentration to immediate left

	Snow	Overland flow (runoff plot 4)			Early snowmelt (water track 7b)			Water track 7a		Water track 7b		Water track 1a		Water track 1b	
		Phase I	Phase II	Phase III	Phase I	Phase II	Phase III	Melt	Post-melt	Melt	Post-melt	Melt	Post-melt	Melt	Post-melt
pH	6.1	5.7	5.7	6.1	nm[a]	nm	nm	5.1	5.5	5.2	5.4	5.8	6.0	5.7	6.0
EC[b]	0.3	1.2*	1.7	0.8*	2.4	2.0	0.8	nm	nm	nm	nm	nm	nm	nm	nm
Ca^{2+}	10.5	33.9*	25.0*	15.0*	54.9	21.0	79.8	38.8	62.9	41.9	41.9	59.9	77.8	70.9	71.9
Mg^{2+}	1.7	14.0*	20.6*	11.5*	41.1	19.7	4.9	23.0	28.0	29.6	33.7	23.0	53.5	50.2	49.4
Na^+	2.2	11.8*	7.4*	3.5*	14.8	6.5	2.2	5.2	10.6	11.8	T[c]	22.2	26.5	27.8	26.1
K^+	0.3	32.0*	25.6*	15.6*	21.7	27.9	13.8	30.7	5.1	22.5	0.0	26.1	4.1	26.3	26.1
$Fe_{(total)}$	0.7	3.9*	3.9	6.1	3.2	1.1	1.1	nm	nm	nm	nm	nm	nm	nm	nm
Mn^{2+}	T[c]	4.7*	3.3*	2.6	6.9	4.0	2.9	nm	nm	nm	nm	nm	nm	nm	nm
Cl^-	T	13.0*	3.4*	1.7*	T	T	T	5.9	0.9	7.1	1.1	5.1	1.7	4.5	2.0
NO_3^-	5.2	4.5	1.3*	1.0	0.0	0.0	0.0	nm	0.0	nm	0.0	nm	0.0	0.0	0.0
PO_4^{3-}	T	0.0	0.0	1.6*	0.0	0.0	0.0	nm	nm	nm	nm	nm	nm	nm	nm
SO_4^{2-}	6.3	5.2	1.7*	0.4*	5.2	1.0	2.6	0.0	0.0	0.0	3.8	1.0	16.0	1.2	3.4
DOC[d]	0.9	22.8*	19.8	13.4*	25.7	18.1	12.9	nm	nm	nm	nm	nm	nm	nm	nm
Estimated negative charge[e] ($mg\,C^{-1}$)	5.2	3.5	4.1	3.8	5.3	4.4	7.9	ce[f]	ce	ce	ce	ce	ce	ce	ce
Suspended particulates[d]	18.5	9.3	37.3	14.6	nm	nm	nm	8.4	4.4	6.1	3.5	6.1	4.6	6.2	3.9

[a] Not measured.
[b] Electrical conductivity ($\mu S\,cm^{-1}$).
[c] Trace.
[d] $mg\,l^{-1}$.
[e] In $\mu Eq\,mg^{-1}$.
[f] Cannot evaluate.

noted by Oswood et al. (Chap. 13, this Vol.), which appear to be important in the ionic balance. The negative charge estimated for measured concentrations of DOC to balance an otherwise excess of cations ranges from 3.5 to 5.2 μEq mg^{-1}, thus being near literature values for soluble humic acids (Stevenson 1982; Thurman 1985). Dissolved inorganic carbon was measurable only in trace amounts. During the later phases of overland flow, particularly phase III, ionic concentrations generally decreased (Table 9.1). Cadle et al. (1987) reported similar trends during snowmelt in northern Michigan.

9.3.2.2 Water Track Flow

Data on chemical concentrations in water-track samples during the three phases of snowmelt (water track 7b in Table 9.1) indicated an enrichment in ionic concentrations during phase I. As more of the contribution to water-

Table 9.2. Water chemistry of snowmelt flow for Imnavait Creek in 1989. Phases as described in section 9.3.1. Mean ion concentrations in μEq l^{-1}; other units as noted

	Snowpack ablation	Phase I	Phase II	Phase III	Postmelt
Duration (h)	349	41	93	254	
Mean discharge[a]		0.13	0.66	0.07	0.02
pH	6.1	6.1	5.9	6.3	6.3
EC[b]	0.3	0.2	0.1	0.1	0.1
H$^+$	0.8	0.9	1.3	0.5	0.5
Ca^{2+}	10.5	42.4	32.4	20.5	39.2
Mg^{2+}	1.7	32.1	23.0	14.0	23.0
Na$^+$	2.2	22.2	12.2	7.4	8.7
K$^+$	0.3	28.1	18.4	7.9	0.3
Fe$_{(total)}$	nm	2.5	2.1	2.9	2.9
Mn^{2+}	nm[c]	1.8	2.2	1.1	0.4
Cl$^-$	T[d]	0.1	2.0	0.6	T
PO$_4^{3-}$	T	T	T	T	T
SO$_4^{2-}$	3.1	10.0	3.1	0.3	1.9
DOC[e]	0.9	15.4	14.6	10.6	13.2
Estimated negative charge[f] (mg C^{-1})	13.8	7.8	5.9	5.0	5.5
Suspended particulates[g]	18.5	11.6	2.7	18.5	5.1

[a] m^3 s^{-1}.
[b] Electrical conductivity (μS cm^{-1}).
[c] Not measured.
[d] Trace.
[e] mg l^{-1}.
[f] In μEq mg^{-1}.
[g] mg l^{-1}.

track flow is derived from late melt (phase III) in tussock tundra, rather than the snowpack in the water track itself, concentrations decreased (with the exception of Ca^{2+}, which is not understood). Contrasts in water chemistry were apparent for water track 7 (SW-facing slope) and water track 1 (East-facing slope). The differences were greatest during postmelt flow. The abundance of cations in samples from water track 1 likely reflected the greater exposure of mineral soil observed on the East-facing slope.

9.3.2.3 Imnavait Creek Flow

During phase I of snowmelt, subnival flow increases for an extended period and the floodplain snowpack slowly becomes saturated. Open channel flow may begin precipitously as a threshold is exceeded and the saturated mass breaks up, rapidly increasing basin discharge (Fig. 9.3). Characteristics for snowmelt water chemistry in Imnavait Creek during 1989 are shown in Table 9.2. The highest ionic concentrations are observed during phase I. During phase II, the pH of the stream reached a seasonal low. The concentrations of other constituents decreased due to dilution and to reductions in the ionic load of the meltwater from the slopes. By this time, most of the ions that are desorbed early have been flushed from the near-surface soil layer. In most cases, ionic concentrations continue to decrease during phase III. During snowpack ablation and phase I, DOC appears to have a more fulvic acid character with high negative charge (13.8 and 7.8 μEq mg^{-1}), whereas DOC later with a negative charge of 5.0 to 5.9 μEq mg^{-1} appears to have a humic-acid character (for charge values compare Stevenson 1982 and Thurman 1985).

9.4 Post Snowmelt Period

9.4.1 Atmospheric Inputs

When concerns about the global movement of industrial pollutants began to receive worldwide attention in the early 1970s, little was known about airborne trace substances in the polar regions (Heintzenberg 1989). In 1971, the National Oceanic and Atmospheric Administration (NOAA) established a remote monitoring site in Arctic Alaska, near Barrow (Quakenbush and Bodhaine 1986). This site – considered representative of the arctic atmosphere under prevailing winds – has provided a continuous record of background atmospheric conditions since 1976 as part of US Global Monitoring for Climate Change (GMCC). Only sporadic short-term analyses of precipitation, commonly only pH, are available from other arctic sites in Alaska. The R4D program (Chap. 1, this Vol.) offered an opportunity to augment long-term sampling at least at one site in the foothills of the Brooks Range.

We placed an AiroChem-Metric wetfall–dryfall collector in the watershed, 80 m off the northeast ridge crest at an elevation of 945 m asl (Fig. 9.2). The sampler, equipped with a weighing rain gauge and event recorder, is the same as those used in the National Atmospheric Deposition Project program (NADP 1988), and its positioning and timing of sample collection followed NADP procedures (Bigelow 1984). Sample treatment followed Standard Methods (Franson 1985). The collector at Imnavait Creek was operated from mid-May through the first week of September, hereafter referred to as the "summer" period. In the late spring and early fall, rime commonly formed on stems of the woody plants in the area and this was sampled as well.

9.4.1.1 Rainfall

Rainfall patterns and water balance for the Imnavait Creek watershed are described in Chap. 6 (this Vol.). Rainfall is an acidic (pH 4.7–5.3; H^+ of 5–19 $\mu Eq\,l^{-1}$), ionically weak solution. The pH values are near the acid end of the normal pH range for precipitation (Pickering 1987). The equivalent ratios indicate that H_2SO_4 accounted for between 53 and 100% of the free acidity. This may reflect inputs from either southern fires, emissions at Prudhoe Bay, or long-range transport (Nriagu et al. 1991). Dayan et al. (1985) reported that storms of northern origin have SO_4^{2-} concentrations much higher than H^+, and suggested neutralization of H_2SO_4 by salts such as $CaCO_3$ or $MgCO_3$. Although the predominant surface winds during the summer months are from the south and southwest, north and northeast winds, which are associated with additional input of HNO_3, do occur. The role of organic acids in the measured free acidity has not been studied at Imnaviat Creek, but they are likely to be significant. Organic anions (oxalate, formate, acetate, and pyruvate) constituted 42% of the measured anions (Harriss, pers. comm.) in lower tropospheric precipitation over the Alaskan Coastal Plain. Dayan et al. (1985), citing analysis by Keene et al. (1984), indicate organic acids contributed as much as 33% of the anion content of precipitation in east–central Alaska.

Average concentrations for selected chemical parameters in summer rain during 1986 through 1991 are shown in Table 9.3. The high annual variability suggests that extra-regional sources play a significant role in determining the ionic composition. The origin of the various ions remains somewhat equivocal. No clear seasonal patterns were observed that might identify particular sources through correlation with weather cycles. It appears that much of the chloride and sulfate is imported with rainfall associated with marine air masses, whereas cations are augmented from local sources, e.g., calcium and probably potassium from road dust (Chap. 15, this Vol.) or from sources south of the Brooks Range such as forest fires. Fires are an important aperiodic source of ions in rainfall north of the Brooks Range (Shaw et al. 1982; see also Breysse 1983 and Cook et al. 1978).

Several peaks were commonly observed in calcium concentration in rainfall during each summer. These were often inversely proportional to rainfall

Table 9.3. Rainfall weighted mean concentrations (μEq l^{-1}) of ions contained in summer precipitation at Imnavait Creek between May and September in 1986 through 1991

	H$^+$	pH	Ca^{2+}	Mg^{2+}	Na$^+$	K$^+$	NH$_4^+$	Cl$^-$	SO$_4^{2-}$	NO$_3^-$
1986	7.7	5.1	4.0	0.8	4.8	4.6	2.2	4.9	7.2	3.5
1987	5.3	5.3	2.0	1.7	1.7	2.3	5.5	1.4	4.0	3.4
1988	10.0	5.0	13.6	5.0	11.7	3.8	4.4	12.7	11.8	8.7
1989	10.3	5.0	7.5	1.3	1.8	0.3	Trace	9.0	2.9	Trace
1990	19.0	4.7	12.5	4.1	7.8	12.8	12.8	41.0	15.9	4.0
1991	5.8	5.2	12.5	3.3	14.4	5.4	12.2	19.0	12.7	9.2

volume. Ca^{2+} is in much higher concentration relative to Mg^{2+} than can be expected from a natural loess of the region. The most likely source is the Dalton Highway adjacent to the trans-Alaskan pipeline. Calcium and potassium originating in road dust are also often important components of dry deposition. Studies over more than a decade have shown that large volumes of calcium-rich dust are generated from the highway each year (Everett 1980; Chap. 15, this Vol.). Dust loads of $1.2\,\mathrm{g\,m^{-2} \cdot yr^{-1}}$ have been measured at 1 km distance from the road (Everett 1980), and trace amounts may reach 5 km. (The entire catchment is within this area of influence.)

9.4.1.2 Dry Deposition

Dry deposition is an important source of ion input. Uncertainty with regard to mechanisms of deposition and the potential for contamination in passive collectors have limited our understanding and interpretation of dry deposition estimates. Data for 1986 and 1988 – years in which there appeared to be no bucket contamination – are presented in Table 9.4. These are the first dry deposition data that have been collected in northern Alaska, except for air chemistry data at the Barrow GMCC site (Quakenbush and Bodhaine 1986). Substantial differences were observed between the 2 years in element concentrations. It is likely in both years that road dust was a principal component with regard to calcium and potassium deposited. In 1988 the large inputs of NH$_4^+$, SO$_4^{2-}$, and NO$_3^-$ may have originated from southern forest fires, a speculation that is reinforced by the elevated levels of these ions in rainfall during that summer. Dry deposition may be responsible for significantly larger inputs of Ca^{2+}, K$^+$, and other ions than rainfall. The dry deposition rate for SO$_4^{2-}$ was nearly double the deposition rate in rainfall. Dry deposition must be considered a significant component in the geochemical mass balance of the Imnavait Creek watershed, and must be further studied in tundra regions in general.

Table 9.4. Dryfall and rime elemental composition ($\mu Eq\,l^{-1}$) during the summer season at Imnavait Creek watershed

	Dryfall elemental composition[a]		Partial ionic composition of rime
	1986[b]	1988[c]	1988
pH	7.6	6.2	5.9
EC[d]	1.6	0.98	3.6
H^+	0.02	0.7	1.3
Ca^{2+}	137.7	26.9	10.5
Mg^{2+}	44.4	9.1	2.5
Na^+	7.0	2.6	2.6
K^+	10.6	14.1	0.3
NH_4^+	1.7	43.2	–
Cl^-	2.8	5.4	Trace
F^-	0.53	1.05	0.0
NO_3^-	1.1	6.9	9.4
PO_4^{3-}	0.9	17.7	–
SO_4^{2-}	2.7	10.8	11.9

[a] For the wash liquor in years free from contamination.
[b] June 3–September 1.
[c] May 12–September 18.
[d] Electrical conductivity ($\mu S\,cm^{-1}$).

9.4.1.3 Rime

The collection and analysis of rime at Imnavait Creek was undertaken in 1988 to examine its input potential, particularly with regard to sulfate and nitrate (Table 9.4). The formation of rime takes place primarily in late May and in late August in the Imnavait Creek area, when surface and air temperature differences are large. The number of rime events varies greatly from year to year. Our observations suggest that rime should not be ignored, because the quantity of SO_4^{2-} and NO_3^- amounted to ca. $0.5\,kg\,ha^{-1}$ or 8% of the total summer input for both SO_4^{2-} and NO_3^-. These estimates are based only on deposition on woody shrubs and, hence, under-represent the total input.

9.4.2 Water Chemistry

9.4.2.1 Overland Flow

During sustained or intense rains with high antecedent moisture, the soil and vegetation above permafrost may saturate, and short episodes of overland flow may occur during the summer. The chemistry of water during such periods of overland flow has been examined during only one period (see

Table 9.5), due to the unpredictable occurrence of the required sequence of events.

9.4.2.2 Active Layer Flow

Rainfall that produces flow in the active layer and in water tracks is common. The ionic contribution of rainfall is variable (see Sect. 9.4.1), and as it passes into the soil surface horizons, its acidity is partially neutralized (Table 9.5). For the period described by Table 9.5, the contribution of K^+ by rainwater was approximately sufficient to account for K^+ concentration in surface flow; the high K^+ concentration in soil solution is probably derived through leaching and cell lysis. Cl^- was the only other ion contributed by rainfall in significant amounts to all flow compartments. Although Ca^{2+} and Mg^{2+} are deposited with most rain events, these ions are also major components of dry deposition (Everett 1989; Everett et al. 1989); both are released to surface flow through leaching of the vegetation and acid dissolution of airborne carbonates.

$Fe_{(total)}$ and Mn^{2+}, as well as silica (Marion and Everett 1989), were measurable in nearly every soil solution sample. The solubilities of both Fe and Mn depend strongly on the redox state and pH of the soil. Flow at the surface and near surface of the watershed contained dissolved oxygen levels of 9.0–11.5 mg l^{-1} (M. Oswood, pers. comm.), and both $Fe_{(total)}$ and Mn^{2+} are largely in insoluble forms. Relatively high concentrations in soil solution at 40 cm depth attest to widespread anoxic conditions, where both $Fe_{(total)}$ and Mn^{2+} are reduced and possibly complexed with organic acids, e.g., oxalic acid in the organic-rich soils.

Both SO_4^{2-} and NO_3^- appear only sporadically in water track and Imnavait Creek flow after the onset of biological activity after snowmelt. Concentrations of these ions increase in late August (Table 9.5) as senescence occurs (Chaps. 10 and 11, this Vol.). In the upper 20 cm, SO_4^{2-} concentration may depend on the oxidation of cell proteins, whereas NO_3^- concentration originates from mineralization and nitrification of organic materials. If exchange takes place between the 20- and 40-cm depth in the soil at all, it is very slow, and the concentrations of SO_4^{2-} and NO_3^- are proportional to their amounts in the solid phase. Concentrations of DOC are similar to those of Imnavait Creek, constituting a significant component of total carbon transport from the catchment (see Chap. 13, this Vol.) and a significant contribution to maintenance of the ionic balance.

9.4.2.3 Imnavait Creek Flow

After snowmelt, flow in Imnavait Creek may decrease to essentially zero for protracted periods (e.g., during summer 1986 as shown in Fig. 17.7, this Vol.). Subsequent flow is determined by the frequency of major rainstorms, either single events that produce a distinct three-phase response hydrograph or

Table 9.5. Comparison of chemical composition of rain and water flowing at surface and subsurface locations as indicated for the period August 2–24, 1989. Soil solution samples taken along transect shown in Fig. 9.2 (values in $\mu Eq\,l^{-1}$)

	H^+	Ca^{2+}	Mg^{2+}	Na^+	K^+	$Fe_{(total)}$	Mn^{2+}	Cl^-	NO_3^-	SO_4^{2-}	DOC^f	Negative charge $(mg\,C^{-1})$	Suspended particulates[f]
Rainfall	12.6	5.0	0.8	1.3	0.57	nm[g]	nm	5.9	0.0	0.94	nm	ce[h]	nm
Overland flow[a]	0.8	31.4	13.2	9.6	0.77	7.9	0.36	11.0	9.4	7.8	15.4	2.3	5.9
Soil solution[b]	nm	79.8	33.7	30.0	4.09	7.9	4.73	3.9	8.9	3.7	nm	ce	37.7
Soil solution[c]	nm	112	60.0	33.5	4.35	29.3	10.2	11.6	16.1	3.7	nm	ce	37.7
Water-track flow[d]	0.8	21.5	30.4	8.3	0.26	4.0	T[i]	13.0	2.6	5.6	13.0	3.4	4.4
Imnavait Creek[e]	0.8	40.9	24.7	2.6	0.51	4.6	T	14.4	3.7	5.3	12.9	3.9	5.0

[a] Includes shallow subsurface flow lining vegetation mat and fibrous organic soil horizon at runoff plot 4 ($n = 3$).

[b] 20 cm depth; $n = 21$.

[c] 40+ cm depth; $n = 21$.

[d] $n = 16$.

[e] $n = 13$.

[f] $mg\,l^{-1}$.

[g] Not measured.

[h] Cannot evaluate.

[i] Trace.

multiple events spaced close enough that the recessional phase is not complete before the next discharge begins. Time-dependent changes in ion concentrations of single events show patterns similar to those described for snowmelt (Sect. 3).

The mean concentrations for all ions were relatively constant, although unusual situations with regard to individual ions are found in particular years (Table 9.6). Further research is required to explain long-term variations. A strong consistency was found in the seasonal pattern for Ca^{2+} concentration in water of Imnavait Creek. Ca^{2+} increased during the low discharge period in mid-summer and decreased again in the fall. A mirror image of this pattern was observed for $Fe_{(total)}$. An extremely large increase in Ca^{2+} was observed during summer 1990, due apparently to unusually low discharge during that year.

9.5 Conclusions

Ion input from the snowpack comprises ca. 20–40% of the total annual ion input, whereas rainfall accounted for 60–80%. Snow was an important source for the addition of NO_3^- and SO_4^{2-}. Input of specific ions by precipitation varies considerably among sampling periods within a given year and between years. During summers with major forest fires (e.g., 1990) many ions and organic acids appear to be imported.

Snowmelt constituted the most important hydrological and geochemical event of any year, accounting for between 50 and 75% of the seasonal discharge of water, and between 60 and 100% of the total seasonal efflux of individual ions, much of this efflux taking place in the event's early phases. The composition of snowmelt runoff differs from the composition of snow because of chemical fractionation. During overland and shallow subsurface flow, composition is further modified by sorption and desorption of solutes from the substrate and leaching of substrate materials. The concentration of cations in discharge decreases progressively during the course of snowmelt. Only dissolved organic carbon and suspended solids increase, which are probably derived from the channel and floodplain.

After the recessional phase of melt, flow may fall to very low levels or cease altogether until protracted rainfall reestablishes flow. Ionic enrichment of the water increases with depth in the active layer, being influenced by contact with the organic substrates. Heavy rains initiate short-duration peak flows during which the patterns in water track and stream chemistry resemble that during snowmelt. Nitrate, and to a lesser degree, sulfate, are retained by the vegetation and thawing soil; neither they nor chloride are significant in stream flow after the first several days of snowmelt.

Geochemical exchange between and among landscape units is most extensive during the snowmelt event, but this takes place at or near the surface. As

Table 9.6. Mean concentration (μEq l^{-1}) of selected ions and other variables measured in Imnavait Creek discharge during 1985 through 1990

Year	pH	EC[a]	H⁺	Ca²⁺	Mg²⁺	Na⁺	K⁺	Fe(total)	Mn²⁺	Cl⁻	NO₃⁻	SO₄²⁻	DOC[b]	Negative charge (mgC⁻¹)	Suspended particulates[b]
1985	5.8	0.10	1.4	54.4	27.2	8.7	3.32	nm[c]	nm	16.6	0.0	0.00	nm	ce[d]	nm
1986	5.9	0.15	1.4	65.9	35.4	4.4	1.79	nm	nm	3.67	1.3	4.16	10.5	9.5	4.1
1987	5.8	0.13	1.7	58.9	31.3	7.9	2.29	15.7	1.45	4.80	0.0	0.42	9.8	11.6	4.7
1988	5.9	0.24	1.2	49.4	23.0	7.8	3.58	19.6	1.18	14.4	0.65	1.56	12.2	7.3	6.8
1989	6.3	0.10	0.54	38.9	21.4	7.4	4.09	3.2	0.34	T[e]	2.1	2.19	12.0	6.0	6.9
1990	5.6	0.15	2.7	71.9	29.6	13.5	md[f]	32.5	36.4	13.8	7.0	12.7	11.9	12.9	9.7

[a] Electrical conductivity (μS cm^{-1}).
[b] mg l^{-1}.
[c] Not measured.
[d] Cannot evaluate.
[e] Trace.
[f] Missing data.

summer progresses this near-surface exchange no longer occurs, but geochemical isolation of communities occurs, except at locations directly adjacent to drainage channels. Precipitation events that are able to fill and exceed soil storage, reestablish hydrological transport among communities for varying lengths of time. The tundra at Imnavait Creek is capable of retaining almost all inorganic anions added from the atmosphere, but a consistent loss of cations occurs during melt discharge and throughout postmelt flows. (The only exception is a net retention of potassium during postmelt.) The loss of cations in summer discharge is offset by organic anions derived mostly from decomposition of organic material. Thus, in most years the Imnavait Creek watershed accumulates imported anions, but exports considerable quantities of calcium, magnesium, iron, and manganese together with carbon.

Acknowledgments. Support for these studies was made possible by grants from the U.S. Department of Energy under the R4D program. B. Ostendorf, D. Luchini, R. Gieck, D. Marrett, and J. Hambrook assisted with sample collection and preparation. Laboratory analyses at Ohio State University was carried out by L. Everett. R. Artz supplied precipitation chemistry for the Barrow GMCC site. D. Barry of the Alaska Fire Service provided forest fire data for the summer of 1988.

References

Bigelow DS (1984) NADP-NTN instruction manual site selection and installation. NADP/NTN Coordination Office, Natl Res Ecol Lab, Colorado State Univ, Fort Collins, 23 pp

Bormann FH, Likens GE (1979) Pattern and processes in a forested ecosystem. Springer, Berlin Heidelberg New York, 253 pp

Breysse PA (1983) Smoke inhalation is hazardous to your health. Soc Am For, Portland, Oregon, pp 170–172

Brimblecombe P, Tranter M, Abrams PW, Beachwood J, Davies TP, Vincent JE (1985) Relocation and preferential elution of acid solute through the snowpack of a small, remote, high latitude Scottish catchment. Ann Glaciol 7: 141–147

Cadle SM, Dasch JM, Kopple RC (1987) Composition of snowmelt and runoff in northern Michigan. Environ Sci Tech 21: 295–299

Chapin SF III (1974) Morphological and physiological mechanisms of temperature compensation in phosphate absorption along a latitudinal gradient. Ecology 55: 1180–1198

Chapin FS III, Fetcher N, Kielland K, Everett K, Linkins AE (1988) Productivity and nutrient cycling of Alaskan tundra: enhancement by flowing soil water. Ecology 69: 693–702

Cook JD, Himel JH, Moyer RH (1978) Impact of forestry burning upon air quality: a state-of-the knowledge characterization in Washington and Oregon. US Environ Protect Agency Rep 910/9-78-052: 58–82

Dayan U, Miller JM, Keene WC, Galloway N (1985) An analysis of precipitation chemistry data from Alaska. Atmos Environ 19: 651–657

Everett KR (1980) Distribution and properties of road dust along the northern portion of the haul road (Dalton Highway), chap 3. In: Brown J, Berg R (eds) Environmental Engineering and Ecological Baseline Investigations Along the Yukon River–Prudhoe Bay Haul Road US Army Cold Regions Res and Eng Lab (CRREL) Rep 80-19: 101–118

Everett KR (1989) Seasonal precipitation chemistry at Imnavait Creek, Alaska. Proc 40th Arct Sci Conf, Inst Arct Biol, Fairbanks, Alaska, 10 pp

Everett KR, Ostendorf B (1988) Hydrology and geochemistry of a small drainage basin in upland tundra, northern Alaska. In: Sennesett K (ed) Proc 5th Int Conf Permafrost, Tapir, Trondheim, Norway, pp 574–579

Everett KR, Marion GM, Kane DL (1989) Seasonal geochemistry of an arctic tundra drainage basin. Holarct Ecol 12: 279–289

Flanagan PW, Bunnell FL (1980) Microflora activities and decomposition. In: Brown J, Miller PC, Tieszen LL, Bunnell FL (eds) An arctic ecosystem: the coastal tundra at Barrow, Alaska. Dowden, Hutchinson and Ross, Stroudsburg, pp 291–334

Franson E (ed) (1985) Standard methods for the examination of water and wastewater, 16th edn. Am Publ Health Assoc, Washington, DC, 1268 pp

Gauch HC (1972) Inorganic plant nutrition. Dowden, Hutchinson and Ross, Stroudsburg, 488 pp

Heintzenberg J (1989) Arctic haze: air pollution in polar regions. Ambio 18: 50–55

Johannessen M, Henricksen A (1978) Chemistry of snow meltwater: changes in concentration during melting. Water Resour Res 14: 615–619

Kane DL, Hinzman LD, Benson CS, Liston GE (1989) Snow hydrology of a headwater arctic basin. 1. Physical measurements and process studies. Water Resour Res 27: 1111–1121

Keene WC, Galloway JN, Holden JD (1984) Organic acidity in precipitation from remote areas of the world. Proc Symp Recent Advances in Pollution Monitoring of Ambient Air and Stationary Sources, Rep EPA 600/9-84-006, US Environ Protect Agency, Research Triangle Park, NC

Liston GE (1988) Seasonal snowcover of the foothills region of Alaska's Arctic Slope: a survey of properties and processes. MS Thesis, Univ Alaska, Fairbanks, 123 pp

Marion GM, Everett KR (1989) The effect of nutrient and water additions on elemental mobility through small tundra watersheds. Holarct Ecol 12: 317–323

Molenaar JG de (1987) An ecohydrological approach to floral and vegetational patterns in arctic landscape ecology. Arct Alp Res 19: 414–424

Nriagu JO, Coker RD, Barrie LA (1991) Origin of sulphur in Canadian Arctic haze from isotope measurements. Nature 349: 142–145

Oswood MW, Everett KR, Schell DM (1989) Some physical and chemical characteristics of an arctic beaded stream. Holarct Ecol 12: 290–295

Pickering RJ (1987) Acid Rain. Water Fact Sheet. US Geol Surv Open File Rep 87-399

Quakenbush TK, Bodhaine BA (1986) Surface aerosols at the Barrow GMCC Observatory: data from 1976 through 1985. NOAA Data Rep ERL-ARL-10, NOAA, pp 1–13

Shaw RW Jr, Binkowski FS, Courtney WJ (1982) Aerosols of high chlorine concentration transported into central and eastern United States. Nature 296: 229–231

Stevenson FJ (1982) Humus chemistry. Wiley, New York, 395 pp

Swanson FJ, Kratz TK, Caine N, Woodmansee RG (1988) Landform effects on ecosystem patterns and processes. Bioscience 38: 92–98

Thurman EM (1985) Organic geochemistry of natural waters. Nijhoff Junk, Dordrecht, 476 pp

Tranter M, Abrahams PW, Blackwood I, Davies TD, Brimblecombe P, Thompson JP, Vincent CE (1987) Changes in streamwater chemistry during snowmelt. In: Jones HG, Orville-Thomas WJ (eds) Seasonal snowcovers: physics, chemistry, hydrology. Reidel, Dordrecht, pp 575–597

Tsiouris S, Vincent CE, Davies TD, Brimblecombe P (1985) The evolution of ions through field and laboratory snowpacks. Ann Glaciol 7: 196–201

Woo M-K (1986) Permafrost hydrology in North America. Atmosphere-Ocean 24: 201–234

Zukowski MD (1990) A study of northern Alaskan snow chemistry. MS Thesis, Univ Alaska Fairbanks, 125 pp

10 Nutrient Availability and Uptake by Tundra Plants

J. P. Schimel, K. Kielland, and F. S. Chapin III

10.1 Introduction

Although tundra in the Imnavait Creek watershed is exposed to low temperature, a short growing season, and in many cases, anaerobic soils (see Chaps. 4,
6, and 11, this Vol.), nutrient availability is the factor that most strongly limits
plant growth and productivity (Billings et al. 1984; Chapin and Shaver 1985).
Nitrogen (N) is the most common limiting element in tundra communities
(Barsdate and Alexander 1975), but phosphorus (P) may be either a sole or co-
limiting nutrient (McKendrick et al. 1980; Shaver and Chapin 1986). Nitrogen
fixation is slow (Alexander and Schell 1973), decomposition and mineralization are limited by cold soils (Nadelhoffer et al. 1991), and N and P immobilization are rapid (Kielland 1990), suggesting that competition from soil
microorganisms may limit nutrient availability to plants.

To understand plant nutrient uptake, studies must focus on the release of
nutrients into available forms (organic and inorganic) and the partitioning of
those nutrients between plants and microbes. In this chapter we examine the
processes controlling nutrient supply and uptake in tundra plants. We discuss
how these processes in the Imnavait Creek watershed vary among communities distributed along the toposequences described by Walker and Walker (Fig.
4.9 in Chap. 4, this Vol.), and how these processes are affected by various
disturbances.

10.2 Controls on Mineralization and Nutrient Supply

10.2.1 Patterns of Nutrient Supply in the Soil

In tundra and bog soils, N and P are usually rapidly released after the soil thaws
(Barel and Barsdate 1978; Gersper et al. 1980; Whalen and Cornwell 1985;
Kielland 1990; A. E. Giblin, pers. comm.). This spring flush represents a large
proportion of the annual nutrient flux and is a result of mineralization – rather
than of nutrient release from the melting snowpack – because nutrients in the
precipitation account for only ca. 5% of the total N flux in the soil (Whalen and
Cornwell 1985; Kielland 1990; Chap. 9, this Vol.). At breakup most of the
nutrients in the snowpack run off into streams and lakes (Hinzman et al. 1991;

J.F. Reynolds and J.D. Tenhunen (Eds.)
Ecological Studies, Vol. 120
© Springer-Verlag Berlin Heidelberg 1996

Chap. 9, this Vol.). Nutrient concentrations are lowest during midsummer, a period when both plants and microbes grow rapidly. Soil phosphate concentrations increase again in the fall (Barel and Barsdate 1978), probably because of reduced microbial uptake, microbial death, leaching of plant litter, or some combination of these processes.

10.2.2 Patterns of Mineralization

Net N mineralization rates in tundra soils are generally low relative to those in temperate soils (Nadelhoffer et al. 1991). Kielland (1990) and Giblin et al. (1991) measured nutrient accumulation in soil cores contained in plastic bags and incubated in situ (the buried bag method) to assess N dynamics in tundra communities, ranging from upland dry heath to wet sedge meadows. Both studies found major differences in the patterns of N cycling across the landscape. Giblin et al. (1991) found the greatest annual mineralization in the driest and the wettest sites: hilltop heath and wet sedge meadow (Fig. 10.1). The only site where N mineralization occurred during the growing season was the hilltop heath; in all other sites N was immobilized during the summer. Kielland (1990), in contrast, found relatively rapid mineralization during the growing season in midslope tussock and shrub communities, whereas mineralization was slow in heath tundra. These differences may have resulted from different parent materials in each site, because Kielland's site was more acid (tussock-

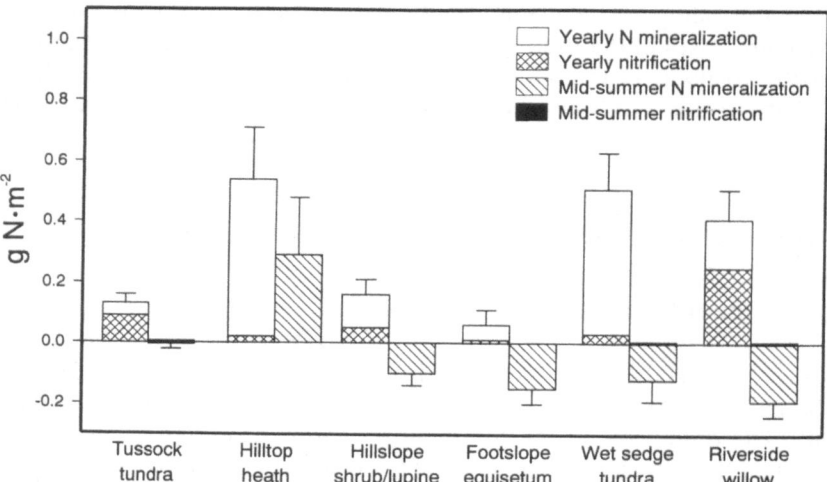

Fig. 10.1. Annual and seasonal patterns of net N mineralization and nitrification in soils along a toposequence of the Sagavanirktok River valley from upland tussock tundra, through heath communities at the top of the slope, down to wet meadow sedge and riverside willow communities. Yearly mineralization and nitrification are represented by means of 4 years at each site (3 years at wet sedge tundra). Midsummer estimates are based on soil cores incubated during 1985 from 9 July to 27 August (7 weeks). (From Giblin et al. 1991)

tundra soil pH 3.7) than the one described by Giblin et al. (tussock tundra soil pH = 5.5).

Seasonal patterns of net mineralization were generally similar to those of inorganic N accumulating on ion-exchange resin bags. Nitrogen accumulated on the resins in all sites during the summer (Giblin et al. 1991), indicating that significant N turnover occurs during the summer, despite the net immobilization measured by the buried bags. Nitrogen also accumulated on the bags over the winter, suggesting nutrient mineralization during the fall freeze-up or the spring thaw. Marion et al. (1982) found that in midslope tussock tundra of the Imnavait Creek watershed ^{15}N appeared to move from soil into plant pools over the growing season, indicating N mineralization and plant uptake during the growing season. These results indicate that while gross mineralization continues, microbial demand for N can immobilize all the mineralized N.

Phosphorus availability may be even more of a function of the winter nutrient release. In the soils in which P accumulated on resin bags, most accumulated from August through May. Microbial biomass contains a very large proportion of total soil P (Giblin et al. 1991); P release from biomass is probably an important source for plant uptake. The large microbial P store also suggests that microbial competition for P may be intense.

10.2.3 Controls on N and P Mineralization

Two major nutrient pools exist in soil: (1) small soluble compounds and polymers, such as protein, that are taken up and metabolized by microorganisms, and (2) large polymers that must be broken down by extracellular enzymes before microbial assimilation. To predict and model nutrient availability and mineralization in soil (Chap. 16, this Vol.) one must consider both pools. The dynamics of the labile pool may be closely associated with the dynamics of microbial biomass (Myrold 1987). Nutrient mineralization from complex soil organic matter, however, may be controlled primarily by depolymerization, which is carried out by extracellular enzymes (Linkins et al. 1984).

Microbial biomass rarely constitutes more than 3–4% of total soil organic matter (Smith and Paul 1986), but it often makes up a significant portion of the active nutrient pool (Myrold 1987) and can act as a major source or sink of nutrients (Paul and Voroney 1984). Tundra soils, such as those at Imnavait Creek, may have a similar proportion of their carbon (C) and N in microbial biomass (Cheng and Virginia 1993). This biomass may vary by an order of magnitude during the year, increasing in early summer and dropping in the fall. The associated nutrient flush could provide as much as 90% of the vegetation's annual P demand (Chapin et al. 1978; Bunnell et al. 1980), and a smaller percentage of the N demand (Flanagan and Van Cleve 1977).

10.2.4 Controls on Decomposition and Mineralization

10.2.4.1 Temperature

Tundra soils are cold and undergo regular freeze–thaw cycles. The low temperatures reduce evapotranspiration and produce permafrost. The result is soils that are often wet, except on ridgetops (Chaps. 5 and 11, this Vol.). The combination of cold and wet leads to the accumulation of organic matter and the slow turnover characteristic of tundra nutrient cycling. The complex effects of temperature on nutrient availability in tundra ecosystems may be grouped into four categories: (1) direct effects on exoenzymatic breakdown of polymers, (2) direct effects on microbial activity and use of enzymatic breakdown products, (3) effects of freeze – thaw events on nutrient release, and (4) indirect effects mediated by effects on thaw depth, soil moisture, and vegetation.

10.2.4.1.1 Enzyme Activities

Enzymatic breakdown of polymers is a critical step in converting nutrients to available forms. Many exoenzymes are active at low temperature, retaining as much as 50% of their activity at 15 °C down to 0 °C (Linkins et al. 1984; McClaugherty and Linkins 1990). In addition to being relatively temperature insensitive, most soil enzymes retain their activity after repeated freeze–thaw cycles (Sinsabaugh and Linkins 1989). Endocellulase, which is responsible for the primary cleavage of cellulose, may be an exception to this pattern: In tussock-tundra soils this enzyme has a high activation energy below 10 °C, inhibiting cellulose decomposition (Linkins et al. 1984). For exoenzymes it is possible that the balance of enzymatic production of monomers and their use by microbes shifts with temperature. Protease activity in tussock-tundra soils, for example, is an order of magnitude more rapid than net N mineralization (Chapin et al. 1988), perhaps accounting for the high concentration of free amino acids and other simple N compounds found in these soils (Kielland 1990).

10.2.4.1.2 Microbial Activity at Low Temperatures

Only a few studies have examined microbial activity at temperatures near or below freezing. Microbial activity often continues as long as soils are unfrozen, and some activity can continue down to −6.5 °C (Coxson and Parkinson 1987; Flanagan and Bunnell 1980; Clein and Schimel 1995). Decomposition at <5 °C can be nearly 50% of decomposition at 10–20 °C in both terrestrial (Ivarson 1974; Taylor and Jones 1990) and aquatic systems (Tam et al. 1983). Below 10 °C, respiration and N mineralization in tussock-tundra soils show little response to temperature (Marion and Miller 1982; Nadelhoffer et al. 1991; see also Chap. 11, this Vol.), and Q_{10} values (relative rate increase for each 10 °C increase) below 2 have also been reported for wet sedge meadow soils (Flanagan and Bunnell 1980; Nadelhoffer et al. 1991). Psychrophilic microbes (those that thrive at low temperatures) are reasonably common in arctic soils (Widden and Parkinson 1978). One can conclude from these studies that low temperature *per se* – even when below zero – does not prevent microbial activity; either freezing itself, or reduced substrate diffusion brought about by freezing, may be more influential. Studies on tundra nutrient cycling should consider winter activity and the amount of free water remaining in the soil.

10.2.4.1.3 Freeze-Thaw Events

A single freeze–thaw event can kill 25–50% of viable microbes in the soil (Soulides and Allison 1961; Skogland et al. 1988), thereby releasing substrates including amino acids and sugars (Taylor and Parkinson 1988). These substrates are used by the surviving microbes (Skogland et al. 1988) resulting in a pulse of N mineralization (DeLuca et al. 1992). Additional freeze–thaw events do not appear to reduce microbial populations further as frost-tolerant microbes continue to survive. As the soil freezes the concentration of solutes in solution increases, which decreases water potential, and this decrease causes microbes to increase their concentration of soluble components (Schimel et al. 1989b). Bacteria accumulate amino acids. Thus, freezing should increase the concentration of N compounds that could be mineralized during a thaw, but these same compounds might act as an antifreeze that limits the proportion of the microbial population killed by freeze–thaw. Freeze–thaw cycles can occur in surface soils throughout the growing season, but may be more common in drier heath and shrub communities than in wetter tussock and wet meadow tundras, because soil moisture acts as a thermal buffer. On the other hand, thaw depth in wet soils can vary by 1–2 cm daily, producing a band of regular freeze–thaw events.

10.2.4.2 Effects of Low Oxygen on Microbial Activity and Mineralization

Anaerobic conditions found in many tundra soils are most common in wet meadow communities, and are less common in midslope soils such as those in tussock and shrub communities (Chap. 11, this Vol.). The soils in tussock and shrub communities are often wet, but not saturated, and low microbial oxygen (O_2) consumption rates may mitigate low O_2 diffusion into the organic soil material. The effects of anaerobiosis in controlling plant nutrient availability are unclear. First, anaerobic conditions reduce the rates of microbial activity and nutrient mineralization, particularly at low temperatures (Flanagan and Bunnell 1980). Counterbalancing these reductions is decreased microbial carbon-use efficiency (Smith et al. 1989), which increases net mineralization as a proportion of gross nutrient turnover. If soil microbes are nutrient-limited under aerobic conditions, anaerobiosis may enhance plant nutrient supply particularly for aerenchymous plants that transport oxygen through their roots. Anaerobiosis may also increase phosphate availability, because iron, which complexes with P in these soils, is more soluble at low redox potentials.

10.2.4.3 Substrate Quality

Few studies have explicitly addressed the role of organic matter quality in controlling nutrient dynamics in tundra soils. The standard indices of organic matter quality, such as C:N ratio, lignin, and lignin:N ratio, correlate poorly with C and N mineralization in tundra soils from across the landscape (Kielland 1990). Shrub tundra soils, for example, have high potentials for N mineralization, but also high lignin content (Kielland 1990). Traditional indices of organic matter quality (e.g., lignin) are limited, because they may relate more to long-term patterns of nutrient dynamics than to the short-term dynamics examined by Kielland (1990). More useful indices for evaluating soil C and nutrient dynamics may be the relative size of the "active," or readily available fraction, of soil organic matter (Parton et al. 1987).

10.3 Fate of Available Nutrients

Soil nutrients can have several possible fates. They may be taken up by microbes or plants, or they may be lost from the system by leaching or conversion to a gas (e.g., N_2O). Nutrient loss is generally limited in tundra soils. N losses to the atmosphere are limited by low rates of nitrification, which is a prerequisite for N_2O production. Leaching of inorganic nutrients is limited by availability, and permafrost blocks vertical percolation. Water flow within the soil can, however, move nutrients laterally between communities along a topographic sequence in a watershed. For example, Giblin et al. (1991) found higher concentrations of NO_3^- in downslope sites than nitrification could account for, and inferred that NO_3^- had been produced in upslope sites (which had relatively high rates of nitrification) and was then transported by flowing soil water. Downslope movement probably accounts for only a small proportion of the nutrients mobilized in upslope communities, but this movement may be important in landscape dynamics over long time scales (decades to centuries).

Competition for nutrients between plant roots and soil microbes is a major contol on plant nutrient uptake. Microbial N uptake exceeds plant uptake in a wide variety of ecosystems (Marion et al. 1982; Jackson et al. 1989; Zak et al. 1990), and competition may also control plant PO_4^{3-} uptake (Jonasson and Chapin 1991). Microbial uptake may therefore be the most critical control on the outcome of this competition.

10.3.1 Microbial Nutrient Uptake and Competition with Plants

Microbial nutrient uptake is controlled by nutrient availability, microbial demand for nutrients, and the carbon available for assimilation. Actual nutrient availability is a function of both the quantity and the form of the nutrients present, particularly for N, because the physical processes (i.e., cation exchange) affecting NO_3^-, NH_4^+, and amino acids differ from one another, as do the relative plant and microbial preferences for these N forms. Ammonium has been considered the critical N pool in plant–microbe competition, because it is the form produced in mineralization, and soil heterotrophs outcompete plants for it (Jackson et al. 1989). Heterotrophs also outcompete nitrifiers for NH_4^+ (Johnson and Edwards 1979), greatly reducing NO_3^- availability in N-limited systems. Soil microbes have a low affinity for NO_3^- (Jackson et al. 1989) and, consequently, plants compete more effectively for NO_3^- when it is present. Competition for NO_3^- may be significant in shrub tundra where nitrification rates are substantial, and in footslope communities where NO_3^- is supplied by downslope movement (Giblin et al. 1991).

Amino acids may be particularly important in tundra N cycles. Although amino acid concentrations vary in tundra soils, they are among the highest measured in terrestrial systems (Kielland 1990), sometimes measuring higher than NH_4^+ concentrations. Amino acid turnover has not been well studied in soils, but it is generally assumed that the amino acid N must be mineralized before it can be used by plants. Most tundra plants can, however, take up amino acids directly (Chapin et al. 1993). Mycorrhizal fungi also allow infected plants to take up amino acids (Read 1991). Tundra sedges compete for amino acid-N in soil as well or better than they compete for NH_4^+ (Schimel and Chapin, unpubl. data). The importance of amino acids in tundra N cycling adds an extra dimension to the role of N form in competition for N, and requires us to consider organic N dynamics more thoroughly.

The extent of microbial nutrient limitation and demand in tundra soils is poorly understood. Microbes are probably N-limited in most tundra soils, perhaps most extremely in tussock tundra where N concentrations are low and immobilization is rapid (Giblin et al. 1991; Schimel 1995). Net immobilization occurs during the growing season in sites across the landscape (Chapin et al. 1988; Giblin et al. 1991), and increases in net mineralization resulting from N additions (Kielland 1990) strongly suggest microbial N limitation. Marion et al. (1982) showed that immobilization appeared to account for ca. 80% of the ^{15}N they applied to tussock-tundra soils at Imnavait Creek. As they added more N the proportion of ^{15}N recovered in plants increased, supporting the idea that microbes were N-limited and out-competed plants for N. Cornstarch additions had no significant effect on N immobilization or plant growth in tussock tundra, suggesting that microbes are not C-limited in this community (Marion et al. 1982; Shaver et al. 1986). The patterns of nitrification also support the idea of extensive microbial N limitation. Although nitrification occurs in tundra communities, including tussock tundra, riverside willow, and dry upland shrub tundra (Kielland 1990; Giblin et al. 1991), the rates are usually stimulated by NH_4^+ additions, suggesting that NH_4^+ availability limits nitrification (Kielland 1990). Microbes in dry shrub tundra may not experience much N limitation, as evidenced by relatively extensive nitrification in this community (Kielland 1990; Giblin et al. 1991). Microbial N availability in wet meadow soils may be enhanced by soil anaerobiosis and the resulting low efficiency of C use. These soils also have very low potentials for N assimilation (Schimel 1995), and this low potential suggests that microbes in wet meadows may be more limited by P than by N (Flanagan and Bunnell 1980).

The effects of C on microbial nutrient uptake are determined by the amount of C immediately available. This is in turn, controlled by the amount of soil C, its quality, and its spatial distribution. The overall quality of soil organic matter controls the size of the active carbon fraction, and thus, the rates of microbial C use and turnover. Even if C and N are plentiful, if they are not simultaneously accessible to a microbe, growth and nutrient uptake will be curtailed.

Spatial compartmentalization of C, N, and P exercise an important control on the cycles of these elements. Spatial compartments – defined as areas in which different nutrients limit microbial activity – can occur on scales from microns to meters, and the most important compartments may vary between communities. They may also exist in a hierarchy of small compartments within larger ones. Rooting and soil stratification are likely to produce important small-scale compartments in all communities. The rhizosphere may be particularly important in moist and wet tundra where it is an area of both increased C (from exudation) and O_2 availability (from transport through roots). The rhizosphere may therefore be N- or P-limited, while the bulk-soil community might be more C-limited. Larger-scale compartments may be defined by topography, hydrology, and plant distribution. Thus, critical large-scale compartments, such as water tracks and tussocks, may greatly affect nutrient dynamics in tundra regions (Chapin et al. 1988).

10.3.2 Plant Uptake

Nutrient acquisition in tundra species involves a variety of factors, including the architecture and longevity of the root systems, the timing and rate of root growth, soil horizons explored, and N sources tapped. Considering the regulatory role that nutrient availability plays in tundra ecosytems, these factors contribute to the structural and functional diversity of tundra communities and their interactions in the landscape.

For the most limiting nutrients in arctic tundra (N and P), absorption is controlled by the rate of nutrient supplied to the root surface, rather than by plant physiological characteristics (Kielland and Chapin 1992; Leadley et al. 1996; see Sect. 14.3.3 in Chap. 14, this Vol.). The diversity of plant uptake capacities and rooting strategies, however, suggests that plant adaptations have a significant bearing on nutrient acquisition. Plant roots encounter ions by diffusion, mass flow, and interception. To a limited extent plants can manipulate these processes. Increased root growth, for example, increases root surface area, enhancing nutrient uptake by both interception and diffusion. Despite the close correlation between P availability and P uptake (e.g., Chapin et al. 1978), mechanistic models of plant nutrient uptake indicate that root growth rate and biomass exercise the most control over nutrient uptake (Silberbush and Barber 1983). Root growth is especially important, because it gives roots access to areas of relatively high nutrient availability.

10.3.2.1 Soil Factors Controlling Nutrient Absorption

Water flow, augmented by spring snowmelt and topographic variation, can substantially modify nutrient cycling and plant community structure. For instance, flow in water tracks may be nearly an order of magnitude faster than the rates at which N and P are estimated to diffuse (Chapin et al. 1988). Increased mass flow alleviates the limitations of nutrient uptake imposed by ion diffusion and enhances nutrient availability downslope. Such flow enhances nutrient supply to the deeply rooted sedge, *Eriophorum vaginatum*, and can generate as much as a tenfold increase in aboveground nutrient stocks and total net annual production (Fig. 10.2), a production that rivals that of temperate forests, despite a much shorter growing season (Waring and Schlesinger 1985). Extensive nutrient movement across the landscape outside water tracks is dictated by topography and is limited to periods after rainfall and snowmelt (Chap. 9, this Vol.).

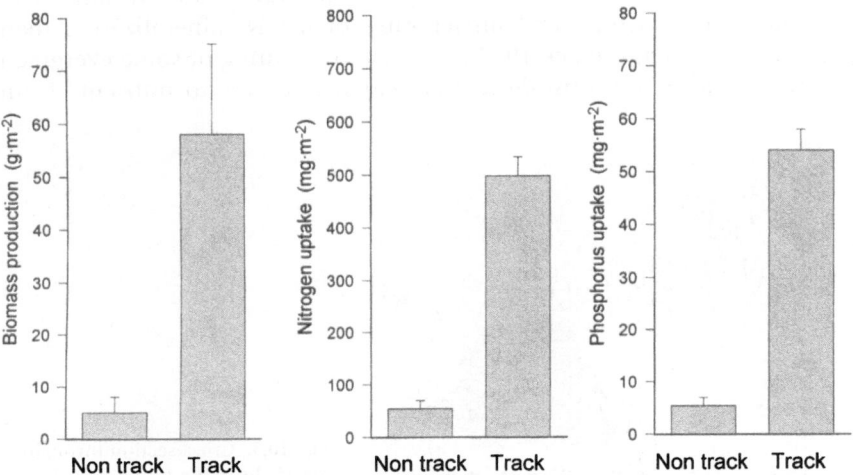

Fig. 10.2. Effect of water tracks on annual biomass production, nitrogen uptake, and phosphorus uptake of *Eriophorum vaginatum*. (From Chapin et al. 1988)

10.3.2.2 Rooting Strategies

Because of the importance of spatial controls over soil nutrient arailability, root system architecture and phenology play key roles in plant nutrient acquisition. Tundra plant taxa differ widely in biomass allocation and rooting architecture. Generally, most tundra plant biomass (60–90%) is belowground in roots, rhizomes, and subsurface stems (Shaver and Cutler 1979), but root system architecture is highly variable (Fig. 10.3). For example, *Eriophorum vaginatum*, common in moist tussock tundra, has thick (up to 1 mm in diameter) unbranched roots that grow straight down into the soil. This enables uptake from freshly thawed soil and perhaps from water flow immediately above the frozen layer (Chap. 9, this Vol.), the latter of which is below the rooting zone of most other species. In contrast, the evergreen shrub *Ledum palustre* has thin (0.24 mm), highly branched roots that allow exploitation of a large soil volume (Kummerow et al. 1983), a trait that is particularly important for acquiring slowly diffusing ions such as ammonium and phosphate. Fine-rooted plants, such as *Ledum*, are common from ridgetop dry heath and shrub communities through midslope tussock tundra, but disappear at lower slope positions. The different rooting strategies found in the dominant vegetation of different communities produces distinctly different vertical profiles of root distribution. For example, Miller et al. (1982) found that tussock tundra had a high proportion of roots in the top 10 cm, and a lower, but uniformly distributed proportion, from 10–50 cm. In a deciduous shrub community, on the other hand, few roots grew in the top 5 cm, but roots did not extend below 25 cm.

The roots of *Ledum* and other evergreen shrubs are concentrated in the uppermost soil horizon throughout the growing season (Kummerow et al. 1983). Such species are exposed to warmer soil with greater variations in temperature and moisture, and higher rates of net N mineralization than deeper soil (Marion and Black 1987). The shallow rooting of some evergreen shrubs (e.g., *Vaccinium vitis-idaea*) may enhance access to nutrients from

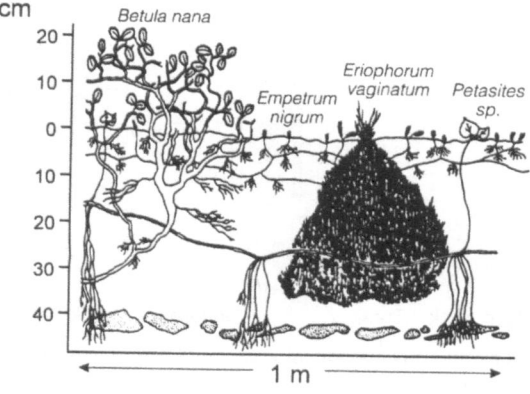

Fig. 10.3. Cross section through dwarf shrub tundra near Eagle Summit, Alaska. (From Miller et al. 1982)

freshly leached leaf litter. Moreover, because of this shallow rooting depth, the roots of *Ledum palustre* and *Vaccinium vitis-idaea* may become active earlier in the season than the more deeply rooted deciduous shrubs such as *Salix* and *Betula*. Because nutrients tend to be more available early in the season (Barsdate and Alexander 1975; Kielland 1990; Giblin et al. 1991), simultaneous with peak plant nutrient uptake (Chapin and Bloom 1976; Marion and Kummerow 1990), root phenology becomes a critical control over plant nutrient budgets in tundra ecosystems.

The timing of root growth in many tundra plants is controlled not by nutrient supply, but by the availability of plant carbon (Kummerow and Russel 1980). Root growth begins when foliage is adequate to support the C drain, and stops when photosynthesis slows in August (Kummerow and Russell 1980; Kummerow et al. 1983). The timing of root growth may therefore have little connection to the timing of root activity. This asynchrony is particularly apparent in species with perennial root systems, such as *Carex* sp. or *Dupontia fischeri*, in which roots may be produced only in the first year of an individual tiller's lifetime (Shaver and Billings 1975). The interactions of root architecture, phenology, and activity pose an important challenge to our understanding of tundra nutrient dynamics.

10.3.2.3 Uptake Characteristics of Tundra Plants

Nutrient uptake by arctic species is less temperature-sensitive than nutrient uptake by temperate plants (Fig. 10.4; Chapin and Bloom 1976). Tundra plants can maintain higher absorption rates at the low temperatures of tundra soils (0–10 °C) in comparison with temperate species, in which adsorption is frequently suppressed below 10 °C. Root growth in tundra species is also insensitive to low temperatures (Shaver and Billings 1977), and roots stay active down to 0 °C. This insensitivity to cold is particularly true for tundra graminoids, such as *Eriophorum* and *Carex*, which grow in wet, organic-rich tussock tundra

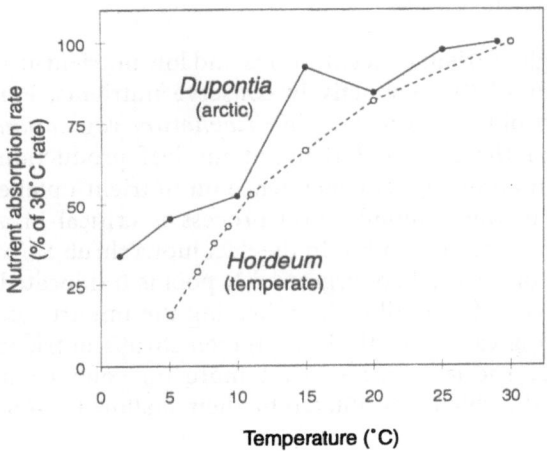

Fig. 10.4. Temperature sensitivity of nutrient absorption exhibited by an arctic (*Dupontia fisherii*) and a temperate (*Hordeum vulgare*) species. (From Chapin 1983)

and wet sedge meadow soils (Bliss 1956). Among arctic plants, deep-rooted species from cold soils tend to be less temperature-sensitive than shallower-rooted species (Chapin and Tryon 1982). Uptake capacity among growth forms correlates loosely with growth and photosynthetic rates, which regulates the demand for nutrients and, consequently, determine the absorption capacity (Clarkson and Hanson 1980). Thus, deciduous shrubs tend to have a higher absorption capacity than graminoids, which in turn have a higher absorption capacity than evergreen shrubs.

Differences in root turnover rates can affect nutrient absorption indirectly. Species such as *Eriophorum vaginatum* and *E. angustifolium* have a nearly annual root system, whereas root turnover in *Carex* requires 6–8 years (Shaver and Billings 1975). These species have different patterns of root suberization, and these differences affect nutrient absorption. Root suberization in *Carex* reduces nutrient uptake less than it does in *Eriophorum* (Chapin and Tryon 1982). Unlike the perrenial roots of *Carex*, *Eriophorum* roots do not fully suberize until late in the season and may be a better competitor for nutrients early in the season, whereas *Carex* may be better able to capitalize on nutrients released from dying microbial populations and senescing plants in the autumn.

Increased N and P absorption by mycorrhizal roots has been recognized for a long time (e.g., Hatch 1937), but the common view held that mycorrhizae primarily enhanced inorganic nutrient uptake. Recent studies have demonstrated that mycorrhizae are able to use amino acids, proteins, and other organic N sources (Read 1991). Findings in Alaskan tundra communities suggest that organic N (amino acids) may supply a substantial proportion of N to tundra plants (Kielland 1990; Chapin et al. 1993; Schimel and Chapin, unpubl. data). Differences among species in their amino acid absorption potential may provide a basis for niche differentiation with regard to N absorption.

10.3.2.4 Retranslocation vs Current Uptake

Because tundra plants have high nutrient concentrations and low nutrient-use efficiencies (NUE; e.g., Chapin 1980), they actively conserve nutrients. For example, retranslocation from old leaves to new in *Eriophorum vaginatum* accounted for at least 85% of the N and P required for leaf production (Jonasson and Chapin 1985), minimizing its dependence on nutrient uptake from the soil. The efficiency of this remobilization process is critical in a low-nutrient environment such as arctic tundra. In the deciduous shrub *Salix pulchra* the equivalent of 12% of the total aboveground N pool is tranlocated into new leaves by the first week after budbreak, indicating the importance of stored reserves for supporting early growth. In evergreen shrubs nutrient movement is slower, so these species must depend more on concurrent uptake to support spring growth, which is facilitated by their shallow rooting depth.

Because much nutrient absorption serves to replenish reserves, rather than to support current growth (Chapin and Bloom 1976; Chapin and Shaver 1985), the timing of nutrient uptake is temporally decoupled from growth in many species. By maintaining physiological readiness of its root system into the fall season, *E. vaginatum* capitalizes on the late-season nutrient flush after aboveground tissue has senesced (Chapin and Bloom 1976), maximizing its annual nutrient gain.

10.4 Disturbances

10.4.1 Vehicle Tracks

The controls over nutrient uptake by plants and microbes provide a basis for understanding many of the impacts of disturbance on tundra communities. Physical disturbance, such as tracked vehicles moving over the tundra, compacts the soil and vegetation, increasing thermal conductance and warming the soil (Chapin and Shaver 1981; Fig. 10.5; Chap. 3, this Vol.). Warmed soils presumably enhance decomposition, which should explain the higher concentrations of available NH_4^+ (Challinor and Gersper 1975) and PO_4^{3-} (Chapin and Shaver 1981) in the soil solution. Because of soil compaction and ground ice melting, the surface of vehicle tracks and other disturbed surfaces generally subsides below the surrounding soil. Water then flows into these tracks and creates wet, poorly oxygenated soils (Challinor and Gersper 1975; Chapin and Shaver 1981). Such water channels have the potential to substantially alter hillslope hydrology and influence plant community dynamics across the landscape.

No one has yet studied microbial responses to tundra disturbances, but microbial processes in undisturbed tundra soils lead us to expect that increased temperature would increase the rates of most microbial processes. Because less oxygen is available in disturbed areas, more microbial activity might be anaerobic.

Plants respond to the 0.8-2.0-fold increase in soil nutrient concentration in vehicle tracks with a 2-to 15-fold increase in production and in N and P pools (Chapin and Shaver 1981). This disproportionately large increase in plant nutrient uptake could reflect any of several processes. Microbes might be less nutrient-limited in disturbed sites, and therefore less effective competitors with plants for nutrients; the warmer soils might promote greater root growth and nutrient uptake; or rapid mass flow of nutrients in track soils may stimulate uptake. Of these possibilities, increased mass flow and reduction in microbial competition seem most probable. Although soils thaw more deeply in vehicle tracks, plant roots do not grow deeper to exploit this new soil volume (Chapin and Shaver 1981), making it difficult to explain the large increases in plant nutrient uptake without invoking changes in plant–micro-

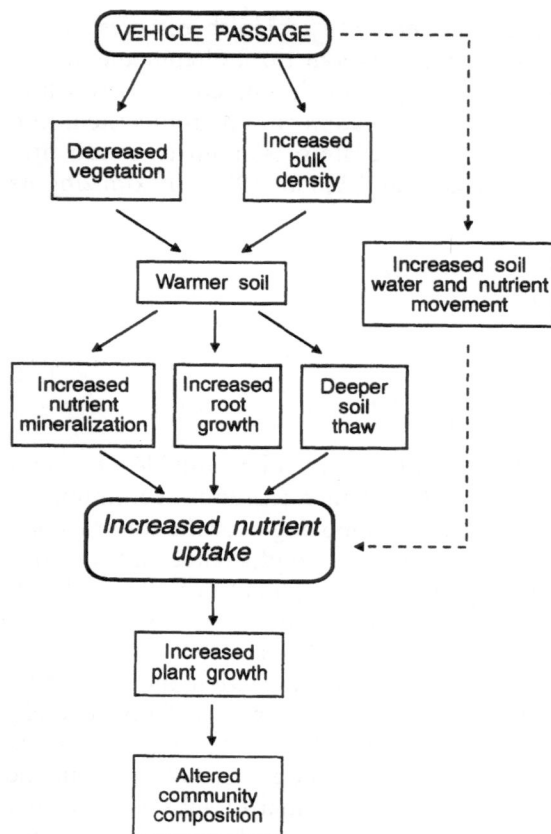

Fig. 10. 5. Effects of vehicle tracks on tundra ecosystem functioning. (From Chapin and Shaver 1981)

bial competition for nutrients or increases in mass flow of nutrients to plant roots.

Plant community structure and productivity change substantially in response to the altered environment of vehicle tracks in upland tundra. Species richness declines in vehicle tracks, reflecting the replacement of many shrub species by a few species of graminoids, which may be favored in wetter soils because their roots are aerenchymous. A few species of sedges dominate wet meadow communities disturbed by vehicles. Large increases in productivity and standing crops of nutrients in vehicle tracks reflect increased leaf nutrient concentrations, increased shoot density, and increased shoot weight of graminoids (Wein and Bliss 1974; Babb 1977; Chapin and Shaver 1981).

10.4.2 Road Dust

Road traffic produces a substantial amount of dust that is deposited on nearby tundra vegetation (Chaps. 3 and 15, this Vol.). In northern Alaska road gravel contains calcareous material and roads are sporadically treated with calcium carbonate to reduce dust; consequently, road dust has a high pH. Tundra receiving large inputs of dust exhibit the following trends: (1) an increase in soil pH, (2) a replacement of *Sphagnum* (which typically grows in acid bogs) by *Aulacomnium* (a more minerotrophic moss), and (3) vigorous growth of *Eriophorum vaginatum*, which tends to displace evergreen shrubs. The enhanced growth and increased tissue nutrient concentrations of *Eriophorum* undoubtedly reflect improved nutrient availability in response to increased pH. The effects of dust on soil microbial processes have not been examined in detail. Increasing pH might be expected to increase nitrification and nutrient mobility, thus potentially improving plant competitiveness for N (Schimel et al. 1989a). Alkaline dust deposition may also enhance bacterial growth and change the composition of the bacterial community (Bääth et al. 1992).

10.4.3 Gray Water

One consequence of increased human activity and industrialization in tundra areas is the production of massive amounts of waste water that is extremely expensive to process. As a result, the impact of gray water on tundra vegetation holds considerable interest. Fertilizer plots demonstrate that tundra responds to nutrient inputs by increased production and a shift in growth-form composition, first to graminoids and then to deciduous shrubs (Shaver and Chapin 1986). As discussed previously, natural areas of high water flow (water tracks) also exhibit high productivity and nutrient uptake. The addition of nutrient-rich gray water may well promote nutrient uptake and productivity in tundra. Experimental addition of water to upland tundra increased both productivity and nutrient mobility (Marion and Everett 1989; Oberbauer et al. 1989). In tussock tundra, which has very high potential N incorporation rates, much of the added N would probably be immobilized initally, but might increase the rate of N turnover and cycling over several years.

10.4.4 Climate Change

Although human disturbances in the Arctic have modified a relatively small proportion of the total land surface, human activity now threatens to modify the entire Arctic through changes in global climate. Temperature increases are expected to be the largest and to occur the earliest in polar regions (Maxwell 1992). Indeed, evidence of climatic warming in the Arctic already exists (Lachenbruch and Marshall 1986; Oechel et al. 1993). Chapin et al. (1992) have discussed many of the likely changes in arctic ecosystems in response to this climatic warming.

Relevant to plant nutrition and plant–microbial interactions, climatic warming will likely lead to a longer growing season, warmer soil temperatures, and deeper thaw. Although changes in soil moisture are more difficult to forecast, predicted increases in precipitation may be concentrated in winter, with predicted increases in evapotranspiration in summer; the result will probably be drier soils. Warmer, drier soils would enhance microbial activity, stimulating respiration and decomposition (Chaps. 16 and 17, this Vol.). This could enhance net N mineralization and increase the N supply to plants. When soils are heated in tundra wet meadow (Chapin and Bloom 1976) or boreal forest (Oechel and Van Cleve 1986), nutrient availability, plant nutrient status, and plant production increase, increases that lend credence to this scenario. In communities in which microbes are N-limited, however, the increased decomposition might not immediately increase N availability to plants, because microbial immobilization would tie up the nutrients. These nutrients would have to be mobilized out of the microbial biomass to allow plants access to them. Stress events such as freezing and thawing could accomplish this as discussed previously.

If these responses occur, we expect upland community composition to shift to vegetation dominated by deciduous shrubs. These species have higher nutrient uptake on a per-root and per-area basis than other growth forms, and they produce litter with a higher nutrient concentration and a lower lignin content than the evergreen shrubs now covering much of the tundra. Enhanced litter quality and microbial activity should result, providing a positive feedback toward increased nutrient availability and nutrient uptake by vegetation.

References

Alexander V, Schell DM (1973) Seasonal and spatial variation on nitrogen fixation in the Barrow, Alaska, tundra. Arct Alp Res 5: 77–88

Bääth E, Frostegård Å, Fritze H (1992) Soil bacterial biomass, activity, phospholipid fatty acid pattern, and pH tolerance in an area polluted with alkaline dust deposition. Appl Environ Microbiol 58: 4026–4031

Babb TA (1977) High arctic disturbance studies associated with the Devon Island Project. In: Bliss LC (ed) Truelove Lowland, Devon Island, Canada: a high arctic ecosystem. Univ Alberta Press, Edmonton, pp 647–654

Barel D, Barsdate RJ (1978) Phosphorus dynamics of wet coastal tundra near Barrow, Alaska. In: Adriano AC, Brisbin I (eds) Environmental chemistry and cycling processes. US DOE Symp Ser. NTIS, Washington, pp 516–537

Barsdate RJ, Alexander V (1975) The nitrogen balance of arctic tundra: pathways, rates, and environmental implications. J Environ Qual 4: 111–117

Billings WD, Peterson KM, Luken JO, Mortensen DA (1984) Interaction of increasing atmospheric carbon dioxide and soil nitrogen on the carbon balance of tundra microcosms. Oecologia 65: 26–29

Bliss LC (1956) A comparison of plant development in microenvironments of arctic and alpine plants. Biol Rev 43: 481–530

Bunnell FL, Miller OK, Flanagan PW, Benoit RE (1980) The microflora: composition, biomass, and environmental relations. In: Brown J, Miller PC, Tieszen LL, Bunnell FL (eds) An arctic ecosystem. Dowden, Hutchinson and Ross, Stroudsburg, pp 255–290

Challinor JL, Gersper PL (1975) Vehicle perturbation effects upon a tundra soil-plant system. II. Effects on the chemical regime. Soil Sci Soc Am Proc 39: 689–695

Chapin FS III (1980) The mineral nutrition of wild plants. Annu Rev Ecol Syst 11: 233–260

Chapin FS III (1983) Direct and indirect effects of temperature on arctic plants. Polar Biol 2: 47–52

Chapin FS III, Bloom A (1976) Phosphate absorption: Adaptation of tundra graminoids to a low temperature, low phosphorus environment. Oikos 26: 111–121

Chapin FS III, Shaver GR (1981) Changes in soil properties and vegetation following disturbance of Alaskan arctic tundra. J Appl Ecol 18: 605–617

Chapin FS III, Shaver GR (1985) Arctic. In: Chabot BF, Mooney HA (eds) Physiological ecology of North American plant communities. Chapman and Hall, New York, pp 16–40

Chapin FS III, Tryon PR (1982) Phosphate absorption and root respiration of different growth forms from northern Alaska. Holarct Ecol 5: 164–171

Chapin FS III, Barsdate RJ, Barel D (1978) Phosphorus cycling in Alaskan coastal tundra: a hypothesis for the regulation of nutrient cycling. Oikos 31: 189–199

Chapin FS III, Fetcher N, Kielland K, Everett KR, Linkins AE (1988) Productivity and nutrient cycling of Alaskan tundra: enhancement by flowing water. Ecology 69: 693–702

Chapin FS III, Jeffries RL, Reynolds JF, Shaver GR, Svoboda J (eds) (1992) Arctic ecosystems in a changing climate. Academic Press, San Diego

Chapin FS III, Moilanen L, Kielland K (1993) Preferential use of organic nitrogen for growth by a non-mycorrhizal arctic sedge. Nature 361: 150–153

Cheng W, Virginia RA (1993) Measurement of microbial biomass in arctic tundra soils using fumigation-extraction and substrate-induced respiration procedures. Soil Biol Biochem 25: 135–141

Clarkson DT, Hanson JB (1980) The mineral nutrition of higher plants. Annu Rev Plant Physiol 31: 239–298

Clein JS, Schimel JP (1995) Microbial activity of tundra and taiga soils at sub-zero temperatures. Soil Biol Biochem 27: 1231–1234

Coxson DS, Parkinson D (1987) Winter respiratory activity in aspen woodland forest floor litter and soils. Soil Biol Biochem 19: 49–59

DeLuca TH, Keeney DR, McCarty GW (1992) Effect of freeze–thaw events on mineralization of soil nitrogen. Biol Fertil Soils 14: 116–120

Flanagan PW, Bunnell FL (1980) Microflora activities and decomposition. In: Brown J, Miller PC, Tieszen LL, Bunnell FL (eds) An arctic ecosystem. Dowden, Hutchinson and Ross, Stroudsburg, pp 291–334

Flanagan PW, Van Cleve K (1977) Microbial biomass, respiration and nutrient cycling in a black spruce taiga ecosystem. Ecol Bull (Stockholm) 25: 261–273

Gersper PL, Alexander V, Barkley SA, Barsdate RJ, Flint PS (1980) The soils and their nutrients. In: Brown J, Miller PC, Tieszen LL, Bunnell FL (eds) An arctic ecosystem. Dowden, Hutchinson and Ross, Stroudsburg, pp 219–254

Giblin AE, Nadelhoffer KJ, Shaver GR, Laundre JA, McKerrow AJ (1991) Biogeochemical diversity along a riverside toposequence in arctic Alaska. Ecol Monogr 61: 415–435

Hatch AB (1937) The physical basis of mycotrophy in the genus Pinus. Black Rock For Bull 6, 168 pp

Hinzman LD, Kane DL, Gieck RE, Everett KR (1991) Hydrologic and thermal properties of the active layer in the Alaskan arctic. Cold Regions Sci Tech 19: 95–110

Ivarson KC (1974) Comparative survival and decomposing ability of four fungi isolated from leaf litter at low temperatures. Can J Soil Sci 54: 245–253

Jackson LE, Schimel JP, Firestone MK (1989) Short-term partitioning of ammonium and nitrate between plants and microbes in an annual grassland. Soil Biol Biochem 21: 409–415

Jonasson S, Chapin FS III (1985) Significance of sequential leaf development for nutrient balance of the cotton sedge, Eriophorum vaginatum L. Oecologia 67: 511–518

Jonasson S, Chapin FS III (1991) Seasonal uptake and allocation of phosphorus in Eriophorum vaginatum L., measured by labelling with [32]P. New Phytol 118: 349–357

Johnson DW, Edwards NT (1979) The effects of stem girdling on biogeochemical cycles within a mixed deciduous forest in eastern Tennessee: II. Soil nitrogen mineralization and nitrification rates. Oecologia 40: 259–271

Kielland K (1990) Processes controlling nitrogen release and turnover in arctic tundra. PhD thesis, Univ Alaska, Fairbanks

Kielland K, Chapin FS III (1992) Nutrient absorption and accumulation in arctic tundra. In: Chapin FS III, Jeffries RL, Reynolds JF, Shaver GR, Svoboda J (eds) Arctic ecosystems in a changing climate. Academic Press, San Diego, pp 321–335

Kummerow J, Russell M (1980) Seasonal root growth in the Arctic tussock tundra. Oecologia 47: 196–199

Kummerow J, Ellis BA, Kummerow S, Chapin FS III (1983) Spring growth of shoots and roots in shrubs of an Alaskan muskeg. Am J Bot 70: 1509–1515

Lachenbruch AH, Marshall BV (1986) Changing climate: geothermal evidence from permafrost in the Alaskan arctic. Science 234: 689–696

Leadley PW, Reynolds JF, Chapin FS III (1996) A model of ammonium, nitrate, and glycine uptake by *Eriophorum vaginatum* roots in the field: ecological implications. (submitted)

Linkins AE, Melillo JM, Sinsabaugh RL (1984) Factors affecting cellulase activity in terrestrial and aquatic ecosystems. In: Klug MJ, Reddy CA (eds) Current perspectives in microbial ecology. Am Soc Microbiol, Washington DC, pp 572–579

Marion GM, Black CH (1987) The effect of time and temperature on nitrogen mineralization in arctic tundra soils. Soil Sci Soc Am J 51: 1501–1508

Marion GM, Everett KR (1989) The effect of nutrient and water addition on elemental mobility through small tundra watersheds. Holarct Ecol 12: 317–323

Marion GM, Kummerow J (1990) Ammonium uptake by field grown *Eriophorum vaginatum* under laboratory and simulated field conditions. Holarct Ecol 13: 50–55

Marion GM. Miller PC (1982) Nitrogen mineralization in a tussock tundra soil. Arct Alp Res 14: 287–293

Marion GM, Miller PC, Kummerow J, Oechel WC (1982) Competition for nitrogen in a tussock tundra ecosystem. Plant Soil 66: 317–327

Maxwell B (1992) Arctic climate: potential for change under global warming. In: Chapin FS III, Jeffries RL, Reynolds JF, Shaver GR, Svoboda J (eds) Arctic ecosystems in a changing climate. Academic Press, San Diego, pp 11–34

McClaugherty CA, Linkins AE (1990) Temperature responses of enzymes in two forest soils. Soil Biol Biochem 22: 29–33

McKendrick JD, Batzli GO, Everett KR, Swanson JC (1980) Some effects of mammalian herbivores and fertilization on tundra soils and vegetation. Arct Alp Res 12: 565–578

Miller PC, Mangan R, Kummerow J (1982) Vertical distribution of organic matter in eight vegetation types near Eagle Summit, Alaska. Holarct Ecol 5: 117–124

Myrold DD (1987) Relationship between microbial biomass nitrogen and a nitrogen availability index. Soil Sci Soc Am J 51: 1047–1049

Nadelhoffer KJ, Giblin AE, Shaver GR, Laudre JA (1991) Effects of temperature and substrate quality on element mineralization in six arctic soils. Ecology 72: 242–253

Oberbauer SF, Hastings SJ, Beyers JL, Oechel WC (1989) Comparative effects of downslope water and nutrient movement on plant nutrition, photosynthesis, and growth in Alaskan tundra. Holarct Ecol 12: 324–334

Oechel WC, Van Cleve K (1986) The role of bryophytes in nutrient cycling in the taiga. In: Van Cleve K, Chapin FS III, Flanagan PW, Viereck LA, Dyrness CT (eds) Forest ecosystems in the Alaskan Taiga: a synthesis of structure and function. Springer, Berlin Heidelberg New York, pp 121–137

Oechel WC, Hastings SJ, Vourlitis G, Jenkins M, Riechers G, Grulke N (1993) Recent change of Arctic tundra ecosystems from a net carbon dioxide sink to a source. Science 361: 520–523

Parton WJ, Schimel DS, Cole CV, Ojima DS (1987) Analysis of factors controlling soil organic matter levels in Great Plains grasslands. Soil Sci Sco Am J 51: 1173–1179

Paul EA, Voroney RP (1984) Field interpretation of microbial biomass activity measurements. In: Klug MJ, Reddy CA (eds) Current perspectives in microbial ecology. Am Soc Microbiol, Washington DC, pp 509–514

Read DJ (1991) Mycorrhizas in ecosystems. Experimentia 47: 376–391

Schimel JP (1995) Ecosystem consequences of microbial diversity and community structure. In: Chapin FS III, Körner C (eds) Arctic and Alpine biodiversity. Ecological Studies 115. Springer, Berlin Heidelberg New York, pp 239–254

Schimel JP, Jackson LE, Firestone MK (1989a) Spatial and temporal effects on plant – microbial competition for inorganic nitrogen in a California annual grassland. Soil Biol Biochem 21: 1059–1066

Schimel JP, Scott W, Killham K (1989b) Changes in cytoplasmic carbon and nitrogen pools in a soil bacterium and a fungus in response to salt stress. Appl Envir Microbiol 55: 1635–1637

Shaver GR, Billings WD (1975) Root production and root turnover in a wet tundra ecosystem, Barrow, Alaska. Ecology 56: 401–409

Shaver GR, Billings WD (1977) Effects of day length and temperature on root elongation in tundra graminoids. Oecologia 28: 57–65

Shaver GR, Chapin FS III (1986) Effect of fertilizer on production and biomass of tussock tundra, Alaska, USA. Arct Alp Res 18: 261–268

Shaver GR, Cutler JD (1979) The vertical distribution of live vascular phytomass in cottongrass tussock tundra. Arct Alp Res 11: 335–342

Shaver GR, Chapin FS III, Gartner BL (1986) Factors limiting seasonal growth and peak biomass accumulation in Eriophorum vaginatum in Alaskan tussock tundra. J Ecol 74: 257–278

Silberbush M, Barber SA (1983) Prediction of phosphorus and potassium uptake by soybeans with a mechanistic mathematical model. Soil Sci Soc Am J 47: 262–265

Sinsabaugh RL, Linkins AE (1989) Natural disturbance and the activity of *Trichoderma viride* cellulase complexes. Soil Biol Biochem 21: 835–839

Skogland T, Lomeland S Goksoyr J (1988) Respiratory burst after freezing and thawing of soil: experiments with soil bacteria. Soil Biol Biochem 20: 851–866

Smith JL, Paul EA (1986) The role of soil type and vegetation on microbial biomass and activity. In: Megusar F, Ganar M (eds) Perspectives in microbial ecology. Slovene Soc Microbiol, Ljubljana, pp 460–466

Smith JL, Norton JM, Paul EA (1989) Decomposition of ^{14}C- and ^{15}N- labeled organisms in soil under anaerobic conditions. Plant Soil 116: 115–118

Souldes DA, Allison FE (1961) Effect of drying and freezing soils on carbon dioxide production, available mineral nutrients, aggregation, and bacterial population, Soil Sci 91: 291–298

Tam T-Y, Mayfield CI, Inniss WE (1983) Microbial decomposition of leaf material at 0 °C. Microb Ecol 9: 355–362

Taylor BR, Jones HG (1990) Litter decomposition under snow cover in a balsam fir forest. Can J Bot 68: 112–120

Taylor BR, Parkinson D (1988) Does repeated freezing and thawing accelerate decay of leaf litter? Soil Biol Biochem 20: 657–665

Waring RH, Schlesinger WH (1985) Forest ecosystems. Academic Press, New York

Wein RW, Bliss LC (1974) Primary production in arctic cottongrass tussock tundra communities. Arct Alp Res 6: 261–274

Whalen SC, Cornwell JC (1985) Nitrogen, phosphorus, and organic carbon cycling in an arctic lake. Can J Fish Aquat Sci 42: 797–808

Widden P, Parkinson D (1978) The effects of temperature on growth of four high arctic soil fungi in a three phase system. Can J Microbiol 24: 415–421

Zak DR, Groffman PM, Pregitzer KS, Christensen S, Tiedje JM (1990) The vernal dam: plant–microbe competition for nitrogen in northern hardwood forests. Ecology 71: 651–656

11 Landscape Patterns of Carbon Dioxide Exchange in Tundra Ecosystems

S. F. Oberbauer, W. Cheng, C. T. Gillespie, B. Ostendorf, A. Sala, R. Gebauer, R. A. Virginia, and J. D. Tenhunen

11.1 Introduction

Ecosystem carbon dioxide (CO_2) exchange is important because it is an indicator of energy captured by the system, it is related via decomposition to nutrient turnover, and long-term carbon storage and release affect atmospheric CO_2 concentrations and the global carbon budget. The large amount of carbon stored in northern soils – and the potential for its release to the atmosphere with climate warming (Miller 1981) – has stimulated much research on ecosystem gas exchange in the Arctic. This work has focused primarily on the potential response of these systems to shifts in climate factors (Oechel and Billings 1992). Because of its stature, tundra is one of the few natural ecosystems for which whole-system CO_2 fluxes have been determined (e.g., Grulke et al. 1990; Oechel et al. 1992, 1993; Tenhunen et al. 1995), because relatively small chambers can be placed over "representative patches" of tundra vegetation (Grulke et al. 1990; Vourlitis et al. 1993).

However, measured CO_2 exchange rates for apparently similar tundra patches vary significantly, suggesting the need for studies on the spatial variability of ecosystem CO_2 fluxes. Important determinants of uptake capacity include the leaf area index (LAI) of the plant community along with the relative frequency of different growth forms present, the stage of phenological development, and radiation input. In the case of long-term vegetation development, LAI is strongly influenced by nutrient availability, which is in turn largely controlled by topography as it affects the hydrological regime and substrate (Chapin and Shaver 1985a; D. A. Walker et al. 1989; M. D. Walker et al. 1989; Ostendorf and Reynolds 1993; Chap. 10, this Vol.). Our studies in the Imnavait Creek watershed were designed to clarify aspects of both spatial and temporal variation of CO_2 exchange, and to document landscape patterns in gas fluxes (Oberbauer et al. 1991, 1992, 1996). Within the Imnavait Creek watershed the topographic gradient from ridgetop to streamside provides a range of community types that are subjected to differences in belowground physical factors including depth of thaw, depth to water table, soil aeration, soil moisture, and soil temperature, and have large differences in root biomass, soil organic matter content and quality, and in soil bulk density (Walker et al. 1989a,b; Oberbauer et al. 1991, 1992; Chap. 5, this Vol.).

In this chapter, we present data from the Imnavait Creek watershed on both daily and seasonal variation in (1) leaf-level CO_2 exchange rates for the

J.F. Reynolds and J.D. Tenhunen (Eds.)
Ecological Studies, Vol. 120
© Springer-Verlag Berlin Heidelberg 1996

dominant plant species, as influenced by light, temperature, phenology, water availability, and nutrition; and (2) ecosystem CO_2 efflux rates, as influenced by plant biomass, soil quality, depth of thaw and the water table, soil moisture, and soil temperature. These fine-scale process data (*sensu*; Rathstetter et al. 1992) provide the basis for a semimechanistic simulation model (GAS-FLUX; Tenhunen et al. 1994) that we used to examine the complex response of tundra ecosystem gas exchange processes in plant communities along a strong environmental gradient.

11.2 Methods

11.2.1 Community Types

Species composition and structure of the community types at Imnavait Creek watershed are described in detail in Walker and Walker (Chap. 4) and Hahn et al. (Chap. 5, this Vol.). The results presented here focus on communities along an idealized toposequence, from dry ridge crests in the watershed to wet footslope locations (see Figs. 4.1 and 4.8 in Chap. 4, this Vol.). Observations were carried out in upslope lichen heath and wet and dry variants of *Cassiope*-dwarf shrub tundra or heaths (cf. Chap. 5); on the upper and lower backslope with tussocks of *Eriophorum vaginatum*, as well as in intertussock areas and water tracks; and in footslope shrub communities and riparian sedge meadows dominated by either *Carex aquatilus* or *Eriophorum angustifolium*. To examine the effect of road dust deposition on ecosystem CO_2 efflux, transects were also established in tussock tundra perpendicular to the Dalton Highway at a point several kilometers from the Imnavait Creek watershed (cf. Chap. 7, this Vol.).

11.2.2 Leaf Photosynthesis

Leaf photosynthesis in response to temperature and light was measured in the field with a PACS 9900 steady-state gas exchange system with CO_2, temperature, and humidity control (Data Design Group, La Jolla, California). Temperature responses were observed at $800\,\mu mol\,m^{-2}\,s^{-1}$ or higher photon flux density, which was above light saturation. For temperature responses of photosynthesis, we did not maintain constant vapor pressure deficit as temperatures were increased, because the responses were intended to match those that might occur under natural conditions when temperatures rise without an increase in absolute humidity.

Measurements to define seasonal trends were made at light saturation with both PACS 9900 and Li-6200 gas exchange systems (Li-Cor Inc., Lincoln, Nebraska). Changes in leaf photosynthesis in response to fertilization and water addition, diurnal courses of net photosynthesis, maximum photosynthesis rates along a toposequence (Fig. 4.1, this Vol.), and leaf and stem respiration were also measured using Li-Cor portable gas exchange systems. Temperature responses of net photosynthesis at CO_2 saturation to define RuBP regeneration capacity were obtained with an ADC Model 225 infrared gas analyzer (Analytical Development Corporation, Hoddesdon, England) in an open gas exchange system similar to that described by Limbach et al. (1982).

Leaf area photosynthetic rates were based on estimates of projected leaf area. Although projected area underestimates the actual leaf area for species with rolled or cylindrical leaves, these estimates are appropriate when combined with total LAI (Hastings et al. 1989; Oberbauer et al. 1989) to calculate light interception and estimate ecosystem flux rates. In the case of the evergreen *Cassiope tetragona.* which has folded, tightly appressed leaves, projected leaf area was estimated at twice the projected surface area of the intact shoots.

11.2.3 Ecosystem Efflux

Ecosystem CO_2 efflux was measured using a dark-chamber technique (Peterson and Billings 1975), which provides a measure of potential CO_2 losses including respiration of aboveground plant biomass and belowground stems and roots along with microbes. Chambers were made from polyvinyl chloride or acrylic tubes of $50\,cm^2$ area and were inserted into the soil several days prior to conducting measurements. During measurements the tubes were capped and the change in CO_2 concentration inside the tube was measured using an Li-6200 gas exchange system (detailed description of the methods is given in Oberbauer et al. 1991, 1992). Soil microbial basal respiration and biomass were measured with an open gas exchange system following procedures of Cheng and Coleman (1989) and Cheng and Virginia (1993).

11.2.4 Ecosystem Net CO_2 Exchange

To determine net ecosystem exchange of CO_2, an Li-6200 portable gas exchange system was used to monitor the change in CO_2 concentration within stirred 0.25- × 0.25-m Plexiglas chambers. These chambers, a modification of those used by Whalen and Reeburgh (1988), were temporarily sealed to aluminum bases that were permanently installed in the soil (Tenhunen et al. 1995).

11.3 CO_2 Uptake

In this section we examine how leaf photosynthesis in the dominant vascular plants is affected by light, nutrition, water availability, temperature, and phenology. We also consider how each of these variables might vary spatially within the watershed.

11.3.1 Factors Affecting CO_2 Uptake

11.3.1.1 Light

Our field measurements of net photosynthesis rate with natural and artificial light over 24-h cycles indicate that tundra species are capable of near maximal rates of photosynthesis at all times of the day (Gebauer, unpubl. data; Oberbauer, unpubl. data). Diurnal rhythms in stomatal conductance that might restrict CO_2 uptake during nighttime hours were not observed in four species (two evergreens and two decidous shrubs) maintained under constant illumination. Semikhatova et al. (1992) reported that as much as 26% of the carbon uptake of tundra plants occurs between 2200 and 0400 h during mid-summer. This proportion may be even higher for lichens, due to dew condensation (Tenhunen et al. 1992; Hahn et al. 1993).

Although Johnson and Tieszen (1976) reported that photosynthesis in leaves of *Betula nana* and *Salix planifolia* ssp. *pulchra* from Meade River, Alaska, did not light-saturate until $1500 \mu mol\, m^{-2} s^{-1}$, light-saturation of the dominant vascular plant species in the Imnavait Creek watershed occurred between 500 and $1000 \mu mol\, m^{-2} s^{-1}$ (Fig. 11.1a). This is similar to species at other tundra sites (Shvetsova and Voznesensky 1970; Tieszen 1973, 1975, 1978; Tieszen and Johnson 1975; Johnson and Tieszen 1976; Limbach et al. 1982; Semikhatova et al. 1992). Bryophytes and lichens light-saturate below $500 \mu mol\, m^{-2} s^{-1}$ (Harley et al. 1989; Tenhunen et al. 1992). Because of low LAI, light-satuation of whole communities of tussock and intertussock at Imnavait Creek (Tenhunen et al. 1995), and tussock at nearby Toolik Lake, (Grulke et al. 1990) are similar to leaf-level values occurring between 500 and $1000 \mu mol\, m^{-2} s^{-1}$.

11.3.1.2 Temperature

Temperature responses of leaf photosynthesis for the important vascular plant species at Imnavait Creek are fairly broad with optima near 15–20 °C and more than 65% of maximum capacity retained between 10 and 25 °C (Fig. 11.1b). These results are similar to those from previous studies of tundra species (Johnson and Tieszen 1976; Limbach et al. 1982; Semikhatova et al. 1992). Seasonal measurements of photosynthetic temperature responses indicate that the magnitude of temperature acclimation in these species is relatively small (Oberbauer and Oechel, unpubl. data). Despite this broad range, photosynthesis may be frequently temperature-limited, because average daily air temperatures during the growing season are generally below 20 °C (Chap. 6, this Vol.). Semikhatova et al. (1992) reported air temperatures below

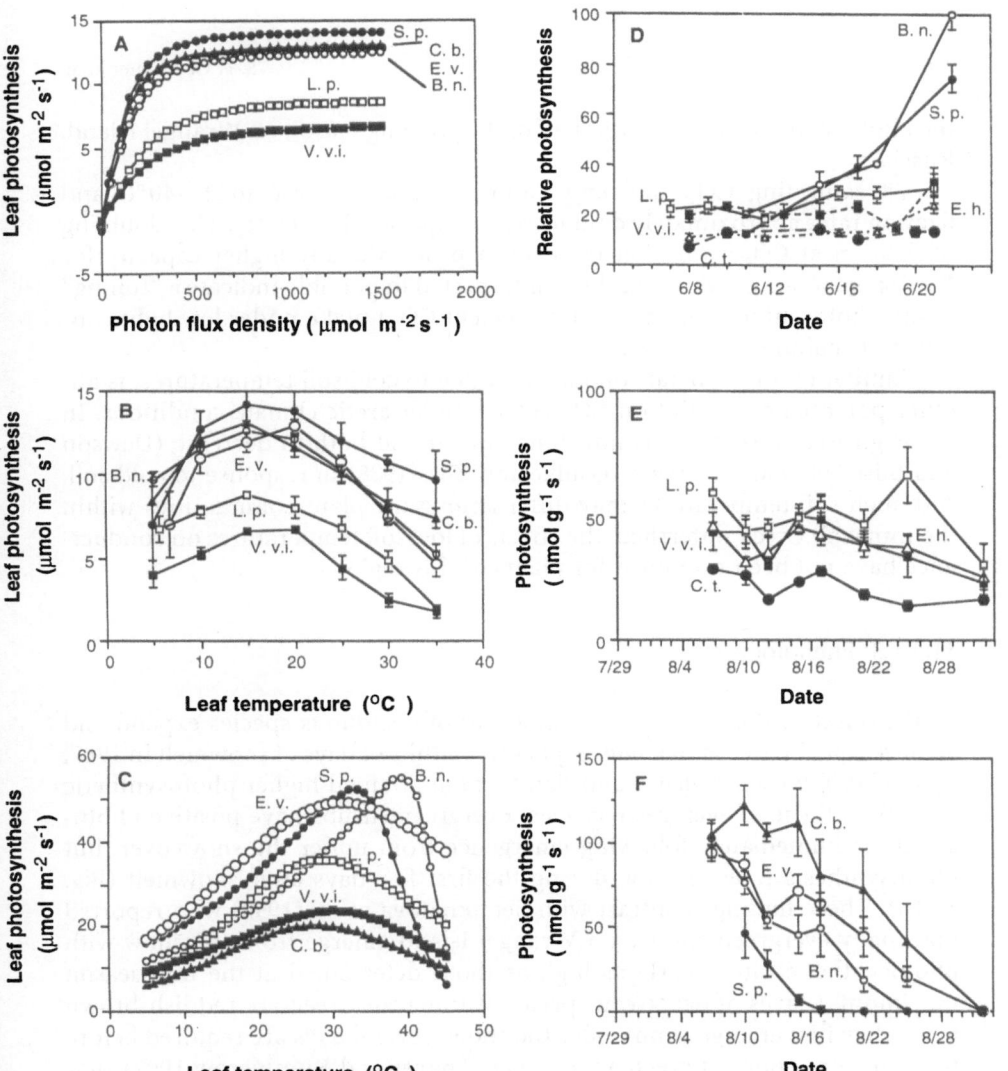

Fig. 11.1A–F. Light, temperature, and seasonal influences on net photosynthesis for dominant vascular plant species at Imnavait Creek watershed growing in situ. **A** Least-squares regressions for photosynthetic light response at 15 °C, 345 μl l^{-1} CO_2, and 35% relative humidity in 1987 fit to the Smith equation (Smith 1937). **B** Temperature responses (± standard error) determined in early August 1987 at 345 μl l^{-1} CO_2, 35% relative humidity and saturating light. **C** Least-squares regressions for CO_2-saturated temperature response according to the function described by Johnson et al. (1942). **D** Relative changes during early-season in maximum photosynthesis rate of deciduous shrubs and evergreens after snowmelt in June 1987 at saturating light. Average conditions during measurements were (mean ± one SD): leaf temperature 12.5 ± 1.7 °C, relative humidity 31.1 ± 7.5%, and CO_2 343 ± 3 μl l^{-1}. **E** Late-season maximum photosynthesis rate of evergreens in 1987 at saturating light and ambient temperature, relative humidity, and CO_2. **F** Late-season maximum photosynthesis rate of deciduous shrubs and graminoid species in 1987 as in **E**. Average conditions during late-season measurements (**E, F**) were (mean ± one SD): leaf temperature 16.6 ± 4.0 °C, relative humidity 48.7 ± 10.3%, and CO_2 327 ± 17 μl l^{-1}. Species symbols are for Evergreens: *L. p.*, *Ledum palustre*; *V. v.i.*, *Vaccinium vitis-idaea*; *C. t.*, *Cassiope tetragona*; *E. h.*, *Empetrum hermaphroditicum*. Deciduous shrubs: *S. p.*, *Salix planifolia* ssp. *pulchra*, *B. n.*, *Betula nana*. Graminoids: *E. v.*, *Eriophorum vaginatum*; *C. b.*, *Carex bigelowii*. (Oberbauer et al., unpubl. data)

the photosynthetic optimum for half of the growing season on Wrangel Island, Russia.

At saturating CO_2, the temperature optima increase to 25–40°C and fixation rates approximately double in most species (Fig. 11.1c). This doubling of fixation at CO_2 saturation indicates a proportionally higher capacity for RuBP regeneration, which has been interpreted to possibly indicate a "tuning" of the photosynthetic apparatus for efficient CO_2 uptake under low light conditions (Tenhunen et al. 1987).

Limitation of stomatal conductance, due to cold soil temperatures, is another potential constraint on CO_2 uptake under arctic climate conditions. In the High Arctic, stomatal conductance was found both to decrease (Dawson and Bliss 1989) and increase (Grulke and Bliss 1988) in response to cold soil. Although soil temperatures may differ among the plant communities within the Imnavait Creek watershed, the effects of low soil temperatures on conductance have not been examined for plants of this region.

11.3.1.3 Phenology

At the onset of the growing season, leaves of deciduous species expand and achieve a positive carbon balance quickly. Within 14 days of snowmelt in 1987, *Salix planifolia* ssp. *pulchra* and *Betula nana* attained higher photosynthetic rates than those of evergreen species. Evergreen shrubs have positive photosynthesis immediately following emergence from under the snowcover, but photosynthesis rates are low during the first few days after snowmelt (Fig. 11.1d). These findings contrast with Semikhatova et al. (1992) who reported that many evergreen species on Wrangel Island emerge from the snow with photosynthetic rates nearly as high as those determined at the mid-season maximum. Leaves of evergreen species at Imnavait Creek are reddish-brown when they first emerge from under the snow. Several days are required before they take on a normal green appearance (Shaver and Kummerow 1992) suggesting that physiological adjustments occur that may affect photosynthetic capacity. The color change is due to reductions in reddish pigments, whereas extractable chlorophyll concentration remains relatively constant during the green-up period (Oberbauer and Oechel, unpubl. data). Restriction of water uptake in the cold or frozen upper soil layers early in the season may contribute to the low photosynthesis rates observed at this time.

Maximum observed net photosynthesis rates of evergreen and deciduous species decline from mid-season values as winter approaches (Fig. 11.1e,f), but evergreens clearly retain full photosynthetic capacity longer than deciduous and graminoid species. Deciduous species exhibited decreases in photosynthesis early in August at Imnavait Creek (Fig. 11.1f). Tieszen and Johnson (1975) also observed a strong seasonality in the photosynthetic capacity of graminoids at Barrow, Alaska. Defoliart et al. (1988) found a similar late-summer reduction in photosynthetic capacity of the sedge *Eriophorum vaginatum* in populations from interior Alaska, and speculated that the response might be a consequence of cold hardening.

11.3.1.4 Water Availability

We tested the possibility that water availability might limit gas exchange rates and productivity at Imnavait Creek by conducting water addition experiments over two growing seasons. In response to irrigation, growth of some species was promoted, but the growth increase seemed to indirectly result from additional supply of nutrients, rather than to improved water status (Oberbauer et al. 1989). Diurnal observations of photosynthesis in dry shrub and tussock tundra did not indicate significant stomatal closure in response to humidity deficits or with reduced water potentials (Gebauer 1994; Gebauer et al., unpubl. data). Surprisingly, *Eriophorum angustifolium*, which grows in waterlogged soil in the riparian zone, appears sensitive to humidity deficits (Gebauer 1994). Water additions to lichen-heath during most of a growing season did not produce a noticeable increase in growth of vascular plant species (Oberbauer, unpubl. data). Similar results have been obtained in lichen-heath communities for other tundra sites in Alaska (McGraw 1985; Fox 1992). Although Murray et al. (1989a) and Tenhunen et al. (1992) reported strong effects of water content (below ca. 800% of dry weight) on the maximum net photosynthesis rate of terminal shoots of *Sphagnum* mosses from tussock tundra, water content of *Sphagnum* in midslope and riparian communities usually remains high, due to the capillary rise of water to the surface. *Sphagnum* growth was not increased by irrigation, but appeared limited by other factors (Murray et al. 1989b), especially by exposure to high light intensity, which leads to photoinhibition (Murray et al. 1993).

Carbon isotope ratios of leaf material do change along the water availability gradient from ridgetop to riparian communities (B. Barnett, pers. comm.), suggesting that upslope communities may have lower stomatal conductances. Matthes-Sears et al. (1988) concluded that during an exceptionally dry period in 1985, upslope individual plants of *Betula nana* and *Salix planifolia* ssp. *pulchra* had lower stomatal conductances and net photosynthesis rates when compared with downslope individuals. Variation in carbon isotope ratios could result from local water stress, but might also occur as a result of lower humidity and exposure to stronger winds at ridgetop locations.

11.3.1.5 Nutrition

Tundra plant growth and production are in most cases enhanced by increased nutrient availability, at least where relatively large additions of nitrogen (N) and phosphorus (P) have been made (Haag 1974; Ulrich and Gersper 1978; Shaver and Chapin 1980; Miller 1982; Chapin and Shaver 1985b, 1989; Kummerow et al. 1987; Chapin et al. 1988; Chap. 10, this Vol.). However, tundra plants growing in situ, unlike many C$_3$ species (Field and Mooney 1986), do not show large increases in photosynthesis with increased leaf N concentrations. At Toolik Lake, increased leaf N and P, in response to heavy fertilization, did not increase photosynthesis of the deciduous shrubs, *Betula nana* and *Salix planifolia* ssp. *pulchra* (Matthes-Sears et al. 1988). Although growth of graminoids, evergreen shrubs, and deciduous shrubs was enhanced by NPK addition at Imnavait Creek (Kummerow et al. 1987), photosynthetic capacity of the measured species in tussock tundra increased only slightly (Oberbauer et al. 1989). Therefore, at the leaf level, characteristics of the photosynthetic apparatus seem strongly fixed and whole plant response in nutrient-rich situations was manifested by changes in allocation and increased leaf area development. Nevertheless, where long-term observations of fertilization plots have been conducted, species with higher photosynthetic capacity on a leaf area basis replace those species originally at the local site.

Results obtained within the Imnavait Creek watershed are consistent with previous studies (Tieszen 1978; Bigger and Oechel 1982). In fact, Bigger and Oechel (1982) reported reductions in leaf photosynthetic rate in response to N fertilization in tussock tundra at Eagle Creek, Alaska. They suggested that essential micronutrients were diluted in the plant biomass, due to the increased growth in response to N addition. Micronutrient limitation of photosynthesis was tested in tussock tundra at Imnavait Creek, but responses of the magnitude reported by Bigger and Oechel (1982) were not found for the addition of any single micronutrient (Oberbauer et al., unpubl. data).

11.3.2 Landscape Patterns in Leaf Photosynthesis

Given the process-level differences in leaf photosynthesis of the species described previously, landscape patterns in leaf photosynthesis vary as a function of spatial and temporal distributions of the important abiotic driving variables in the watershed. This includes light, temperature, water, and nutrient availability. Landscape patterns of CO$_2$ flux also depend on the biochemical characteristics of the species, their leaf area, and phenological status of the individual plants. Species biochemistry and nutritional status determine the baseline from which photosynthesis responds to the current microenvironment. In this section we discuss general landscape patterns of the abiotic driving variables and examine spatial patterns in leaf photosynthetic potential, both within species and with regard to community distribution (Oberbauer and Oechel, unpubl. data) for the dominant species at mid-season in each of six communities along a 600-m transect from ridgetop to stream edge in the Imnavait Creek watershed (see Fig. 11.2).

The low angle of direct-beam radiation combined with meso- and microtopography results in a complex mosaic of radiation regimes on the tundra surface. The contrast in microclimate between illuminated and shaded areas is particularly strong during late-evening, through the night period, and

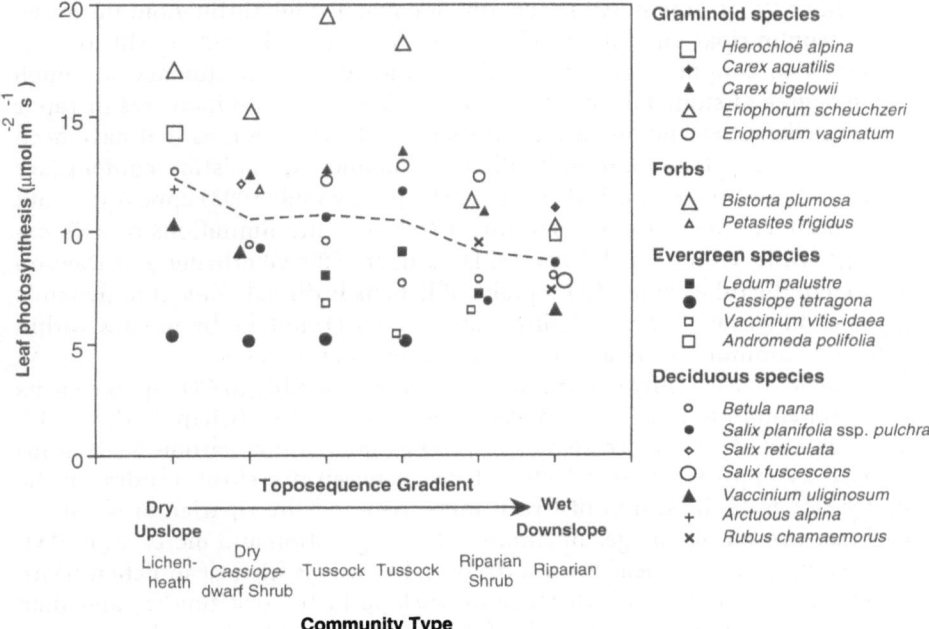

Fig. 11.2. Light-saturated photosynthesis rate at mid-season of the eight most abundant vascular plant species at each of six sites along a toposequence from ridgetop to stream edge within the Imnavait Creek watershed in 1987. Average conditions during measurement were (mean ± one SD): leaf temperature 19.8 ± 4.0 °C, relative humidity 52.2 ± 10.8%, and CO_2 317 ± 10 µl l^{-1}. *Dashed line* indicates the mean of all species rates observed. (Oberbauer and Oechel, unpubl. data)

in the early-morning hours when diffuse light levels are low. These patterns of incident solar flux over the landscape may play a role in determining community characteristics and distribution.

Variations in air temperature during the growing season within the Imnavait Creek watershed are small, due to relatively small topographic variation. Large temperature differences may occur on calm nights when cold air settles into the basin, or during summer snow storms, when elevational differences of a few meters determines whether the vegetation becomes snow-covered for a few days. In both cases light levels are usually low during these periods, suggesting that the cumulative effects of air temperature on photosynthetic production are small. Rather, the large variation in leaf temperatures observed depends on net radiation, latent heat exchange, and convective cooling. The combination of high winds and a short-statured canopy on ridgetops may result in substantially lower leaf temperatures within upslope communities relative to those of downslope communities.

In most plant communities in the watershed, reduced water availability or water stress does not usually limit CO_2 uptake of vascular plants. Similar conclusions have been drawn for tussock tundra in interior Alaska (Oberbauer and Miller 1979). Soil moisture remains adequately high throughout the growing season in riparian, water track, tussock, intertussock, and dry shrub communities to support transpirational demand with open stomata, although some degree of stomatal closure may occur in response to increases in vapor pressure deficit at midday in certain species. The correlations that have been found between plant community distribution and soil moisture content (see Chaps. 14, 17, and 18, this Vol.; Ostendorf and Reynolds 1993) appear to result from indirect effects; rather than direct soil moisture limitations on CO_2 exchange (D. A. Walker et al. 1989; M. D. Walker 1989; Oberbauer and Dawson 1992). On the other hand, CO_2 uptake of lichens is directly linked to dynamic changes in surface layer moisture status as is CO_2 uptake by mosses within ridgetop communities (Hahn 1991; Tenhunen et al. 1992).

Nutrient availability determines landscape variability in CO_2 uptake via its effect on community biomass development, i.e., on LAI (Chap. 5, this Vol.). Tenhunen et al. (1994; see also Sect. 11.5) estimated that carbon fixation increased by approximately a factor of two between dry shrub tundra on the upper slopes of the Imnavait Creek watershed and the riparian areas of the basin as a result of changes in community composition and increases in LAI. Interestingly, average leaf N on a dry-weight basis is highest in lichen heath vegetation, decreases with distance downslope in tussock tundra, and then increases in riparian vegetation. Leaf P on a dry-weight basis declines consistently with distance downslope; levels in the riparian vegetation are only 65% of those in the lichen-heath (Oberbauer and Oechel, unpubl. data).

Carbon dioxide uptake is strongly determined by the phenological status of plants within the community (Tenhunen et al. 1994). Phenological development is related to the timing of snow melt (Chap. 6, this Vol.) and the cumulative exposure to above-freezing air temperatures. Viewed from a landscape perspective, onset of development of the vegetation canopy is to a large degree a function of patterns in snow accumulation (Sorenson 1941; Murray and Miller 1982; Shaver and Kummerow 1992). For example, ridgetops are typically snow-free earlier and are, thus, phenologically advanced as compared with communities further downslope. The progression of snowmelt and community development is additionally influenced by topography, elevation, and effects of flowing water.

Despite large differences in the photosynthetic capacity of the individual component species, we found that the average maximum photosynthetic capacity was similar among the communities (Fig. 11.2). Although community CO_2 fluxes depend on the proportional distribution of leaf area operating at various photosynthetic rates – not the average maximum rate of the community – it is interesting that the upslope communities are composed of a mixture of species with high and low maximum photosynthetic rates, whereas the riparian community is composed of species with similar maximum rates. It is paradoxical that heath communities have the lowest productivity, yet contain species with the highest photosynthetic rates. The two species present at the widest range of sites, *Betula nana* and *Salix planifolia* ssp. *pulchra*, showed reductions in light-saturated photosynthesis with distance downslope, although such patterns were not apparent for *Eriophorum vaginatum, Bistorta plumosa*, or *Cassiope tetragona*.

Large differences in photosynthetic capacity on a leaf area basis did not occur for most species within the range of habitats found in the Imnavait Creek watershed. Significant differences did occur, however, in the photosynthetic capacity of different growth forms. Forbs had the highest rates, deciduous shrubs and graminoids had intermediate rates, and evergreens had the lowest rates. From the perspective of modeling landscape patterns of carbon uptake, it is preferable to consider photosynthetic rates at the level of growth form, rather than at the species level. Limbach et al. (1982) and Oberbauer and Oechel (1989) concluded that evergreen species have uniformly low rates. However, considerable differences were found among graminoid and deciduous species suggesting that grouping species by growth form for the purposes of modeling carbon flux may be unwarranted.

11.4 CO$_2$ Efflux

Landscape patterns of potential CO$_2$ efflux have been extensively examined based on darkened-chamber techniques (Billings et al. 1977). Although small, respiration from aboveground biomass under artificially darkened conditions is included in these estimates. Studies of ecosystem efflux from arctic sites have in most cases focused on the role of abiotic factors, such as soil temperature, soil moisture, and water table, although the extent that organic matter quality determines efflux has recently received greater attention (Nadelhoffer et al. 1991, 1992; Cheng and Virginia 1993). Our studies of CO$_2$ efflux from communities of the Imnavait Creek watershed examined the relative importance of variation in the soil physical environment and biotic properties of the soil communities (substrate quality, root density and biomass, and microbial population dynamics; see also Chap. 10, this Vol.).

11.4.1 Factors Affecting CO$_2$ Efflux

With the exception of CO$_2$ released by physical processes, such as dissolved CO$_2$ or CO$_2$ displaced by moving water (Peterson and Billings 1975), the amount of living biomass and the amount and quality of organic matter determine the efflux rate for a given site. Because tundra ecosystems usually have greater biomass belowground (as much as 85–98% of the biomass in wet sedge tundra is belowground; Billings et al. 1977; Chapin and Shaver 1985a), the major portion of CO$_2$ efflux is derived from this source. Billings et al. (1977) suggested that roots account for 50–90% of the belowground efflux. Time-varying factors that modify belowground respiration are (1) depth to which

Table 11.1. Mid-season dark respiration rates of selected tundra species. Rates for leaves measured in situ at 15 °C and expressed on a leaf-weight and leaf-area basis. Rates include the contribution of supporting stem tissue in cuvette, except for *Carex bigelowii and Eriophorum vaginatum*, which are for leaves only ($n = 6$). Rates for stems measured in situ at 17 °C and expressed on a surface area and stem dry-weight basis ($n = 5$)

	Leaves		Stems	
	$\mu mol\,m^{-2}\,s^{-1}$	$nmol\,g^{-1}\,s^{-1}$	$\mu mol\,m^{-2}\,s^{-1}$	$nmol\,g^{-1}\,s^{-1}$
Deciduous shrubs				
Salix planifolia ssp. *pulchra*	1.61	16.0	0.73	2.07
Betula nana	0.96	12.8	0.81	2.42
Evergreens				
Ledum palustre	1.21	8.7	0.83	3.88
Vaccinium vitis-idaea	0.80	5.3		
Graminoids				
Eriophorum vaginatum	1.27	12.8		
Carex bigelowii	0.77	10.9		

soil is thawed, (2) depth to water table (volume of aerated soil), (3) soil temperature, and (4) soil moisture content.

11.4.1.1 Live Plant Biomass

Leaf respiratory rates of evergreen species are lower than those of graminoids and deciduous species, and woody stems have relatively low, dark respiration rates (Table 11.1). Respiratory rates of vascular plants are generally greater than those of lichens and mosses (Tenhunen et al. 1992). Temperature is the primary environmental factor determining vascular plant respiration, whereas the metabolism of lichens and mosses is directly correlated with short- (hours) and long-term (days, weeks) changes in water content (Harley et al. 1989; Murray et al. 1989a; Hahn 1991; Tenhunen et al. 1992). Because dewfall and rainfall usually occur when temperatures are low, and moisture content is lowest when temperatures are high, the physiological activity of lichens and dry-site mosses is often out of phase with that of vascular plants. Vascular plants also have strong seasonal changes in above- and belowground activity that influence the CO_2 efflux rates (McNulty and Cummins 1987). Moore (1989) suggested that such phenological trends were responsible for strong seasonal changes in CO_2 efflux in fen sites.

11.4.1.2 Soil Quality

Although arctic soils may be generally characterized as cold, wet, and rich in organic matter, large differences exist at local and regional scales, due to the

Table 11.2. Soil characteristics of six communities in the Imnavait Creek watershed

Community	N (%)	C (%)	C:N	Bulk density (g cm⁻³)
Lichen heath	0.93	16.8	19.4	0.379
Cassiope-dwarf shrub heath				
Dry	0.50	11.5	25.6	0.310
Moist	1.66	35.1	22.0	0.113
Tussock	0.48	43.1	91.5	0.106
Intertussock	0.69	39.0	60.7	0.059
Water track	1.43	34.6	26.7	0.089
Wet sedge				
Carex aquatilis	1.06	29.8	31.1	0.078
Eriophorum angustifolium	1.20	20.15	16.79	0.152

N, nitrogen; C, carbon.

effects of topography, hydrology, underlying parent materials, and microclimate (D. A. Walker et al. 1989; Giblin et al. 1991). In terms of hydrology, tundra types may be coarsely classified as dry, moist, and wet (Nadelhoffer et al. 1992), although within each class, many distinctive soil types are recognized that differ with regard to organic carbon content, microbial biomass, total nitrogen content, bulk density, etc. The entire range of soil types may be found in the Imnavait Creek watershed (Table 11.2; D. A. Walker et al. 1989; Cheng and Virginia 1993).

In a laboratory incubation study of six arctic soils, Nadelhoffer et al. (1991) found that the quality of soil organic matter was more important than soil temperature in controlling rates of CO_2 efflux. Cheng et al. (1996) found that carbon availability for microbial utilization varied significantly and systematically among soils collected along a toposequence in the watershed (see Chap. 4, this Vol.). They reported high carbon availability in wetter sites and lower carbon availability in the relatively dry sites.

Soil basal respiration is an integrative measure of soil respiration potential. Differences in basal respiration reflect differences in both microbial biomass and microbial metabolic activity. In a study of seven communities during the 1990 growing season, Cheng et al. (1996) found that (1) basal respiration rate (expressed on a per-gram-soil basis) was inversely proportional to soil bulk density, (2) soil organic carbon content was negatively related to soil bulk density, and (3) lichen-heath soil had the lowest basal respiration rate, whereas water track soil has the highest basal respiration rate (Fig. 11.3). Variation in basal respiration rate correlated strongly with that of ecosystem CO_2 efflux from these different communities measured with darkened-chamber techniques (Oberbauer et al. 1991, 1992, 1996). Long-term changes in total microbial respiration appears to strongly determine the seasonal pattern of ecosystem CO_2 efflux at Imnavait Creek (Oberbauer et al. 1992), apparently in direct response to changes in soil temperature and depth

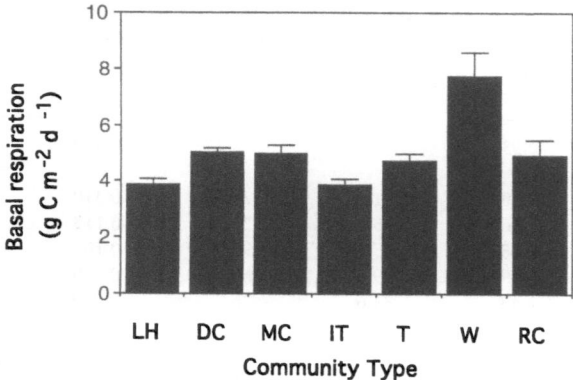

Fig. 11.3. Basal respiration rates of upper 5 cm of soil from six communities along a toposequence from ridgetop to stream edge within Imnavait Creek watershed incubated in the laboratory at 22 °C under aerobic conditions. Community designations are: *LH*, lichen heath; *MC*, moist *Cassiope*-dwarf shrub; *DC*, dry *Cassiope*-dwarf shrub; *IT*, intertussock areas at mid-slope; *T*, tussocks of *Eriophorum vaginatum* at mid-slope; *W*, water track; *RC*, riparian wet sedge meadow (*Carex aquatilus*). Communities described in Walker and Walker (Chap. 4, this Vol.) and Hahn et al. (Chap. 5, this Vol.)

of aeration, rather than to any changes in microbial biomass and basal respiration (Cheng et al. 1996).

11.4.1.3 Thaw Depth and Depth to Water Table

Variations in seasonal patterns in the depth of the water table and in the depth of thaw in seven communities along a toposequence in Imnavait Creek watershed is shown in Fig. 11.4. Thaw depth and depth to water table affect CO_2 efflux via their influence on the total volume of aerated soil. Efflux of CO_2 from thawed soil below the water table is reduced essentially to zero. Oxygen microelectrode measurements indicate that arctic soils are aerated down to the water table (Chap. 12, this Vol.). Billings et al. (1982, 1983), using tundra microcosms, showed that depth to water table was a stronger control on net carbon exchange via soil respiration than nutrient status. Moore (1989) found large differences in CO_2 efflux for fen sites with different water tables, and Gorham (1991) identified depth to water table as one of the factors most likely to affect carbon balance of northern systems in response to climate change.

At Imnavait Creek effluxes from two riparian communities remained relatively low until water table depth increased near the end of the season (Oberbauer et al. 1992). We attributed the differences in efflux between and within these communities to variations in the water table depth. The water table in these two communities may change rapidly enough to affect the daily pattern of CO_2 efflux, as occurred following a summer thundershower, efflux

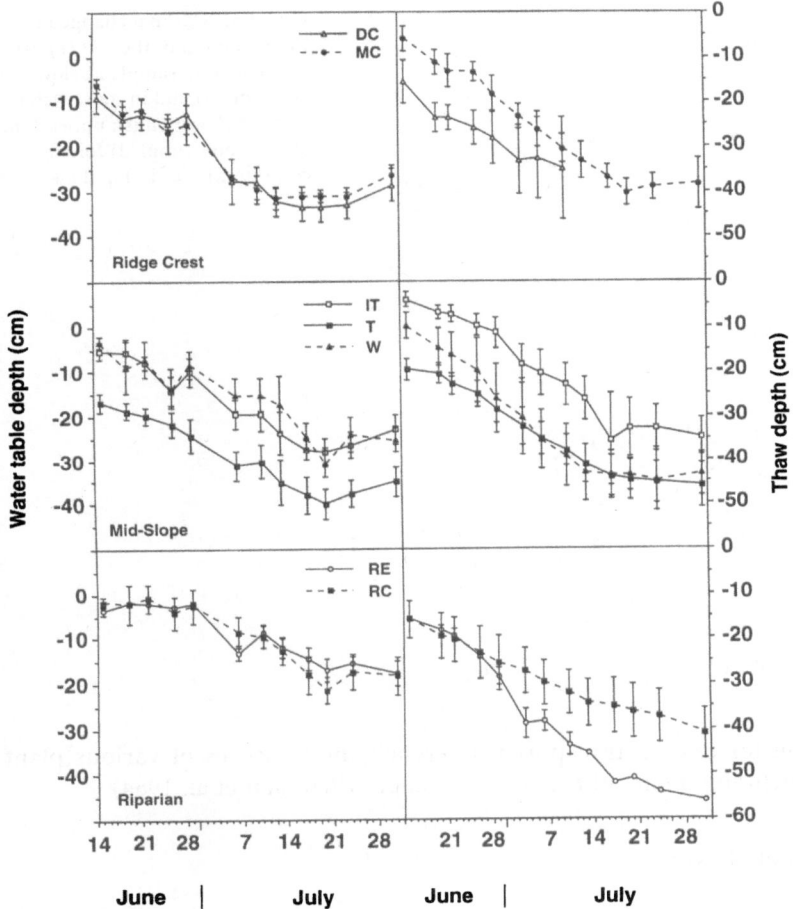

Fig. 11.4. Seasonal patterns in depth to water table and in depth of thaw in soil measured from the moss surface for seven communities along a toposequence from ridgetop to stream edge within the Imnavait Creek watershed during 1990. With some variation in sampling, symbol indicates the average value for two replicates collected from five plots six times daily. Frequent rain occurred until 4 July followed by continuous drying (see Fig. 11.6). Community designated as *RE* refers to pure stands of *Eriophorum angustifolium*; other designations as in Fig. 11.3

decreasing as the water table rose. In contrast, Billings et al. (1977) reported a sharp rise in efflux following a rainfall. The rain increased soil temperature; it was not possible to determine if the increase in respiration was due to increased temperature or displacement of CO_2 trapped in soil pores.

Our results suggest that for the purpose of modeling tundra ecosystem carbon balance, soil beneath the water table may be treated as nonrespiring, but water table depth is of critical importance (Oberbauer et al. 1992; Tenhunen et al. 1994; Chap. 17, this Vol.). Refinements might consider differences in aeration that may occur between standing and flowing water and by

238   S. F. Oberbauer et al.

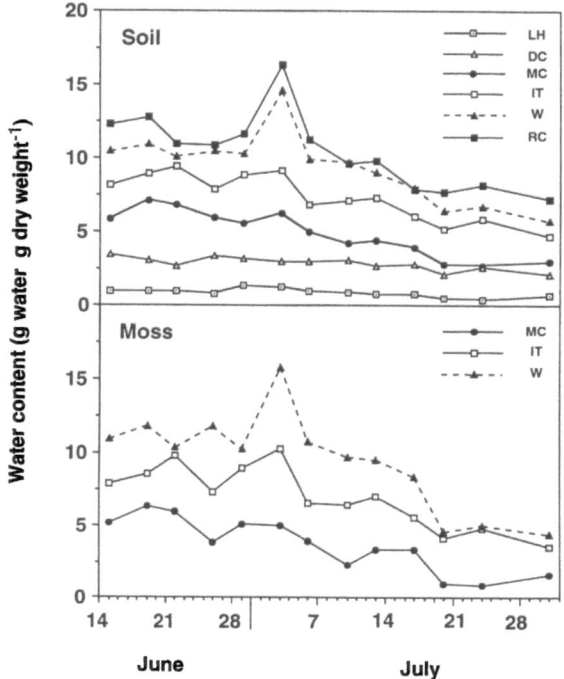

Fig. 11.5. Seasonal change in water content of the soil (*upper panel*: 0–5 cm sampled below green moss) and in moss water content (*lower panel*: upper 2 cm of cushions sampled) for six communities as in Fig. 11.4

accounting for oxygen transport via aerenchymous tissues of various plant species, including many of the riparian sedges (Peterson et al. 1984)

11.4.1.4 Soil Moisture

Luken and Billings (1985) reported a negative effect of water content on CO_2 release from a subarctic bog. However, Svensson (1980) found a positive relationship between soil CO_2 release and soil water content. Decker (1981) and Siegwolf (1987) reported parabolic responses with a defined water content optimum for respiration in alpine heath soils as has generally been suggested for decomposition. (Waring and Schlesinger 1985). We (Oberbauer et al. 1991) similarly identified a parabolic response for intertussock and water-track tundra, although our database was not sufficiently large to adequately separate the effects of depth to water table from the direct effects of soil moisture.

The range of water content change that occurs in soils of various tundra communities is difficult to define. Even considerations of where to measure water content is a challenge. This is illustrated in Fig. 11.5 by biweekly water content observations of soil (0–5 cm below green moss) and moss (upper 2 cm of cushions) for the summer of 1990 in various community types in the watershed. The season was characterized by warm, moist weather in the spring, which led to rapid and early development of the vegetation (full canopy expansion by mid-June). Moist conditions with frequent rain prevailed until 4 July, after which rain was absent and the watershed continued to dry until August. The dry-down effect and its variation in the communities at different topographic locations are apparent.

Our laboratory studies and field observations of moist tundra soils (*Carex* meadow, water track, midslope intertussock, and moist *Cassiope*-shrub in Fig. 11.5) suggest that respiration is insensitive to moisture across a broad range of water content (essentially the range of conditions observed for these communities in Fig. 11.5, although late season water content below $5\,g\,g^{-1}$ in dry shrub may have an effect). We conclude that soils in tussock tundra do not usually dry sufficiently enough to limit CO_2 efflux, whereas in the well-drained upland tundra sites (dry *Cassiope*-dwarf shrub tundra and lichen heath) this is not the case. For example, lichen heath soils at Imnavait Creek were below $1\,g\,H_2O\,g^{-1}$ soil ($0.4\,g\,H_2O\,cm^{-3}$ soil) for most of the growing season in 1990. In comparison, soils in the wet sedge community near Imnavait Creek remained above $6\,g\,H_2O\,g^{-1}$ soil even after prolonged drought. Late-season CO_2 efflux rates from ridgetop sites in 1990 were considerably lower than might be expected based on soil temperatures, suggesting that soil moisture content had indeed decreased to limiting levels. Soil moisture data suggest that dewfall changes the hydration of the soil surface layer (0–2 cm depth) at these sites on a diurnal basis (Oberbauer et al. 1996). Such changes could be responsible for diurnal changes in microbial physiological activity and in CO_2 efflux that are similar to patterns observed for lichen respiration.

11.4.1.5 Soil Temperature

Most field studies of CO_2 efflux in arctic and alpine sites suggest that soil temperature is a primary controlling factor (Billings et al. 1977, 1978; Decker 1981; Moore 1986; Luken and Billings 1985; Siegwolf 1987; Grulke et al. 1990; Oberbauer et al. 1991, 1992). As an enzymatically regulated process, the relationship between CO_2 efflux and soil temperature might be expected to be exponential (Waring and Schlesinger 1985) as reported by Decker (1981) and Siegwolf (1987) for alpine soils. Whereas Peterson and Billings (1975) found an exponential response in CO_2 efflux to long-term seasonal temperature changes, they found a linear response to diurnal changes. Other studies also report both linear and nonlinear relationships between CO_2 efflux and soil temperature (Luken and Billings 1985; Moore 1986). Our field and laboratory results for Imnavait Creek soils have been consistent with an exponential response (Oberbauer et al. 1991, 1992).

Soil temperature variations are primarily responsible for diurnal changes in CO_2 efflux. Defining appropriate measures of soil temperature to relate to the vertical CO_2 efflux at the soil surface measured in darkened chambers is difficult, because the temperature change with increasing depth in the soil is very strong. The best correlations for soils at Imnavait Creek were obtained between CO_2 efflux and soil "surface" temperature measured at 1 cm below the moss surface. Over the course of the day soil temperatures in deeper layers lag behind the changes in soil temperature in the upper soil horizon where most of the respiratory activity takes place (Billings et al. 1977; Nadelhoffer et al. 1991).

At drier sites in the watershed, such as found in lichen heath, the interactive effects of temperature and water content determine soil CO_2 efflux. At very low soil moisture contents respiration is relatively independent of temperature. Sensitivity of mineralization to soil temperature increases with increasing soil moisture (Nadelhoffer et al. 1991). Our studies indicate that correlations between CO_2 efflux and soil temperature were much stronger for wet sites than for the drier ridgetop sites, although drier sites showed much greater diurnal variation in soil temperature than wetter sites (Oberbauer et al. 1991, 1992).

11.4.2 Landscape Patterns of CO_2 Efflux

Rates of CO_2 efflux from riparian sites are primarily determined by the effects of water table and depth of thaw on the volume of aerated active soil, and secondarily by soil temperature (Fig. 11.6). Soil moisture does not directly limit CO_2 efflux. Our field observations further indicated that the effect of water table was negligible at depths below 10–11 cm, i.e., that there is an effective soil layer ca. 10 cm thick from which CO_2 may diffuse to the surface. The dependency of CO_2 efflux on water table depth and soil temperature is described by a model that calculates the volume of soil available as a source for CO_2, and respiratory activity dependent on temperature (Oberbauer et al. 1992). Efflux of CO_2 at midslope tussock and intertussock sites is similarly regulated by depth of thaw, water table, and soil temperature. At these sites direct soil moisture effects may occasionally be limiting (Fig. 11.7; Oberbauer et al. 1991). In these communities abiotic factors explain similar amounts of the variation in observed rates as found for the riparian vegetation.

For well-drained sites on ridgetops, the importance of the depth of thaw and the water table in controlling CO_2 efflux is greatly diminished. Soil temperature and soil moisture appear to best explain seasonal variation in efflux rates in both lichen-heath vegetation and *Cassiope*-dwarf shrub tundra, although in the latter case some influence of thaw depth and water table depth is discernable (Fig. 11.8).

Based on site quality and abiotic differences in the watershed, we expected substantial differences in average seasonal CO_2 efflux (Fig. 11.9). Highest efflux was found for water track tundra and lowest efflux for lichen heath vegetation. These measurements include the seasonal period with greatest biological activity, but do not include observations immediately following snowmelt or during extreme late-season. In the case of riparian tundra, our study terminated when carbon losses were at their maximum due to high temperatures and low water

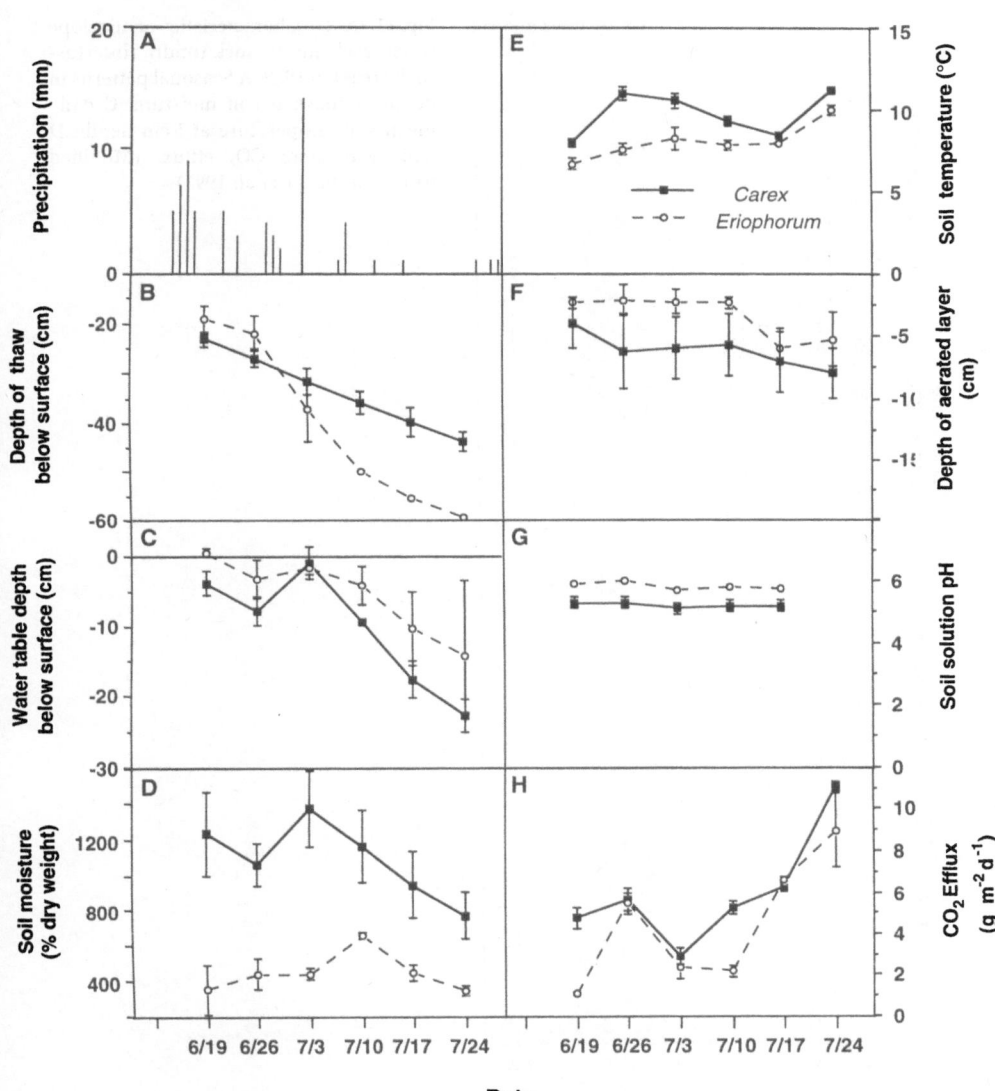

Fig 11.6A–H. Characteristics of riparian communities in 1990. A Precipitation; B depth of thaw; C depth to water table; D soil moisture content at 0–5 cm below green moss; E daily mean soil temperature at 5 cm depth under *Carex aquatilus* and *Eriophorum angustifolium*; F depth of soil aerated layer estimated via oxygen diffusion resistance; G soil solution pH; H daily ecosystem CO_2 efflux based on integration of measurements at 4-h intervals. Values are means ± one standard error; six plots for *Carex* and two plots for *Eriophorum*. (Modified from Oberbauer et al. 1992)

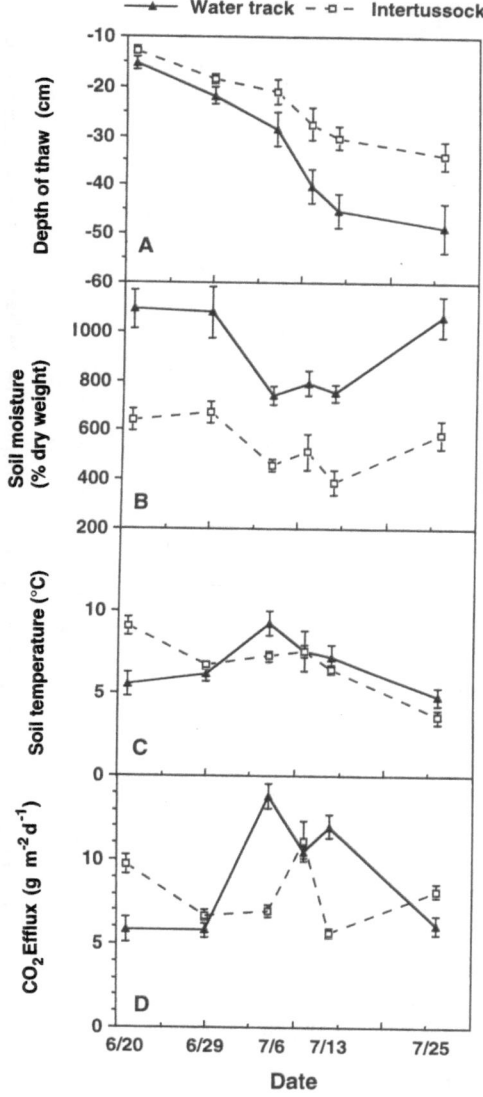

Fig. 11.7A–D. Characteristics of midslope water track and tussock tundra (intertussock areas) in 1989. **A** Seasonal patterns in depth of thaw; **B** soil moisture; **C** daily mean soil temperature at 5 cm depth; **D** daily ecosystem CO_2 efflux. (Modified from Oberbauer et al. 1991)

tables. The high efflux rates probably persisted for some time, and the average rates may be underestimates.

The average efflux rate among communities was very similar to that found for soil basal respiration (Figs. 11.3 and 11.8), supporting the idea that microbial respiratory potential exerts a strong control on CO_2 efflux variation at the landscape level. These results are consistent with the suggestion by Poole and Miller (1982) that carbon losses increase with community productivity. For example, on the ridgetop the moist and dry *Cassiope*-dwarf shrub communi-

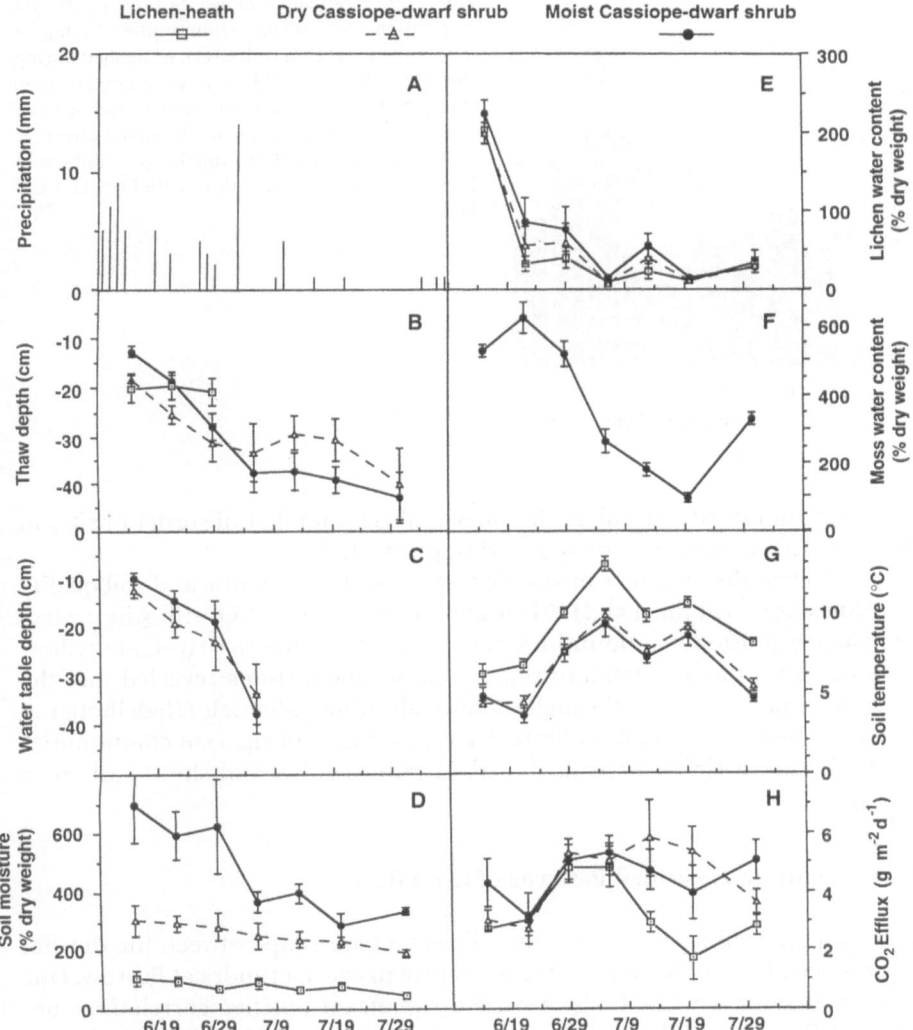

Fig. 11.8A–H. Characteristics of three upland tundra heaths in 1990. **A** Precipitation; **B** thaw depth measured from moss surface; **C** depth to water table; **D** soil moisture content at 0–5 cm below green moss; **E** daily average water content of lichen *Masonhalea richardsonii* observed at 4-h intervals; **F** daily average water content of mosses observed at 4-h intervals; **G** daily mean soil temperature at 5 cm depth; **H** daily ecosystem CO_2 efflux based on integration of measurements at 4-h intervals. Values are means ± 1 SE ($n = 6$). (Oberbauer and Tenhunen, unpubl. data)

ties had higher CO_2 efflux rates than the lichen heath (Oberbauer et al. 1992), tussocks had higher efflux rates than intertussocks (Tenhunen et al. 1995), and water tracks had greater effluxes than intertussock (Oberbauer et al. 1991). In each case the differences in productivity are similarly ordered. Moore (1986),

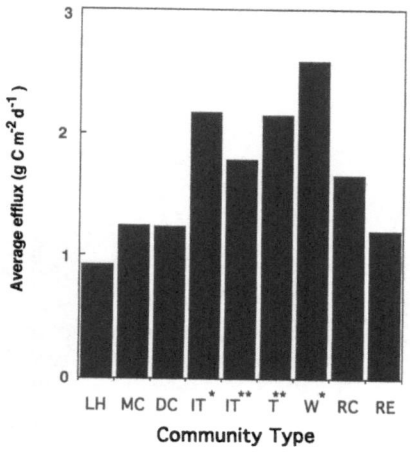

Fig. 11.9. Seasonal average carbon efflux via respiration for seven communities along a toposequence at Imnavait Creek watershed during 1989 (sampled from 20 June to 25 August) or 1990 (sampled from 15 June to 31 August). *, data from 1989 using stirred chamber; **, stirred chamber 1990. All other data from unstirred chambers in 1990. Community designations as in Figs. 11.3 and 11.4

in a study of five sites in subarctic Quebec, also found that the order of sites in terms of CO_2 evolution corresponded to productivity.

However, the influence of abiotic factors may mask biotic and soil quality considerations. Giblin et al. (1991) found highest rates of CO_2 efflux in the less productive hillslope communities and lower efflux for the riverside willow community, although standardized laboratory incubations revealed that the riverside willow soils had the highest mineralization potentials (Nadelhoffer et al. 1991). In this case, high water table, as in the case of riparian communities of the Imnavait Creek watershed, acts to reduce efflux and shunt carbon to storage.

11.4.3 Daily and Seasonal Patterns of CO_2 Efflux

Peterson and Billings (1975) found a linear relationship between the diurnal course of soil CO_2 efflux and soil temperature in coastal tundra at Barrow. Our studies at Imnavait Creek also revealed significant positive correlations between CO_2 efflux and soil temperature for most sampling dates, although correlations over the day were stronger for wet sites than dry sites, suggesting that other factors operate over the daily time scale at dry sites that are not well understood.

Efflux of CO_2 for most sites within the Imnavait Creek watershed was maximal at mid-season as soil surface temperatures also reached their maxima (Figs. 11.6 and 11.7). Billings et al. (1977) found that maximum ecosystem respiration from coastal tundra at Barrow occurred in conjunction with the highest mean soil temperature. Moore (1989) described strong seasonal changes in CO_2 efflux with maxima near midsummer in subarctic patterned fen. Poole and Miller (1982) compared mid- and late-season, but again reported lower efflux rates for the late-season measurements. Grulke et al. (1990)

found higher rates at mid-season for tussock tundra than during late-season, which they attributed to differences in soil temperature. Giblin et al. (1991) measured CO_2 efflux at early, mid-, and late season in six community types, and found the highest rates in mid-season for all but two riparian sites close to the nearby Sagavanirktok River, where soil temperatures were relatively high early in the season. However, CO_2 efflux in two riparian communities we sampled in the Imnavait Creek watershed – wet sedge meadows with *Carex aquatilus* and stands of *Eriophorum angustifolium* – did not reach maximum until after the seasonal maximum in temperature when the water tables fell. The same pattern was found for tussock and intertussock sites in 1990 (Tenhunen et al. 1995). Thus, seasonal and diurnal soil temperature variation must be generally recognized as a major factor determining CO_2 efflux with additional interactive effects such as those of water table.

Difficulties encountered in describing the exact causes for variation in ecosystem CO_2 efflux may be attributed to the complex interactive effects of thaw depth, depth to water table, soil moisture, water flow, and soil temperature gradients on both root and microbial metabolism. Depending on the particular season, microbial basal respiration may remain relatively stable (Sect. 11.4.1.2; Cheng et al. 1996) or may in fact respond to changes in carbohydrate, nitrogen, or phosphorus availability (Luken and Billings 1985; Nadelhoffer et al. 1992). Freeze–thaw events provide one example of short-term events that are known to release carbohydrates (Luken and Billings 1985; Chap. 10, this Vol.) and stimulate mineralization. Such effects could be overlaid on an otherwise consistent and stable functioning of the soil microbes, leading to changes in respiratory capacity.

11.4.4 Dust Deposition Effects on CO_2 Efflux

Although our knowledge of factors regulating CO_2 efflux from undisturbed tundra soil has grown considerably in recent years, much less is known about changes in ecosystem processes, especially soils processes, under the influence of disturbance (Chap. 10, this Vol.). Odum (1985) suggested that one of the responses to be expected in stressed ecosystems is an immediate increase in community respiration. To estimate the potential magnitude of such an effect in tundra, we studied the response of intertussock community CO_2 efflux to an obvious and easily quantifiable disturbance (Walker and Everett 1987; Chap. 3, this Vol.), that of long-term deposition of high pH, calcium-containing dust from the Dalton Highway (see also Chaps. 15 and 18, this Vol.). In accordance with previous studies (Toolik Uplands Road Effects Site of Walker and Everett 1987), our observations were carried out along transects perpendicular to the Dalton Highway and extending to a distance of 800 m from the road in tussock tundra (as described in Chap. 4, this Vol.), a point at which dust input becomes small. Nevertheless, atmospheric inputs of calcium in the Imnavait Creek watershed are still substantial (Chap. 9, this Vol.), although it lies ca. 2 km from the highway.

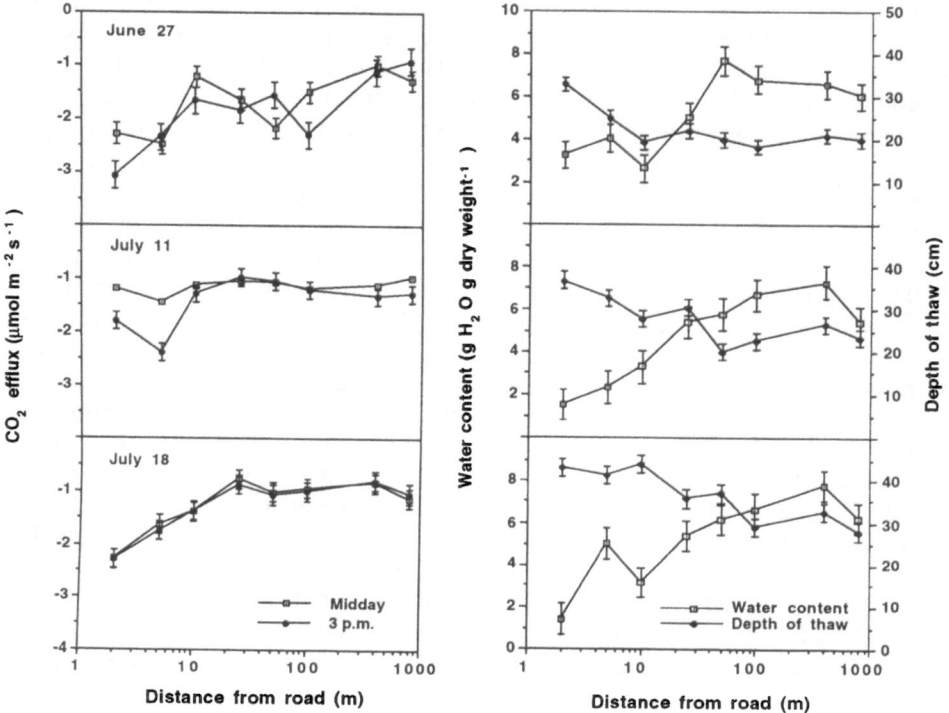

Fig. 11.10. Observed CO_2 efflux rate, soil water content, and depth of thaw on 3 days during summer 1990 along transects perpendicular to the Dalton Highway at the Toolik Uplands Road Effects Site near Imnavait Creek watershed

Three transects 40 m apart were sampled with plots at 2, 5, 10, 25, 50, 100, 400, and 800 m from the road. The terrain over which the transects lay is relatively flat with a slight downhill gradient between the 400- and 800-m plots. Plots up to 10 m from the road were obviously disturbed. Moss biomass decreased by 50% in these plots, whereas lichens disappeared completely in the 2- and 5-m plots (see also Spatt 1978; Meininger and Spatt 1988). Vascular plant biomass, primarily *Eriophorum vaginatum*, was not affected or even increased slightly in plots nearest the road.

Community CO_2 efflux was monitored in the plots along with abiotic factors on three days during summer 1990, at midday and at ca. 3 P.M. (Fig. 11.10). Measurements of soil temperature indicated that there were no differences along the transects at 1, 5, and 10 cm depth. In contrast, soil moisture near the road decreased to less than 1 g g^{-1} dry weight of soil, whereas soil moisture at 50- to 800-m from the road was similar to that observed for tussock tundra at the intensive research site (Fig. 11.5). Plots closest to the road also exhibited much deeper thaw, despite having similar temperature profiles at the observational times to those plots farthest from the road. Although the soils were very dry, CO_2 efflux was found to be approximately twice as large near the road as in those plots further than 25 m from the road. Efflux gradually decreased to the apparently "normal" level as distance from the road increased.

Soil texture is modified and soil density increases near the road. The natural communities most similar in terms of soils characteristics to those found near the road may be the dry *Cassiope* and lichen heath types in the watershed, although these are not similar from the standpoint of vegetation. Nevertheless, it is clear from Fig. 11.8 that high rates of CO_2 efflux may occur at very low soil water content also in these communities. A simple explanation for the increase in community CO_2 efflux near the road is not apparent from the observations of abiotic factors. We must draw the conclusion that dust deposition has led to a rather complex shift in ecosystem function that is obvious in CO_2 efflux rates, but requires much more intensive study at the organismic level. The response shown in Fig. 11.10 cannot be predicted based on the extensive observations of CO_2 exchange in undisturbed communities, although it may be related to the controls that have been investigated. This is of course suggested by the observed changes in community structure and composition, which probably extends to the microbial component as well.

11.5 Landscape Patterns in Net CO_2 Exchange

We examined the complexity of interactions affecting tundra net ecosystem gas exchange throughout the Imnaviat Creek watershed with a computer simulator (GAS-FLUX) (Tenhunen et al. 1994; Chap. 14, this Vol.). GAS-FLUX utilizes the information on leaf photosynthesis, stem respiration, and CO_2 efflux presented in Sections 11.3 and 11.4, and data related to the physiological response of *Sphagnum* mosses. Predicted values for net CO_2 exchange in tussock tundra for three dates are shown in Fig. 11.11. For each date 36 diurnal simulations were conducted in order to extensively sample the range of environmental conditions occurring within the watershed (based on long-term microclimate monitoring; Tenhunen, unpubl. data). The zero point in the figure depends on soil CO_2 efflux, which was held constant (ca. 120 mmol m^{-2} day^{-1}).

The sensitivity of daily carbon fixation to increasing radiation input is evident. Low carbon fixation potential occurs before leaves expand and the moss cushions grow (15 June). As both structural (LAI increase; see Chap. 17, this Vol.) and physiological change occur with development of the vegetation (cf. Fig. 11.1d), carbon fixation potential increases to the maximum observed at mid-season (15 and 30 July). Later, fixation potential decreases due to the onset of senescence (cf. Fig. 11.1f). Scatter of the symbols of each type on a particular date and at a constant light value is the result of changes in aboveground temperature and moss water content.

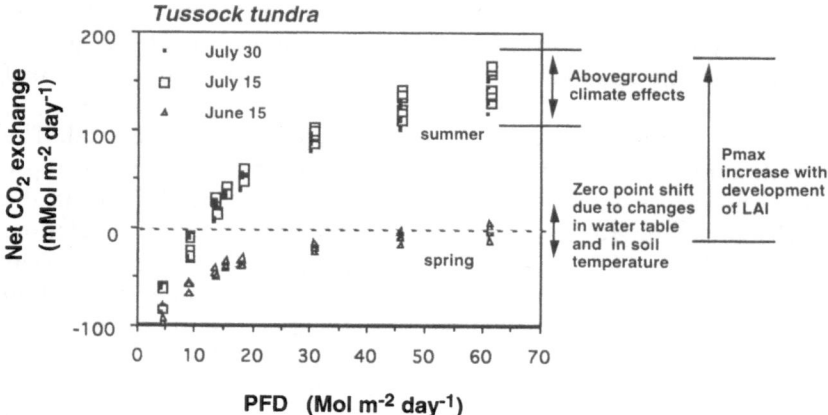

Fig. 11.11. Simulated values for integrated ecosystem diurnal net gas exchange vs total integrated photon flux density incident on plant canopy. Results shown for tussock tundra at three times during the summer growing season as described in the text. Shifts in ecosystem carbon balance (the shift in the location of the zero point on the *y*-axis) that result from soil temperature or water table changes are indicated by *arrows*

The effect of variation in soil CO_2 efflux on diurnal tundra carbon balance may be visualized as a shift in the zero point of the *y*-axis in Fig. 11.11. Diurnal soil CO_2 efflux may vary from 0 to >200 mmol m^{-2} day^{-1}. With a fully aerated profile in tussock tundra, diurnal soil CO_2 efflux was approximately 170 mmol m^{-2} day^{-1} with a mean soil temperature of 4 °C, and increased to 210 mmol m^{-2} day^{-1} at 8.5 °C. Increasing the soil temperature raises the zero or break-even point and reduces net uptake, whereas increasing the water table has the opposite effect. Thus, overall diurnal carbon balance is negative or positive depending on the interplay of incident light, soil temperature, water table, and aboveground phenological development.

To obtain a landscape perspective of gas exchange in Imnavait Creek watershed, the driving variables in GAS-FLUX may be defined as a function of both time and space. We linked GAS-FLUX to routines providing the driving variables on an hourly basis for the summer season of 1986 and parameterized the simulator in terms of the structure and physiology of three generalized vegetation zones: (1) riparian (wet meadow), (2) tussock tundra, and (3) upslope heath tundra. As a simplification, the characteristics of moist CDS heath (Fig. 5.8, this Vol.) were ascribed to other heath types as well. Although the carbon fixation in upslope areas will be overestimated, the effect of this simplification at the watershed level should be small, due to the small surface cover by dry heath types. Consequences of this simplification are that all vegetation zones are viewed as having organic soil profiles, and soil CO_2 efflux is similarly affected in all zones by variation in water table.

The simulated 1986 seasonal course of net gas exchange for the three vegetation zones is shown in Fig. 11.12. The season was characterized by dry

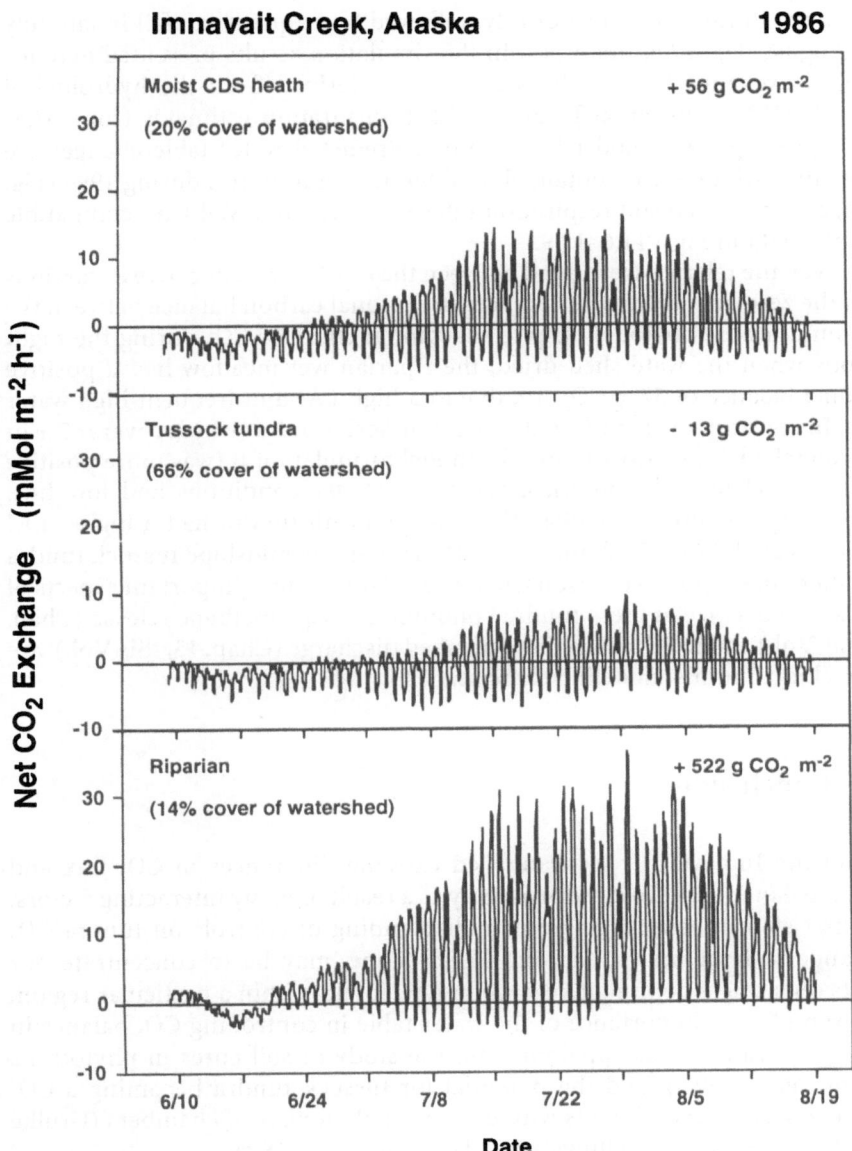

Fig. 11.12. Simulated time courses of ecosystem net carbon dioxide exchange during the dry summer of 1986 for three generalized vegetation types along the toposequence from ridgetop to stream edge within the Imnavait Creek watershed. The predicted CO_2 exchange is based on the linked models GAS-FLUX and TOPMODEL (model formulation B; Chap. 17, this Vol.). Cover values for vegetation types in the watershed are indicated *at the left* in each panel; total integrated CO_2 balance for the simulation period indicated *at the right*. For details for vegetation parameterization, see Tenhunen et al. (1994) and Chaps. 14 and 17 (this Vol.)

conditions in June and most of July, followed by frequent rainfall in late July and August (Fig. 17.6, this Vol.). In the simulation results presented here for net CO_2 exchange, water table depths were derived from the hydrological model, TOPMODEL, linked hourly with transpiration estimates from GAS-FLUX (see Fig. 17.7, model B, this Vol.). Predicted water table changes are compatible with the data obtained in different communities during 1990 (Fig. 11.4), and predicted soil respiration rates (Fig. 17.7, this Vol.) are compatible with the data in Figs. 11.6–11.9.

Given the relative area found between the positive and negative excursions from the zero line, relative differences in seasonal carbon balance between the communities are obvious. Despite substantial efflux of CO_2 during the night periods when the watershed dried, the riparian wet meadow had a positive seasonal balance of $522\,g\,CO_2\,m^{-2}$, due to a high LAI and frequent high water table. In contrast, the net CO_2 balance for tussock tundra (which covers 67% of the watershed) was $-13\,g\,CO_2\,m^{-2}$ for model formulation B (was more positive in the model formulation A), a reflection of dry conditions and low LAI. Despite dry conditions in 1986, the upslope heath tundra had a higher LAI (Hastings et al. 1989; Tenhunen et al. 1994) than the midslope tussock tundra and, therefore, a positive seasonal balance of $56\,g\,CO_2\,m^{-2}$. Important aspects of the carbon balance of these tundra communities, e.g., methane release (Chap. 12, this Vol.) and DOC loss in the watershed discharge (Chap. 13, this Vol.), are not included in this modeling exercise.

11.6 Conclusions

Within the Imnavait Creek watershed extreme differences in CO_2 flux and carbon balance occur in close proximity as a result of many interacting factors. The first step in generalizing our understanding of controls on tundra CO_2 exchange to large scales, e.g., the North Slope, may be to concentrate our efforts on a single dominant vegetation type found within a particular region. For example, the importance of the water table in controlling CO_2 balance in wet coastal tundra was apparent from the study of soil cores in phytotrons (Billings et al. 1983), and the potential for tussock tundra becoming a CO_2 source is seen in experiments with even a small number of chambers (Grulke et al. 1990; Oechel and Billings 1992; Tenhunen et al. 1995).

Nevertheless, measured CO_2 exchange rates for apparently similar tundra "patches" vary greatly. This is shown in Table 11.3 where we compare estimates of seasonal and diurnal net CO_2 exchange obtained from field, laboratory, and computer experiments. The estimates of CO_2 loss from tussock tundra reported by Oechel et al. (1993) are very large. Based on the models and data presented, these large negative fluxes require further scrutiny and explanation. The large variation in rates shown in Table 11.3 suggest that we must

Table 11.3. Estimates of seasonal and diurnal net ecosystem CO_2 exchange for wet, moist, and dry tundra communities. Simulations represent 1986 70-day season shown in Fig. 11.12 using the linked GAS-FLUX/TOPMODEL in the Imnavait Creek watershed

Community	Net CO_2 exchange rate	Reference
Seasonal ($g\,C\,m^{-2}\,year^{-1}$)		
Wet coastal tundra	40	Coyne and Kelley (1978)
Wet coastal tundra	−34	Oechel et al. (1993)
Riparian	142	Coupled model (this study)
Tussock tundra	−50	Grulke et al. (1990)
Tussock tundra	−53 to −286	Oechel et al. (1993)
Tussock tundra	−3.5	Coupled model (this study)
Upslope "heath" (see text)	15	Coupled model (this study)
Diurnal ($g\,CO_2\,m^{-2}\,day^{-1}$)		
Wet coastal tundra	−2 to 10	Billings et al. (1982)
Salix stand LA1 = 1	15	Eckardt et al. (1982)
CO_2 uptake only		
Salix stand LAl = 2	20	Eckardt et al. (1982)
CO_2 uptake only		
Riparian		Coupled model (this study)
average	7.5	
maximum	17	
minimum	−2.9	
Tussock tundra	−7 to 3	Tenhunen et al. (1995)
Tussock tundra		Coupled model (this study)
average	−0.3	
maximum	3.1	
minimum	−3.1	
Upslope "heath"		Coupled model (this study)
average	0.8	
maximum	5.7	
minimum	−3.9	

question the extent of our knowledge of "tundra" gas exchange. Although an oversimplification, we must conclude that the average CO_2 exchange rate for tundra depends *at a minimum* on the relative cover of dry, moist, and wet community types composing the landscape. By weighting the seasonal CO_2 balances in Fig. 11.12 according to their overall cover, we estimated an overall positive uptake of $76\,g\,CO_2\,m^{-2}$ for the Imnavait Creek watershed during the dry summer of 1986. Some of this excess is stored in peat in wet locations, some is exported (Chap. 13, this Vol.), and some leaves as methane gas (Chap. 12, this Vol.).

We conclude the following: net seasonal CO_2 loss occurs in tussock tundra under dry conditions and with low LAI development. However, CO_2 exchange will vary strongly from year to year depending on rainfall, irradiance, and temperature. Difficult to predict are the indirect effects of environmental factors on interannual variation in phenological development and the adaptive use of carbon and nutrient stores by the vegetation to support CO_2 fixation capacity. Furthermore, net seasonal CO_2 gains may occur in upslope communities even in dry years and organic carbon is transported between communities (Chap. 13, this Vol.), leading to shifts in process rates and directions along the toposequence. In the model simulation shown for the dry year 1986, carbon fixation by heath located higher on the slopes was adequate to compensate for tussock tundra located below it. Thus, carbon transport must be examined more closely in future studies. Although, as in the case of riparian sedge meadows, the extent of areal coverage by certain communities may be small, the impact of these sites may be of considerable significance at the landscape level; due to intense biological activity. Studies of ecosystem gas exchange rates at the landscape or regional levels must carefully consider variations in biological activities on the land surface that are not constant, but dynamically regulated.

In arctic tundra long-term studies are required to obtain an adequate assessment of the hydrological balance (Chaps. 6 and 9, this Vol.). As a result of the strong coupling of hydrology to CO_2 efflux, and possibly to phenological development, long-term observations of CO_2 balance, spatial variation in water table, and carbon transport are necessary before general conclusions can be made about the average carbon balance of a particular landscape. The temporal variation in CO_2 balance must be appreciated and understood before integrative signals, such as seasonal and annual changes in atmospheric CO_2 concentration above the tundra surface, can be interpreted. Our results demonstrate conclusively that spatial perspectives are required in models designed to assess climate change or disturbance effects on tundra ecosystems (Chap. 14, this Vol.).

Acknowledgments. This work was supported by the U.S. Department of Energy, Office of Health and Environmental Research R4D program, contract no. DE-FG03-84ER60250, and German Federal Ministry for Science and Technology (grant no. BEO51-0339476A). Logistic support and maintenance of Toolik Lake Field Station by the Institute of Arctic Biology is appreciated. Field assistance by Chris Lesieutre, Deborah Forster, Lisa Balduman, Joseph Lumianski, and Margaret Zalejko is gratefully acknowledged.

References

Bigger CM, Oechel WC (1982) Nutrient effect on maximum photosynthesis in arctic plants. Holarct Ecol 5: 158–163

Billings WD, Peterson KM, Shaver GR, Trent AW (1977) Root growth, respiration, and carbon dioxide evolution in an arctic tundra soil. Arct Alp Res 9: 129–137

Billings WD, Peterson KM, Shaver GR (1978) Growth, turnover, and respiration rates of roots and tillers in tundra communities. In: Tieszen LL (ed) Vegetation and production ecology of an Alaskan arctic tundra. Springer, Berlin Heidelberg New York, pp 415–434

Billings WD, Luken JO, Mortenson DA, Peterson KM (1982) Arctic tundra: a source or sink for atmospheric carbon dioxide in a changing environment. Oecologia 53: 7–11

Billings WD, Luken JO, Mortenson DA, Peterson KM (1983) Increasing atmospheric carbon dioxide: possible effects on arctic tundra. Oecologia 58: 286–289

Chapin FS III, Shaver GR (1985a) Arctic. In: Chabot BF, Mooney HA (eds) Physiological ecology of North American plant communities. Chapman and Hall, New York, pp 16–40

Chapin FS III, Shaver GR (1985b) Individualistic growth response of tundra plant species to environmental manipulations in the field. Ecology 66: 564–576

Chapin FS III, Shaver GR (1989) Differences in growth and nutrient use among arctic plant growth forms. Funct Ecol 3: 73–80

Chapin FS III, Fetcher N, Kielland K, Everett K, Linkins AE (1988) Productivity and nutrient cycling of Alaskan tundra: enhancement by flowing soil water. Ecology 69: 693–702

Cheng W, Coleman DC (1989) A simple method for measuring CO_2 in a continuous air-flow system: modifications to the substrate-induced respiration technique. Soil Biol Biochem 21: 385–388

Cheng W, Virginia RA (1993) Measurement of microbial biomass in arctic tundra soils using fumigation-extraction and substrate-induced respiration procedures. Soil Biol Biochem 25: 135–141

Cheng W, Virginia RA, Oberbauer SF, Tenhunen JD, Gillespie CT, Reynolds JF (1996) Spatial and temporal variation in soil nitrogen, microbial biomass, and respiration in an arctic catena (submitted)

Coyne PI, Kelley JJ (1978) Meteorological assessment of CO_2 exchange over an Alaskan arctic tundra. In: Tieszen LL (ed) Vegetation and production ecology of an Alaskan arctic tundra. Springer, Berlin Heidelberg New York, pp 299–319

Dawson TE, Bliss LC (1989) Intraspecific variation in the water relations of Salix arctica, an arctic-alpine dwarf willow. Oecologia 79: 322–331

Decker P (1981) Respiratorische Kohlenstoffverluste und Kohlenstoffbilanz einer alpinen Grasheide (Caricetum curvulae). PhD Thesis, Univ Innsbruck, Innsbruck, Austria

Defoliart LS, Griffith M, Chapin FS III, Jonassons S (1988) Seasonal patterns of photosynthesis and nutrient storage in Eriophorum vaginatum L., an arctic sedge. Funct Ecol 2: 185–194

Eckardt FE, Heerfordt L, Jørgenson HM, Vaag P (1982) Photosynthetic production in Greenland as related to climate, plant cover, and grazing pressure. Photosynthetica 16: 71–100

Field C, Mooney HA (1986) The photosynthesis–nitrogen relationship in wild plants. In: Givnish TJ (ed) On the economy of plant form and function. Cambridge Univ Press, Cambridge, pp 25–55

Fox JF (1992) Responses of diversity and growth-form dominance to fertility in Alaskan tundra fellfield communities. Arct Alp Res 24: 233–237

Gebauer RLE (1994) The effect of waterlogging on photosynthesis, nutrient status and growth of the arctic species Eriophorum vaginatum and Eriophorum angustifolium. PhD Thesis, San Diego State Univ, San Diego and Univ California, Davis

Giblin AE, Nadelhoffer KJ, Shaver GR, Laundre JA, McKerrow AJ (1991) Biogeochemical diversity along a riverside toposequence in arctic Alaska. Ecol Mongr 61: 415–435

Gorham E (1991) Northern peatlands: role in the carbon cycle and probable responses to climatic warming. Ecol Appl 1: 182–195

Grulke NE, Bliss LC (1988) Comparative life history characteristics of two high arctic grasses, Northwest Territories. Ecology 69: 484–496

Grulke NE, Riechers GH, Oechel WC, Hjelm U, Jaeger C (1990) Carbon balance in tussock tundra under ambient and elevated atmospheric CO_2. Oecologia 83: 485–494

Haag RG (1974) Nutrient limitations to plant production in two tundra communities. Can J Bot 52: 103–116

Hahn S (1991) Photosynthese und Wasserhaushalt von Flechten in der Tundra Alaskas: Gaswechselmessungen unter natürlichen Bedingungen und experimentelle Faktorenanalyse. PhD Thesis, Univ Würzburg, Würzburg

Hahn SC, Tenhunen JD, Popp PW, Meyer A, Lange OL (1993) Upland tundra in the foothills of the Brooks Range, Alaska: diurnal CO_2 exchange patterns of characteristic lichen species. Flora 188: 125–143

Harley PC, Tenhunen JD, Murray KJ, Beyers J (1989) Irradiance and temperature effects on photosynthesis of tussock tundra *Sphagnum* mosses from the foothills of the Philip Smith Mountains, Alaska. Oecologia 79: 251–259

Hastings SJ, Luchessa SA, Oechel WC, Tenhunen JD (1989) Standing biomass and production in water drainages of the foothills of the Philip Smith Mountains, Alaska. Holarct Ecol 12: 304–311

Johnson DA, Tieszen LL (1976) Aboveground biomass allocation, leaf growth, and photosynthesis patterns in tundra plant forms in arctic Alaska. Oecologia 24: 159–173

Johnson F, Eyring H, Williams R (1942) The nature of enzyme inhibitions in bacterial luminescence: sulfanilamide, urethane, temperature, and pressure. J Cell Comp Physiol 20: 247–268

Kummerow J, Mills JN, Ellis BA, Hastings SJ, Kummerow A (1987) Downslope fertilizer movement in arctic tussock tundra. Holarct Ecol 10: 312–319

Limbach WE, Oechel WC, Lowell W (1982) Photosynthetic and respiratory responses to temperature and light of three Alaskan tundra growth forms. Holarct Ecol 5: 150–157

Luken JO, Billings WD (1985) The influence of microtopographic heterogeneity on carbon dioxide efflux from a subarctic bog. Holarct Ecol 8: 306–312

Matthes-Sears U, Matthes-Sears W, Hastings SJ, Oechel WC (1988) Biomass production, photosynthesis and tissue nutrient concentrations of two dwarf deciduous shrub species along a natural slope in arctic Alaska. Arct Alp Res 20: 342–351

McGraw JB (1985) Experimental ecology of *Dryas octopetala* ecotypes. III. Environmental factors and plant growth. Arct Alp Res 17: 229–239

McNulty AK, Cummins WR (1987) The relationship between respiration and temperature in leaves of the arctic plant, *Saxifraga cernua*. Plant Cell Environ 10: 319–325

Meininger CA, Spatt PD (1988) Variations of tardigrade assemblages in dust-impacted arctic mosses. Arct Alp Res 20: 24–30

Miller PC (1981) Carbon balance in northern ecosystems and the potential effect of carbon dioxide-induced climatic change. CONF-8003118, Natl Tech Info Serv Springfield, Virgina

Miller PC (1982) Environmental and vegetational variation across a snow accumulation area in montane tundra in central Alaska. Holarct Ecol 5: 85–98

Moore TR (1986) Carbon dioxide evolution from subarctic peatlands in eastern Canada. Arct Alp Res 18: 189–193

Moore TR (1989) Plant production, decomposition, and carbon efflux in a subarctic patterned fen. Arct Alp Res 21: 156–162

Murray C, Miller PC (1982) Phenological observations of major plant growth forms and species in montane and *Eriophorum vaginatum* tussock tundra in central Alaska. Holarct Ecol 5: 109–116

Murray KJ, Harley PC, Beyers J, Walz H, Tenhunen JD (1989a) Water content effects on photosynthetic response of *Sphagnum* mosses from the foothills of the Philip Smith Mountains, Alaska. Oecologia 79: 244–250

Murray KJ, Tenhunen JD, Kummerow J (1989b) Limitations on *Sphagnum* growth and net primary production in the foothills of the Philip Smith Mountains, Alaska. Oecologia 80: 256–262

Murray KJ, Tenhunen JD, Nowak RS (1993) Photoinhibition as a control on photosynthesis of *Sphagnum* mosses. Oecologia 96: 200–207

Nadelhoffer KJ, Giblin AE, Shaver GR, Laundre JA (1991) Effects of temperature and substrate quality on element mineralization in six arctic soils. Ecology 72: 242–253

Nadelhoffer KJ, Giblin AE, Shaver GR, Linkens AE (1992) Microbial processes and plant nutrient availability in arctic soils. In: Chapin FS III, Jeffries RL, Reynolds JF, Shaver GR, Svoboda J (eds) Arctic ecosystems in a changing climate. Academic Press, San Diego, pp 281–300

Oberbauer SF, Dawson TE (1992) Water relations of vascular plants. In: Chapin FS III, Jeffries RL, Reynolds JF, Shaver GR, Svoboda J (eds) Arctic ecosystems in a changing climate. Academic Press, San Diego, pp 259–280

Oberbauer SF, Miller PC (1979) Plant water relations in montane and tussock tundra vegetation types in Alaska. Arct Alp Res 11: 69–81

Oberbauer SF, Oechel WC (1989) Maximum CO_2 assimilation rates of vascular plants along an Alaskan arctic tundra slope. Holarct Ecol 12: 312–316

Oberbauer SF, Hastings SJ, Beyers JL, Oechel WC (1989) Comparative effects of downslope water and nutrient movement on plant nutrition, photosynthesis and growth in Alaskan tundra. Holarct Ecol 12: 324–334

Oberbauer SF, Tenhunen JD, Reynolds JF (1991) Environmental effects on CO_2 efflux from water track and tussock tundra in Arctic Alaska, USA. Arct Alp Res 23: 162–169

Oberbauer SF, Gillespie CT, Cheng W, Gebauer R, Sala Serra A, Tenhunen JD (1992) Environmental effects on CO_2 efflux from riparian tundra in the northern foothills of the Brooks Range, Alaska, USA. Oecologia 92: 568–577

Oberbauer SF, Gillespie CT, Cheng W, Sala A, Tenhunen JD (1996) Diurnal and seasonal patterns of carbon efflux from upland tundra communities of the northern foothills of the Brooks Range, Alaska, USA (in preparation)

Odum EP (1985) Trends expected in stressed ecosystems. BioScience 35: 419–422

Oechel WC, Billings WD (1992) Effects of global change on the carbon balance of arctic plants and ecosystems. In: Chapin FS III, Jeffries RL, Reynolds JF, Shaver GR, Svoboda J (eds) Arctic ecosystems in a changing climate. Academic Press, San Diego, pp 139–168

Oechel WC, Riechers GH, Lawrence WT, Prudhomme TI, Grulke N, Hastings SJ (1992) "CO_2LT" an automated, null-balance system for studying the effects of elevated CO_2 and global climate change on unmanaged ecosystems. Funct Ecol 6: 86–100

Oechel WC, Hastings SJ, Jenkins M, Riechers G, Grulke N, Vourlitis G (1993) Recent change of arctic tundra ecosystems from a net carbon dioxide sink to a source. Nature 361: 520–523

Ostendorf B, Reynolds JF (1993) Relationships between a terrain-based hydrologic model and patch-scale vegetation patterns in an arctic tundra landscape. Landscape Ecol 8: 229–239

Peterson KM, Billings WD (1975) Carbon dioxide flux from tundra soils and vegetation as related to temperature at Barrow, Alaska. Am Midl Nat 94: 88–98

Peterson KM, Billings WD, Reynolds DN (1984) Influence of water table and atmospheric CO_2 concentration on the carbon balance of arctic tundra. Arct Alp Res 16: 331–335

Poole DK, Miller PC (1982) Carbon dioxide flux from three arctic tundra types in North-Central Alaska, USA. Arct Alp Res 14: 27–32

Rathstetter EB, King AW, Cosby BJ, Hornberger GM, O'Niell, Hobbie JE (1992) Aggregating fine-scale ecological knowledge to model coarser-scale attributes of ecosystems. Ecol Appl 2: 55–70

Semikhatova OA, Gerasimenko TV, Ivanova TI (1992) Photosynthesis, respiration, and growth of plants in the Soviet Arctic. In: Chapin FS III, Jeffries RL, Reynolds JF, Shaver GR, Svoboda J (eds) Arctic ecosystems in a changing climate. Academic Press, San Diego, pp 169–192

Shaver GR, Chapin FS III (1980) Response to fertilization by various plant growth forms in an Alaska tundra: nutrient accumulation and growth. Ecology 61: 662–675

Shaver GR, Kummerow J (1992) Phenology, resource allocation, and growth of arctic vascular plants. In: Chapin FS III, Jeffries RL, Reynolds JF, Shaver GR, Svoboda J (eds) Arctic ecosystems in a changing climate. Academic Press, San Diego, pp 193–212

Shvetsova VM, Voznesensky VL (1970) Diurnal and season variation in the rate of photosynthesis in some plants of western Taimyr. Bot Zh 55: 66–76

Siegwolf R (1987) CO_2-Gaswechel von Rhododendron ferrugineum L. im Jahresgang an der alpinen Waldgrenze. PhD Thesis, Univ Innsbruck, Innsbruck, Austria

Smith E (1937) The influence of light and carbon dioxide on photosynthesis. Gen Physiol 20: 807–830

Sorenson T (1941) Temperature relations and phenology of the northeast Greenland flowering plants. Meddr Gronland 125: 1–305

Spatt PD (1978) Seasonal variation of growth conditions in a natural and dust impacted Sphagnum (Sphagnaceae) community in northern Alaska. MS Thesis, Univ Cincinnati, Cinncinati, Ohio, p 103

Svensson BH (1980) Carbon dioxide and methane fluxes from the ombotrophic parts of a subarctic mire. In: Sonesson M (ed) Ecology of a subarctic mire. Swed Nat Sci Res Counc, Stockholm. Ecol Bull 30: 235–250

Tenhunen JD, Beyschlag W, Lange OL, Harley PC (1987) Changes during summer drought in leaf CO_2 uptake rates of macchia shrubs growing in Portugal: limitations due to photosynthetic capacity, carboxylation efficiency, and stomatal conductance. In: Tenhunen JD, Catarino F, Lange OL, Oechel WC (eds) Plant response to stress: functional analysis in mediterranean ecosystems. Springer, Berlin Heidelberg New York, pp 305–327

Tenhunen JD, Lange OL, Hahn S, Siegwolf R, Oberbauer SF (1992) The ecosystem role of poikilhydric tundra plants. In: Chapin FS III, Jeffries RL, Reynolds JF, Shaver GR, Svoboda J (eds) Arctic ecosystems in a changing climate. Academic Press, San Diego, pp 213–238

Tenhunen JD, Siegwolf RA, Oberbauer SF (1994) Effects of phenology, physiology, and gradients in community composition, structure, and microclimate on tundra ecosystem CO_2 exchange. In: Schulze E-D, Caldwell MM (eds) Ecophysiology of photosynthesis. Springer, Berlin Heidelberg New York, pp 431–460

Tenhunen JD, Gillespie CT, Oberbauer SF, Sala A, Whalen SC (1995) Climate effects on the carbon balance of tussock tundra in the Philip Smith Mountains, Alaska. Flora 190: 273–283

Tieszen LL (1973) Photosynthesis and respiration in arctic tundra grasses: field light intensity and temperature responses. Arct Alp Res 5: 239–251

Tieszen LL (1975) CO_2 exchange in the Alaskan arctic tundra: seasonal changes in the rate of photosynthesis of four species. Photosynthetica 9: 376–390

Tieszen LL (1978) Photosynthesis in the principal Barrow, Alaska, species: a summary of field and laboratory responses. In: Tieszen LL (ed) Vegetation and production ecology of an Alaskan arctic tundra. Springer, Berlin Heidelberg New York, pp 241–268

Tieszen LL, Johnson DA (1975) Seasonal pattern of photosynthesis in individual grass leaves and other plant parts in arctic Alaska with a portable $^{14}CO_2$ system. Bot Gaz 136: 99–105

Ulrich A, Gersper PL (1978) Plant nutrition limitations of tundra plant growth. In: Tieszen LL (ed) Vegetation and production ecology of an Alaskan arctic tundra. Springer, Berlin Heidelberg New York, pp 457–481

Vourlitis GC, Oechel WC, Hastings SJ, Jenkins MA (1993) A system for measuring CO_2 and CH_4 flux in unmanaged ecosystems: an Arctic example. Funct Ecol 7: 369–375

Walker DA, Everett KR (1987) Road dust and its environmental impact on Alaskan taiga and tundra. Arct Alp Res 19: 479–489

Walker DA, Binnian E, Evans BM, Lederer ND, Nordstrand E, Webber PJ (1989) Terrain, vegetation, and landscape evolution of the R4D research site, Brooks Range Foothills, Alaska. Holarct Ecol 12: 238–261

Walker MD, Walker DA, Everett KR (1989) Wetland soils and vegetation, Arctic Foothills, Alaska. US Fish Wildlife Serv, Biol Rep 89(7)

Waring RH, Schlesinger WH (1985) Forest ecosystems concepts and management. Academic Press, Orlando

Whalen SC, Reeburgh WS (1988) A methane flux time series for tundra environments. Global Biogeochem Cycles 2: 399–409

12 Control of Tundra Methane Emission by Microbial Oxidation

S. C. WHALEN, W. S. REEBURGH, and C. E. REIMERS

12.1 Introduction

The recent global increase of 1% per year in the concentration of atmospheric methane (CH_4) is well documented (Rasmussen and Khalil 1984; Steele et al. 1987; Blake and Rowland 1989). This increase causes concern because CH_4 is an important trace gas in the earth's atmosphere. Greenhouse warming from CH_4 is 25% of CO_2-induced warming, and together these gases account for 75% of the radiative trapping from atmospheric gases (Rodhe 1990).

Major terms of the atmospheric CH_4 budget have been identified, but considerable uncertainty remains over their magnitude (Cicerone and Oremland 1988). This uncertainty stems from a scarcity in data on regional CH_4 flux (Fung et al. 1991), errors in scaling up from "representative" sites to a regional and global basis (Matson et al. 1989), and a lack of information on climate-induced biogeochemical feedbacks in the global CH_4 cycle (Mooney et al. 1987). Nevertheless, emission from northern peatland and tundra is clearly a major component of the atmospheric CH_4 budget. Recent estimates of source strength for high-latitude (>50° N) environments range from $14-106\,Tg\,CH_4\,yr^{-1}$ (Sebacher et al. 1986; Matthews and Fung 1987; Whalen and Reeburgh 1988; Aselmann and Crutzen 1989; Moore et al. 1990; Whalen and Reeburgh 1990a; Fung et al. 1991; Reeburgh and Whalen 1992), or about 3–20% of the total atmospheric budget of $540\,Tg\,CH_4\,yr^{-1}$ (Cicerone and Oremland 1988).

Global climate models (Grotch 1988; Mitchell 1989) predict a worldwide increase in surface air temperature over the next 50 years because of increasing atmospheric concentrations of radiatively active gases such as CO_2 and CH_4. The most pronounced warming is expected at high latitudes where doubling of atmospheric CO_2 concentration may increase the mean surface air temperature 8–10 °C in winter and 2–4 °C in summer (Grotch 1988). Climate models also forecast an increase in the precipitation, evaporation, and duration of the thaw at high latitudes, and most predict a decrease in summer soil moisture content. Variability on a scale of years to decades makes it difficult to detect a greenhouse-driven warming trend from records of surface air temperature (Wigley and Raper 1990), but temperature distributions in permafrost point to a recent multidecade warming trend for the Arctic (Lachenbruch and Marshall 1986).

J.F. Reynolds and J.D. Tenhunen (Eds.)
Ecological Studies, Vol. 120
© Springer-Verlag Berlin Heidelberg 1996

Understanding the carbon balance for high-latitude environments is essential to predicting regional response to perturbations in climate. Taiga and coastal plain tundra have accumulated carbon since the last Glacial Maximum (Adams et al. 1990), and are now considered a carbon sink (Billings 1987; see also Chap. 11, this Vol.). These regions store 27% of the earth's soil carbon (Post et al. 1982). Thawing of permafrost is expected to alter the carbon balance. Emission of CO_2 and CH_4 to the atmosphere may increase because of direct discharge of sequestered gas (Post 1990), microbial decomposition in previously frozen soil, and enhanced peat oxidation in upper soil horizons (Gorham 1991).

The role of biogeochemical feedbacks in high-latitude climate change is uncertain. Khalil and Rasmussen (1989) and Lashof (1989) suggest that an increase in soil temperature will increase methanogenesis and result in a positive climate feedback. Whalen and Reeburgh (1990b), however, argue that a lowered water table in highly permeable soil will enhance CH_4 oxidation and lead to lower fluxes and a negative feedback through consumption of atmospheric CH_4.

Relationships have been reported between net CH_4 flux and soil temperature (e.g., Crill et al. 1988) and water table position (e.g., Moore and Knowles 1989). These relationships have been important in placing bounds on the high-latitude response to climate change (Post 1990; Gorham 1991). The atmospheric CH_4 flux, however, results from the balance between CH_4 production (methanogenesis) and consumption (oxidation), so that an understanding of the controls on these processes is required to predict the influence of climate change on atmospheric CH_4 flux and biogeochemical feedbacks. In this chapter, we synthesize results of our field and laboratory investigations of CH_4 oxidation and the physiological properties of CH_4-oxidizing bacteria (methanotrophs) from environments ranging from arctic tundra and taiga to landfills. We develop a conceptual model of the environmental controls on CH_4 oxidation to better predict the impact of climate change on CH_4 emission from some arctic tundra soils.

12.2 Sampling Procedure

Many of the methods and sites used in this study have been described in detail elsewhere (Whalen and Reeburgh 1988; Whalen et al. 1990; Whalen and Reeburgh 1990b). We established permanent sampling stations in a subarctic muskeg and along a pond margin in the University of Alaska Fairbanks Arboretum (64° N). Floristic and physical characteristics of these sites are similar to those of arctic tundra. Additional arctic tundra sampling stations were established near the Kuparuk River (68° N) on Alaska's North Slope (Chaps. 1 and 4, this Vol.). Taiga sampling stations were established in the Bonanza Creek Experimental Forest (BCEF) at 64° N, near Fairbanks, Alaska. Subarctic sampling stations were visited weekly throughout the thaw season;

arctic stations were usually sampled three times. Soil temperatures, water table depths, and maximum thaw depths varied widely during the thaw seasons over the study period. The mean arboretum soil temperature to permafrost varied from 1.0 °C (intertussock depression) to 11.2 °C (*Carex* site) relative to the soil surface, the water table depth varied from −40 cm (moss) to +65 cm (*Carex*), and the maximum thaw depth was −96 cm (*Eriophorum* tussock) in the arboretum (Whalen and Reeburgh 1992). Mean soil temperatures to 13 cm varied from 8.9 °C (black spruce) to 12.9 °C (south-facing aspen), and the water table depth was consistently deeper than 1 m at stations in BCEF (Whalen et al. 1990). Arctic sites had mean soil temperatures to 13 cm varying from 2.1 °C (moss) to 13.5 °C (intertussock depression). Water table depths ranged from −23 cm (hummock) to 20 cm (intertussock depression), and thaw depths ranged from −25 to −50 cm.

We determined net CH_4 fluxes using static chambers. Fluxes were calculated from the linear regression of chamber CH_4 concentration vs time. Methane concentration in syringe samples was measured by gas chromatography using a flame ionization detector. Each chamber consisted of a 0.075-m^2 skirted aluminum base permanently implanted in the soil plus lucite vertical sections and lids. A water-filled channel provided a seal between components. Soil CH_4 distributions were determined by analyzing air pumped from known depths in moist soil or water removed from equilibration samplers in waterlogged soil (Hesslein 1976). Supporting data were collected when fluxes were measured. Soil temperature profiles were determined with portable or permanently placed thermistor probes. Water table depths were monitored with shallow wells, and thaw depths were measured with a steel probe. Other ancillary data were occasionally collected.

Some CH_4 consumption rate measurements were made on soil cores (10- to 12-cm long ×6.7 cm diameter) placed in mason jars. Results of jar and chamber experiments are comparable. The cores are retained in plastic core tubes with a sealed bottom. The chamber and jar area: volume ratios agree to within 5%. Chambers and jars were sometimes amended with CH_4 to study the capacity for soil CH_4 oxidation or allowed to "run down" under a free-air atmosphere (1.7 ppm CH_4) to determine the threshold for CH_4 consumption. Biological consumption was measured over a 0.5- to 24-h time course, depending on the season or the experiment, by collecting syringe samples and analyzing them for CH_4. Equilibration or relaxation experiments were used to establish equilibration times of gases added to the headspace in jar experiments. These experiments involved adding CH_4 to acetylene-poisoned soil cores placed in mason jars. Headspace samples were collected and analyzed at 1-min intervals until changes in headspace CH_4 could not be resolved (6–8 min).

Soil O_2 concentrations were estimated from measurements made with micro-Clark electrodes constructed according to the design of Revsbech and Ward (1983). The microelectrodes were lowered into the substrate at each study site using a micromanipulator attached to a section of aluminum angle that had been pushed firmly into the ground. To calibrate the microelectrode readings, background currents evaluated from readings within the anaerobic soil layers at depth were subtracted from each recorded signal. These corrected values were referenced against a background-corrected value recorded in air-saturated distilled water. The latter measurement was made in water that had been placed in a glass petri dish at the soil surface before profiling was started. Background readings were generally 4–12% of the 100% air-saturated output. O_2 concentrations for the 100% air-saturated water samples were calculated using the temperature of the water and the barometric pressure. These estimates agreed with O_2 concentrations measured for the same water by small-volume Winkler techniques to within ±5%.

Physiological and environmental controls on CH_4 oxidation were studied on pure and enriched microbial cultures, bulk soils, and cores collected from taiga, tundra, and the surface of a retired landfill (Berkeley North Waterfront Park, Berkeley, California). Bulk soils were homogenized and the temperature and moisture dependence of CH_4 oxidation were assessed by adjusting the temperature or moisture content of soil aliquots in jars and following the change in headspace CH_4 concentration over a 12-h time course. The end products of CH_4 oxidation were determined for cores and bulk soils in jars with $^{14}CH_4$-amended headspaces. Jar headspaces were sampled over a 2- to 12-h time course, and syringe samples were injected into a stripping/oxidation line, where $^{14}CO_2$ was trapped directly, and $^{14}CH_4$ was trapped as $^{14}CO_2$ after combustion. Soil from these experiments was assayed for ^{14}C by dry combustion at the end of the experiment. The substrate dependence of CH_4 oxidation was determined in initial velocity experiments where culture aliquots, bulk soils, or soil cores were exposed to different headspace CH_4 concentrations. Half-saturation constants (K_s) and maximum uptake rate (V_m) for CH_4 oxidation were calculated by a least-squares fit (untransformed data) of a rectangular hyperbola to the rate vs concentration data.

12.3 Results and Discussion

12.3.1 Methane Flux and Environmental Variables in Tundra and Taiga

Characteristics, processes, and components of the development and function of arctic tundra and taiga ecosystems have been reviewed by Bliss et al. (1981), Tieszen (1978), Van Cleve et al. (1986), and Chapin et al. (1992). We consider controls on CH_4 flux in both tundra and taiga environments. This broad approach will likely lead to a more general understanding of controls on trace gas emissions.

The soils of the two regions show sharply contrasting physical characteristics. A thaw depth of less than 0.5 m in arctic tundra soils often limits internal drainage and maintains the water table position at or near the soil surface

(Chap. 11, this Vol.). Bliss (1975) reported that arctic soils on Devon Island remained saturated below the uppermost few centimeters for the entire summer, and Rieger (1975) estimated that up to 90% of tundra soils are at or near saturation throughout the thaw season. Hinzman et al. (1991) reported that subsurface mineral soils in an Alaskan arctic watershed remained saturated throughout the thaw period while the moisture content of surficial organic soils varied between 10 and 90% by volume (Chaps. 6, 11, this Vol.). Permafrost in taiga soils is confined to black spruce sites where an understory of feather moss provides insulation. Soils are well drained except in low-lying black spruce bogs. Moisture content of the highly organic surface to 10 cm of soil at our taiga sites averaged 72% (w/w) over the thaw season (Whalen et al. 1991), and moisture content of the mineral soil from 30 to 60 cm (less frequently determined) averaged 22%.

The magnitudes and seasonal variations in CH_4 fluxes from arboretum (arctic tundra surrogate) and taiga environments also differed. Waterlogged arboretum soils emit CH_4 throughout the thaw season (Fig. 12.1a). Fluxes range over a factor of 10^4 and exhibit a pronounced seasonality. As soils thaw, rates of CH_4 emission track the increase in thaw depth and soil temperature until the maximum thaw depth is reached (Fig. 12.1a–c). Methane emission rates then decrease with falling soil temperature until freeze-up. Well-drained taiga soils rarely emit CH_4 in to the atmosphere; instead, these soils show consumption of atmospheric CH_4 (negative flux; Fig. 12.1d). Methane consumption rates in taiga soils show less variability than CH_4 emission rates in arboretum soils, ranging over a factor of about three. No seasonal trend appears in CH_4 consumption, although subsurface taiga soil temperatures show a distinct seasonality and greater variation (0–12 °C; Fig. 12.1e) than arboretum soil temperatures (−1 to 5 °C; Fig. 12.1b). The methane fluxes and physical characteristics shown in Fig. 12.1 are representative of several permanent sampling sites. But waterlogged taiga soils, such as spruce bogs, show CH_4 emission and serve as regional point sources for atmospheric CH_4, whereas well-drained alpine and moist meadow tundra consume atmospheric CH_4 or show no emissions (Whalen and Reeburgh 1990a; Whalen et al. 1991).

Soil CH_4 distributions in waterlogged arboretum and moist taiga soils differ as well. Arboretum soils show an increase in porewater CH_4 concentration with increasing depth in the first few centimeters below the water table surface and nearly uniform CH_4 concentrations of 200–800 μM at greater depths (Fig. 12.2a,b). The CH_4 concentration in the air-filled pore space of taiga soils decreases from atmospheric levels (1.7 ppm) at the soil surface to the detection limit (0.1 ppm) at 40 cm (Fig. 12.3). Soil CH_4 concentrations are approximately 10^5- to 10^6-fold higher in waterlogged arboretum soils than in moist taiga soils (solubility of CH_4 is 2 nM in an atmosphere of 1.7 ppm CH_4).

Oxygen distributions determined with microelectrodes in arboretum soils show a rapid depletion of O_2 at or within 1 cm below the water table. Moist arboretum soils appear supersaturated (350 μM) with O_2 at the air–soil inter-

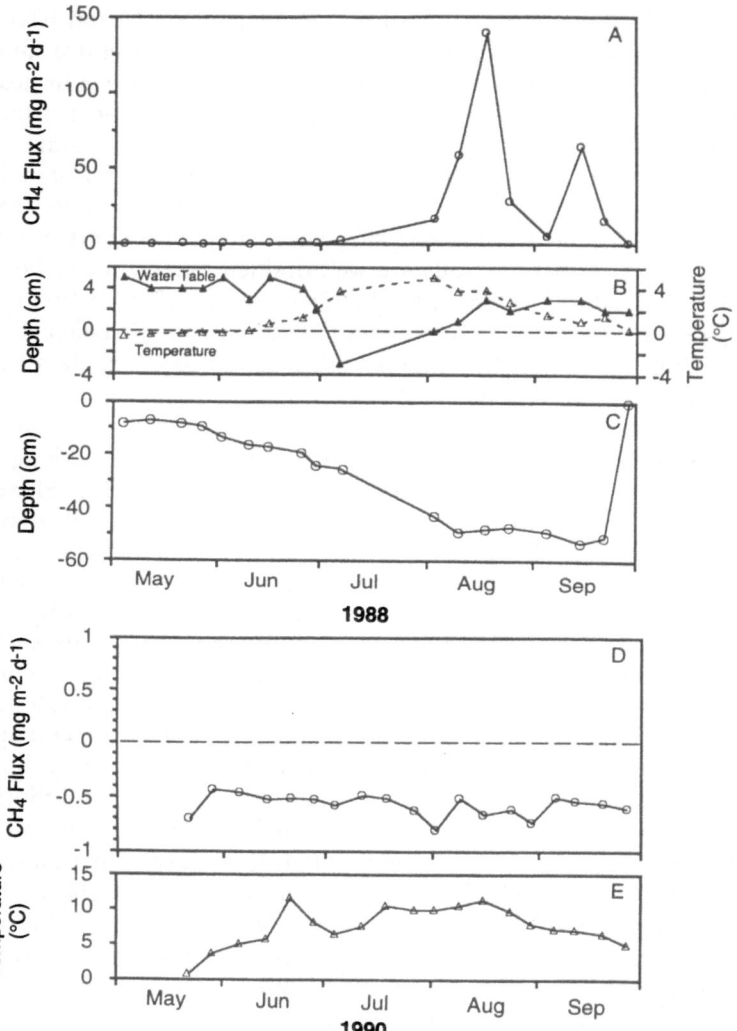

Fig. 12.1. A Seasonal changes in CH$_4$ flux; **B** water table depth and soil temperature at 12 cm depth; **C** thaw depth at an intertussock depression (site BH1) in the University of Alaska Fairbanks arboretum; **D** seasonal changes in CH$_4$ flux; and **E** soil temperature at 13 cm depth at site UP3A (white spruce) in the Bonanza Creek Experimental Forest

face (Fig. 12.4a), but they are probably only air-saturated. The higher electrode readings appear to be caused by a higher rate of oxygen supply across the membrane of the O$_2$ microelectrodes on drying (Revsbech and Ward 1983). Moist soils remain oxygenated to the water table, although O$_2$ concentrations decrease gradually with depth. Elevation of the water table to the soil–air interface was observed to consistently restrict the oxygenated soil zone to a 4–7 mm surface layer (Fig. 12.4b). This latter profile is similar to the O$_2$ distri-

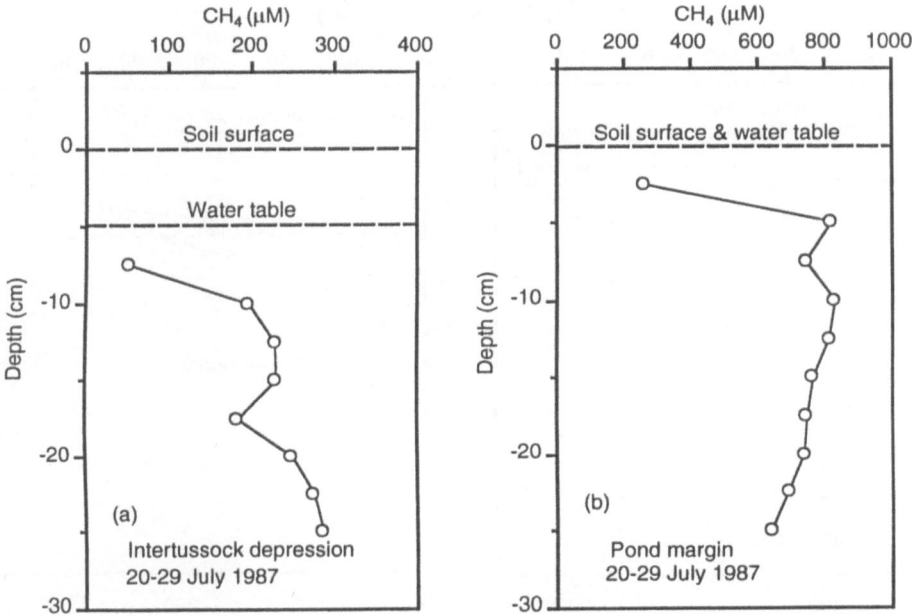

Fig. 12.2a,b. Depth distributions of CH_4 concentration in waterlogged soils in the University of Alaska Fairbanks arboretum. The intertussock depression site (**a**) is adjacent to an *Eriophorum* tussock, and the pond margin site (**b**) was covered with moss and detritus. An equilibration sampler, which was deployed 2 weeks before sample collection, was used to sample these profiles

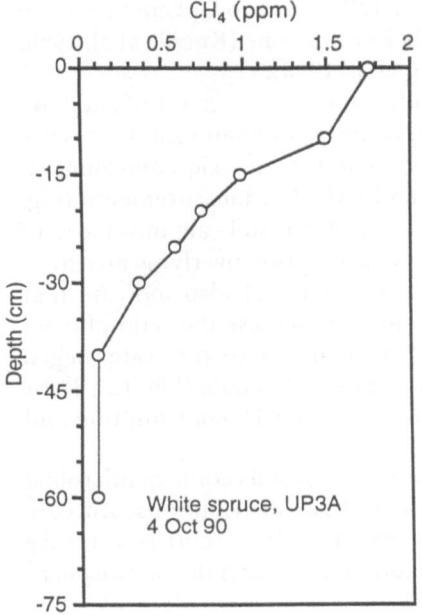

Fig. 12.3. Depth distribution of CH_4 in moist soil of a mature white spruce stand, site UP3A, in the Bonanza Creek Experimental Forest. Samples were collected by pumping gas through a probe inserted to appropriate depths

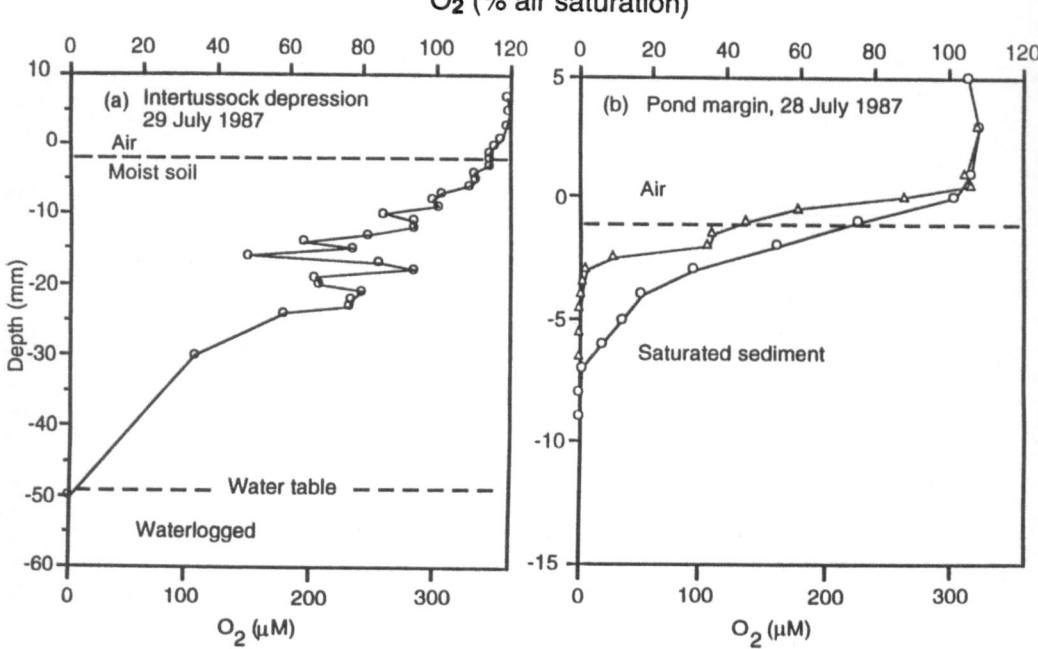

Fig. 12.4a,b. Representative depth distributions of microelectrode-determined dissolved O_2 in waterlogged intertussock depression (**a**) and pond margin (**b**) sites in the University of Alaska Fairbanks arboretum. Sites are identical to those in Fig. 12.3. Note the millimeter depth scale

butions in surficial lake sediments, which generally show no O_2 1 cm below the surface and CH_4 concentrations increasing below this zone (Kuivila et al. 1988; Remsen et al. 1989; Frenzel et al. 1990; King et al. 1990).

Microbial CH_4 production and consumption have been reviewed by Oremland (1988) and Hanson (1980), respectively. Methanogenesis occurs under anaerobic conditions; CH_4 oxidation requires an oxic environment. Profiles of CH_4 (Fig. 12.2), O_2 (Fig. 12.4), and CH_4 flux measurements (Fig. 12.1a) for Arboretum sites with the water table at the soil–air interface are consistent with a shallow surface zone of CH_4 consumption overlying an extensive soil zone of CH_4 production. Figures 12.2 and 12.4 also indicate that lowering the water table position in these soils will increase the ratio of oxic/ anoxic soil if thaw depth is held constant. In contrast to the waterlogged arboretum soils, soil CH_4 distributions (Fig. 12.3) and flux data (Fig. 12.1d) for moist taiga environments suggest an extensive zone of CH_4 consumption and no region of CH_4 production.

Jar experiments with moist arctic tundra and taiga soils confirm microbial CH_4 oxidation as the cause of differences in soil gas distributions. All core sections taken from the soil surface to the water table (12 cm) at a mossy hummock from arctic tundra showed CH_4 oxidation in amended atmosphere experiments (Fig. 12.5a). Core sections to depths of 60 cm in taiga soils also showed oxidized CH_4 under similar conditions (Fig. 12.5b). Oxygen availability has been demonstrated as a control on CH_4 oxidation in surficial lake sediment (King 1990; King et al. 1990). Figures 12.4 and 12.5 suggest no O_2 limitation of CH_4 oxidation above the water table in arboretum and taiga soils.

12.3.2 Physiology, Controls, and Potential for Microbial CH_4 Oxidation

End-member (extreme) environments are useful in placing boundaries on rates of microbial activity, environmental controls on rates, and physiological differences among bacterial communities and isolated organisms. Our CH_4 oxidation studies included microbial communities and isolates from arctic tundra, taiga, and landfill surface soils, thus covering populations naturally exposed to a wide range of CH_4 concentrations (see Figs. 12.2 and 12.3). Also, landfill soil CH_4 concentrations may reach 47% by volume (Schumacher 1983).

Microbial communities and isolates from these diverse environments showed remarkable physiological similarity. Values for K_s varied form 6.2 to 9.3 μM CH_4 for isolated methanotrophs and from 1.7 to 6.2 μM for soil commu-

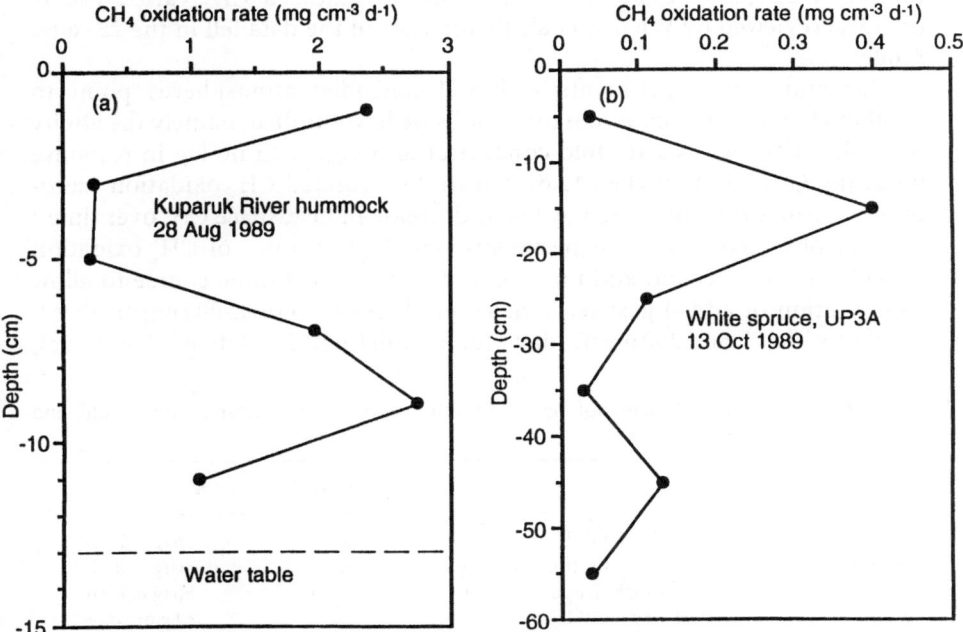

Fig. 12.5a,b. Depth distributions of CH_4 consumption in soils from an arctic hummock adjacent the Kuparuk River (a) and a white spruce site (UP3A) in BCEF (b). These amended atmosphere experiments illustrate zones of consumption; CH_4 oxidation rates are enhanced, but not maximum. Additional data for b are given in Fig. 12.1d,e and Fig. 12.3

nities (Table 12.1). Directly comparable (cellular) values of V_m were also similar, varying from 80 to 98 fg cell^{-1}h^{-1} (see Table 12.1) or from 140–172 mg CH$_4$ g cell^{-1}h^{-1}. No differences were apparent among organisms from a variety of environments when all K_s or culture V_m values were ranked. Furthermore, the median threshold for CH$_4$ oxidation was subambient in all cases, with a range of only 0.14 ppm (Table 12.2), and the mean percentage of ^{14}CH$_4$ incorporated into microbial biomass also showed a restricted range, from 40 to 58% (Table 12.2). Differences in V_m among soil samples (Table 12.1) likely relate to microbial density. Qualitatively, a larger population of methanotrophs in a landfill cover soil than in moist arctic tundra soil is reasonable. Remsen et al. (1989) argue that V_m provides a comparative measure of population size, because values for cultures vary only over a four- to fivefold range.

Comparison of kinetic constants, thresholds, and incorporation of oxidized CH$_4$ in cells points to a fundamental similarity between the methanotrophic communities studied here and most isolates and natural communities from terrestrial and aquatic environments. Values for K_s varied from 0.8 to 66 µM (surveyed by Jørgensen and Degn 1983; Whalen et al. 1990); approximately 75% of the observations fell between 2 and 10 µM. Values of V_m (reviewed by Bedard and Knowles 1989) varied from 160 to 496 mg CH$_4$g cell^{-1}h^{-1}. The CH$_4$ oxidation threshold for a temperate forest soil was 0.3 ppm (Yavitt et al. 1990). Cellular incorporation of oxidized CH$_4$ varied widely (2–73%; reviewed by Whalen et al. 1990); most of the data fell in the 25–60% range.

Jar and core experiments with CH$_4$-amended atmospheres point to another characteristic common to the soils we have studied, namely the ability to oxidize CH$_4$ over a >10^3-fold concentration range with no lag in response following CH$_4$ addition (Fig. 12.6). Substrate-saturated CH$_4$ oxidation (zero-order consumption; indicated by linear decrease in headspace CH$_4$ over time) was not observed in these experiments. The highest rates of CH$_4$ oxidation (calculated from second and third observations of each time course to allow equilibration of added gas; see Whalen et al. 1990) were 45 000 mg m^{-2}day^{-1}, 720 mg m^{-2}day^{-1}, and 290 mg m^{-1}day^{-1} for landfill (Fig. 12.6a), taiga (Fig. 12.6c),

Table 12.1. Kinetics of CH$_4$ oxidation by soil composites, cores, and cultures from landfill and arctic tundra environments

Sample	Location	K_s(µM)	V_m
Culture	Berkeley landfill	9.3	96 fg cell^{-1}h^{-1}
Culture[a]	Lichen heath, Imnavait Creek	6.3	80 fg cell^{-1}h^{-1}
Culture[a]	Hummock, Kuparuk River	6.2	98 fg cell^{-1}h^{-1}
Core E	Berkeley landfill	6.2	61 g m^{-2}day^{-1}
Soil composite	Berkeley landfill	2.5	60 µg(g$_{soil}$)$^{-1}$day^{-1}
Core K2	Moss, Kuparuk River	1.7	489 mg m^{-1}day^{-1}
Core K3	Moss, Kuparuk River	3.2	866 mg m^{-1}day^{-1}

[a] Mean of two experiments from same culture.

Table 12.2. Median threshold for CH_4 oxidation and percent of oxidized $^{14}CH_4$ incorporated into microbial biomass

Experiment	Location	Median threshold (ppm)	$^{14}CH_4$ in biomass (%)
Jar	Moist arctic tundra, Kuparuk River	0.37(10)[a]	58 ± 2(3)
Jar	Taiga, site UP3A, BCEF	0.25(6)	40 ± 2(4)
Jar	Berkeley landfill	0.23(8)	42 ± 26(2)
Chamber	Moist arctic tundra, Dalton Hwy transect	0.27(2)	
Chamber	Taiga, site UP3A, BCEF	0.36(12)	

BCEF, Bonanza Creek Experimental Forest.
[a] Number of observations shown in parentheses; ± values represent standard error of the mean.

and arctic tundra (Fig. 12.6b) soils, respectively. A high population of methanotrophs is a plausible explanation for the high rate of CH_4 oxidation in landfill soil. Methane oxidation rates for taiga and arctic tundra soils are remarkable when put in perspective. The observed (290 mg m^{-2} d^{-1}) and calculated (489–866 mg m^{-2} day^{-1}; Table 12.1) CH_4 consumption potentials for arctic tundra cores are ca. three to ten times the mean CH_4 emission rate of 90 mg m^{-2} day^{-1} for wet tundra (Whalen and Reeburgh 1990a). A CH_4 consumption rate of 720 mg m^{-2} day^{-1} for taiga soil is at least 1500 times that observed in a free-air atmosphere (Fig. 12.1d). The data suggest first-order (concentration-dependent) CH_4 consumption in moist taiga soils, consistent with the observed lack of seasonality in CH_4 consumption rates (Fig. 12.1d).

Relaxation experiments for arctic tundra and landfill cores gave first-order equilibration constants of −0.7 and −1.54 min^{-1}, respectively, indicating that equilibration of added CH_4 in amended atmosphere experiments was rapid. These data suggest that moist surface soils of arctic tundra and taiga offer little resistance to the diffusion of gases, ensuring a continuous supply of atmospheric O_2 and CH_4 for methanotrophic communities in moist taiga and arctic tundra soils. Unlike taiga soils, however, the principal microbial CH_4 supply in arctic tundra soil is upward diffusion through water-saturated soil. A high capacity for CH_4 oxidation is essential for soil microbes to arrest gas diffusing from waterlogged soils through the moist soil zone to the atmosphere.

The apparent physiological similarity among methanotrophs from extreme environments suggests that physical controls on CH_4 oxidation may be responsible for differences in CH_4 flux between taiga and arboretum soils (Figs. 12.1a,d). Analysis of environmental influences on substrate-saturated CH_4 oxidation in bulk landfill soils points to soil moisture content as a powerful control on CH_4 oxidation (Whalen et al. 1990). Soils at an initial gravimetric moisture content of 11% showed a CH_4 oxidation rate of 116 mg l^{-1} day^{-1}. Increasing the soil moisture in 15% increments decreased the CH_4 oxidation rate to a soil moisture content of 41% (Whalen et al. 1990); soils were visibly

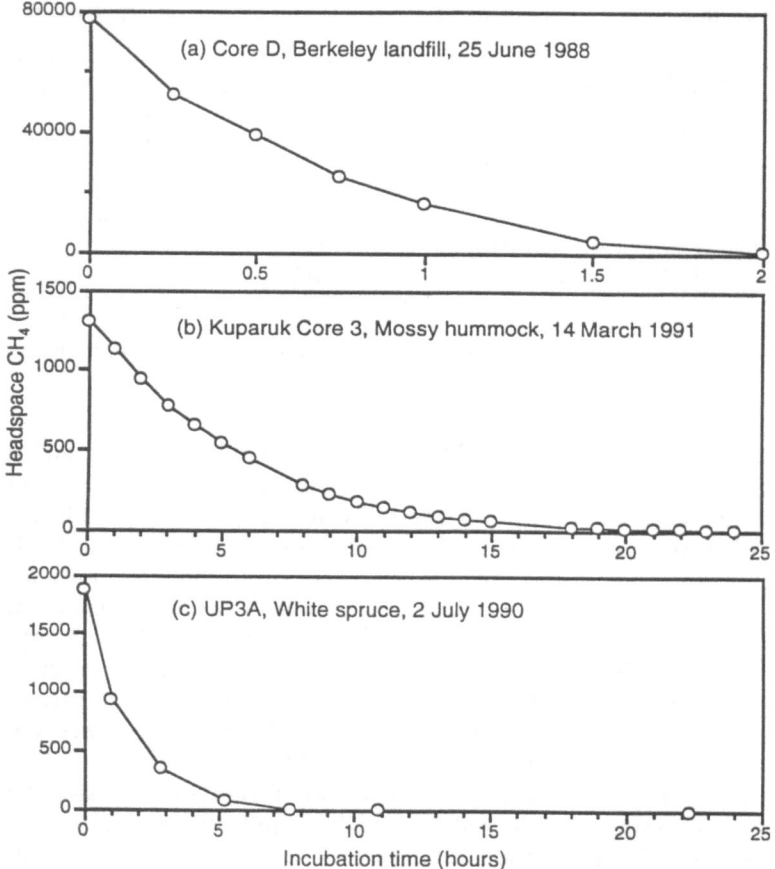

Fig. 12.6a–c. Methane-consumption measurements in amended atmosphere experiments involving cores in jars (**a,b**) and an in situ soil chamber (**c**)

waterlogged at that point. The CH_4 oxidation rate was reduced 20-fold to 6.1 mg l^{-1}day^{-1}, and further increases in soil moisture had no influence on this oxidation rate. These observations are consistent with transition from gas phase to aqueous molecular diffusion (10^4-fold less rapid) in CH_4 supply to the cell membrane, suggesting that oxidation rate is limited by CH_4 supply to the cell.

A water table positioned at the soil surface has a profound effect on CH_4 oxidation. The oxidized zone in arboretum soil is water-saturated and restricted to approximately the first centimeter of soil (Fig. 12.4b). Methane oxidation at typical substrate levels (Fig. 12.2b) proceeds at the aqueous phase V_m of about 90 fg cell^{-1}h^{-1} (Table 12.1). We calculate a CH_4 oxidation rate of about 108 mg m^{-2}day^{-1} (11 mg l^{-1}day^{-1}), assuming a 1 cm-thick oxidized zone and 0.5- $\times$ 10^4 methanotrophs cm^{-3} peat (Williams and Crawford 1983). If we assume a mean atmospheric emission rate of 90 mg m^{-2}day^{-1} (Whalen and Reeburgh 1990a), 55% of the CH$_4$ diffusing to the soil surface in wet tundra is oxidized. This amount is only 12–22% of V$_m$ for moist arctic tundra cores (Table 12.1), and is a direct consequence of aqueous molecular diffusion of CH$_4$ into a thin oxidized zone.

Our calculated rate of substrate-saturated CH$_4$ oxidation in waterlogged arctic tundra compares reasonably well with area-based CH$_4$ oxidation rates for freshwater sediment. King (1990) reported CH$_4$ oxidation rates from 0.5 to 76 mg m^{-2}day^{-1} for pond sediment in the Florida Everglades. Kuivila et al. (1988) and Frenzel et al. (1990) give CH$_4$ oxidation rates of 7.2 and 4.8 mg m^{-2}day^{-1} for Lake Constance and Lake Washington, respectively. The volume-based rate of CH$_4$ oxidation we calculate for tundra also lies within the V$_m$ range reported for lake sediment. King (1990) reports V$_m$ varying from 254 to 553 mg l^{-1}day^{-1} for Danish pond sediment; Lidstrom and Somers (1984) and Kuivila et al. (1988) give values from 0.2 to 15.9 mg l^{-1}d^{-1} for Lake Washington sediment. Moreover, the data also agree with a V$_m$ of 6.1 mg l^{-1}day^{-1} for waterlogged landfill soil (Whalen et al. 1990). Again, differences in rates may reflect variations in bacterial numbers. Collectively, the data from the arboretum, arctic tundra, freshwater lakes, and manipulated landfill soils point to a fundamental similarity among waterlogged environments: CH$_4$ concentrations are sufficient to support CH$_4$ oxidation rates at or approaching aqueous phase V$_m$, but area-based rates will be as much as an order of magnitude lower than for drained soils exposed to similar CH$_4$ concentrations.

The depth to which the water table must be lowered for CH$_4$ oxidation to result in zero net CH$_4$ emission can be estimated from waterlogged arboretum soil CH$_4$ concentrations (Fig. 12.2) and V$_m$ for arctic tundra cores (Table 12.1). Methane fluxes across the water table into the moist soil zone were estimated using a stagnant film gas transfer model (Broecker and Peng 1974), which assumes that the gas phase is well-mixed and that transfer from the aqueous phase is limited by molecular diffusion across a laminar surface layer. The thickness of the stagnant surface layer is controlled by agitation of the aqueous phase. We assume a stagnant film thickness of 200μ for the water table surface. This value is typical for a still, free-water surface (Kling et al. 1991), and probably results in an over-estimate of CH$_4$ flux. The consumption-zone thickness required to balance CH$_4$ flux across the water table surface was 10 cm and 42 cm for CH$_4$ profiles in Fig. 12.2a and b, respectively. Figure 12.2a is more representative of soil CH$_4$ distributions in moist arboretum soils; pond margin sediment CH$_4$ concentrations are consistently four- to fivefold higher. This preliminary estimate shows that these soils have the capacity to oxidize large quantitites of CH$_4$ with only a slight lowering of water table; increasing the extent of the surface oxidized zone and the resulting shift in supply of CH$_4$ to cell membranes from aqueous to gaseous molecular diffusion result in an effective control on CH$_4$ emissions.

Figure 12.7 illustrates CH_4 oxidation and its controls in arctic tundra soils without vascular plants. Vascular transport of CH_4 to the atmosphere by aquatic plants (Cicerone and Shetter 1981; Seiler et al. 1984; Sebacher et al. 1985) bypasses the surface CH_4 oxidizing zone and may limit CH_4 oxidation in tundra soils vegetated by *Eriophorum* (Tenhunen et al. 1995) and *Carex*. Methane oxidation occurs from the soil surface to about 1 cm below the water table. A water table at the soil surface restricts CH_4 oxidation to a shallow surface zone where aqueous-phase molecular diffusion transports CH_4 from the underlying zone of production. Soils with a subsurface water table have a more extensive zone of CH_4 oxidation. The aqueous-phase CH_4 flux from the water table and atmospheric CH_4 are consumed in this situation. Taiga represents the extreme case: since there is no subsurface CH_4 source, all CH_4 must diffuse through the soil from the atmosphere.

12.3.3 Methane Oxidation by Tundra Soils in a Warmer Climate

These data permit an estimate of the future role of CH_4 oxidation in the fraction of arctic tundra vegetated by nonvascular plants, which includes about 10–20% of wet meadow tundra and 50–65% of tussock and low-shrub

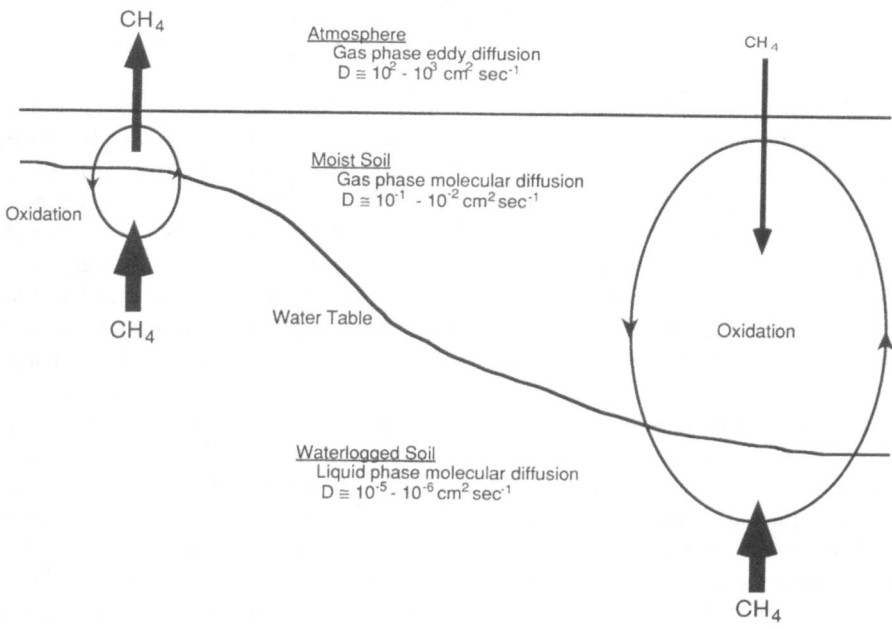

Fig. 12.7. Diagram showing the effect of changes in water table level on CH_4 oxidation and CH_4 emission and consumption. Characteristic diffusivities are given for free air, moist soil, and waterlogged soil

tundra (Whalen and Reeburgh 1988). General circulation models (GCMs; Mitchell 1989) all predict warmer mean air temperatures for the Arctic, but they disagree in the predicted influence of this temperature increase on the hydrological regime. Specifically, some GCMs project an unchanged water table, but most forecast a reduced water table during the thaw period. Initial models of biogeochemical feedbacks to warmer climate predict increased CH_4 emission from northern wetlands because of CH_4 released from the permafrost and more rapid methanogenesis in an increased active zone during an extended thaw (Khalil and Rasmussen 1989; Lashof 1989). These models do not consider CH_4 oxidation and assume no shift in water table position.

Our data also suggest that warmer soils and an unchanged water table will favor increased CH_4 emission. Methane oxidation will remain restricted to a surface zone of waterlogged soil. In the more likely instance that warmer soils are accompanied by a lowered water table, CH_4 oxidation will be enhanced. The oxic/anoxic boundary will drop with the water table, increasing the vertical extent of the oxidized zone at the expense of the anoxic zone. These low-resistance tundra soils permit rapid gas-phase molecular transport of both O_2 and CH_4 to soil microbial communities with a demonstrated high capacity for CH_4 oxidation. Also, soil warming will probably increase CH_4 consumption more than production, because surface soils where oxidation occurs will undergo earlier and greater warming than the subsurface sites of methanogenesis. Lashof (1989) assumed a Q_{10} of three for methanogenesis, which is equivalent to our measured Q_{10} of 1.9 for CH_4 oxidation (Whalen et al. 1990), given the sensitivity of calculated values to the temperature range examined (Ahlgren 1987).

Two factors beyond enhanced CH_4 oxidation will contribute to reduced CH_4 emission from a warmer, drier tundra. Firstly, a drop in the water table will promote aerobic decomposition of near-surface labile material previously decomposed anaerobically in waterlogged soils. Moore and Knowles (1989) showed that CO_2 emission increased linearly and CH_4 evolution decreased logarithmically as water tables were lowered in laboratory columns packed with fen, bog, and swamp soils. Secondly, the organic material made available for microbial decomposition by an increased thaw depth is probably recalcitrant and will result in little additional CH_4 production. Clymo (1983), Farrish and Grigal (1988), and Yavitt et al. (1988) have demonstrated a decreasing potential for decomposition with increasing peat depth.

12.4 Conclusions

Tundra soils show a remarkable potential for CH_4 oxidation that might result in a negative feedback to atmospheric CH_4 increases under appropriate conditions. Results given here show the physiological similarities among diverse communities of methanotrophs and the strong control of CH_4 oxidation by soil moisture. These results can be extended to predict the impact of an altered water table on CH_4 emission for any environment. Clearly, CH_4 oxidation must be included as a biogeochemical feedback in climate change models.

Three lines of future research will advance this work. Firstly, large-scale manipulation experiments (tens of m^2) can provide a seasonal verification of results of the day-length jar and chamber experiments performed here. Secondly, methanotrophic bacteria can be quantified in natural habitats and bacterial numbers correlated with CH_4 fluxes and soil characteristics;. RNA signature probes (Topp and Hanson 1991) are a promising approach in this area. Finally, the focus of future work should be expanded to include net emissions and feedbacks of all radiatively active gases. Conditions that lead to reduced CH_4 emissions also result in enhanced CO_2 emissions (Jenkinson et al. 1991; Chap. 11, this Vol.).

Acknowledgments. Contribution number 888 from the Institute of Marine Science, University of Alaska, Fairbanks. This work was sponsored by NASA grant no. IDP 88-032.

References

Adams JM, Faure H, Faure-Denard L, McGlade JM, Woodward FI (1990) Increases in terrestrial carbon storage from the Last Glacial Maximum to the present. Nature 348: 711–714

Ahlgren G (1987) Temperature functions in biology and their application to algal growth constants. Oikos 49: 177–190

Aselmann I, Crutzen PJ (1989) Global distribution of natural freshwater wetlands and rice paddies: their net primary productivity seasonality and possible methane emissions. J Atmos Chem 8: 307–358

Bedard C, Knowles R (1989) Physiology biochemistry and specific inhibitors of CH_4, NH_4^+ and CO oxidation by methylotrophs and nitrifiers. Microbiol Rev 53: 68–84

Billings WD (1987) Carbon balance of Alaskan tundra and taiga ecosystems: past, present and future. Quat Sci Rev 6: 165–177

Blake DR, Rowland FS (1989) Continuing worldwide increase in tropospheric methane, 1978–1987. Science 239: 1129–1131

Bliss LC (1975) Devon Island. In: Rosswall T, Heal OW (eds) Structure and function of tundra ecosystems. Swed Natl Res Counc, Stockholm. Ecol Bull 20: 17–60

Bliss LC, Heal OW, Moore JJ (1981) Tundra ecosystems: a comparative analysis. Cambridge Univ Press, Cambridge

Broecker WS, Peng T-H (1974) Gas exchange rates between air and sea. Tellus 26: 21–35

Chapin FS III, Jeffries RL, Reynolds JF, Shaver GR, Svoboda J (1992) Arctic ecosystems in a changing climate. Academic Press, San Diego

Cicerone RJ, Oremland RS (1988) Biogeochemical aspects of atmospheric methane. Global Biogeochem Cycles 2: 299–327

Cicerone RJ, Shetter JD (1981) Sources of atmospheric methane: measurements in rice paddies and a discussion. J Geophys Res 86: 7203–7209

Clymo RS (1983) Peat. In: Gore AJP (ed) Mires: swamp, bog, fen and moor. Elsevier, New York, pp 159–224

Crill PM, Bartlett KB, Harriss RC, Gorham E, Verry ES, Sebacher DI, Mazdar L, Sanner W (1988) Methane flux from Minnesota peatlands. Global Biogeochem Cycles 3: 371–384

Farrish KW, Grigal DF (1988) Decomposition in an ombrotrophic bog and a minerotrophic fen in Minnesota. Soil Sci 145: 353–358

Frenzel P, Thebrath B, Conrad R (1990) Oxidation of methane in the oxic surface layer of a deep lake sediment (Lake Constance). FEMS Microbiol Ecol 73: 149–158

Fung I, John J, Lerner J, Matthews E, Prather M, Steele LP, Fraser PJ (1991) Three-dimensional model synthesis of the global methane cycle. J Geophys Res 96: 13,033–13,065

Gorham E (1991) The role of subarctic and boreal peatlands in the global carbon cycle and their probable responses to "greenhouse" climate warming. Ecol Appl 1: 182–195

Grotch SL (1988) Regional intercomparison of general circulation models, predictions and historical climate data. US Dept Energy (DOE/NBB-0084 TR 041), Washington DC

Hanson RS (1980) Ecology and diversity of methylotrophic organisms. Adv Appl Microbiol 26: 3–39

Hesslein RH (1976) An in situ sampler for close-interval interstitial water studies. Limnol Oceanogr 21: 912–914

Hinzman LD, Kane DL, Gieck RE, Everett KR (1991) Hydrologic and thermal properties of the active layer in the Alaskan Arctic. Cold Reg Sci Technol 19: 95–110

Jenkinson DS, Adams DE, Wild A (1991) Model estimates of CO_2 emissions from soils in response to global warming. Nature 351: 304–306

Jørgensen L, Degn H (1983) Mass spectrometric measurements of methane and oxygen utilization by methanotrophic bacteria. FEMS Microbiol Ecol 20: 331–335

Khalil MAK, Rasmussen RA (1989) Climate-induced feedbacks for global cycles of methane and nitrous oxide. Tellus 41(B): 554–559

King GM (1990) Dynamics and controls of methane oxidation in a Danish wetland sediment. FEMS Microbiol Ecol 74: 309–324

King GM, Roslev P, Skovgaard H (1990) Distribution and rate of methane oxidation in sediments of the Florida Everglades. Appl Environ Microbiol 56: 2902–2911

Kling GW, Kipphut GW, Miller MC (1991) Arctic lakes and streams as gas conduits to the atmosphere: implications for tundra carbon budgets. Science 251: 298–301

Kuivila KM, Murray JW, Devol AH, Lidstrom ME, Reimers CE (1988) Methane cycling in the sediments of Lake Washington. Limnol Oceanogr 33: 571–581

Lachenbruch AH, Marshall BV (1986) Changing climate: geothermal evidence from permafrost in the Alaskan arctic. Science 234: 689–696

Lashof DA (1989) The dynamic greenhouse: feedback processes that may influence future concentrations of atmospheric trace gases and climate change. Clim Change 14: 213–242

Lidstrom ME, Somers L (1984) Seasonal studies of methane oxidation in Lake Washington. Appl Environ Microbiol 47: 1255–1260

Matson PA, Vitousek PM, Schimel DS (1989) Regional extrapolation of trace gas flux based on soils and ecosystems. In: Andreae MO, Schimel DS (eds) Exchange of trace gases between terrestrial ecosystems and the atmosphere. Wiley, New York, pp 97–108

Matthews E, Fung I (1987) Methane emission from natural wetlands: global distribution area and environmental characteristics of sources. Global Biogeochem Cycles 1: 61–86

Mitchell JFB (1989) The greenhouse effect and climate. Rev Geophys 27: 115–139

Mooney HS, Vitousek PM, Matson PA (1987) Exchange of material between terrestrial ecosystems and the atmosphere. Science 238: 926–932

Moore TR, Knowles R (1989) The influence of water table levels on methane and carbon dioxide emissions from peatland soils. Can J Soil Sci 69: 33–38

Moore TR, Roulet N, Knowles R (1990) Spatial and temporal variations of methane flux from subarctic/northern boreal forest fens. Global Biogeochem Cycles 4: 29–46

Oremland RS (1988) Biogeochemistry of methanogenic bacteria. In: Zehnder AJB (ed) Biology of anaerobic microorganisms. Wiley, New York, pp 641–706

Post WM (1990) Report of a workshop on climate feedbacks and the role of peatlands, tundra and boreal ecosystems in the global carbon cycle. Oak Ridge National Laboratory (ORNL/TM-11457), Oak Ridge, Tennessee

Post WM, Emanuel WR, Zinke PJ, Stangenberger AG (1982) Soil carbon pools and world life zones. Nature 298: 156–159

Rasmussen RA, Khalil MAK (1984) Atmospheric methane in recent and ancient atmospheres: concentrations, trends and the interhemispheric gradient. J Geophys Res 89(D7): 11 599–11 605

Reeburgh WS, Whalen SC (1992) High latitude ecosystems as CH_4 sources. Ecol Bull (Copenhagen) 42: 62–70

Remsen CC, Minnich EH, Stephens RS, Buchholz L, Lidstrom ME (1989) Methane oxidation in Lake Superior sediments. J Great Lakes Res 15: 141–146

Revsbech NP, Ward DM (1983) Oxygen microelectrode that is insensitive to medium composition: use in an acid microbial mat dominated by *Cyanidium calderium*. Appl Environ Microbiol 45: 755–759

Rieger S (1975) Arctic soils. In: Ives JD, Barry RG (eds) Arctic and alpine environments. Methuen, London, pp 749–769

Rodhe H (1990) A comparison of the contribution of various gases to the greenhouse effect. Science 248: 1217–1219

Schumacher MM (1983) Landfill methane recovery. Noyes Data Corp, Park Ridge, New Jersey

Sebacher DI, Harriss RC, Bartlett KB (1985) Methane emission to the atmosphere through aquatic plants. J Environ Qual 14: 40–46

Sebacher DI, Harriss RC, Bartlett KB, Sebacher SM, Grice SS (1986) Atmospheric methane sources: Alaskan tundra, an alpine fen and a subarctic boreal marsh. Tellus 38(B): 1–10

Seiler W, Holzapfel-Pschorn A, Conrad R, Scharfee D (1984) Methane emission from rice paddies. J Atmos Chem 1: 241–268

Steele LP, Fraser PJ, Rasmussen RA, Khalil MAK, Conway TJ, Crawford AJ, Gammon RH, Masarie KA, Thoning KW (1987) The global distribution of methane in the troposphere. J Atmos Chem 5: 125–171

Tenhunen JD, Gillespie CT, Oberbauer SF, Sala A, Whalen SC (1995) Climate effects on the carbon balance of tussock tundra in the Philip Smith Mountains, Alaska. Flora 190: 273–283

Tieszen LL (1978) Vegetation and production ecology of an Alaskan arctic tundra. Springer, Berlin Heidelberg New York

Topp E, Hanson RS (1991) Metabolism of radiatively important trace gases by methane-oxidizing bacteria. In: Rogers JE, Whitman WB (eds) Microbial production and consumption of greenhouse gases: methane, nitrogen oxides and halomethanes. Am Soc Microbiol, Washington DC, PP 71–90

Van Cleve K, Chapin FS III, Flanagan PW, Viereck LA, Dyrness CT (1986) Forest ecosystems in the Alaskan taiga. Springer, Berlin Heidelberg New York

Whalen SC, Reeburgh WS (1988) A methane flux time series for tundra environment. Global Biogeochem Cycles 2: 399–409

Whalen SC, Reeburgh WS (1990a) A methane flux transect along the trans-Alaska pipeline haul road. Tellus 42(B): 237–249

Whalen SC, Reeburgh WS (1990b) Consumption of atmospheric methane by tundra soils. Nature 346: 160–162

Whalen SC, Reeburgh WS (1992) Interannual variations in tundra methane emission: a 4-year time series at fixed sites. Global Biogeochem Cycles 6: 139–152

Whalen SC, Reeburgh WS, Sandbeck KA (1990) Rapid methane oxidation in a landfill cover soil. Appl Environ Microbiol 56: 3405–3411

Whalen SC, Reeburgh WS, Kizer KS (1991) Methane consumption and emission by taiga. Global Biogeochem Cycles 5: 261–273

Wigley TML, Raper SCB (1990) Natural variability of the climate system and detection of the greenhouse effect. Nature 344: 324–327

Williams RT, Crawford RL (1983) Microbial diversity in Minnesota peatlands. Microbiol Ecol 9: 201–214

Yavitt JB, Lang GE, Downey DM (1988) Potential methane production and methane oxidation rates in peatland ecosystems of the Appalachian Mountains, United States. Global Biogeochem Cycles 2: 253–268

Yavitt JB, Downey DM, Lang GE, Sexstone AJ (1990) Methane consumption in two temperate forest soils. Biogeochemistry 9: 39–52

13 Dynamics of Dissolved and Particulate Carbon in an Arctic Stream

M. W. Oswood, J. G. Irons III, and D. M. Schell

13.1 Introduction

There are large stores of soil organic carbon in arctic tundra (Schlesinger 1977; Post et al. 1982), which has stimulated much research on understanding the net carbon balance in these terrestrial ecosystems (e.g., Shaver et al. 1992; Oechel et al. 1993; Tenhunen et al. 1995; see also Chaps. 11 and 17, this Vol.). However, aquatic ecosystems also play an important role in tundra carbon budgets. Kling et al. (1991) found that freshwater ecosystems in arctic Alaska showed net positive fluxes of CO_2 to the atmosphere, and thus serve as sources, rather than sinks, of atmospheric carbon. Much of the CO_2 in arctic aquatic systems is ultimately derived from terrestrial sources – eroding peat, soil dissolved organic matter, and inorganic carbon. Higher soil temperatures or a longer ice-free season at high latitudes may result in increased terrestrial decomposition and thus increased fluxes of organic materials to streams (Forsberg 1992; Oswood et al. 1992; Chap. 10, this Vol.).

Streams at high latitudes may serve as especially sensitive indicators of anthropogenic disturbances – including climate change – that directly or indirectly affect terrestrial ecosystems (Hammar 1989; Oswood et al. 1992). As part of the R4D project (Chap. 1, this Vol.), we initiated carbon balance studies in the headwaters of Imnavait Creek, a small stream typical of zero – order drainages in the Arctic Foothills. Tundra streams generally have characteristics that are intermediate between spring-fed streams (stable discharge and temperature, very high densities of benthic invertebrates) and mountain streams (intermittent flow, very low densities of benthic invertebrates; Craig and McCart 1975). Although lentic ecosystems in the northern arctic and subarctic regions have been intensively studied (e.g., Hobbie 1980, 1984), this is not the case for high-latitude running waters (Harper 1981; Peterson et al. 1986; Oswood et al. 1992). In this chapter, we present seasonal dynamics of both particulate and dissolved organic carbon concentrations over a 2-year period. We also estimated spatial variability in stream carbon and total export of carbon from the watershed.

J.F. Reynolds and J.D. Tenhunen (Eds.)
Ecological Studies, Vol. 120
© Springer-Verlag Berlin Heidelberg 1996

13.2 Site Description

13.2.1 Imnavait Creek Watershed

The Imnavait Creek watershed encompasses a 210-ha (2860-m stream length) headwater section of an arctic tundra stream (Oswood et al. 1989a,b). The stream is hydrologically bounded by continuous permafrost. Slopes in the watershed are dominated by tussock tundra with occasional exposed mineral soil (Chap. 4, this Vol.). Soils are highly stratified, with 20-cm living and dead mosses, and peaty organic and organic-mineral material overlying glacial till (Hinzman et al. 1991; Chap. 4, this Vol.). Details of the summer weather conditions are given in Chap. 6 (this Vol.). Depth of thaw reaches ca. 25–100 cm. During the summer air temperatures in the watershed can drop precipitously – approaching freezing – and snowstorms can occur.

13.2.2 Description of Imnavait Creek

Imnavait Creek originates in a broad, relatively flat, sedge-dominated string bog with numerous, often poorly defined channels. Through much of its length, the stream is relatively straight and characterized by narrow, steep-sided channel segments bordered by peat. It is a typical "beaded" stream, composed of relatively large rounded pools (<1–3 m deep, and up to ca. 6 m diameter) formed by ice-wedge erosion and connected by very shallow, narrow channels, giving the appearance of beads on a string (Hopkins et al. 1955; Bird 1967; Oswood et al. 1989a; Fig. 13.1). Substrates of both pools and channels are

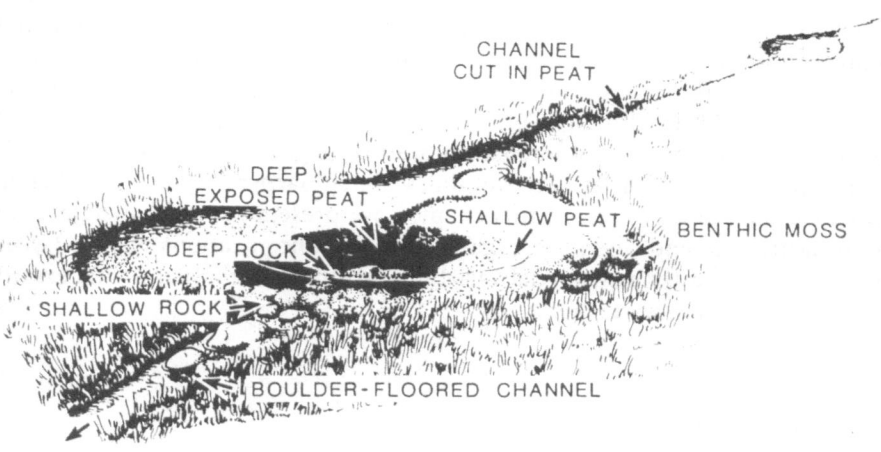

Fig. 13.1. View of beaded stream showing typical benthic substrates. "Deep" indicates water depth >1 m. (From Oswood et al. 1989a)

dominated by peat with occasional boulder, rock, or live moss (Bird 1967; Oswood et al. 1989a).

Snowmelt runoff occurs in mid-May to early June (Kane et al. 1991; Chap. 6, this Vol.) with annual peak discharge associated with breakup (Fig. 13.2). Low flow is common in June and much of July. Late summer storms produce highly variable flow until the end of the surface flow season in mid-September. Stream water is acidic (pH of 5.3–6.1) with very low alkalinity ($\leqslant 3\,mg\,l^{-1}$ as $CaCO_3$).

The extreme dichotomy between channels and pools gives Imnavait Creek characteristics of both lentic and lotic systems (Oswood et al. 1989a). During low summer flows, pools thermally stratify, so that temperatures in surface waters and channels may be very high (ca. 20 °C) (Oswood et al. 1989a; Irons and Oswood 1992). Similarly, benthic macroinvertebrate communities in Imnavait Creek contain taxa typical of both lentic systems, e.g., dytiscid beetles and phryganeid caddisflies, and lotic systems, e.g., simuliid (black fly) larvae and nemourid stoneflies (Viavant 1989).

13.3 Field and Laboratory Procedures

Water samples for particulate organic carbon (POC) and dissolved organic carbon (DOC) analyses were obtained from a 3.7-m² pool immediately upstream of the flume (see Fig. 9.2 in Chap. 9, this Vol.). These "grab samples" were taken approximately three times weekly during low flow and as often as three times daily in 1987 or daily in 1986 during high-discharge events associated with snowmelt and storms. In 1986, samples were not collected during the first 3 days of snowmelt; we used DOC and POC concentrations (noted hereafter as [DOC] and [POC], respectively) for the first 3 days of the 1987 hydrological season as estimates of carbon concentration for this period. Beginning on 19 June 1987, we also took samples from a larger (17.5 m²) pool. On two dates (one with low flow and the other with high flow), we randomly sampled 27 pools of different size classes and spatial position (see below). Supplementary samples of large particulate organic material were obtained with a drift net (filtered through a 1-mm mesh sieve) emptied as necessary (infrequently during low flow and more often during high-discharge events).

Grab samples (two replicates) of stream water for analysis of organic carbon were collected in acid-washed 1-liter Nalgene bottles. Samples were refrigerated until processed. Stream water samples were first filtered through a 1-mm Nitex mesh screen to remove coarse particulate organic material (CPOM) then filtered through precombusted (500 °C) glass fiber filters (ca. 0.45 µm). The filtrate was stored frozen at −20 °C in 125 ml acid washed bottles for DOC analyses and the filter was placed in a precombusted scintillation vial and stored at −20 °C for POC analysis. The DOC analysis was performed on a Technicon Autoanalyzer II system (Industrial Method 45176W/A) equipped

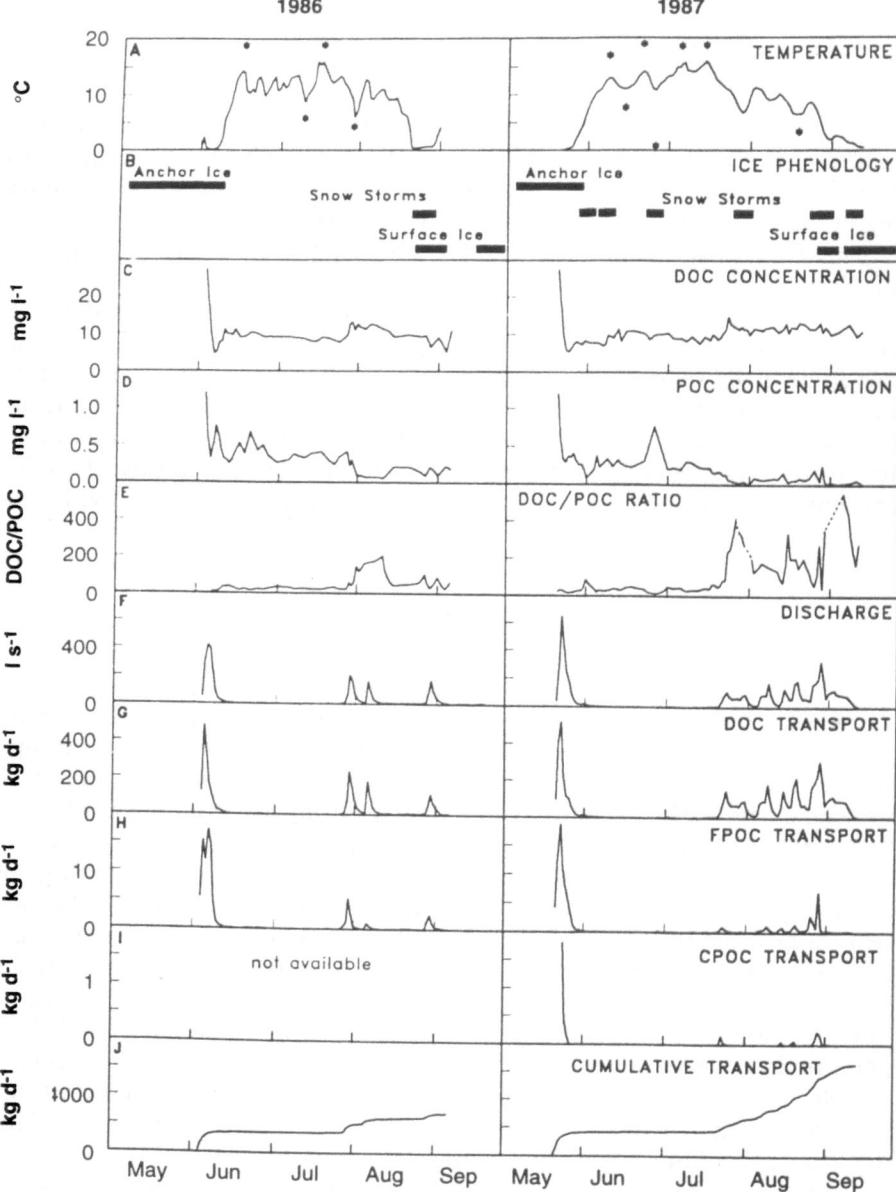

Fig. 13.2A–J. Characteristics of Imnavait Creek, 1986 and 1987. **A** Surface water temperature (5-day running means). Daily maxima and minima (from hourly data) are denoted by asterisks where substantially different from 5-day mean; **B** snow and ice phenology. Anchor ice is attached to the substrate, often with stream water flowing on top of it; **C** dissolved organic carbon (DOC); **D** particulate organic carbon (POC); **E** ratio of DOC to POC. Dates when POC <20 µg/l eliminated and interpolated between the remaining points (*dashed lines*); **F** discharge of Imnavait Creek (L. Hinzman and D. Kane, pers. comm.); **G** daily means of DOC transport (< 0.45 µm); **H** daily means of FPOC transport (0.45 µm < FPOC < 1 mm); **I** daily means of CPOC transport (>1 mm); **J** cumulative transport (DOC + FPOC + CPOC). For **C** and **D**, daily means based on $n = 2–6$; linear interpolation for missing samples

with 5-cm colorimeter cells. Aspirated samples were acidified and stripped of inorganic carbon dioxide by liquid film exposure to a stream of carbon-dioxide-free nitrogen in a large bore coil. The sample stream was aliquoted and segmented, mixed with acid-persulfate solution and subjected to shortwave UV irradiation. The CO_2 generated from the oxidation of the organic carbon was diffused through a silicone membrane and reacted with a buffered phenolphthalein solution. The change in color was proportional to the concentration of organic carbon in the sample. Standards consisted of known concentrations of potassium hydrogen phthalate. Precision was ca. $\pm 0.2 \, mg \, l^{-1}$ with a lower limit of ca. $0.4 \, mg \, l^{-1}$. Precombusted glass fiber filters were analyzed for POC using a Perkin-Elmer CHN Analyzer (Perkin-Elmer 1981) with acetanilide (S & J Reagents) standards. Filter blanks (filtered stream water or distilled water) were used to correct POC values. The CPOM sampled via drift net was filtered through a 1-mm mesh screen and dried at $50 \, °C$, then ashed at $500 \, °C$ to obtain ash-free dry weight (AFDW) of CPOM. The AFDW was converted to carbon assuming a 50% carbon content.

Average [DOC] and [POC] were calculated as both discharge-weighted and time-averaged concentrations. Discharge-weighted concentrations were computed as (Concentrations x Discharge)$_{daily}$/Discharge$_{total \, seasonal}$. Daily transport was calculated as mean daily concentration x mean daily discharge, with carbon concentration values between sampling dates interpolated linearly and summed for annual transport.

All pools and channels were numbered sequentially and the surface area was determined. The reach of Imnavait Creek was divided longitudinally into three "valley" positions: headwaters, mid-valley, and lower. Three categories of pool sizes were established to provide approximately an equal number of pools in each size category. Pool size (surface area) ranges were: small pools = $1-3.4 \, m^2$; medium = $3.7-10.4 \, m^2$; large = $11.8-35.3 \, m^2$. Two replicate 1-l samples for determination of [DOC] and [POC] were collected from three randomly selected pools in each pool size and valley position category. At low flow, pools are nearly isolated and function as small ponds (i.e., thermal stratification occurs); at high flows thermal isolation is destroyed (Oswood et al. 1989a). We sampled once at low water flow (9 July 1987; discharge $0.024 \, l \, s^{-1}$) and once at high water flow (27 July 1987; discharge $65.42 \, l \, s^{-1}$) to examine differences in [DOC] and [POC] among pool sizes. A two-way analysis of variance (ANOVA; valley position x pool size) was carried out for each date. Sample sizes (each date) were as follows: 3 pool sizes x 3 valley positions x 3 replicate pools = 27, with mean values of water samples from each pool used.

13.4 Physical Regime

The exaggerated pool-riffle (beaded) morphology and high-latitude climate of Imnavait Creek combine to produce a highly eccentric physical regime (Fig.

13.2). From September through late May surface ice is continuous, with free water remaining only beneath the deep ice cover of larger pools; flow between pools is highly unlikely. Initial runoff from snowmelt flows over anchor ice and frozen pool and channel substrates (Fig. 13.2b,f) with later snowmelt flowing over unfrozen substrates. The onset of spring melt differed by about 2 weeks in 1986 and 1987 (Fig. 13.2f). A period of very low water flow commences immediately following snowmelt runoff and lasts until mid- to late July. During this period, pools are nearly isolated hydrologically, and flow trickles through ill-defined channels containing *Carex aquatalis*. Riparian vegetation provides little shading when long day lengths produce high temperatures (ca. 20 °C) in surface waters of pools and in channels (Fig. 13.2a). Low water flow through pools contributes to thermal stratification in pools (Oswood et al. 1989a).

Daily minimum temperatures approach freezing during the occasional summer snowstorm (Fig. 13.2a). Convective storms from mid-July to late August are associated with irregular discharge peaks. The timing of these storm flows and total discharge derived from rainfall differed substantially between 1986 and 1987 (Fig. 13.2f). During storm flows, pools become hydrologically connected, and stratification in pools is rapidly destroyed (Oswood et al. 1989a). Surface ice formation begins as early as mid- to late August. The ice-free season can thus be divided into three relatively discrete phases: (1) a brief period of high discharge (over frozen substrates) from snowmelt, (2) very low flow, during which the stream takes on a lentic character, and (3) irregular flows generated by rainstorms, during which the stream alternates between lentic and lotic characteristics (e.g., 1986) or maintains primarily a lotic character (e.g., 1987).

13.5 Carbon in Imnavait Creek

13.5.1 Concentrations

Organic carbon in the stream is dominated by the dissolved fraction (Table 13.1.). Average [DOC] differed slightly between 1986 and 1987, ranging from ca. 9–13 mgl^{-1}, with little difference between discharge-weighted and time-weighted concentrations. Concentrations of fine particulate organic carbon [FPOC] ranged from ca. 0.2–0.5 mgl^{-1}, about two orders of magnitude less than [DOC]. The [CPOC]s were even lower, averaging 0.01 mgl^{-1}. Thurman (1985) reported that [DOC] in small streams in arctic and alpine environments ranged from 1–5 mgl^{-1}, with an average of 2 mgl^{-1}. Meybeck (1988; summarized in Spitzy and Leenheer 1991) also found 2 mgl^{-1} as a typical [DOC] for tundra rivers. In a nonbeaded tundra stream adjacent to Imnavait Creek, [DOC] averaged 6.4 mgl^{-1} (Peterson et al. 1986). Based on our sampling in 1986 and 1987, [DOC]s in Imnavait Creek averaged ca. 14 and 11 mgl^{-1}, respectively (Table 13.1.), which is considerably greater than the 2 mgl^{-1} average cited

Table 13.1. Concentrations and transport of organic carbon in Imnavait Creek

	Concentration (mg l^{-1}; time-weighted)		Concentration (mg l^{-1}; discharge-weighted)		Transport (kg year^{-1})	
	1986	1987	1986	1987	1986	1987
DOC	9.73	10.55	13.77	11.07	2727.0	6447.0
FPOC	0.30	0.20	0.48	0.17	94.2	99.2
CPOC	–	–	–	0.01	–	3.2
Total	10.03	10.75	14.24	11.25	2821.2	6549.4

DOC, dissolved organic carbon, FPOC, fine particulate organic carbon; CPOC, coarse particulate organic carbon.

above. However, these values are within the 8–25 mg l^{-1} range of average [DOC] found for taiga streams (Thurman 1985).

Beneath the ice cover during winter, [DOC]s are very high (>30 mg l^{-1}; Oswood, unpubl. data), but drop rapidly during snowmelt discharge in late May (Figs. 13.2c and 13.3a). This steady and rapid decline occurs across both the rising and falling limbs of the snowmelt discharge peak. During the remainder of the ice-free season, the daily mean [DOC]s vary from ca. 5–15 mg l^{-1}. More frequent sampling during mid- to late summer (18 July to 7 August) storm flows in 1987 (Fig. 13.3c,d) allowed examination of carbon concentrations in relation to short-term variations in discharge. In 1987, the first stormflow following the early-season "dry period" produced a moderate increase in [DOC] associated with the peak and falling limb of the storm hydrograph, but subsequent stormflows did not produce corresponding increases in [DOC] although [DOC] remained somewhat higher over the remainder of the ice-free season compared with the "dry" season (Figs. 13.2c and 13.3c).

Like [DOC], [POC]s were high early in snowmelt runoff (Fig. 13.2d) and declined rapidly across both the rising and falling periods of the snowmelt discharge (Fig. 13.3b). However, [POC] differed from [DOC] in relation to summer storm flows. In 1987 the first storm flow following the early season "dry" period produced a sharp decrease in [POC] in the rising limb of the hydrograph, then a sharp increase associated with peak flow, followed by a rapid decline in concentration (Fig. 13.3d). As for [DOC], later storm-flow hydrographs produced no clearcut patterns in associated [POC] (Fig. 13.3d).

Although [DOC]/[POC] ratios (based upon discharge-weighted averages) were consistently high – 28 in 1986 and 63 in 1987 (Table 13.1.) – they showed distinct seasonal variation. Ratios were relatively low both during the snowmelt period and during the early-summer low-flow period (Figs. 13.2e). During the later-summer storm-flow period, the [DOC]/[POC] ratios showed irregular peaks, approaching 200 in 1986 and 600 in 1987 (Fig. 13.2e).

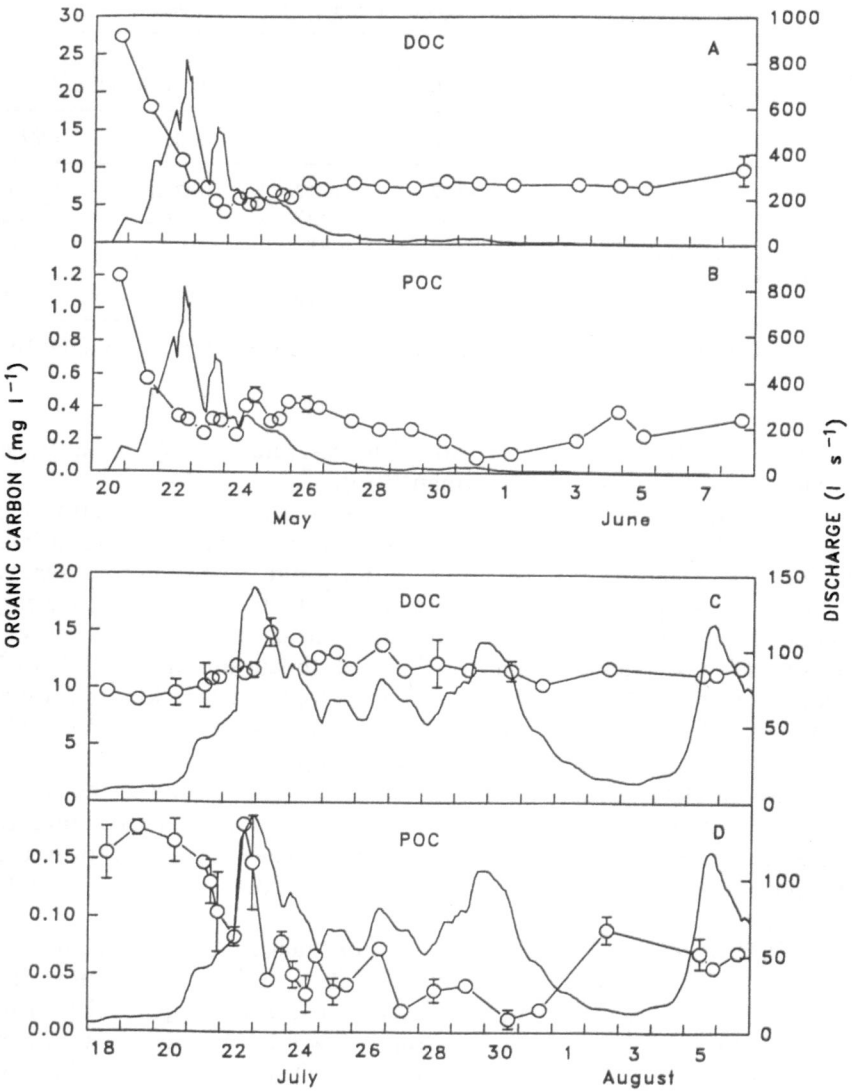

Fig. 13.3A–D. DOC, POC, and stream discharge for two 20-day periods in 1987. Discharge data are hourly (*continuous line*). Carbon samples (*circles*) are means of two samples collected every other day at low flow and two to three times daily at high flow

13.5.2 Transport

The Imnavait Creek watershed exported an estimated 2821 kg C in 1986 and 6549 kg C in 1987. Expressed as specific transport, i.e., carbon transported per unit area of drainage basin, the export was 1.34 g C m⁻² in 1986 and

3.12 g C m^{-2} in 1987. The smaller amount in 1986 appears to reflect the much lower discharge during both snowmelt and the late-summer storm runoff periods (Figs. 13.2f,j). The DOC contributed more than 96% (1986) to 98% (1987) of the total carbon loss (Table 13.1.). Carbon transport (both DOC and POC) occurs in two distinct phases: a very brief pulse (several days) during early-season snowmelt and during multiple stormflow events in late summer. Between these phases there is a period of little transport. Cumulative transport over the entire running-water season indicates that much of the year's carbon transport (especially in 1987) occurred during late-season storm-flow events (Fig. 13.2j).

13.5.3 Spatial Variability

Individual pools (the "beads") vary greatly in size. Over the reach of Imnavait Creek contained in our study site, pool surface areas ranged from ca. 1–35 m^2 with corresponding depths ranging from <0.5–2 m. Smaller pools tend to have a greater littoral area per unit volume, greater contact between the water column and the benthos, and less buffering of physical variability (e.g., temperature or hydraulic extremes). Similarly, subtle upstream-to-downstream differences in soil, vegetation, or permafrost characteristics along the Imnavait Creek valley or cumulative upstream effects (e.g., Vannote et al. 1980) could possibly give rise to differences in organic carbon inputs to Imnavait Creek.

There were no significant differences in [DOC] among pool sizes or valley position at either high or low discharge (Fig. 13.4). This indicates that the choice of the pool for water sampling should have little effect on subsequent calculations of carbon transport. In contrast, [POC]s differed significantly among pool sizes at low water flow: [POC]s were substantially higher in small pools (Fig. 13.4). At high flow, [POC]s did not differ among pool sizes or valley position. Based upon observations of zooplankton on FPOC filters and in CPOC (drift net) samples, at low discharge we suspect that small pools quickly develop substantial phytoplankton and zooplankton populations.

13.5.4 Seasonal Dynamics

Beaded streams are an extreme development of the pool-riffle dichotomy in streams. This dichotomy at Imnavait Creek is reflected in the mix of lentic and lotic macroinvertebrates (Viavant 1989), thermal stratification of pools during low summer flows (Oswood et al. 1989a), and by the development of higher [POC] at low flow, especially in small pools (Fig. 13.4). The [DOC]s showed no differences among pools sizes at either low or high flows.

We propose a simple qualitative model of seasonal dynamics of carbon concentrations and transport in Imnavait Creek, incorporating watershed-stream interations and stream sources of carbon. The organic carbon budget of

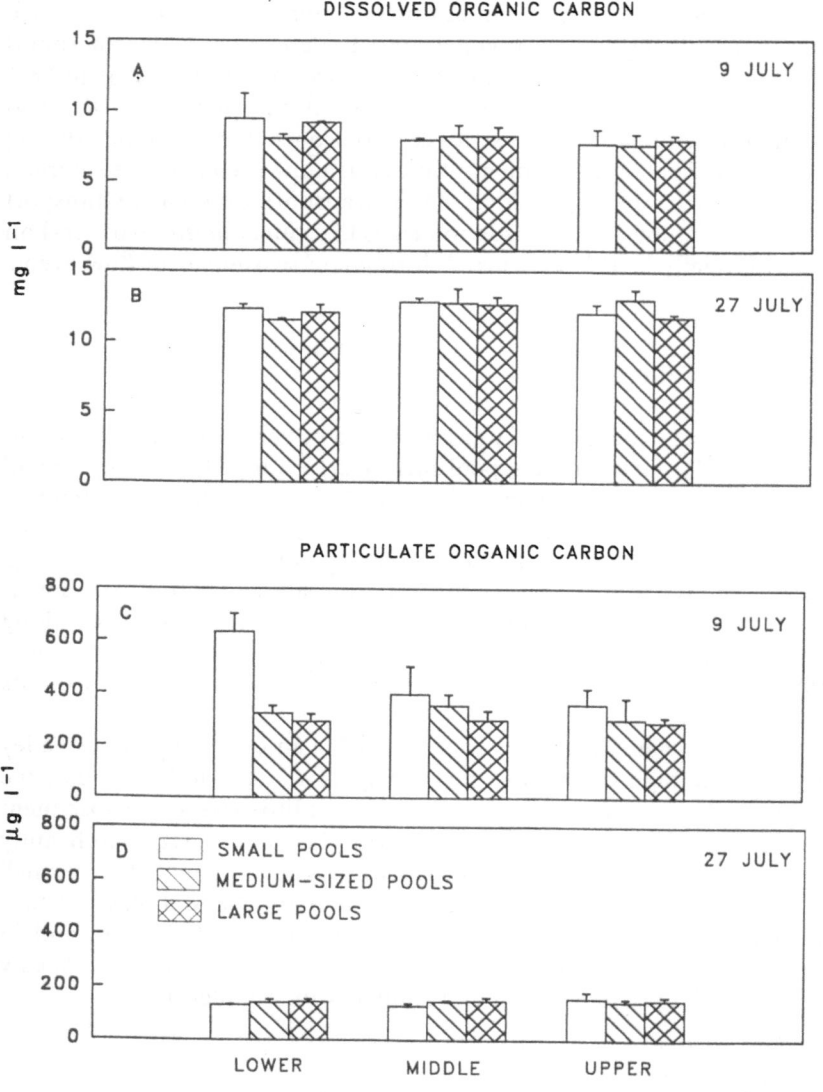

Fig. 13.4A–D. DOC and POC concentrations in Imnavait Creek at low flow (9 July 1987) and high flow (27 July 1987) at different valley locations

Imnavait Creek can be divided into three distinct temporal phases. During breakup, high concentrations of both DOC and POC, coupled with high discharge, transport substantial carbon from the watershed. The source of this carbon is likely senescent leaves from the previous fall and particles of peat entrained in ice and snow and transported to the stream via overland flow. The second phase coincides with a period of low flow, relatively low [DOC]s, relatively high [POC]s (Fig. 13.2d), relatively low [DOC]/[POC] ratios, and little carbon export. During this period, [POC]s rise – likely due to the buildup of phyto- and zooplankton populations – while [DOC]s remain relatively low, probably because little water (either surface or subsurface flow) enters the stream from the watershed. At the beginning of the third phase (associated with summer storm flows), [POC] drops sharply with the rising limb of the hydrograph (Fig. 13.3d), apparently as plankton is flushed from the pools, then rises again as overland flow begins to add peat particles to the stream. As these particles are flushed and depleted, [POC] falls again. The [DOC] levels increase as subsurface flow from the organic soil increases and remain high while flow is high. Because high flow coincides with high [DOC], most carbon is exported during this period. The timing and magnitude of each of these phases may vary from year to year, depending on precipitation patterns; nevertheless, we believe that the overall pattern likely remains the same.

13.6 Conclusions

Estimates of bulk transport of riverine carbon from various ecoregions or biomes to the oceans are essential inputs to calculations of global carbon budgets. Because these are based on estimates of average discharge and average discharge-weighted carbon concentrations (e.g., Mulholland and Watts 1982), it is important that these averages be based on representative streams encompassing the physiographic diversity inherent to an ecoregion, e.g., mountain, tundra, and spring-fed streams for the Alaskan arctic (Craig and McCart 1975). Unfortunately, logistical difficulties and research costs have limited studies at high latitudes (Harper 1981), and even a very recent summary of carbon and mineral transport by North American and Soviet rivers indicates that "The fluxes . . . reported for the Alaskan rivers must be considered very tentative" (Telang et al. 1991).

How does the carbon transport (i.e., transport per unit area of drainage basin) from Imnavait Creek compare to other systems? Our values for specific transport of 1.34 and $3.12\,g\,C\,m^{-2}$ for 1986 and 1987, respectively, are nearly identical to the 1.9 and $3.2\,g\,C\,m^{-2}$ reported by Peterson et al. (1986) in 1979 and 1980, respectively, for the adjacent Kuparuk River. Our values also fall within the range $1.2–4.6\,g\,C\,m^{-2}$ found by Mulholland and Watts (1982) for seven large Alaskan rivers and by Moore (1988), who reported a range of $1.3–4.8\,g\,C\,m^{-2}$ for several subarctic fens and bogs in northern Quebec. Over a broader geographic scale. Mulholland and Watts (1982) summarized specific transport for 82 large rivers of North America, and reported values of $1–6\,g\,C\,m^{-2}$ for boreal or temperate forests. Hope et al. (1994) reviewed carbon export from temperate and boreal watersheds (including wetlands) in North America, New Zealand, Russia, and Europe, and found that most catchments were between 1 and

$10\,g\,C\,m^{-2}$. Thus, Imnavait Creek appears to have about average specific transport compared with rivers of temperate and boreal regions.

Regression equations relating specific transport vs annual runoff (cm) have been developed by Brinson (1976) for upland watersheds, Mulholland and Kuenzler (1979) for swamp-draining watersheds, and by Mulholland and Watts (1982) for large rivers in North America. When annual runoff from Imnavait Creek during 1986 and 1987 is used in these regressions, we found that observed exports from Imnavait Creek were greater than that predicted from upland streams and large rivers, and nearly as great as export from swampy watersheds. This has several potential causes.

Soils in the Imnavait Creek watershed consist of a shallow layer of live and dead vegetation (moss and other plants) overlying a layer of peaty decomposing plant materials, which in turn overlies mineral soils. Summer precipitation apparently does not significantly affect the deeper mineral soils, and the upper organic soils saturate and drain rapidly (Hinzman et al. 1991). Several researchers have noted the sharp decline in DOC concentration with passage through deeper soil profiles (Wallis et al. 1981; McDowell and Wood 1984; Cronan 1990; Qualls and Haines 1991). Meyer (1990) and Koprivnjak and Moore (1992) suggest that mineral horizons can decouple upland DOC sources from streams, but in situations where mineral horizons are bypassed by water flow or have low absorptive capacity, water may have higher [DOC]s. Water moving downslope to the Imnavait Creek channel moves through carbon-rich surficial soils with little contact with mineral soils. Studies of a New England bog (McKnight et al. 1985) and a peatland boreal forest river in Finland (Heikkinen 1989) indicate that the major source of DOC is the upper layer of living and dead *Sphagnum* and peat. At Imnavait Creek, runoff from the low arctic rainfall is largely confined to the carbon-rich upper soil layer of plants and peat, likely giving rise to high rates of carbon transport in relation to the limited runoff. Thus, patterns of soil drainage and carbon density in soils appear to produce high watershed carbon outputs

A more mechanistic understanding of seasonal changes in organic carbon concentrations in stream water requires an understanding of terrestrial (especially soil) organic carbon stores, hydrological transport of organic carbon to the stream, and in-stream sources and sinks of carbon. Some relatively simple empirical models (e.g., Grieve 1984) have been proposed to use commonly available data (e.g., discharge) to predict organic carbon concentrations. In general, DOCs increase with storms; "flushing" of the soil or hyporheic (deep stream bed) organic stores is likely a major source of DOC (Thurman 1985; Tipping et al. 1988; Heikkinen 1989 Sedell and Dahm 1990). This is the pattern in Imnavait Creek, which shows flushing of both DOC and POC with increased storm discharge (Fig. 13.3). However, identical storm flows can produce very different effects depending on factors such as the length of time from previous flushing (Foster and Grieve 1982; Tate and Meyer 1983). Such complex "historical" changes in availability of stores of both DOC and POC are apparently involved in the different responses of Imnavait Creek DOC and POC to a series of summer storms in 1987 (Fig. 13.3). Similarly, the lower runoff in 1986 may have resulted in relatively lower [DOC]s and higher [POC]s in 1986 than in 1987 (Table 13.1).

Many characteristics of streams and rivers are determined by their surrounding terrestrial ecosystems (Hynes 1975; Firth and Fisher 1992). Water flow through watershed soils transports dissolved organic and inorganic materials to streams. Dissolved organic materials enter stream food webs via bacteria and fungi, and dissolved inorganic materials – especially phosphorus and nitrogen – affect stream productivity. Hammar (1989) and Walker (Chap. 3, this Vol.) note many potential direct threats to aquatic ecosystems in polar regions including acidification, oil spills, and eutrophication. However, changes in carbon concentrations and transport in arctic streams will serve significantly as sensitive indicators of changes in terrestrial ecosystem processes (Forsberg 1992). Many of the disturbances noted by Walker (Chap. 3, this Vol.) resulting in changes in terrestrial primary productivity, decomposition, and species composition, for example, will be detected in the carbon and nutrient chemistry of streams, because streams are a "fingerprint" of what has transpired physically, biologically, and chemically in their watersheds.

Acknowledgments. We thank L. Hinzman for providing the hydrology data. Financial support was provided by the US Department of Energy. Field and laboratory assistance was provided by B. Barnett, B. Bond, B. Ostendorf, A. Porchet, T. Viavant, and J. Vohden.

References

Bird JB (1967) The physiography of arctic Canada. Hopkins Press, Baltimore

Brinson MM (1976) Organic matter losses from four watersheds in the humid tropics. Limnol Oceanogr 21: 572–582

Craig PC, McCart PJ (1975) Classification of stream types in Beaufort Sea drainages between Prudhoe Bay, Alaska, and the Mackenzie Delta, NWT, Canada. Arct Alp Res 7: 183–198

Cronan CS (1990) Patterns of organic acid transport from forested watersheds to aquatic ecosystems. In: Perdue EM, Gjessing ET (eds) Organic acids in aquatic ecosystems. Wiley, New York, pp 245–260

Firth P, Fisher SG (eds) (1992) Global climate change and freshwater ecosystems. Springer, Berlin Heidelberg New York

Forsberg C (1992) Will an increased greenhouse impact in Fennoscandia give rise to more humic and coloured lakes? Hydrobiologia 229: 51–58

Foster IDL, Grieve IC (1982) Short-term fluctuations in dissolved organic matter concentrations in streamflow draining a forested watershed and their relation to the catchment budget. Earth Surface Proc Landforms 7: 417–425

Grieve IC (1984) Concentrations and annual loading of dissolved organic matter in a small moorland stream. Freshwater Biol 14: 533–537

Hammar J (1989) Freshwater ecosystems of polar regions: vulnerable resources. Ambio 18: 6–22

Harper PP (1981) Ecology of streams at high latitudes. In: Lock MA, Williams DD (eds) Perspectives in running water ecology. Plenum Press, New York pp 41–68

Heikkinen K (1989) Organic carbon transport in an undisturbed boreal humic river in northern Finland. Arch Hydrobiol 117: 1–19

Hinzman LD, Kane DL, Gieck RE, Everett KR (1991) Hydrologic and thermal properties of the active layer in the Alaskan arctic. Cold Regions Sci Technol 19: 95–110

Hobbie JE (1980) Limnology of tundra ponds: Barrow, Alaska. Dowden, Hutchinson and Ross, Stroudsburg

Hobbie JE (1984) Polar limnology In: Taub FT (ed) Lakes and reservoirs: ecosystems of the world, no 23. Elsevier, Amsterdam, pp 63–105

Hope D, Billett MF, Cresser MS (1994) A review of the export of carbon in river water: fluxes and processes. Environ Pollut 84: 301–324

Hopkins DM, Karlstrom TNV, Black RF, Williams JR, Péwé TR, Fernald AT, Muller EH (1955) Permafrost and ground water in Alaska. US Geol Surv Prof Pap 264-F. US Gov Printing Office, Washington DC

Hynes H B N (1975) The stream and its valley. Proc Int Assoc Theor Appl Limnol 19: 1–15

Irons JG III, Oswood MW (1992) Seasonal temperature patterns in an arctic and two subarctic Alaskan (USA) headwater streams. Hydrobiologia 237: 147–157

Kane DL, Hinzman LD, Benson CS, Liston GE (1991) Snow hydrology of a headwater arctic basin 1. Physical measurements and process studies. Water Resour Res 27: 1099–1109

Kling GW, Kipphut GW, Miller, MC (1991) Arctic lakes and streams as gas conduits to the atmosphere: implications for tundra carbon budgets. Science 251: 298–301

Koprivnjak J-F, Moore TR (1992) Sources, sinks, and fluxes of dissolved organic carbon in subarctic fen catchments. Arct Alp Res 24: 204–210

McDowell WH, Wood T (1984) Podzolization: soil processes control dissolved organic carbon concentrations in stream water. Soil Sci 137: 23–32

McNight D, Thurman EM, Wershaw RL (1985) Biogeochemistry of aquatic humic substances in Thoreau's Bog, Concord, Massachusetts. Ecology 66: 1339–1352

Meybeck M (1988) How to establish and use world budgets of riverine materials. In: Lerman A, Meybeck M (eds) Physical and chemical weathering in geochemical cycles. Kluwer, Dordrecht, pp 247–272

Meyer JL (1986) Dissolved organic carbon dynamics in two subtropical blackwater rivers. Arch Hydrobiol 108: 119–134

Meyer JL (1990) Production and utilization of dissolved organic carbon in riverine ecosystems. In: Perdue EM, Gjessing ET (eds) Organic acids in aquatic ecosystems. Wiley, New York, pp 281–299

Moeller JR, Minshall GW, Cummins KW, Petersen RC, Cushing CE, Sedell JR, Larson RA, Vannote RL (1979) Transport of dissolved organic carbon in streams of differing physiographic characteristics. Organic Geochem 1: 139–150

Moore TR (1988) Dissolved iron and organic matter in northern peatlands. Soil Sci 145: 70–76

Mulholland PJ, Kuenzler EJ (1979) Organic carbon export from upland and forested wetland watersheds. Limnol Oceanogr 24: 960–966

Mulholland PJ, Watts JA (1982) Transport of organic carbon to the oceans by rivers of North America: a synthesis of existing data. Tellus 34: 176–186

Oechel WC, Hastings SJ, Vourlitis G, Jenkins M, Richhers G, Grulke N (1993) Recent change of Arctic tundra ecosystems from a net carbon dioxide sink to a source. Nature 361: 520–523

Oswood MW, Everett KR, Schell DM (1989a) Some physical and chemical characteristics of an arctic beaded stream. Holarct Ecol 12: 290–295

Oswood MW, Irons JG III, Hilgert JW, Slaughter CW (1989b) Effects of riparian vegetation removal on an Alaskan subarctic stream In: Ashton WS (ed) Groundwater: Alaska's hidden resource. Inst Water Resources, Univ Alaska, Faribanks, Report IWR 112: 3–13

Oswood MW, Milner AM, Irons JG III (1992) Climate change and Alaskan rivers and streams. In: Firth P, Fisher SG (eds) Global climate change and freshwater ecosystems. Springer, Berlin Heidelberg New York, pp 192–210

Perkin-Elmer Instruments (1981) Model 240C elemental analyzer instruction manual. Norwalk, CT

Peterson BJ, Hobbie JE, Corliss TL (1986) Carbon flow in a tundra stream ecosystem. Can J Fish Aquat Sci 43: 1259–1270

Post WM, Emanuel WR, Zinke PJ, Stangenberger AG (1982) Soil carbon pools and world life zones. Nature 298: 156–159

Qualls RG, Haines BL (1991) Geochemistry of dissolved organic nutrients in water percolating through a forest ecosystem. Soil Sci Soc Am J 55: 1112–1123

Schlesinger WH (1977) Carbon balance in terrestrial detritus. Annu Rev Ecol Syst 8: 51–81

Schlesinger WH, Melack JM (1981) Transport of organic carbon in the world's rivers. Tellus 33: 172–187

Sedell JR, Dahm CN (1990) Spatial and temporal scales of dissolved organic carbon in streams and rivers. In: Perdue EM, Gjessing ET (eds) Organic acids in aquatic ecosystems. Wiley, New York, pp 281–299

Shaver GR, Billings WD, Chapin FS III, Giblin AE, Nadelhoffer KJ, Oechel WC, Rastetter EB (1992) Global change and the carbon balance of arctic ecosystems. BioScience 42: 433–441

Spitzy A, Leenheer J (1991) Dissolved organic carbon in rivers. In: Degens ET, Kempe S, Richey JE (eds) Biogeochemistry of major world rivers. Wiley, New York, pp 213–232

Tate CM, Meyer JL (1983) The influence of hydrologic conditions and successional state on dissolved organic carbon export from forested watersheds. Ecology 64: 25–32

Telang SA, Pocklington R, Naidu AS, Romankevich EA, Gitelson II, Gladyshev MI (1991) Carbon and mineral transport in major North American, Russian arctic, and Siberian rivers: the St Lawrence, the Mackenzie, the Yukon, the Arctic Alaskan rivers, the arctic basin rivers in the Soviet Union, and the Yenisei In: Degens ET, Kempe S, Richey JE (eds) Biogeochemistry of major world rivers. Wiley, New York, pp 75–104

Tenhunen JD, Gillespie CT, Oberbauer SF, Sala Serra A, Whalen SC (1995) Climate effects on the carbon balance of tussock tundra in the Philip Smith Mountains, Alaska. Flora 190: 273–283

Thurman EM (1985) Organic geochemistry of natural waters. Martinus Nijhoff/Dr. W. Junk, Dordrecht

Tipping E, Hilton J, James B (1988) Dissolved organic matter in Cumbrian lakes and streams. Freshwater Biol 19: 371–378

Vannote RL, Minshall GW, Cummins KW, Sedell JR, Cushing CE (1980) The river continuum concept. Can J Fish Aquat Sci 37: 130–137

Viavant TR (1989) Community structure, trophic relationships, and habitat ecology of the benthic macroinvertebrates in an Alaskan arctic tundra beaded stream. MS Thesis, Univ Alaska, Fairbanks

Wallis PM, Hynes HBN, Telang SA (1981) The importance of groundwater in the transportation of allochthonous dissolved organic matter to the streams draining a small mountain lake. Hydrobiologia 79: 77–90

IV Modeling Landscape Function

14 Patch and Landscape Models of Arctic Tundra: Potentials and Limitations

J. F. REYNOLDS, J. D. TENHUNEN, P. W. LEADLEY, H. LI, D. L. MOORHEAD, B. OSTENDORF, and F. S. CHAPIN III

14.1 Introduction

A synthesis and integration of data gathered at the Imnavait Creek catchment in the form of up-to-date simulation models was a major goal of the R4D program sponsored by the Department of Energy (Chap. 1, this Vol.). Models were proposed as research tools to test our basic understanding of the structure and function of arctic ecosystems, as a means for providing initial management assessments of potential response to energy-related development, and as a vehicle for extrapolation of research results to other arctic sites and landscapes (NRC 1982). Similar desires for the use of models as natural resource management tools have been expressed in ecosystem science for over 20 years (Hammond 1972; Cooper 1976; Loucks 1985; Slocombe 1993; Kaufmann et al. 1994). While significant progress has been made, much work is still required before truly integrated models will be able to effectively summarize and illustrate the complex interactions affecting physical-biological-social systems, particularly within a management framework (e.g., Pickett et al. 1994).

In this chapter, we present a summary of some of the models that were developed as part of the R4D program. We review progress made on models at a variety of scales, e.g., from nutrient uptake by individual roots to nutrient availability within arctic landscapes, and examine the potentials and limitations of these models for providing insight on patch and landscape level function in tundra regions.

Our approach to modeling arctic systems has been to focus in each instance on a single phenomenon and scale (both temporal and spatial) in order to clearly define the system of interest (O'Neill 1988). We agree with Starfield and Bleloch (1991) that models should be constructed for a particular purpose, based on assumptions and data relevant for that purpose, rather than attempting to serve "all purposes" and include everything known about a subject. Different types of models are needed, ranging from simple phenomenology to detailed mechanisms. Each model type is used at the scale for which it is best suited (e.g., whole plant growth, ecosystem productivity, watershed discharge, etc.). Knowledge generated from a model at one level may contribute to our total system understanding and, thus, help build or alter a model at another level (Reynolds and Leadley 1992; Reynolds et al. 1993).

J.F. Reynolds and J.D. Tenhunen (Eds.)
Ecological Studies, Vol. 120
© Springer-Verlag Berlin Heidelberg 1996

14.2 Modeling Framework

14.2.1 Spatial Simulation Units

The arctic landscape is a rich mosaic of soil types, topography, lakes, creeks and vegetation (Chap. 4, this Vol.). Such heterogeneity must be considered when modeling the effects of disturbance on landscape function (DeAngelis et al. 1985; Turner 1989; Turner and Dale 1991; Hannsson et al. 1994). In this context – and consistent with our theme of avoiding complex, "all purpose models" – three distinct tasks were addressed by our modeling efforts: (1) the description of processes that occur at *specific* locations or patches in the landscape; (2) characterization of energy and material transport *between* patches; and (3) interfacing of patch and transport processes in a *landscape* framework. These modeling tasks were accomplished using the spatial "simulation units" defined in Table 14.1. This scheme illustrates important spatial and temporal linkages in arctic landscapes and provides a framework for integrating structural, functional, and transport linkages within and across ecosystem boundaries in the watershed.

The smallest simulation unit is at the scale of an individual plant. Plants are coupled to their environment via numerous plant-soil-atmosphere-animal vectors (Table 14.1). We define "patch" ecosystems as relatively discrete and internally homogeneous units of land, e.g., the biotic community residing on a particular soil type. While true homogeneity is rarely observed in nature (Kotliar and Wiens 1990), we view patches as parcels of land that have similar responses to energy and water inputs, regardless of their spatial location. These physically and biologically homogeneous areas may be identified on the basis of similar soil, vegetation, slope, and aspect and are conceptually identical to "hydrologic response units," which are land areas that respond similarly throughout when subjected to hydrologic events, e.g., precipitation (Battaglin et al. 1993).

A series of connected and interrelated patches form a flow path (Woodmansee 1988, 1990). An example is Walker's toposequence in the Imnavait Creek watershed (see Fig. 4.8, this Vol.), which identifies five patches – crest, shoulder, upper backslope, lower backslope, and footslope – on the basis of differences in soils, vegetation physiognomy, geologic surface forms, and plant community types. Another example is the toposequence defined by Schimel et al. in the nearby Sagavanirktok River valley, (see Fig. 10.1, this Vol.), which consists of six ecosystem types running from tussock tundra on upland sites, down through a series of floodplain terraces, to the river. Typically, water, atmospheric turbulence, and animals are the main vectors of transport of materials and energy between patches. Field studies at the level of flow paths correspond to examining the degree of "connectedness" between patches (examples given in Chaps. 5, 8, 13, this Vol.). Understanding flow paths is important since ecosystem processes are directly affected by transfers of mate-

Table 14.1. Spatial simulation units

Spatial simulation unit			Typical coupling vectors	
Name	Spatial scale[a] (m^2)	Structural components	Water	Energy/nutrients
Plant	10^{-4}–10^{-1}	Leaves Stems Roots Soil volume	Soil water Transpiration Water uptake	Photosynthesis Respiration Nutrient uptake Herbivory
Patch	10^0–10^4	Plants Soil Microbes Animals	Precipitation Infiltration Discharge/run-on Evapotranspiration Soil water balance	Net carbon balance Nutrient cycling Decomposition Trace gas flux
Flow path	10^2–10^5	Patch ecosystems Toposequences or soil catenas	Soil water flux and discharge [Hillslope Darcian flow dominant]	Mass flux of sediment Mass flux of dissolved nutients Aeolian transport
Landscape	10^6–4×10^6	Flow-path ecosystems Groundwater; channel storage	Channel flow [Turbulent flow dominant]	Dispersal Trace gas flux
Region[b]	10^7–10^{10}	Landscapes (= integrative flow systems) Lakes; rivers		Hydrologic transport of sediments Hydrologic transport of dissolved nutrients Aeolian transport Migration

[a] Examples of "typical" values for arctic tundra; based on Osmond et al. (1980), Woodmansee (1988, 1990) and Walker and Walker (1991).
[b] "Mesoscale" in Walker and Walker (1991), which includes 2nd-order watersheds.

rials across system boundaries (Forman and Gordon 1986) and these boundaries may vary in response to disturbance (Wiens et al. 1985).

The Imnavait Creek watershed consists of an assemblage of flow path ecosystems that deliver mass and energy (mainly by gravitational processes) to the stream. The sum total of these flow paths corresponds to our use of the term "landscape." This simple definition is based on the concept of an "integrated flow system," which is analogous to a specific drainage basin associated with a water gauging station (Li et al. 1977). Thus, at the landscape scale of interest, the entire Imnavait Creek watershed is the simulation unit.

The largest spatial unit is a region (Table 14.1). Regional studies involve a mixture of integrated flow systems or landscapes that make up the scale of interest, e.g., the 22-km^2 R4D region (see Fig. 1.1, this Vol.) or the entire North Slope (Walker and Walker 1991).

Table 14.2. Types of models developed in the R4D program. Numbers 1–9 correspond to those shown in Fig. 14.1 [chapters cited are in this volume]

Types of models	References
Mechanistic	
① Photosynthesis, respiration (GAS-FLUX)	Chap. 5 (5.5); Chap. 14 (14.3.1); Tenhunen et al. (1992, 1994)
Decomposition and mineralization (GENDEC)	Chap. 16 (16.3.4, 16.5); Chap. 14 (14.3.4); Moorhead and Reynolds (1992, 1993)
Plant growth	Chap. 14 (14.3.2); Leadley and Reynolds (1992); Reynolds and Leadley (1992)
Nutrient uptake	Chap. 14 (14.3.3); Leadley et al. (1996)
Soil respiration	Chap. 5 (5.5); Chap. 17; Oberbauer et al. (1991)
Soil thermal regime (TDHC)	Chap. 6 (6.4.1); Hinzman et al. (1991); Kane et al. (1991, 1992)
Surface energy budget	Chap. 6 (6.2.1); Kane et al. (1991, 1992)
Transport	
② Hydrology (discharge, soil water storage; TOPMODEL)	Quinn et al. (1995); Chap. 17
③ Hydrology (discharge; T-HYDRO)	Chap. 14 (14.4.1); Chap. 18 (18.3.1.1); Ostendorf and Reynolds (1993, 1995)
Hydrology (snowmelt)	Chap. 6 (6.4.2); Kane et al. (1992)
Hydrology (runoff; HBV)	Chap. 6 (6.4.3); Hinzman and Kane (1991); Hinzman et al. (1993); Kane et al. (1991, 1992)
④ Dust deposition	Chap. 15
Topographically derived	
⑤ T-VEG	Chap. 18 (18.3.1.2)
⑥ T-NUT, T-PLT	Chap. 18 (18.3.1.3, 18.3.1.4)
⑦ TVM	Chap. 14 (14.4.2); Chap. 17 (17.3.1); Ostendorf and Reynolds (1995)
Landscape	
⑧ GAS-FLUX/TOPMODEL	Chap. 17
⑨ T-MAP	Chap. 18

14.2.2 Types of Models

Reynolds and Leadley (1992) described numerous ecological models developed for arctic tundra. One of these, the arctic tundra simulator (ARTUS), was specifically designed to address effects of disturbance on tussock tundra (Miller et al. 1984). These models – restricted mainly at the plant or patch scale – were largely motivated by research interests in the International Biome Program (IBP) (Wielgolaski 1975; Miller et al. 1975, 1978, 1979, 1984; Chap. 2, this Vol.). Reynolds and Leadley (1992) concluded that most of these models

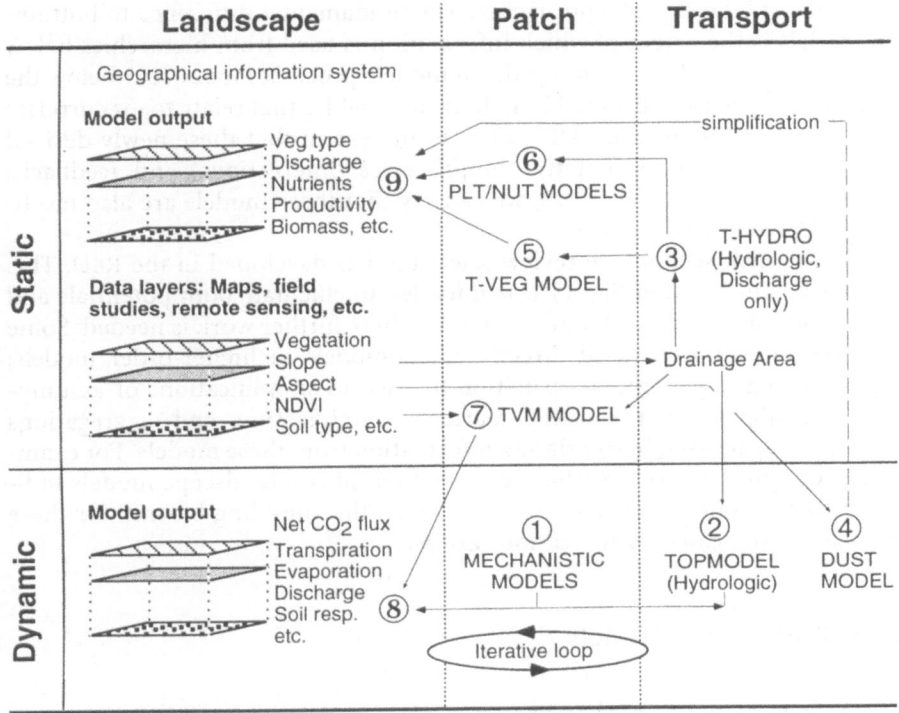

Fig. 14.1. Framework for integrating structural and transport linkages within and across eco-system boundaries. Illustrated are "building blocks" for landscape models described in Chap. 17 (*8*) and Chap. 18 (*9*). Some of these models (i.e., *1–7*) were used directly as submodels; others served to generate information that led to simplifications or assumptions. See Table 14.2 for further details

were either too complex for long-term simulations or were too empirical for extrapolation. Nevertheless, these models provided an excellent baseline for model development in arctic ecosystems.

Examples of the various types of models developed in the R4D program that are based on these previous efforts are given in Table 14.2 and Fig. 14.1. These models are referred to as mechanistic, transport, topographically derived, and landscape. Our mechanistic models may be viewed as "bottom-up" models in that they employ information and relationships from below the scale of focus and are generally based on processes or mechanisms that link or relate various elements to one another. We constructed mechanistic models for a range of arctic phenomena at the scale of organs, individuals, and patches (Table 14.1) as part of our effort to obtain fundamental understanding of biotic and abiotic controls on carbon, nutrient, and water fluxes in arctic ecosystems. Our transport, topographically derived, and landscape models – at the scale of flow paths and landscapes (Table 14.1) – are, on the other hand, best viewed as examples of "top-down" approaches. The fundamental difference to bottom-up models is the extent to which information is used from hierarchies below the level of focus. That is, in top-down models processes that occur below the scale of interest are subsumed into derived variables that relate to, or correlate with, the variable of focus. Of course, to the extent that these newly derived relationships are representative of processes, interactions, and feedbacks among real parts of the system, we can say that these models are also mechanistic (Reynolds et al. 1993).

In the next section, we review select models developed in the R4D. This overview, illustrated in Fig. 14.1, is intended to elucidate both potentials and limitations of these models and to suggest where further work is needed. Some of these models were used directly as submodels in higher-order models; others served to generate information that led to simplifications or assumptions in higher-order models. Various simplification and aggregations schemes may be used in translating information from these models. For example, in Chapters 17 and 18 (this Vol.), we present two landscape models indicated by #8 and #9, respectively, in Fig. 14.1; the "building blocks" for these models are indicated by the various arrows.

14.3 Bottom-Up Models

14.3.1 Ecosystem Gas Exchange

14.3.1.1 Motivation

Considerable effort has been devoted to the examination of ecosystem carbon dioxide exchange in tundra regions (Chap. 11, this Vol.), beginning with investigations near Barrow, Alaska, during the IBP (Chap. 2, this Vol.). In spite of this, a general concept to explain the variation in observed flux rates based on physical and biological control mechanisms has been lacking. Miller et al. (1984) attempted to integrate information on microclimate, species composition, spatial structure, and gas exchange in ARTUS, but their approach was highly empirical and hence limited with respect to modeling the effects of disturbance on ecosystems. We developed the GAS-FLUX model to provide a mechanistic framework for integrating knowledge of tundra gas exchange (Tenhunen et al. 1994). Our ultimate objective is to link this simulator with models of plant growth and nutrient availability in order to examine ecosystem dynamics under a wide range of conditions, including disturbance and climate change.

14.3.1.2 Description

GAS-FLUX considers the gas exchange and the microclimate of three types of landscape patches or vegetation types in the Imnavait Creek watershed: moist

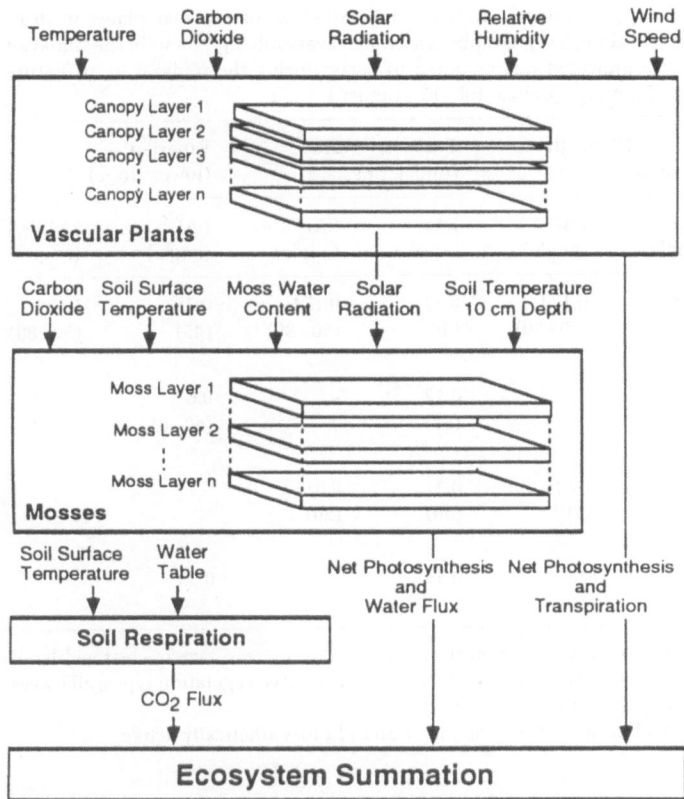

Fig. 14.2. Structure of the GAS-FLUX model as applied to tundra plant communities. Summation of CO_2 and water-exchange processes in vascular plant *canopy layers*, *moss layers*, and *soil* provides estimates of ecosystem water and carbon balance. (Modified from Tenhunen et al. 1994)

Cassiope dwarf shrub (CDS) heath (upper slopes), tussock tundra (midslopes), and riparian (lower slopes) (cf. Chaps. 4, 5, 17, this Vol.). Each patch is considered to be homogeneous, but composed of a mixture of vascular and non-vascular (primarily *Sphagnum* mosses) plants. The vascular plant canopy is divided into a series of layers each containing leaf and stem material for one of three functional groups with a defined physiological response: deciduous shrubs, graminoids, and evergreen shrubs (Fig. 14.2, Table 14.3).

Light interception by both stems and leaves is considered. Deciduous shrubs and graminoids occur in significant quantities in all three vegetation types, with their lowest leaf area index (LAI) in tussock tundra and highest LAI in the riparian meadows. Stem area index (SAI) of the deciduous shrubs plays an important role in light interception in all three vegetation types; evergreen shrubs are equally present in the upper and midslope locations, but disappear in riparian sites. Light incident on the understory moss vegetation is computed as the light leaving the lowest layer of the vascular plant canopy. Sunlit and

Table 14.3. Simplified canopy characteristics used in GAS-FLUX for vascular plants in three community types distributed along a topographic and water-availability gradient in the Imnavait Creek watershed. Structure indicated was assumed to occur during the midseason or mature stage of phenological development from ca. July 15–August 1

	Moist CDS heath (upslope)		Tussock tundra (midslope)		Riparian (lower slope)	
	LAI[a] (angle)	SAI[b] (angle)	LAI (angle)	SAI (angle)	LAI (angle)	SAI (angle)
Deciduous shrub (D)	0.24 (45)	0.174 (50–80)	0.12 (45)	0.174 (50–80)	0.35 (45)	0.151 (50–80)
Graminoid (G)	0.24 (85)	–	0.12 (85)	–	0.6 (60–85)	–
Evergreen shrub (E)	0.2 (30)	0.016 (30)	0.2 (30)	0.016 (30)	–	–
Total vascular plant LAI	0.68		0.44		0.95	

[a] Leaf area index ($m^2 m^{-2}$); average angles from the horizontal in degrees; average leaf widths of $D = 1.0$ cm, $G = 0.3$ cm and $E = 0.3$ cm, considered the same in all three vegetation types; all leaves and stems considered to be alive and non-clustered.

[b] Stem area index ($m^2 m^{-2}$); all stems $\leqslant 6$ mm diameter and photosynthetically active.

shaded leaf area is calculated for each canopy layer as well as sunlit and shaded portions of the ground surface. Leaf level photosynthesis for the vascular plants is described with the Farquhar et al. (1980) equations, which are based on ribulose-1,5-bisphosphate carboxylase-oxygenase (Rubisco) kinetics as mediated by (1) the concentrations of competing gaseous substrates, CO_2 and O_2, and (2) the ratio of ribulose-1,5-bisphosphate (RuBP) concentration to enzyme-active sites. CO_2 assimilation is linked to an empirical stomatal conductance model (Collatz et al. 1991). Average hourly gas exchange rates for each canopy layer (vascular plants) and ground area (mosses) are obtained by weighting the sunlit and shaded area fluxes. Poikilohydric plants such as *Sphagnum* exhibit a dynamic response to slow changes in water content during the summer season, which presents added difficulties (Chap. 11, this Vol.); we derived the parameters for the Farquhar et al. model based on extensive measurements of CO_2 gas exchange over a range of water contents (Tenhunen et al. 1992, 1994).

Ecosystem efflux of CO_2 from tundra communities with deep organic soils varies primarily in response to changes in the depth to water table and/or soil moisture and temperature (Chaps. 11, 17, this Vol.). Soil respiratory flux is determined by using regression equations (Chap. 11, this Vol.); Oberbauer et

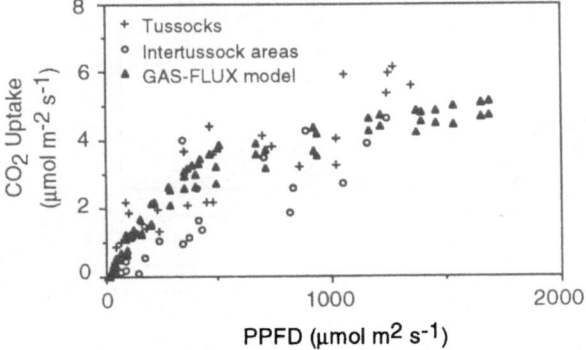

Fig. 14.3. Estimated CO_2 uptake of *Eriophorum vaginatum* tussocks and intertussock areas as a function of PPFD at the time of measurement. Comparison with the mean tussock tundra ecosystem uptake estimated from GAS-FLUX under a variety of weather conditions. (Tenhunen et al. 1994, 1995)

al. 1991, 1992). Net ecosystem gas exchange is computed by the balance of CO_2 flux rates from the aboveground (leaf, stem, moss) and soil components of the model. A comparison of net CO_2 uptake as described by GAS-FLUX and measured with small chambers in tussock tundra of the Imnavait Creek catchment is shown in Fig. 14.3.

14.3.1.3 Potentials and Limitations

With GAS-FLUX, we are able to elucidate various abiotic and biotic interactions that contribute to ecosystem gas exchange. For example, the integrated daily CO_2 flux in tussock tundra on a clear day is $20\,\mu mol^{-2}$ vs. $-20\,\mu mol\,m^{-2}$ if the day is overcast, where the individual contributions of each system component can be quantified (Fig. 14.4). Tenhunen et al. (1992) used GAS-FLUX to examine the effect of similar short-term changes in microclimate vs. the effects of longer-term changes in seasonal soil moisture and in phenology. Since seasonal increases in mean soil temperature are small, the primary effect of temperature is seen in diurnal changes in CO_2 efflux (Chap. 17, this Vol.). As described in Chapters 11 and 17 (this Vol.), the balance between carbon fixation as influenced by phenology, LAI development, and radiation input and CO_2 efflux as determined by water table depth and soil aeration establishes tundra as either a source or sink for CO_2.

GAS-FLUX could be significantly improved with additional field data, including stem respiration, gas exchange of entire moss cushions, and whole-system gas exchange under a variety of conditions. Further work on lichen and ridge-top heath sites are needed to better represent these important arctic vegetation types in the model. Coupling of the abiotic-biotic interactions at the patch scale in GAS-FLUX presently occurs via water table fluctuations, directly

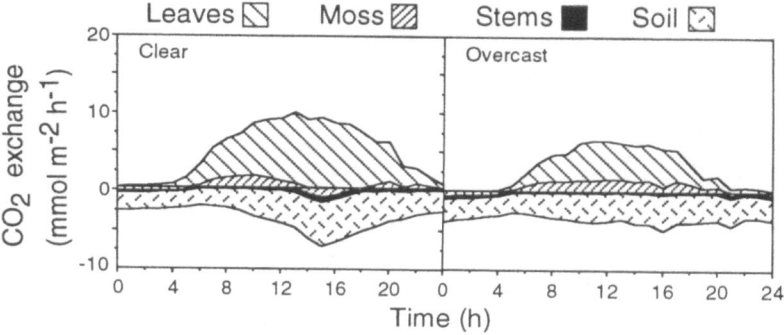

Fig. 14.4. Simulated CO_2 gas exchange of tussock tundra vegetation on a warm, clear day at midseason. The fluxes for leaves, *Sphagnum* moss, twigs, and soil are shown, additively, with carbon fixation by leaves and moss shown as positive fluxes and respiration of moss, twigs, and soil as negative. *Left* Simulations for a clear day at midseason (integrated daily CO_2 flux = $20\,\mu mol\,m^{-2}$); *right* for an overcast day (integrated daily CO_2 flux = $-20\,\mu mol\,m^{-2}$). Soil was considered to be moist and well aerated. *Sphagnum* water content was set at 750% of dry weight. (Modified from Tenhunen et al. 1995)

influencing soil aeration, soil respiration, and thus carbon balance. While we do consider seasonal changes in phenology and LAI, interannual variations in environmental conditions that might alter these parameters and, ultimately, the magnitude of carbon dioxide uptake are not. As we continue to improve this model, our capability to better understand and predict the response of tundra ecosystems to climate change will also improve. In Chapter 17 (this Vol.), we describe an exploratory study to link GAS-FLUX with a watershed model in order to consider interactions between carbon balance and spatial aspects of water table variability.

14.3.2 Plant Growth

14.3.2.1 Motivation

A number of vascular and nonvascular whole-plant growth models have been developed for arctic ecosystems (reviewed in Reynolds and leadley 1992). These models range from: (1) budget-type models where plants are treated as "black boxes", i.e., plant biomass is lumped into a single box and growth is described as the difference between inputs and outputs (usually fixed values); (2) flux-type models that contain a high degree of comprehensiveness in terms of gas exchange processes per se, but have a highly empirical treatment of plant growth processes; and (3) mechanistic models that range greatly in terms of the degree of phenomenology and mechanism used. Because of the emphasis on modeling at the IBP wet meadow sites (especially Barrow, Alaska), most of these models were developed for mesic tundra ecosystems. Recently, Rastetter

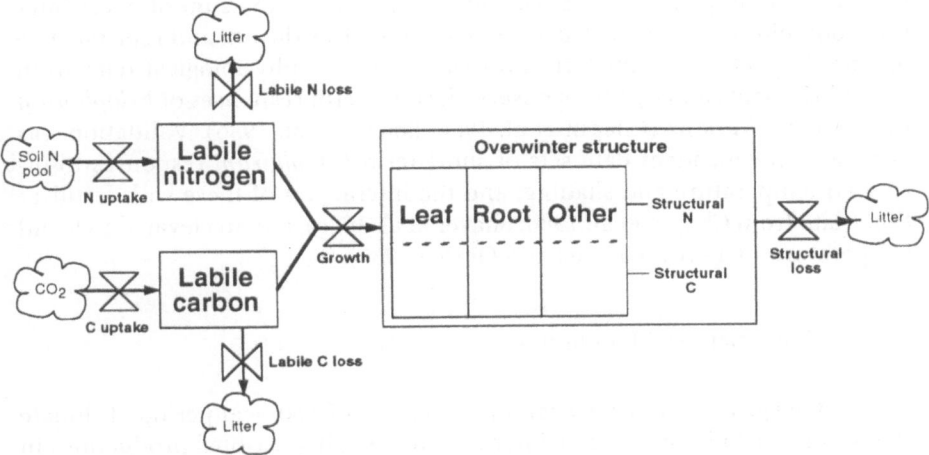

Fig. 14.5. Conceptual diagram of plant growth model. *Boxes* represent state variables, *arrows* are fluxes, *valves* signify control over fluxes, and *clouds* represent infinite sinks or sources. (From Leadley and Reynolds 1992)

et al. (1991) applied a biogeochemical model to arctic tundra; this model is mechanistic but highly aggregated, treating all plants as a single entity.

As an initial step in developing a general framework for modeling arctic plant growth and development, we concentrated efforts on a single, ecologically important species, *Eriophorum vaginatum*. We selected *E. vaginatum* because it is one of the dominant vascular plant species in tussock tundra. It is found in many arctic and subarctic ecosystems, and microclimate and soil processes are profoundly affected by the tussocks that it forms (Chapin et al. 1979; Bliss 1981). Consequently, a large amount of experimental data on its growth and physiology exists, perhaps more than for any other arctic species.

14.3.2.2 Description

A complete description of the model, including validation and behavior, is given in Leadley and Reynolds (1992) and Reynolds and Leadley (1992). It utilizes a mechanistic framework and includes the effects of light, temperature, season length, nitrogen availability, and CO_2 concentration on growth dynamics. The spatial simulation units are plants (Table 14.1) contained in a core in the center of a tussock that is $3.3 \times 10^{-3}\,\text{m}^2$. The temporal resolution is 1 year. A diagram of the model is shown in Fig. 14.5 where the variables represent the overwinter status of the plant, i.e., January 1 of each year. Physiological activity is restricted to the arctic summer, defined as the time between spring snowmelt (late May) and before freezing of the soil (mid-September) in Imnavait Creek watershed. Carbon (C) and nitrogen (N) are divided into overwinter labile and structural components, where overwinter structure is the sum of overwinter leaf, root, plus other plant structures, e.g., stem, leaf sheath, and reproductive organs (Fig. 14.5). Parameterization was based on physiological data from individual plants and by fitting observed, short-term responses of *Eriophorum* to nutrient additions (Chapin et al. 1986; Shaver et al. 1986). Validation was done with independent data sets of short-term *Eriophorum* responses to increased temperature and shading, and the interaction of these with fertilization (data from Chapin et al. 1986; Shaver et al. 1986), and to elevated CO_2 and temperature (data from Tissue and Oechel 1987).

14.3.2.3 Potentials and Limitations

We used the growth model to examine the effects of a 50-year period of climate change on peak biomass (overwintering biomass plus seasonal production) in *E. vaginatum*. Three scenarios (see Fig. 14.6) were simulated. In scenario 1, the model predicts that a simultaneous increase in the direct effects of temperature, season length, and CO_2, with no change in N availability, would result in a slight decrease in peak biomass. This is a result of decreasing overwinter biomass that is not offset by a slight increase in seasonal production. In scenario 2, a simulated long-term doubling of N availability resulted in a ca. 70% increase in peak biomass, whereas with concurrent changes in climate and N availability (scenario 3), the model predicted a slight decline in peak biomass compared to increases in N alone. In essence, the model predicts that climate change will have substantial effects on *E. vaginatum* only indirectly through changes in N availability, which emphasizes the importance of being able to describe any natural or anthropogenic effect on N cycling in tundra ecosystems (see below).

Our model and experimental evidence (e.g., Shaver and Chapin 1986) indicate that N availability is the factor that ultimately limits *E. vaginatum* productivity under a wide range of conditions. In addition, both the model and data suggest that allocation patterns and mortality are much more important than physiological factors such as photosynthesis rates and nutrient uptake kinetics in determining the growth and biomass of *E. vaginatum*. These results are important in the context of future modeling efforts for other major tundra species.

The ability of the model to predict the responses of *E. vaginatum* to climate change or other disturbances is limited by the assumptions (1) that competitive relationships with other plants species remain unchanged, (2) that all of the factors that potentially limit growth are included in the model (e.g., nutrients other than nitrogen are not important), and (3) that there will be no profound changes in environment such as the loss of the permafrost layer. In addition, the model cannot account for within-season changes in environment because it has a very coarse time resolution. As described in Reynolds and Leadley (1992), we are examining this potential error by comparing it with

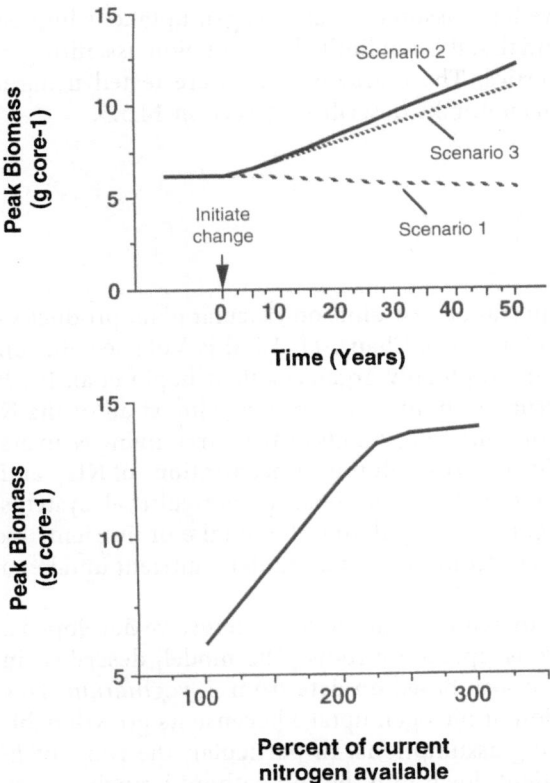

Fig. 14.6. *Above* Response of *E. vaginatum* biomass (core = $3.32 \times 10^{-3} \, m^{-2}$) to long-term climate change simulated by linear increases over a 50-year period in temperature (8–13 °C), season length (100–120 days), atmospheric CO_2 concentration (340–680 µl/l). Nitrogen availability varied from 9–18 g m^{-2} year^{-1}. *Scenario 1* Climate change with no change in N availability; *2* an increase in N availability with no change in climate; *3* climate change with an increase in N availability. *Below* Simulated response of *E. vaginatum* to long-term changes in N availability. (From Leadley and Reynolds 1992)

another model of *E. vaginatum* that uses a daily time step. Another caveat is that we have attempted to model the long-term responses of *E. vaginatum* to environmental variation, but we were only able to validate it using relatively short-term data sets. Thus, the predictions made for the climate change scenarios in Fig. 14.6 must be viewed as extrapolations well outside the range of validation. These long-term simulations are important, however, because they show that the long-term responses of arctic plants to environmental change are potentially very different than those predicted by short-term simulations (see Reynolds and Leadley 1992 for specific examples).

In addition to these limitations of the model, we have also made several important simplifying assumptions about the regulation of nitrogen uptake in

E. vaginatum. In particular, we have assumed that nitrogen uptake is limited only by nitrogen availability and that it is not limited by root biomass, nitrogen uptake kinetics, or soil properties. These assumptions were tested using a highly detailed model of nitrogen uptake described in Section 14.3.3.

14.3.3 Nitrogen Uptake

14.3.3.1 Motivation

Nitrogen is one of the most important constraints on vascular plant productivity in tussock tundra (Shaver et al. 1986; Chaps. 11, 16, this Vol.; see above). Tussock tundra occurs together with highly organic soils (Chapin et al. 1979; Chap. 4, this Vol.). Since external N inputs are very low, almost all of the N comes from the breakdown of organic compounds or from free-living N-fixers (Nadelhoffer et al. 1992). In addition, soil solution concentrations of NH_4^+ and NO_3^- in tussock tundra are far lower than those in agricultural systems (Marion et al. 1989). Therefore, factors regulating the uptake of nutrients by arctic plants may be very different from those that regulate nutrient uptake in mineral soils.

Given the importance of N dynamics in arctic ecosystems, we developed a mechanistic model to simulate N uptake by roots. The model, described in detail by Leadley et al. (1996), is also based on data from *E. vaginatum.* This sedge is well suited to simulation of nitrogen uptake because its growth habit permits a number of simplifying assumptions. In particular, the roots of *E. vaginatum* grow relatively straight down into the soil without branching, are not heavily suberized, are nonmycorrhizal, and experience only intraspecific competition for nitrogen.

14.3.3.2 Description

The root system of a plant is envisioned as a set of roots growing parallel to each other down into soil cylinders. Each soil cylinder is divided into numerous concentric sub-cylinders around an individual root (Fig. 14.7). The model is conceptually similar to the Barber and Cushman (1981) model, but has several technical differences and includes a term for the supply of nutrients from microbial mineralization and immobilization of nutrients ("S" in Fig. 14.7) that is not included in the Barber and Cushman model. The model was parameterized using field measurements of the properties of tussock tundra soils and field and laboratory measurements of *E. vaginatum* root uptake kinetics (details provided in Leadley et al. 1996). The model accounts for the supply, soil flux, and uptake of NH_4^+, NO_3^-, and glycine by *E. vaginatum.* It can be used to examine factors likely to regulate the uptake from each of these sources of N and to estimate the relative contributions of each source to the overall N nutrition of this species.

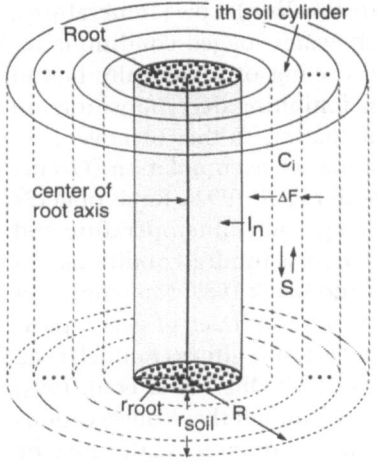

Fig. 14.7. Diagram of root segment and surrounding soil cylinder divided into numerous concentric subcylinders. Net exchange of nutrients in ith subcylinder by diffusion and mass flow (ΔF), C_i is concentration of soil solution, S is supply rate of nutrients (gain or loss) from microbial activity, I_n is flux into root, and R is radius from center of root axis to outer soil cylinder. (Modified from Leadley et al. 1996)

14.3.3.3 Potentials and Limitations

Extensive sensitivity analysis using this model suggests that ammonium, nitrate, and even small molecular weight amino acids can make significant contributions to the nitrogen nutrition of *E. vaginatum*. The sensitivity analysis also suggests that the rate of supply of nitrogen compounds through microbial mineralization is the most important factor in determining the rate of nitrogen uptake by *E. vaginatum* and that soil properties, the intensity of root exploration of the soil, and root uptake kinetics play minor roles. As a consequence, it was concluded that nitrogen uptake in arctic ecosystems is sensitive to factors different to those in fertile agricultural systems. Specifically, root morphology and soil factors are very important in determining nitrogen uptake in agricultural systems (Barber and Siblerbush 1984), but are not important in *E. vaginatum*. Because of this, efforts to predict nitrogen uptake by *E. vaginatum* and other arctic plants based on root uptake kinetics, soil solution concentrations, and root length (e.g., the model of Marion and Everett 1989), puts an inappropriate emphasis on factors that are poorly correlated with nitrogen uptake. These conclusions, based on a detailed, process-level mechanism, were used to justify the simplifications made in the *E. vaginatum* plant growth model (Sect. 14.3.2) and to develop a simplified nutrient availability model for use at the landscape scale (Sect. 14.4.2).

14.3.4 Decomposition

14.3.4.1 Motivation

Nitrogen dynamics in arctic soils, especially the relationships between nutrient mineralization, microbial immobilization, and plant uptake, are still not well understood (Kielland and Chapin 1992; Chap. 10, this Vol.). Low temperatures, permafrost, low nutrient inputs, and frequently water-logged conditions reduce the rate of organic matter turnover and cycling of organically bound nutrients; furthermore, many of the conditions inhibiting decomposition are enhanced by the accumulation of dead organic matter, so that turnover rate may actually decrease with increasing organic matter accumulation (Oechel and Billings 1992; Tenhunen et al. 1992; Nadelhoffer et al. 1992). Rates of net N mineralization seem relatively insensitive to changes in soil temperature and moisture content within the normal range of tussock tundra conditions (2–10 °C, <0.4 bar; Marion and Miller 1982; Marion and Black 1987; Nadelhoffer et al. 1991, 1992). In contrast, respiration in tundra soil (an index of decomposition) is greatly influenced by soil temperature and water regimes (Bunnell et al. 1977; Flanagan and Bunnell 1980; Peterson et al. 1984; Oberbauer et al. 1991; Nadelhoffer et al. 1992; Chaps. 11, 17, this Vol.). Although these observations seem contradictory, nutrients released from decaying organic matter may be immobilized by decomposer microorganisms so that the net mineralization rate is more dependent on litter quality and microbial carbon:nutrient balance than the overall decay rate.

The dynamic physiochemical nature of arctic soils, combined with the uncertainties of biological interactions, makes it difficult to predict patterns of nutrient availability, especially so with regard to potential effects of disturbance. Mathematical models can aid in understanding the nature and behavior of such complex dynamics. We developed a general model to examine interactions between decaying litter, decomposer microorganisms, and C and N soil pools. Our approach was aimed to explore the mechanisms underlying observed patterns of decomposition using a model that balances simplicity with enough detail to suggest the reasons for system behavior.

14.3.4.2 Description

Our model, GENDEC (Moorhead and Reynolds 1991, 1993), considers the effects of soil moisture, soil temperature, and season length on pools of carbon and nitrogen including labile plant compounds, holocellulose (cellulose + hemicellulose), resistant plant compounds (e.g., lignins), dead and live microbial mass, mineral N, and CO_2 (Fig. 14.8). Nitrogen is assumed to balance carbon flows and the effect of N limitation is determined by comparing available N to the amount required for maximum potential decay of a carbon (C) pool based on temperature, moisture, and decay rate coefficient; actual decay of N-limited substrates is reduced in proportion to any existing deficit. Conversely, net N mineralization occurs when the quantities of N released from decomposition exceed microbial demands. The total quantities of carbon and N available for microbial use consist of the sum of all losses from the dead organic matter pools; available N includes mineral forms. Microbial growth and respiration are driven by C availability, i.e., total losses from C substrates

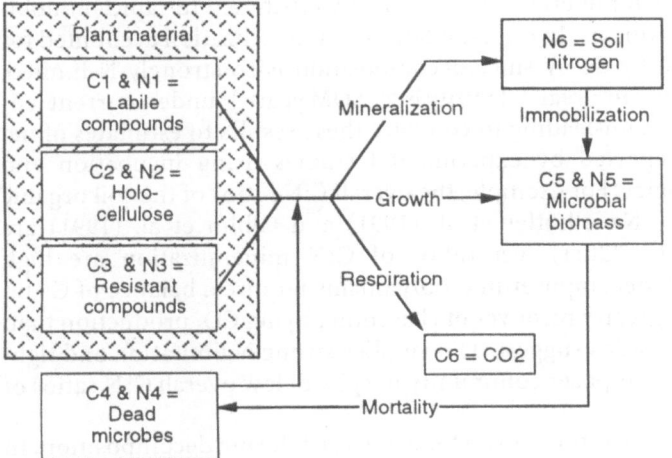

Fig. 14.8. Carbon and nitrogen flows in the GENDEC decomposition model. (From Moorhead and Reynolds 1993)

given fixed assimilation efficiencies. Microbial turnover includes a fraction of the standing biomass, a portion of incremental growth, and a portion of the standing biomass killed by wetting-drying events.

14.3.4.3 Potentials and Limitations

We used GENDEC to explore potential interactions of soil temperatures, season length, and soil moisture regimes on C and N dynamics of a tussock tundra soil, i.e., mineralization and immobilization independent of plant activities. Three major conclusions were reached: (1) the amount of simulated N turnover from decaying organic matter is sufficient to meet plant demands, hence, the model exhibits reasonable behavior under current environmental conditions; (2) previously reported inconsistencies in patterns of C and N mineralization for organic arctic soils likely result from different fates of mineral forms of C (lost from the system as CO_2) and N (rapidly immobilized within the system) – an insight that would be difficult to gain from experiments utilizing common incubation methods; and (3) response of N dynamics to changes in climate regimes is unpredictable, given uncertainties in climate change scenarios for this region (Moorhead and Reynolds 1993).

Under current climate conditions, simulated mineral N pools are very small in tussock tundra soils, maximum values averaging $<26\,\mathrm{mg\,N\,m^{-2}}$, which is equivalent to ca. $1.52\,\mu\mathrm{g\,N\,g^{-1}}$ soil organic matter (SOM) (assuming soil organic matter is ca. 50% carbon) and is slightly lower than extractable nitrate + ammonia concentrations generally reported for tussock tundra soils (Marion and Black 1987; Kielland 1990; Giblin et al. 1991; Nadelhoffer et al. 1991). The reason for such a small mineral N pool is usually attributed to rapid immobilization by microorganisms. In our simulations, immobilization accounted for all of the N released by decay since decomposition is so strongly N-limited, amounting to $5.28 \, g \, N \, m^{-2} year^{-1}$ ($310 \, \mu g \, N \, g^{-1} \, SOM \, year^{-1}$) under current climatic conditions. It is misleading to compare these results to estimates of net N mineralization reported by experimental studies using incubation and leaching methodologies. For example, the overall C:N ratios of the soil organic matter examined by Nadelhoffer et al. (1991) and Giblin et al. (1991) are relatively low (16:1 −220:1), yet ratios of C:N mineralization are high (≥450:1). In general, decomposer microorganisms require a balance of C and N that requires a far greater turnover of N accompanying CO_2 production than these mineralization ratios suggest. This implies strong N limitation and tight cycling within the decomposer community in spite of low overall C:N ratios of soil organic matter.

In field situations, plants compete for N released during decomposition. In our simulations with the *E. vaginatum* growth model (Sect. 3.2), we found that $9-18 \, g \, N \, m^{-2} \, year^{-1}$ (within tussocks) was needed to obtain reasonable results, which converts to about $1.8-7.2 \, g \, N \, m^{-2} \, year^{-1}$ at Imnavait Creek, where 20–40% of the soil surface is covered with tussocks. In comparison, Shaver et al. (1990) estimated that tussock tundra vegetation at the nearby Sagavanirktok River requires about $350 \, mg \, N \, m^{-2} \, year^{-1}$ to realize aboveground production. If belowground production requires twice this amount, as much as $1.15 \, g \, N \, m^{-2} \, year^{-1}$ could be needed to meet total plant demand. The amount of N turnover estimated by GENDEC exceeds this quantity. In contrast, many of the reported annual net N mineralization rates are too small to support observed plant growth (e.g., Giblin et al. 1991; Nadelhoffer et al. 1991), although some species may also obtain amino acids directly from soil solution (Kielland 1990; Keilland and Chapin 1992).

Insight to this apparent paradox can be gained by comparing observed and simulated patterns of C and N mineralization. Efflux of carbon dioxide from tundra soils appears more closely related to decomposition than many estimates of net N mineralization. Our simulations produced $79.5 \, g \, CO_2\text{-}C \, m^{-2} \, year^{-1}$ ($4.7 \, mg \, CO_2\text{-}C \, g^{-1} \, SOM$), which is comparable to observations reported for tussock tundra soils (Poole and Miller 1982; $127 \, g \, CO_2\text{-}C \, m^{-2} year^{-1}$, Giblin et al. 1991; $68 \, g \, CO_2\text{-}C \, m^{-2} year^{-1}$, Nadelhoffer et al. 1991; $8 \, mg \, CO_2\text{-}C \, g^{-1} \, SOM \, year^{-1}$). Ratios of C mineralization:N mineralization in our simulations – based on cumulative CO_2-C efflux and average size of the N-mineral pool – are much higher (ca. 3550:1 to 5000:1) than those reported by Nadelhoffer et al. (1991; 800:1) and Giblin et al. (1991; 450:1), although these field measurements probably represent five separate extractions of a smaller, transient mineral pools. Also, Nadelhoffer et al. (1991) and Giblin et al. (1991) found no correlation between soil respiration and N-mineralization rates. Similarly, there was no constant relationship between the size of the mineral N pool and either CO_2 efflux or net N immobilization in simulations. However, the quantity of C mineralized as CO_2 in GENDEC was directly proportional to net N immobilized because decomposer microorganisms require a balance of C and N to simultaneously meet energetic and nutritional needs. This means that N release from dead organic matter must balance C flux, but such a relationship is difficult to demonstrate by experimentally extracting a small, transient mineral N pool from incubated soils under conditions of rapid N immobilization (e.g., Giblin et al. 1991; Nadelhoffer et al. 1991). Both experimental and simulation data indicate that N is tightly cycled within organic arctic soils. We suggest that N availability to plants is proportionally related to the cumulative turnover of N from decaying organic matter rather than any static measure of soil mineral N pool size.

Results of simulations under various climate change scenarios suggest large potential variability compared to current C and N dynamics, depending on the rates and directions of changes in soil moisture and temperature regimes. Although increasing temperatures increased estimated N turnover, current levels of soil moisture appear close to optimum for decomposition, which indicates that any net changes in soil moisture may decrease C mineralization. More detailed soil profile descriptions and soil climate data are necessary to more accurately characterize both temporal patterns and net change in decomposition and nutrient dynamics in these systems.

14.4 Top-Down Models

14.4.1 Hydrologic Transport

14.4.1.1 Motivation

The strong relationship between water and plant structure/function in the Arctic is well known. Numerous researchers have found strong correlations between both patterns and productivity of tussock tundra vegetation and moisture gradients (e.g., Bliss et al. 1984; Jasieniuk and Johnson 1982; Jorgenson 1984; Peterson and Billings 1980; Webber 1978; Chapin et al. 1988; Matthes-Sears et al. 1988; Murray et al. 1989; Gebauer et al. 1995; Chaps. 4, 5, 11, this Vol.) and an array of chemical, physical, and biological variables, e.g., thaw depth and heat input (Webber 1978; Walker et al. 1989; Hastings et al. 1989; Shaver et al. 1990). The importance of water is illustrated in Webber's (1978) conceptual model of the principal environmental variables affecting vegetation at Barrow, Alaska (Fig. 14.9). While the role of topography on the water regime at the patch scale is explicit in Fig. 14.9, the importance of this relationship may be implied as well when we consider vegetation patterns and landforms at larger spatial and temporal scales. Over longer periods of time, integrated, complex feedbacks among a number of processes often govern the performance of arctic plants under natural circumstances more strongly than do short-term responses of individual processes per se (Chapin et al. 1992).

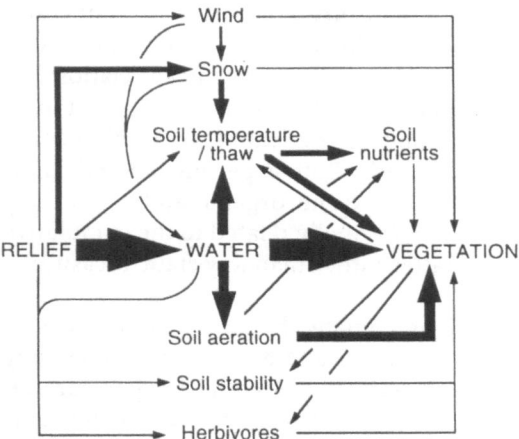

Fig. 14.9. Principal environmental variables that control vegetation at Barrow, Alaska. Strength of control indicated by *thickness of line*. Redrawn from Webber (1978)

Is there a predictive relation between soil water and vegetation type at Imnavait Creek or between soil water and nutrient availability? These questions and others led to the development of simple hydrology models that could estimate soil water status for any pixel in the watershed. This was seen as a first step in developing landscape models for tundra ecosystems. We used two models, TOPMODEL and T-HYDRO (#2 and #3, respectively, Fig. 14.1 and Table 14.2). TOPMODEL, which predicts both water routing and soil water deficits, was linked to the GAS-FLUX model to simulate landscape patterns of ecosystem gas exchange (#8, described in Chap. 17, this Vol.). The water routing algorithm used in TOPMODEL was also used in a vegetation typing scheme (#7). T-HYDRO, which only computes water routing, was used in vegetation typing model (#5) and in landscape models for nutrient availability and growth (#6), the latter as part of the landscape model T-MAPS (#9, described in Chap. 18, this Vol.). TOPMODEL is described in Chapter 17, this Vol.; our work with T-HYDRO is described below.

14.4.1.2 Description

T-HYDRO is a spatially explicit watershed model that utilizes raster-based topographic information to generate a two-dimensional water flow field for the Imnavait Creek watershed. Details are given in Ostendorf and Reynolds (1993). The watershed is subdivided into a grid of 21 250 square pixels with 10-m side length and, for each pixel, we compute the total discharge (volume of water) leaving a pixel per year based on the difference between run-on, precipitation, and evapotranspiration (units of m^3 H_2O pixel^{-1} year^{-1}) (Fig. 14.10). The water-routing algorithm arrived at using this approach differs slightly from that used by Quinn et al. (1991) in TOPMODEL. The result is a "drainage area" map that shows the total upslope area that "drains" into a given pixel. The map for Imnavait Creek is shown in Fig. 18.4 in Chapter 18 (this Vol.).

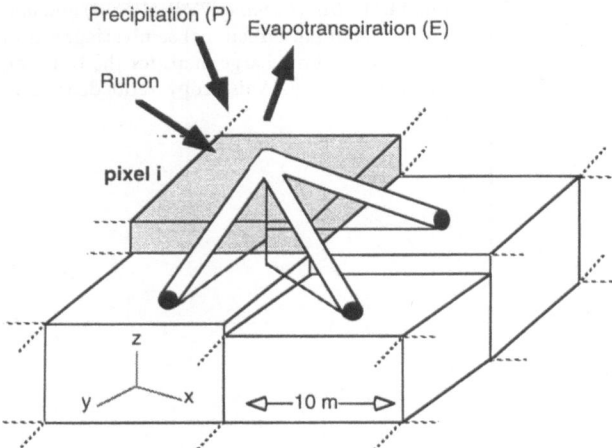

Fig. 14.10. Water-routing scheme used in T-HYDRO. Centers of pixels are connected by a network of nodes or "pipes" of equal diameter filled with soil; the elevational difference between pixels is the gravitational potential between nodes. Darcy's law is used to estimate the relative discharge through the connections and, ultimately, its routing through the landscape. Redrawn from Ostendorf and Reynolds (1993)

14.4.1.3 Potentials and Limitations

We compared the drainage area map produced by T-HYDRO with a normalized difference vegetation index (NDVI) map derived from a SPOT satellite image. NDVI is the difference between the infrared and red bands divided by the sum of the two spectral bands (White and Running 1994) and is linearly related to the amount of photosynthetically active radiation (PAR) intercepted by vegetation (Sellers 1987). NDVI may provide an indirect measure of net primary productivity (NPP) since there is a close relationship between integrated intercepted PAR and NPP in many types of vegetation (Monteith 1981) and NDVI is related to NPP on a biome scale (Goward et al. 1985). Comparing the drainage area map with the NDVI map using regression analysis, we were able to account for ca. 43% of the spatial variance in NDVI (see Fig. 18.8, this Vol.).

To examine the spatial relationships between the NDVI and discharge, we constructed semivariograms of the maps. Since semivariance may not only depend upon the lag distance but also on direction (Cressie 1991), we examined *directional* semivariograms coinciding with the major environmental gradients in the Imnavait Creek watershed – N75°E (northeast to southwest, i.e., perpendicular to the stream channel) and N15°W (northwest to southeast, i.e., parallel to the main stream channel) – as well as those without reference to direction (*omnidirectional*). When two or more directional semivariograms differ significantly, the spatial structure of the system is called *anisotropic*. Anisotropic semivariograms occur when there is a strong gradient in the landscape. We used the anisotropy ratio to characterize the anisotropic structure of the NDVI and discharge maps (Trangmar et al. 1985):

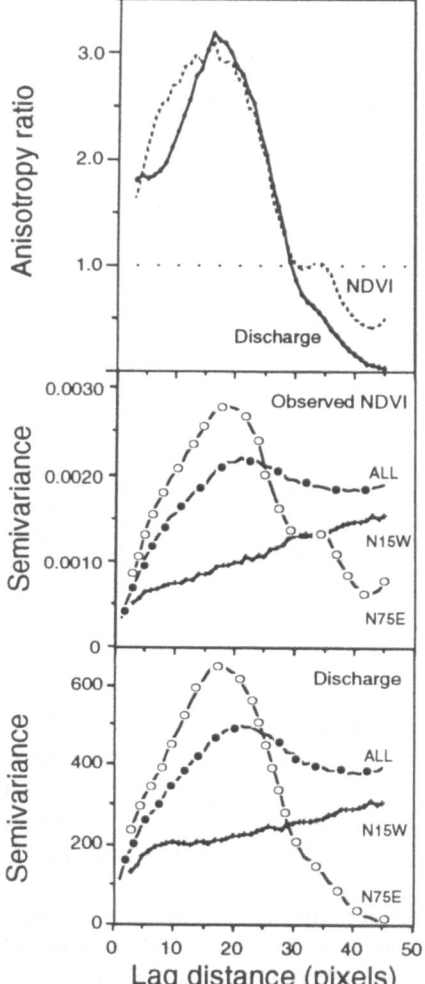

Fig. 14.11. *Directional* (N75°E, N15°W) and *omnidirectional* (all directions) semivariograms of the NDVI and discharge maps for the Imnavait Creek watershed. Anisotropy ratio defined in text

$$k(h) = \frac{\gamma(h,\theta_1)}{\gamma(h,\theta_2)},$$

(1)

where k(h) is the anisotropy ratio at lag distance h, and $\gamma(h, \theta_1)$ and $\gamma(h, \theta_2)$ are semivariance in two directions, θ_1 = N75°E and θ_2 = N15°W, respectively. If k(h) is equal or close to 1, spatial heterogeneity is defined as *isotropic*.

The semivariograms of the maps are quite similar: omnidirectionals peak at ca. 21 pixels, N75°E directionals peak at ca. 18, and N15°W directionals show strong trend effects (Fig. 14.11). Both maps have an apparent anisotropic structure since the anisotropy ratio is above the isotropic line over most spatial scales. Strong anisotropy suggests that the tundra landscape portrayed by the two maps is structured differently in the two directions, which is consistent with our field experience. Anisotropy is expected because the watershed has distinct vegetation zones in the N75°E direction and, generally in the same direction, gradients of elevation and drainage. Thus, the differences between semivariograms in the two directions likely represent real differences in the landscape.

T-HYDRO is parameterized with an estimate of the water budget based on a single year and our regression model is based on a single NDVI scene; a multi-year data set would be valuable in order to test the generality of these relationships. Nevertheless, these results support Webber's conceptual model (Fig. 14.9) and our hypothesis that the influence of topography on the local water regime can be used as a tool to predict vegetation patterns in the landscape. There are of course numerous, complex interactions and feedbacks that exist in arctic ecosystems that contribute to the high variability of the vegetation patterns, including soil properties, slope, radiation, temperature, and the glaciation history or the age and successional state of the evolving landscape (Walker and Walker 1991; Chap. 4, this Vol.). Extensions of this analysis are described below and in Chapters 17–18 (this Vol.).

14.4.2 Topographically Derived Vegetation Model

14.4.2.1 Motivation

As discussed above, we hypothesized that the general pattern of arctic vegetation – and other important long-term processes as shown in Fig. 14.9 – might be explained by topographically derived variables related to soil water. In our efforts to build simple landscape models, we made a number of assumptions directly related to Fig. 14.9. These assumptions included: average soil moisture during the growing season may be derived from local water discharge and slope; vegetation types are predictable from soil moisture and slope; the productivity and biomass of poikilohydric plants are related to soil moisture; and N availability is related to soil moisture. Based on these assumptions, we developed several "topographically derived" models (#5–#7, Fig. 14.1 and Table 14.2), so named since the only input variables are topographically derived, i.e., discharge and/or slope. Topographically derived models of moisture and nutrient availability and plant growth are described in Chapter 18 (this Vol.). We describe a vegetation model (#7) below.

14.4.2.2 Description

Ostendorf and Reynolds (1995) describe a topographically derived model (TVM) for predicting vegetation types based on the relationships between slope ($\tan \beta$), discharge (δ), and plant vegetation. A general overview of the approach is given in Fig. 14.12. The objective is to predict plant vegetation

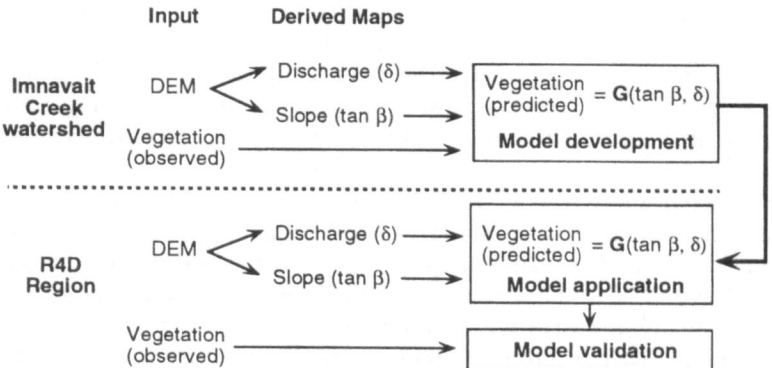

Fig. 14.12. Scheme for TVM based on functional relationships (*G*) between slope (*tan β*), discharge (δ), and plant vegetation types. The model was developed and parameterized using data from the Imnavait Creek watershed and then tested against the larger R4D region. (From Ostendorf and Reynolds 1996)

types at a particular location (pixel) based solely on a function of topographically derived surrogates of soil moisture. Slope determines the gravitational hydrological gradient and hence influences flow velocity; slope angle is inversely related to soil moisture. Discharge is estimated by the water-routing algorithm used in TOPMODEL (Quinn et al. 1991). Slope, discharge, and vegetation maps in the watershed were systematically scanned to determine the frequency of occurrence of each of five different vegetation types associated with particular slope and discharge (an example is shown in Fig. 17.3, this Vol.). The vegetation type associated with each slope and discharge combination was determined by one of two functions (G, Fig. 14.12): (1) randomly selecting a community type based on relative frequencies or (2) selecting the community type with the highest frequency of occurrence. The latter gave slightly better results (see Ostendorf and Reynolds 1996 for details of the search algorithms, functions, and spatial error analysis).

We used five vegetation types: (1) *aquatic types* dominated by sedges, grasses, or forbs; (2) *water track/riparian* tundra; (3) *moist types* (= tussock tundra) dominated by mesic dwarf shrubs, sedges, and mosses; (4) *dry types* (= heath tundra) dominated by ericaceous dwarf shrubs and lichens; and (5) *barren* or partially vegetated areas. Types 2–4 account for more than 95% of the vegetation in the Imnavait Creek and adjacent watersheds (Walker et al. 1989; see Chaps. 4 and 11, this Vol.). The model was developed and parameterized using vegetation data from the Imnavait Creek watershed, an area equal to about 10% of the R4D region (Fig. 1.1, this Vol.). It was then applied it to the total R4D region (202436 pixels) as a test of its validity.

14.4.2.3 Potentials and Limitations

The TVM model gives excellent predictions of the *total* percentage of the five vegetation types in the R4D region (compare measured vs. predicted areas in

Table 14.4. Validation of the TVM model for R4D region (22 km^2). The $\hat{\kappa}$ statistic evaluates the overall accuracy of spatial predictions. Z-values based on a standardized sample number of 457 spatially independent pixels. (Modified from Ostendorf and Reynolds 1996)

Vegetation type	Measured area	Predicted area	Spatial accuracy
	(%)	(%)	(%)
Aquatic types	0.8	0.7	3.0
Water track/riparian	20.8	17.3	57.1
Moist types	58.1	63.6	75.9
Dry types	20.1	18.3	44.4
Barren	0.2	0.1	0.3
		Goodness of fit	59.3
		$\hat{\kappa} =$	0.29
		Z =	7.9[a]

[a] $P < 0.001$ such that $\hat{\kappa} = 0$.

Table 14.4). In terms of spatial accuracy, the strength of the predictions is ca. 60%. While the general patterns of the major community types are reproduced, the absence of certain features in the Imnavait Creek watershed that occur in the larger R4D region, e.g., lakes and certain slope/discharge gradients, is a major source of error.

TVM was inspired by Webber's (1978) model and the "hydroecological approach" of Molenaar (1987). Molenaar distinguishes between operational and conditional functionality of environmental factors. Water is operational in the sense that it serves as a "nutrient." In addition to this "direct, causal effect" of water, Molenaar lists several possible conditional or "physiological noncausal" functions, namely, water as a solvent for nutrients, and as a factor regulating aeration, redox conditions and acidity, phosphorus supply, metal ion availability, humus formation and breakdown, and nutrient cycling and soil formation. These functions are especially important for arctic ecosystems as shown in Webber's (1978) model. They are a result of the long-term interactions of vegetation, soil, and hydrological processes (Fig. 14.9). It appears impossible to separate such multiple factor interactions. However, by quantifying slope and discharge gradients in the TVM model, we lump a suite of interrelated factors: tan β and ∂ are continuous gradients and, as such, they serve as integrators of complex and long-term feedbacks between vegetation and the environment.

The TVM model represents a top-down approach based on topography, but it may be linked to small-scale, bottom-up models of arctic plants (e.g., Leadley and Reynolds 1992). With slope and discharge, we are able to account for a large proportion of the variance in vegetation patterning and, hence, this model could be used to delineate areas where we might expect similar ecosystem behavior at the patch scale. Further work is warranted to examine the generality of Webber's model (Fig 14.9) and the potential of TVM-type models as scaling tools.

14.5 Conclusions

One limitation to developing integrated models for the arctic is that we still do not understand many of the basic processes that control ecosystem responses to multiple stresses. An example is the role of individual species in ecosystem function. Within a given climate regime, different vegetation types exhibit different rates of carbon and nutrient cycling, water balance, and energy balance largely because of the effects of individual species on resource supply, trophic structure, and/or disturbance regime. In the arctic, ecophysiological studies show that plant species differ substantially in growth rate, nutrient uptake, tissue turnover, and litter quality (e.g., Chapin 1980; Chapin et al. 1992), processes that determine productivity and nutrient cycling. However, such detailed species-specific information for all of these processes has never been incorporated into an ecosystem model because of our lack of understanding of the specific linkages and feedback mechanisms involved. Perhaps the best alternative is to identify and employ functional groups – i.e., species that share a similar physiology or growth form, etc. – traits that can have important ecosystem consequences (Chapin 1993). Another limitation is our lack of understanding of how the destruction and fragmentation of natural habitats affect different components of a ecosystem. Individual organisms may modify their behavior, community and population dynamics can shift, and system-level fluxes of carbon, water and nutrients may be impacted in complex ways (Sousa 1984; Goigel-Turner 1987; Pickett et al. 1989; Dunning et al. 1992; DeAngelis and White 1994). Field measurements taken only at one level, e.g., the community, may belie effects on individual species, functional groups, and populations, which in turn may occur at different spatial and temporal scales (see Robinson et al. 1992).

These gaps in our knowledge – as well as many others – suggest that it is still premature to expect fully integrated models of arctic ecosystem function and that the models described here must be understood as "next steps". In the absence of comprehensive understanding, there is a danger that largely untested hypotheses will become incorporated into models and influence our considerations. Such imperfections remain obscure because model testing and validation are often inadequate or impossible, particularly at larger spatial and temporal scales (Oreskes et al. 1994; Reynolds et al. 1996). Similarly, a wholly empirical approach to predicting arctic ecosystem responses to energy-related disturbances is impractical since it is impossible to design and conduct the many different combinations of experiments needed to sort out complex ecosystem interactions. Also, the size and expense of equipment restricts studies to relatively small spatial scales and the majority of experiments and field manipulations are short term. Long-term changes that may lead to new feedbacks, homeostasis, or shifts in species are not apparent.

Models are based on assumptions and conceptualizations about how ecosystems behave and on important factors, relationships, and interactions that must be considered. While simulation modeling has limitations, our limited body of empirical knowledge can be greatly extended when all assumptions and limitations are brought together in a best "state-of-the-art formulation, as attempted in some of the R4D models described in this book. These models have a great potential to provide important information and predictions that can help us link short-term, single-factor experiments to our understanding of long-term, multifactor disturbances.

Acknowledgments. Financial support was provided by the US Dept. of Energy (DE-FG03-84R60250 and DE-FG05-92ER61455), NASA Global Change Fellowship (to P. Leadley), and the German Federal Ministry for Science and Technology (grant BEO51-0339476A).

References

Barber SA, Cushman JH (1981) Nitrogen uptake model from agronomic crops. In: Iskandar IK (ed) Modeling wastewater renovation-land treatment. Wiley, New York, pp 383–409

Barber SA, Silberbush M (1984) Plant root morphology and nutrient uptake. In: Barber SA, Bouldin DR (eds) Roots, nutrient and water influx, and plant growth. Am Soc Agric Spec Publ 49, Madison, Wisconsin, pp 65–88

Battaglin WA, Hay LE, Parker RS, Leavesley GH (1993) Applications of a GIS for modeling the sensitivity of water resources to alterations in climate in the Gunnison River Basin, Colorado. Water Res Bull 25: 1021–1028

Bliss LC (1981) North American and Scandinavian tundras and polar deserts. In: Bliss LC, Heal OW, Moore JJ (eds) Tundra ecosystems: a comparative analysis. Cambridge University Press, Cambridge, pp 8–24

Bliss LC, Svoboda J, Bliss DI (1984) Polar deserts, their plant cover and plant production in the Canadian High Arctic. Holarct Ecol 7: 305–324

Bunnell FL, Tait DEN, Flanagan PW, Van Cleve K (1977) Microbial respiration and substrate weight loss. I. A general model of the influences of abiotic variables. Soil Biol Biochem 9: 33–40

Chapin FS III (1980) The mineral nutrition of wild plants. Annu Rev Ecol Syst 11: 233–260

Chapin FS III (1993) Functional role of growth forms in ecosystem and global processes. In: Ehleringer JR, Field CB (eds) Scaling physiological processes: leaf to globe. Academic Press, San Diego, pp 287–312

Chapin FS III, Van Cleve K, Chapin MC (1979) Soil temperature and nutrient cycling in the tussock growth form of *Eriophorum vaginatum*. J Ecol 67: 169–189

Chapin FS III, Shaver GR, Kedrowski RA (1986) Environmental controls over carbon, nitrogen, and phosphorus fractions in *Eriophorum vaginatum* in Alaskan tussock tundra. J Ecol 74: 167–195

Chapin FS III, Fetcher N, Kielland K, Everett K, Linkins AE (1988) Productivity and nutrient cycling of Alaskan tundra: enhancement by flowing soil water. Ecology 69: 693–702

Chapin FS III, Jefferies RL, Reynolds JF, Shaver GR, Svoboda J (1992) Arctic plant physiological ecology: a challenge for the future. In: Chapin FS III, Jefferies RL, Reynolds JF, Shaver GR, Svoboda J (eds) Arctic ecosystems in a changing climate: an ecophysiological perspective. Academic Press, San Diego, pp 3–8

Collatz GJ, Ball JT, Grivet C, Berry JA (1991) Physiological and environmental regulation of stomatal conductance, photosynthesis and transpiration: a model that includes laminar boundary layer. Agric For Meteorol 54: 107–136

Cooper DF (1976) Ecosystem models and environmental policy. Simulation (May 1976): 133–138

Cressie NAC (1991) Statistics for spatial data. John Wiley & Sons, New York

DeAngelis DL, White PS (1994) Ecosystems as products of spatially and temporally varying forces, ecological processes, and landscapes: a theoretical perspective. In: Davis SM, Ogden JC (eds) Everglades: the ecosystem and its restoration. St Lucie Press, Delray Beach, pp 9–27

DeAngelis DL, Waterhouse JC, Post WM, O'Neill RV (1985) Ecological modelling and disturbance evaluation. Ecol Model 29: 399–419

Dunning JB, Danielson BJ, Pulliam HR (1992) Ecological processes that affect populations in complex landscapes. Oikos 65: 169–175

Farquhar GD, von Caemmerer S, Berry JA (1980) A biochemical model of photosynthetic CO_2 assimilation in leaves of C_3 species. Planta 149: 78–90

Flanagan PW, Bunnell FL (1980) Microflora activities and decomposition. In: Brown J, Miller PC, Tieszen LL, Bunnell FL (eds) An arctic ecosystem: the coastal tundra at Barrow, Alaska. Dowden, Hutchinson and Ross, Stroudsburg, pp 291–334

Forman RTT, Gordon M (1986) Landscape ecology. Wiley and Sons, New York

Gebauer R, Tenhunen JD, Reynolds JF (1995) Gradients of soil moisture and oxygen in an arctic landscape and its effects on plant growth. Plant and Soil (in press)

Giblin AE, Nadelhoffer KJ, Shaver GR, Laundre JA, McKerrow AJ (1991) Biogeochemical diversity along a riverside toposequence in arctic Alaska. Ecol Monogr 61: 425–435

Goigel-Turner M (ed) (1987) Landscape heterogeneity and disturbance. Springer, Berlin Heidelberg New York

Goward SN, Tucker CJ, Dye DG (1985) North American vegetation patterns observed with the NOAA-7 advanced very high resolution radiometer. Vegetatio 64: 3–14

Hammond AL (1972) Ecosystem analysis: biome approach to environmental research. Science 175: 47–48

Hannsson L, Fahrig L, Merriam G (eds) (1994) Mosaic landscapes and ecological processes. Chapman and Hall, London

Hastings SJ, Luchessa SA, Oechel W, Tenhunen JD (1989) Standing biomass and production in water drainages of the foothills of the Philip Smith Mountains, Alaska. Holarctic Ecol 12: 304–311

Hinzman LD, Kane DL (1991) Snow hydrology of a headwater arctic basin 2. Conceptual analysis and computer modeling. Water Resour Res 27: 1111–1121

Hinzman LD, Kane DL, Gieck RE, Everett KR (1991) Hydrologic and thermal properties of the active layer in the Alaskan Arctic. Cold Reg Sci Technol 19: 95–110

Hinzman LD, Kane DL, Everett KR (1993) Hillslope hydrology in an arctic setting. Proc Permafrost 6th Int Conf, Beijing, China. South China Univ Technol Press, pp 267–271

Jasieniuk MA, Johnson EA (1982) Peatland vegetation organization and dynamics in the western subarctic, Northwest Territories, Canada. Can J Bot 60: 2581–2593

Jorgenson T (1984) The response of vegetation to landscape evolution on glacial till near Toolik Lake, Alaska. In: Inventorying forest and other vegetation of the high latitude and high altitude regions. Proc Int Symp, Soc Am Forest Regional Tech Conf, Fairbanks, Alaska, pp 134–141

Kane DL, Hinzman LD, Zarling JP (1991) Thermal response of the active layer in a permafrost environment to climatic warming. Cold Reg Sci Technol 19: 111–122

Kane DL, Hinzman LD, Woo M, Everett KR (1992) Arctic hydrology and climate change. In: Chapin FS III, Jefferies RL, Reynolds JF, Shaver GR, Svoboda J (eds) Arctic ecosystems in a changing climate: an ecophysiological perspective. Academic Press, New York, pp 35–57

Kaufmann MR, Graham RT, Boyce DA, Moir WH, Perry L, Reynolds RT, Bassett RL, Mehlhop P, Edminster CB, Block WM, Corn PS (1994) An ecological basis for ecosystem management. Gen Tech Rep RM-246, US Dept Agric, Ft Collins, Colorado, 22 pp

Kielland K (1990) Processes controlling nitrogen release and turnover in arctic tundra. Thesis, Univ Alaska, Fairbanks, 177 pp

Kielland K, Chapin FS III (1992) Nutrient absorption and accumulation in arctic plants. In: Chapin FS III, Jefferies RL, Reynolds JF, Shaver GR, Svoboda J (eds) Arctic ecosystems in a changing climate: an ecophysiological perspective. Academic Press, San Diego, pp 321–336

Kotliar NB, Wiens JA (1990) Multiple scales of patchiness and patch structure: a hierarchical framework for the study of heterogeneity. Oikos 59: 252–260

Leadley PW, Reynolds JF (1992) Long-term response of an arctic sedge to climate change: a simulation study. Ecol Appl 2: 323–340

Leadley, PW, Reynolds JF, Chapin FS III (1996) A model of ammonium, nitrate, and glycine uptake by Eriophorum vaginatum roots in the field: ecological implications. (submitted)

Li EA, Schanholtz VO, Contractor DN, Carr JC (1977) Generating rainfall excess based on readily determinable soil and land use characteristics. Trans ASAE 20: 1070–1078

Loucks OL (1985) Looking for surprise in managing stressed ecosystems. BioScience 35: 428–432

Marion GM, Black CH (1987) The effect of time and temperature on nitrogen mineralization in arctic tundra soils. Soil Sci Soc Am J 51: 1501–1508

Marion GM, Everett K (1989) The effect of nutrient and water additions on elemental mobility through small tundra watersheds. Holarct Ecol 12: 317–323

Marion GM, Miller PC (1982) Nitrogen mineralization in a tussock tundra soil. Arct Alpine Res 14: 287–293

Marion GM, Hastings SJ, Oberbauer SF, Oechel WC (1989) Soil-plant element relationships in a tundra ecosystem. Holarct Ecol 12: 296–303

Matthes-Sears U, Matthes-Sears WC, Hastings SJ, Oechel WC (1988) The effects of topography and nutrient status on the biomass, vegetative characteristics, and gas exchange of two deciduous shrubs on an arctic tundra slope. Arct Alpine Res 20: 342–351

May RM (1994) Ecological science and the management of protected areas. Biodivers Conserv 3(5): 437–448

Miller PC, Collier BD, Bunnell FL (1975) Development of ecosystem modeling in the tundra biome. In: Patten BC (ed) Systems analysis and simulation in ecology, vol 3. Academic Press, New York, pp 95–115

Miller PC, Stoner WA, Tieszen LL, Allessio M, McCown B, Chapin FS III, Shaver G (1978) A model of carbohydrate, nitrogen, phosphorus allocation and growth in tundra production. In: Tieszen LL (ed) Vegetation and production ecology of an Alaskan arctic tundra. Ecological Studies 29. Springer, Berlin Heidelberg New York, pp 577–598

Miller PC, Billings WD, Oechel WC (1979) A modeling approach to understanding plant adaptation to low temperatures In: Underwood LS, Tieszen LL, Callahan AB, Folk GE (eds) Comparative mechanisms of cold adapation. Academic Press, New York, pp 181–214

Miller PC, Miller PM, Blake-Johnson M, Chapin FS III, Everett KR, Hilbert DW, Kummerow J, Linkins AE, Marion GM, Oechel WC, Roberts SW, Stuart L (1984) Plant-soil processes in Eriophorum vaginatum tussock tundra in Alaska: a systems modeling approach. Ecol Monogr 54: 361–405

Molenaar JG (1987) An ecohydrological approach to floral and vegetational patterns in arctic landscape ecology. Arctic Alpine Res 19: 414–424

Monteith JL (1981) Does light limit crop production? In: Johnson CB (ed) Physiological processes limiting plant productivity, Butterworths, London, pp 23–38

Moorhead DL, Reynolds JF (1991) A general model of litter decomposition in the northern Chihuahuan Desert. Ecol Model 56: 197–219

Moorhead DL, Reynolds JF (1992) Modeling the contribution of decomposer fungi in nutrient cycling. In: Carroll GC, Wicklow DT (eds) The fungal community. Marcel Dekker, Boston, pp 691–714

Moorhead DL, Reynolds JF (1993) Effects of climate change on decomposition in arctic tussock tundra: a modeling synthesis. Arct Alpine Res 25: 403–412

Murray KJ, Tenhunen JD, Kummerow J (1989) Limitations on moss growth and net primary production in tussock tundra areas of the foothills of the Philip Smith Mountains, Alaska. Oecologia 20: 256–262

Nadelhoffer KJ, Giblin AE, Shaver GR, Laundre JA (1991) Effects of temperature and substrate quality on element mineralization in six arctic soils. Ecology 72: 242–253

Nadelhoffer KJ, Giblin AE, Shaver GR, Linkins AE (1992) Microbial processes and plant nutrient availability in arctic soils. In: Chapin FS III, Jefferies RL, Reynolds JF, Shaver GR, Svoboda J (eds) Arctic ecosystems in a changing climate: an ecophysiological perspective. Academic Press, San Diego, pp 281–300

National Resource Council (1982) Arctic Terrestrial Environmental Research Programs of the Office of Energy Research. Department of Energy: evaluation and recommendations. National Academy Press, Washington, DC, 63 pp

Oberbauer SF, Tenhunen JD, Reynolds JF (1991) Environmental effects on CO_2 efflux from water track and tussock tundra in arctic Alaska, USA. Arct Alpine Res 23: 162–169

Oberbauer SF, Gillespie CT, Cheng W, Gebauer R, Sala Serra A, Tenhunen JD (1992) Environmental effects on CO_2 efflux from riparian tundra in the northern foothills of the Brooks Range, Alaska, USA. Oecologia 92: 568–577

Oechel WC, Billings WD (1992) Effects of global change on the carbon balance of arctic plants and ecosystems. In: Chapin FS III, Jefferies RL, Reynolds JF, Shaver GR, Svoboda J (eds) Arctic ecosystems in a changing climate: an ecophysiological perspective. Academic Press, San Diego, pp 139–168

O'Neill RV (1988) Hierarchy theory and global change. In: Rosswall T, Woodmansee RG, Risser PG (eds) Scales and global change. Scientific Committee on Problems of the Environment (SCOPE), John Wiley & Sons, New York, pp 29–45

Oreskes N, Shrader-Frechette K, Belitz K (1994) Verification, validation, and confirmation of numerical models in the earth sciences. Science 263: 641–646

Osmond CB, Bjorkman O, Anderson DJ (1980) Physiological processes in plant ecology. Springer, Berlin Heidelberg New York

Ostendorf B, Reynolds JF (1993) Relationships between a terrain-based hydrologic model and patch-scale vegetation patterns in an arctic tundra landscape. Landscape Ecol 8: 229–237

Ostendorf B, Reynolds JF (1996) A model of arctic tundra vegetation derived from topographic gradients. (submitted)

Peterson KM, Billings WD (1980) Tundra vegetational patterns and succession in relation to microtopography near Atkasook, Alaska. Arct Alpine Res 12: 473–482

Peterson KM, Billings WD, Reynolds DN (1984) Influence of water table and atmospheric CO_2 concentration on the carbon balance of arctic tundra. Arct Alpine Res 16: 331–335

Pickett STA, Kolasa J, Armesto JJ, Collins SL (1989) The ecological concept of disturbance and its expression at various hierarchical level. Oikos 54: 129–136

Pickett STA, Burke IC, Dale VH, Gosz JR, Lee RG, Pacala SW, Shachak M (1994) Integrated models of forested regions. In: Groffman PM, Likens GE (eds) Integrated regional models. Chapman and Hall, New York, pp 120–141

Poole DK, Miller PC (1982) Carbon dioxide flux from three arctic tundra types in north-central Alaska, USA. Arct Alpine Res 14: 27–32

Quinn P, Beven K, Chevallier P, Planchon O (1991) The prediction of hillslope flow paths for distributed hydrological modeling using digital terrain model. Hydrol Processes 5: 59–79

Quinn P, Ostendorf B, Beven K, Tenhunen JD (1995) Spatial and temporal predictions of soil moisture patterns and evaporative dynamics using TOPMODEL and the GAS-FLUX model for an Alaskan catchment. (in preparation)

Rastetter EB, Ryan MG, Shaver GR, Melillo JM, Nadelhoffer KJ, Hobbie JE, Aber JD (1991) A general model describing the responses of the C and N cycles in terrestrial ecosystems to changes in CO_2, climate, and N deposition. Tree Physiol 9: 101–126

Reynolds JF, Leadley PW (1992) Modeling the response of arctic plants to climate change. In: Chapin FS III, Jefferies RL, Reynolds JF, Shaver GR, Svoboda J (eds) Arctic ecosystems in a changing climate. Academic Press, San Diego, pp 413–438

Reynolds JF, Hilbert DW, Kemp PR (1993) Scaling ecophysiology from the plant to the ecosystem: a conceptual framework. In: Ehleringer J, Field C (eds) Scaling processes between leaf and the globe. Academic Press, New York, pp 127–140

Reynolds, JF, Virginia RA, Schlesinger WA (1996) Defining functional types for models of desertification. In: Smith TM, Shugart HH, Woodward FI (eds) Towards the development of a functional classification of plants. Cambridge University Press, Cambridge (in press)

Robinson GR, Holt RD, Gaines MS, Hamburg SP, Johnson ML, Fitch HS, Martinko EA (1992) Diverse and contrasting effects of habitat fragmentation. Science 257: 524–526

Sellers PJ (1987) Canopy reflectance, photosynthesis, transpiration: the role of biophysics in the linearity of their interdependence. Remote Sensing Environ 21: 143–183

Shaver GR, Chapin FS III (1986) Effect of fertilizer on production and biomass of tussock tundra, Alaska, USA. Arct Alpine Res 18: 261–268

Shaver GR, Chapin FS III, Gartner BL (1986) Factors limiting seasonal growth and peak biomass accumulation in *Eriophorum vaginatum* in Alaskan tussock tundra. J Ecol 74: 257–278

Shaver GR, Nadelhoffer KH, Giblin AE (1990) Biogeochemical diversity and element transport in a heterogeneous landscape, the north slope of Alaska. In: Turner MG, Gardner RH (eds) Quantitative methods in landscape ecology. Ecological Studies 82. Springer, Berlin Heidelberg New York, pp 105–125

Slocombe DS (1993) Implementing ecosystem-based management. BioScience 43: 612–622

Sousa WP (1984) The role of disturbance in natural communities. Annu Rev Ecol Syst 15: 353–391

Starfield AM, Bleloch AL (1991) Building models for conservation and wildlife managment, Burgress Int Group, Edina, Minnesota, 253 pp

Tehnunen JD, Lange OL, Hahn S, Siegwolf R, Oberbauer SF (1992) The ecosystem role of poikilohydric tundra plants. In: Chapin FS III, Jefferies RL, Reynolds JF, Shaver GR, Svoboda J (eds) Arctic ecosystems in a changing climate: an ecophysiological perspective. Academic Press, San Diego, pp 213–238

Tehnunen JD, Siegwolf R, Oberbauer SF (1994) Effects of phenology, physiology, and gradients in community composition, structure, and microclimate on tundra ecosystem CO_2 exchange. In: Shulze E-D, Caldwell M (eds) Ecophysiology of photosynthesis. Springer, Berlin Heidelberg New York, pp 431–460

Tenhunen JD, Gillespie CT, Oberbauer S, Sala A, Whalen S (1995) Climate effects on the carbon balance of tussock tundra in the Philip Smith Mountains, Alaska. Flora 190: 273–283

Tissue DT, Oechel WC (1987) Response of *Eriophorum vaginatum* to elevated CO_2 and temperature in the Alaskan tussock tundra. Ecology 68: 401–410

Trangmar BB, Yost RS, Uehara G (1985) Application of geostatistics to spatial studies of soil properties. Adv Agron 38: 44–94

Turner M (1989) Landscape ecology: the effect of pattern on process. Annu Rev Ecol Syst 20: 171–197

Turner MG, Dale VH (1991) Modeling landscape disturbance. In: Turner MG, Gardner RH (eds) Quantitative methods in landscape ecology. Springer, Berlin Heidelberg New York, pp 323–351

Walker DA, Binnian E, Evans BM, Lederer ND, Nordstrand E, Webber PJ (1989) Terrain, vegetation and landscape evolution of the R4D research site, Brooks Range foothills, Alaska. Holarctic Ecol 12: 238–261

Walker DA, Walker MD (1991) History and pattern of disturbance in Alaskan arctic terrestrial ecosystems: a hierarchical approach to analysing landscape change. J Appl Ecol 28: 244–276

Webber PJ (1978) Spatial and temporal variation of the vegetation and its productivity. In: Tieszen LL (ed) Vegetation and production ecology of an Alaskan arctic tundra. Ecological Studies Series 29. Springer, Berlin Heidelberg New York, pp 37–112

White JD, Running SW (1994) Testing scale dependent assumptions in regional ecosystem simulations. J Veg Sci 5: 687–702

Wielgolaski FE (1975) Principles in the use of wide-scale models on tundra data. In: Wielgolaski FE (ed) Fennoscandian tundra ecosystems. Part 2. Animals and systems analysis. Springer, Berlin Heidelberg New York, pp 245–250

Wiens JA, Crawford CS, Gosz JR (1985) Boundary dynamics: a conceptual framework for studying landscape ecosystems. Oikos 45: 421–427
Woodmansee RG (1988) Ecosystem processes and global change. In: Rosswall T, Woodmansee RG, Risser PG (eds) Scales and global change. Scientific Committee on Problems of the Environment, John Wiley and Sons, New York, pp 11–27
Woodmansee RG (1990) Biogeochemical cycles and ecological hierarchies. In: Zonneveld IS, Forman RT (eds) Changing landscapes: an ecological perspective. Springer, Berlin Heidelberg New York, pp 57–71

15 Modeling Dry Deposition of Dust Along the Dalton Highway

R. Lamprecht and W. Graber

15.1 Introduction

The Dalton Highway is a gravel road that was constructed in 1974 for transport of materials required in building the trans-Alaska pipeline. Presently, the road serves as a supply link between Fairbanks and the northern oil fields at Prudhoe Bay. The Dalton Highway is heavily traveled by large vehicles (we counted an average of 130 trucks per day during the summer of 1991). These vehicles generate large dust plumes in their wake (Fig. 15.1) particularly during periods of dry weather. This dust is deposited on the adjacent vegetation where it has detrimental effects on some arctic species (Everett 1980; Santelmann and Gorham 1988; Walker and Everett 1987; Auerbach 1992; see Chaps. 3, 9, 11, and 18, this Vol.).

In this chapter, we present a model to calculate dust dispersal patterns and the amount of dust deposited onto the vegetation along a specific stretch of the Dalton Highway. We compare our results with measured dust deposition loads near Toolik Lake (see Fig. 1.1, this Vol.). This model provides input information for the landscape model presented in Chapter 18 (this Vol.). The deposition algorithm is discussed in light of recent experimental findings on how particles cross the turbulent transition layer immediately above the canopy of plants. Specifically, the dispersal of heavy particles is simulated by a modified Lagrangian random walk model that incorporates a probabilistic scheme for calculating dry deposition and resuspension.

15.2 Model Fundamentals

The following is intended to provide a technical overview of the basic principles of stochastic air quality modeling for passive air constituents and to provide background for Section 15.6 on dry deposition. A more detailed treatment of random walk models for dispersion is given by de Baas (1988). In air pollution applications, random walk techniques are commonly used to model dispersion and deposition processes. The emitted gaseous material is represented by computer "particles" that mathematically consist of an ensemble of many different position vectors that are moving in the computational domain. These particles are moved by the average wind velocity and the random turbu-

J.F. Reynolds and J.D. Tenhunen (Eds.)
Ecological Studies, Vol. 120
© Springer-Verlag Berlin Heidelberg 1996

Fig. 15.1. Photograph showing the Dalton Highway in arctic Alaska and the release of several dust plumes by moving trucks. In the center of the picture the trans-Alaska pipeline crossing the Dalton Highway is visible. (Courtesy of K. Everett)

lent fluctuations of the wind. Hence, the components u_i of the particle velocity vector $\vec{u}$ can be split into two additive terms (see Table 15.1 for definitions of all symbols):

$$u_i = \langle u_i \rangle + \sum_i \Delta_{i,j} u_j' .$$

(1)

The first term represents the advection due to the average Eulerian flow conditions, which is either measured or obtained by a meteorological boundary layer model. The second term denotes the Lagrangian stochastic movement of a particle according to the intensity of turbulent fluctuations in the lower tropospheric boundary layer. The fluctuations u_j' of the wind components are given in a mobile, Lagrangian wind coordinate system in which the x-axis is parallel to the mean wind direction and the two other components are perpendicular to it. Because the average wind components are given in a fixed, Eulerian coordinate system in which the x-axis points towards the east and the y-axis towards the north, the components of the wind fluctuations have to be transformed by an orthogonal rotation $\Delta_{i,j}$ at each time integration step, Δt.

After each time step a new particle position vector $\vec{r}(t + \Delta t)$ is obtained by discrete integration of Eq. (1), i.e. by adding the incremental particle displacement vector $\vec{u}\Delta t$ to the position vector from the previous time step:

$$\vec{r}(t + \Delta t) = \vec{r}(t) + \vec{u}\Delta t.$$

(2)

Thus, the particle location at the initial time t_0 of an air pollution episode is identical to the release position of the emitted air constituents.

Table 15.1. List of mathematical symbols

Symbol	Name	Dimensions	Characteristics
a_d	Vertical drift correction term	$m\,s^{-2}$	
α_c	Conversion factor	$kg\,m^{-2}$	
c_p	Specific heat of air	$J\,kg^{-1}\,K^{-1}$	At constant pressure
$\Delta_{i,j}$	Orthogonal rotation		Matrix for coordinate transformation
$\Delta x \Delta y$	Area of resolution of the receptor grid	m^2	
Δt	Time integration step	s	$0.1 \cdot min\ (T_{Lj})$
$\Delta \tau$	Time interval of constant emissivity	s	
$erf(\gamma)$	Gaussian error function		
F	Deposition flux	$kg\,m^{-2}\,s^{-1}$	
f_c	Coriolis parameter	s^{-1}	$1.46 \cdot 10^{-4} sin(\Phi)$
g	Acceleration of gravity on earth	$m\,s^{-2}$	9.81
H	Sensible heat flux	$J\,m^{-2}\,s$	
h_{PBL}	Mixing height	m	
K_H	Eddy diffusivity of heat	$m^2\,s^{-1}$	
k	von Kármán constant		Dimensionless constant: 0.35–0.4
χ	Airborne concentration	$kg\,m^{-3}$	At a specific height (often 1 m above ground level)
L	Monin-Obukhov length	m	Measure denoting atmospheric stability
λ	Specific dimensionless constant		0.8
$m_{ij,v}$	Cumulative portion of the deposited mass	$kg\,m^{-2}$	For the receptor cell (i, j)
N_p	Total amount of particles released		
p_n	Probability of resuspension		After n reflections at the ground
$\Phi_H(\tfrac{z}{L})$	Dimensionless temperature gradient		Flux-profile relationships in the surface layer
$\Psi(\tfrac{z}{L})$	Dimensionless correction term		For the diabatic wind profile
Q	Source strength	$kg\,m^{-3}\,s^{-1}$	
$R_j(\Delta t)$	Lagrangian autocorrelation function		$exp(-\Delta t/T_{Lj})$
$R_p(\Delta t)$	Lagrangian autocorrelation function		For heavy particles
$\vec{r}(t)$	Particle position vector	m	
ρ	Density of air	$kg\,m^{-3}$	
σ_j	Intensity of turbulence	$m\,s^{-1}$	For the $j - th$ directional coordinate
σ_0	Turbulence at 0.2 m AGL	$m\,s^{-1}$	$\sigma_z(z = 0.2\,m)$
t_0	Starting time of releasing particles		
T_{Lj}	Lagrangian timescale	s	For the $j - th$ directional coordinate
T_{Lp}	Lagrangian timescale	s	For heavy particles
Θ	Potential air temperature	K	
u_*	Friction velocity	$m\,s^{-1}$	
u'_j	Wind velocity fluctuation	$m\,s^{-1}$	
$<\vec{u}>$	Average wind velocity	$m\,s^{-1}$	
v_d	Deposition velocity	$m\,s^{-1}$	$F \cdot \chi^{-1}$
v_g	Gravitational settling velocity	$m\,s^{-1}$	
w_d	Probability of dry deposition		
ξ_j	Gaussian white noise		
z_0	Roughness length	m	Measure of the earth's surface roughness
z	Vertical coordinate	m	Height above ground level

The computer model we created from this scheme is a single particle model, because only one particle is released at a time and follows along a trajectory according to Eq. (2). An ensemble of trajectories is built up by repeating this process under stationary advection conditions in order to improve the dispersion statistics and the subsequent deposition calculation. Because the random fluctuations of the air, which determine the displacements of the different particles, are uncorrelated, each particle is moved on a different trajectory. Perfect reflection of the particles is assumed at the top of the mixing height. At the ground, each incoming particle is reflected or deposited with prescribed respective probabilities.

The basic principle from which the velocity fluctuations u'_j [Eq. (1)] of the wind are derived is the Langevin equation, which is used in statistical mechanics to describe Brownian motion. The Langevin equation for the turbulent motion of a particle due to a retarding force and a random acceleration is

$$\frac{du'_j}{dt} = -\alpha u'_j + \beta \xi_j(t). \tag{3}$$

This is a stochastic differential equation where α and β are coefficients determined by the physical properties of the turbulent flow and ξ_j is Gaussian white noise.

The Langevin equation can be solved numerically with a Eulerian finite difference scheme with time step Δt. Hence, the three-dimensional Markov chain model for the components u'_j of the wind velocity fluctuation vector is obtained:

$$u'_j(t + \Delta t) = R_j u'_j(t) + \sqrt{1 - R_j^2}\, \sigma_j \xi_j + \delta_{jz} T_{Lj}(1 - R_j) a_d. \tag{4}$$

The subscript j of the different vectors involved in Eq. (4) refers to the three-directional components x, y, z, which are defined with respect to the Lagrangian wind coordinate system. ξ_j denotes random numbers representing Gaussian white noise with zero mean and unit variance. $R_j(\Delta t) = \exp(-\Delta t/T_{Lj})$ is the Langrangian autocorrelation function and σ_j the standard deviation of the wind fluctuations for the $j-th$ component, repectively. σ_j is a measure for the intensity of turbulence, and the Lagrangian time-scale T_{Lj} indicates, for the coordinate direction j, the correlation time during which an inertia-less airborne molecule remains suspended within a particular eddy. δ_{jz} denotes the Kronecker delta, i.e., $\delta_{jz} = 1$ if $j = z$, otherwise 0. In order to satisfy the conservation condition of atmospheric correlation between two subsequent model time steps, as indicated by the autocorrelation function $R_j(\Delta t)$, the size of the integration time step was controlled according to $\Delta t = 0.1 \cdot \min(T_{Lj})$ for $j = x, y$, z. A more detailed discussion concerning an estimation of the required parameters, $<u_i>$ [Eq. (1)], σ_j and T_{Lj} [Eq. (4)], necessary to run the random walk model is given in Section 15.4.

For turbulence, where σ_z varies with height, e.g., in crop canopies (Legg and Raupach 1982) or in non-neutral boundary layers, an extension of the

Langevin equation becomes necessary. The equation of motion for a fluid particle in inhomogeneous flows must include a drift correction term to avoid unrealistic particle accumulation in regions where σ_z is small. Different forms for a drift correction a_d [Eq. (4)] have been proposed (Legg and Raupach 1982; Sawford 1985). The model described here makes use of the drift correction term proposed by Sawford:

$$a_d = \frac{1}{2}\left(1 + \frac{u_z'^2}{\sigma_z^2}\right)\partial\sigma_z^2/\partial z. \tag{5}$$

Within the limits of weakly inhomogeneous turbulence, Sawford's correction is equivalent to the drift term introduced by Legg and Raupach, whereas Eq. (5) is still valid in arbitrarily strong inhomogeneous turbulence.

15.3 Modeling Heavy Particle Dispersion

For gaseous diffusion, the turbulence characteristics of fluid particles are assumed to be identical to those of the surrounding air. Heavy particles, such as dust, are transferred in the atmosphere by the same process of turbulent diffusion, except that a few minor modifications are necessary. Firstly, a heavy particle has greater inertia, which causes it to respond to accelerations more slowly than a fluid particle, so that the heavy particle cannot follow exactly the directional changes of fluid elements ("inertia effect", Csanady 1963). Secondly, due to the influence of gravity, heavy particles have mean downward velocities that contribute to an overall faster depletion of a dust plume. As a consequence, a heavy particle does not remain within a particular turbulent eddy, but rather continuously drops out of the influence of the surrounding fluid. The result is that a heavy particle loses correlation more rapidly than a passive tracer, which is known as the "crossing trajectory effect" (Yudine 1959).

Csanady (1963) demonstrated that for particles up to 400–500 µm in diameter, the effect of inertia is negligible. However, the effect of the crossing trajectories has to be considered for heavy particle dispersion. Hashem and Parkin (1991) derived an expression for the vertical component of the Lagrangian time scale between a marked parcel of air and a heavy particle:

$$T_{Lp} = \frac{\sigma_z}{\lambda(\sigma_z + \upsilon_g)}T_{Lz}; \tag{6}$$

and thus yields for the auto-correlation function:

$$R_p(\Delta t) = \exp\left\{-\frac{\Delta t}{T_{Lz}}\lambda\left(1 + \frac{\upsilon_g}{\sigma_z}\right)\right\}, \tag{7}$$

where v_g stands for the gravitational settling velocity and λ is a specific constant. Revising Eq. (4) with regard to these corrections, and furthermore, adding the term $\delta_{jz} \cdot v_g$ for the settling velocity, Hashem reports good agreement between Lagrangian model simulations and dispersion measurements of Snyder and Lumeley (1971) for heavy particles of different sizes.

The most important factors influencing dispersion and dry deposition of particles can be categorized according to characteristics of the atmosphere (Sect. 15.4), properties of the depositing species (Sect. 15.5) and the nature of the surface (Sect. 15.6).

15.4 Estimation of Atmospheric Boundary Layer Parameters

Knowledge of the mean windfield is of primary importance for calculating the transport and dispersion of dust. This information, together with shielded air temperature measurements, was obtained from a meteorological mast operated by the Arctic Long Term Ecological Research site (Chap. 2, this Vol.). The mast is located in the vicinity of the University of Alaska field station near Toolik Lake, approximately 2 km from the site we selected for the dust deposition calculations. This site is slightly slanted along the slope of a small hill where the terrain is relatively flat, so that detailed topographical features could be neglected for the deposition calculation. The wind data obtained at the mast is assumed to be representative for the dust site. In particular, the wind patterns recorded during the summer of 1991 (Fig. 15.2) show a daily cycle that can be explained by the anabatic and katabatic winds of a nearby valley that leads to the Atigun Pass in the Brooks Range.

Unfortunately, the wind and temperature data are measured only at two height levels and relatively close to the ground, i.e., 1 and 5 m. No other wind information representative for this area was available. Another meteorological mast of the University of Alaska, located about 10 km northeast of Toolik Lake, provided additional air temperature data at 3 and 10 m above the ground. Air temperatures measured over relatively flat terrain, such as the North Slope of Alaska, can be assumed to be horizontally homogeneous. In order to satisfy the requirement that the air temperature measurements originate from the atmospheric surface layer, and not from a transition layer below, the readings from this second mast were used.

Using this wind and temperature data, all atmospheric surface layer parameters needed to run the dust model were estimated. For that purpose an iteration was performed by fitting the diabatic wind profile equation, which incorporates the integrated empirical flux-profile relationships of Businger et al. (1971) and Dyer (1974), to the measured wind data, starting the iteration with an arbitrary value for L:

$$u(z) = \frac{u_*}{k}\left\{\ln\left(\frac{z}{z_0}\right) + \Psi\left(\frac{z}{L}\right)\right\}, \tag{8}$$

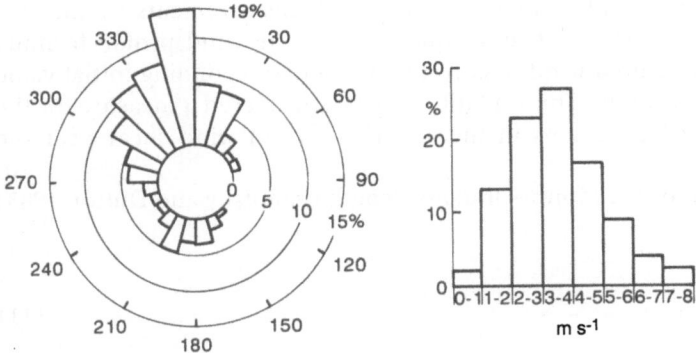

daytime: 0600-1900 hr.

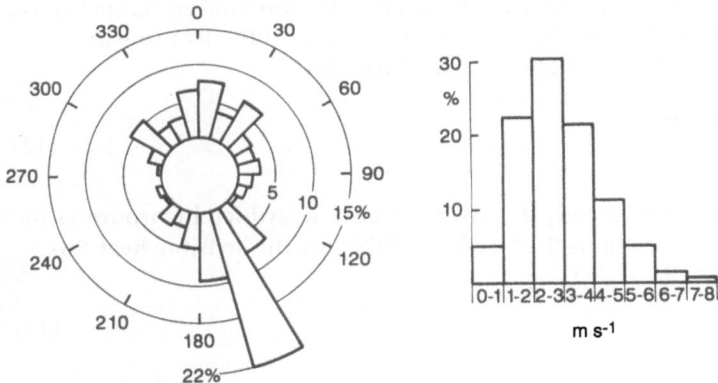

nighttime: 1900-0600 hr.

Fig. 15.2. Diurnal frequency of wind direction (*left*) and velocity (*right*) during the summer of 1991 between 15 June and 19 August

with the correction terms for non-neutral cases

$$\Psi\left(\frac{z}{L}\right) = \begin{cases} -2\ln\left(\frac{x+1}{2}\right) - \ln\left(\frac{x^2+1}{2}\right) + 2\arctan(x) - \frac{\pi}{2} : L < 0 \text{ unstable} \\ 4.7\frac{z}{L} \qquad\qquad\qquad\qquad\qquad\qquad\quad : L > 0 \text{ stable} \end{cases} \tag{9}$$

and

$$x = \left(1 - 15\frac{z}{L}\right)^{1/4} . \tag{10}$$

The Monin-Obukhov length, L, may be considered the height above ground where the production of turbulence by mechanical forces equals the production of turbulence by buoyant forces. L is positive in a stable atmosphere, negative in an unstable atmosphere, and approaches $\pm\infty$ when the atmosphere is neutral. The mean roughness length z_0 was determined to be 0.01 m using the adiabatic wind profile and the wind speed measurements at the two elevations z_1 and z_2. The first least-squares fit of the wind profile formula [Eq. (8)] to the measured wind data resulted in a corresponding initial value for the friction velocity u_*. This quantity may be considered a measure for the turbulent velocity fluctuations in the air arising from wind shear near the ground.

The equation for the Monin-Obukhov length (Panofsky and Dutton 1983) is

$$L = \frac{-u_*^3 \Theta \rho c_p}{kgH},\tag{11}$$

where ρ denotes the density of air, c_p is the specific heat of air at constant pressure, k is the von Kármán constant, taken as 0.4, g is the acceleration of gravity, H is the sensible heat flux and Θ denotes the potential air temperature, taken at a height of 6.5 m, as the average value between the two sensors.

With the flux-gradient relation for heat transfer

$$H = -K_H \frac{\partial(\rho c_p \Theta)}{\partial z}\tag{12}$$

and the expression for the eddy diffusivity of heat, K_H, which incorporates the stability dependent functions Φ_H (see Dyer 1974) for the sensible heat flux

$$K_H = \frac{ku_*z}{\Phi_H}\tag{13}$$

$$\Phi_H = \begin{cases} \left(1 - 16\frac{z}{L}\right)^{-1/2} & : L < 0 \\ 1 + 5\frac{z}{L} & : L > 0 \end{cases}\tag{14}$$

one obtains two quadratic equations for $L > 0$ and $L < 0$, respectively, as functions of $\Theta, \frac{\partial \Theta}{\partial z}$ and u_*. The lapse rate $\frac{\partial \Theta}{\partial z}$ can be derived from the measured air temperature data. Hence, with the initial friction velocity u^* derived from Eq. (8), a refined second estimation of the Monin-Obukhov length L can be calculated according to the above-mentioned formulas, producing yet another value for u_* by recalculating Eq. (8), and so forth, until convergence for all unknown parameters – the Monin-Obukhov length, the friction velocity, as well as wind direction and wind velocity $<\vec{u}>$ (Eq. (1)) in the atmospheric surface layer – is achieved.

With estimates of these parameters, both the turbulence intensities (σ_j) and the intrinsic Lagrangian time scales (T_{Lj}) of the atmosphere can then be determined using the parameterization scheme proposed by Hanna (1982). These two parameters are most influential for the simulation of random turbulent diffusion of airborne material as described by the random walk process of Eq. (4).

The vertical profiles thus applied for wind and turbulence in the atmospheric surface layer are probably not valid in the transition layer immediately above the canopy (Garratt 1980), but errors due to this inadequacy appear to be negligible for this study. Sensitivity tests of the model to different boundary conditions due to variations of the σ_j and T_{Lj} values exhibited no remarkable differences in the results of the deposition calculations. In view of the general uncertainty regarding the magnitude of the various micrometeorological quantities below the canopy top, we assumed that all σ_j and T_{Lj} parameters were constant within the canopy layer ($z \leqslant 0.2\,\text{m}$). A review of turbulence profiles in and above plant canopies has been published by Raupach (1988).

The top of the mixing layer, which is assumed to be impenetrable for any dust particle reaching the upper boundary of the model domain, was estimated according to the following relationship of Zilitinkevich (1972), valid for the stable boundary layer:

$$h_{PBL} \sim 0.4\sqrt{\frac{u_*L}{f_c}}, \tag{15}$$

where f_c denotes the Coriolis parameter. In particular, ground-based inversions (as observed by K. Everett, pers. comm.) over the study area, which can appear under very stable atmospheric conditions in combination with stagnant air, have been taken into account by the model. Combinations of small values both for $L > 0$ and u_*, which yielded mixing layer depths smaller than 30 m according to Eq. (15), occurred on two different days during the study period in the summer of 1991. Conversely, for the cases of unstably stratified atmospheres, a general mixing layer depth of 1000 m was assumed.

15.5 Dust Characterization and Mass Transfer Through the Atmosphere

Gravitational settling is the major mechanism affecting the atmospheric residence time and dry deposition of coarse particles with diameters greater than about 10 μm. The settling velocity v_g is a function of size, density and shape of a particle, and is negligible for small particles when compared with the turbulent vertical velocities of the surrounding fluid. Thus, particle size distribution is of primary importance for an accurate computation of transport and dry deposition of heavy particles.

At the selected study site two transects oriented towards the prevailing wind direction were established to determine the particle size spectra by a dust collection experiment. Along each transect, we distributed 40 dust samplers at five different distances (10, 30, 70, 150 and 300 m) on either side of the road by means of a compass and measuring tapes. At each collector site two dish samplers were installed, mounted about 20 cm above the ground, i.e., slightly

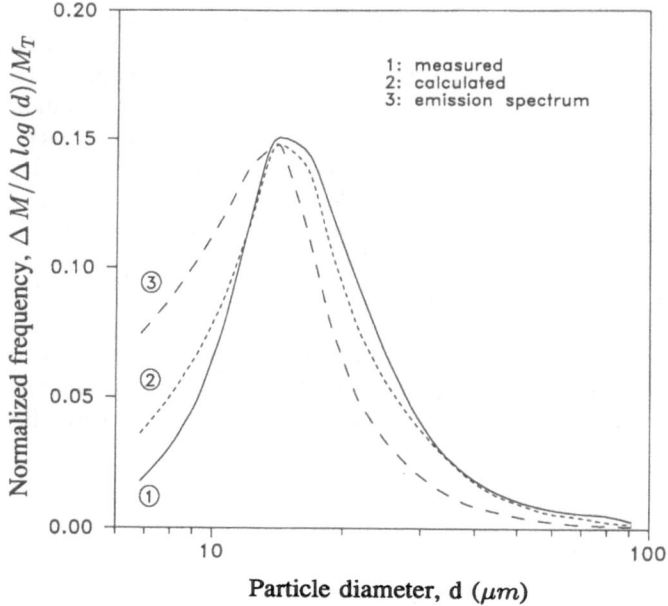

Fig. 15.3. Normalized number spectrum of dust particles as derived from measured (*1*) and computed (*2*) values at the different collector locations. *3* is the corresponding density function of the emitted particle sizes.

above the plant canopy. The period of exposure was approximately 24 h. To prevent blow off the dust samplers were covered with a double-sided self adhesive foil on which the particles were trapped. The collected dust grains were counted by means of an image processing system with regard to logarithmically equal size classes with diameter limits d_i and d_f of each grade. A description of the dust sampling and particle counting technique is given in the VDI guideline 2119 (1991). For each sampling distance, 16 independent particle size distribution measurements were averaged. After combining the results from the two dust transects and considering normalization, the log-normal size distribution pattern was obtained (curve 1, Fig. 15.3).

This size distribution pattern of the dust particles is of the same general nature as the size distribution of wind-blown dust reported by Chepil (1957). The only distinguishing feature between our distribution curve and the one observed by Chepil is in the position of the predominant diameter of the dust particles on the logarithmic scale. Better conformity of our results exists with the particle size spectra measured by Noll and Fang (1989). These authors also report typical mass peaks for coarse dust particles between 10 and 30 μm diameter depending on wind speed. Higher wind speeds, i.e., increasing friction velocities, cause the peak of the predominant diameter of the settled dust particles to shift gradually towards larger values.

The chemical composition of the road's topping material was analyzed by XRD with a Siemens D500 X-ray diffractometer (Siemens, Erlangen, Germany). The coarser fraction of the gravel material was analyzed separately revealing, after the material had been pulverized, the chemical compound $CaCO_3$, which could not be found in the finer dust fraction. Regarding the other components, the two samples appeared to be identical, i.e., quartz (SiO_2), albit ($NaAlSi_3O_8$) and a chlorite compound were analyzed. The mean density of these compounds was $2.65\,g\,cm^{-3}$.

Soil particles and other airborne materials are seldom spherical. Such particles are characterized by their Stokes' diameter, which is the diameter of a sphere with the same density and sedimentation velocity. Large particles that fall under gravity attain a steady sedimentation velocity as soon as the drag force balances the gravitational force. The drag force becomes important with increasing relative motion between the air and the particle. The drag coefficient of a sphere is a function of the Reynolds number, and increases nonlinearly with increasing particle diameter (Fuchs 1964). For intermediate-sized particles the Stokes' formula is applicable, and for particle sizes in the aerosol range, Fuchs (1964) gives analogous expressions for the gravitational settling velocities v_g.

15.6 Theory of Particle Dry Deposition into Vegetation

Dry deposition of gases and particles is complex and affected by a multiplicity of factors. Therefore, it is not surprising that large inconsistencies have been reported in the literature regarding, in particular, the range of published deposition velocities v_d (e.g., Sehmel 1980; Davidson and Yee-Lin 1990). The dry deposition velocity is defined for both gases and particles as the deposition flux F divided by an airborne concentration χ at a certain reference height. Therefore, v_d is the key factor that determines how fast the atmosphere is freed of airborne material by dry deposition.

Transport across the laminar sublayer to the receptor surface can occur by Brownian diffusion, interception, impaction or sedimentation (Chamberlain and Little 1981). In wind tunnel experiments with simple surfaces, several investigators (Chamberlain 1966; Sehmel and Hodgson 1978) demonstrated that particle dry deposition due to these mechanisms depends primarily on particle size, surface characteristics, and air flow conditions above the receptor surface. Interception and impaction become important for diameters greater than about $0.1-1\,\mu m$. Sedimentation is significant for horizontally oriented surfaces and for particles with diameters greater than about $1-10\,\mu m$ (Davidson and Yee-Lin 1990). Figure 15.4 illustrates deposition velocities as a function of particle size. The different regions in the deposition curve demonstrate the relative importance of the different mechanisms in particle dry deposition. The deposition velocity is always greater than – or in the case of coarse particles, equal to – the gravitational settling velocity.

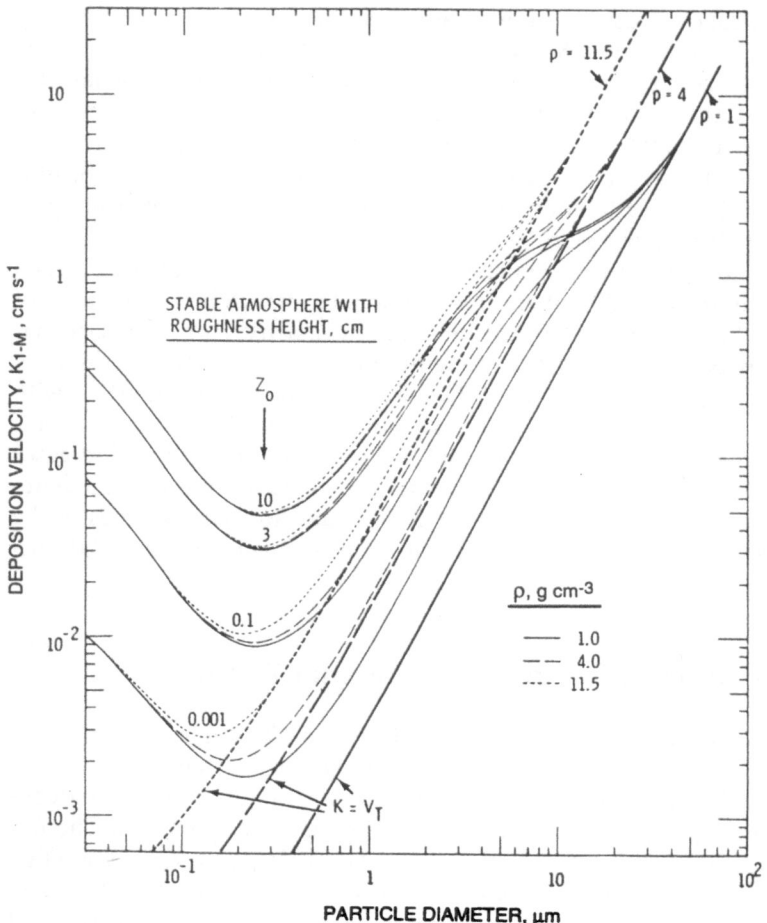

Fig. 15.4. Predicted deposition velocities at 1 m above ground for $u_* = 50 \, \text{cm s}^{-1}$ and different particle densities. V_T denotes terminal settling velocity. (After Sehmel 1980)

Lagrangian particle trajectory models can be easily extended to simulate the particulate mass transfer through the atmosphere and the subsequent dry deposition on the ground. When a particle hits the ground, irreversible deposition can be assumed or, physically more appropriate, a probabilistic method can be used that accounts for the phenomenon of particle resuspension (Zannetti and Al-Madani 1983; Axenfeld et al. 1984; Nicholson 1988; Noll and Fang 1989; Underwood 1991).

This study makes use of the probabilistic approach for dry deposition and resuspension proposed by Axenfeld et al. (1984). According to this method, a relationship between the probability of deposition w_d and the deposition velocity v_d, as a function of particle size, is given. Moreover, it is evident that the

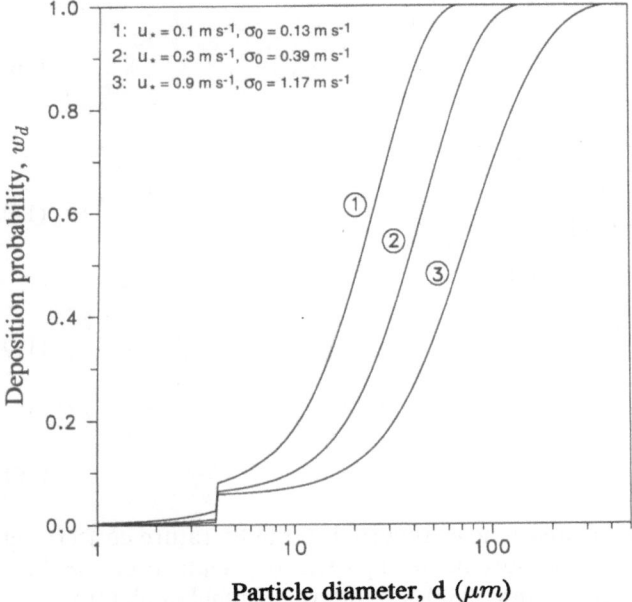

Fig. 15.5. Deposition probability as a function of particle diameter and friction velocity according to Axenfeld et al. (1984)

particle deposition probability must also be related to the magnitude of the turbulent fluctuations in the roughness layer near the ground: $\sigma_0 = \sigma_z(z \simeq 0.2\,\text{m})$. Larger particles that encounter these turbulent eddies are provided with sufficient inertia to cross this sublayer and deposit. Therefore, the increasing settling velocity v_g of larger particles favors progressively higher deposition probabilities as shown in Fig. 15.5. Smaller particles, however, will partly follow the air eddies promoting their ability to remain suspended in the atmosphere. With increasing wind speed, and thus higher turbulence intensity near the ground (higher u_* and σ_0 values), the deposition probability for a particle of any size drops (Fig. 15.5). This is consistent with the experimental findings of Noll and Fang (1989) who observed that higher wind velocities cause the peak of the predominant diameter of the settled dust particles to shift gradually towards larger values. Because air eddies are also assumed to be equally responsible for the resuspension of already-deposited particles, the expression $1 - w_d$ is regarded to be the probability of resuspension.

Following Axenfeld et al. (1984), the Maxwell-Boltzmann velocity distribution for molecules in a turbulent flow, shifted by the corresponding settling velocity for a heavy particle near the ground, constitutes the starting point for the derivation of a relationship for the particle dry deposition probability. The following formalism for the deposition probability can thus be derived:

$$w_d = \frac{\sqrt{2\pi}\,\frac{v_d}{\sigma_0}}{F_g + \sqrt{\frac{\pi}{2}\,\frac{v_d}{\sigma_0}}}, \tag{16}$$

where

$$F_g = \sqrt{\frac{\pi}{2}}\,\frac{v_g}{\sigma_0} + \frac{exp\!\left(-\gamma^2\right)}{1 + erf\!\left(\gamma\right)} \tag{17}$$

with the Gaussian error function:

$$erf\!\left(\gamma\right) = \frac{2}{\sqrt{\pi}} \int_0^\gamma e^{-x^2}\,dx \tag{18}$$

and

$$\gamma = \frac{v_g}{\sqrt{2}\,\sigma_0} \tag{19}$$

In view of the large inconsistencies reported in the literature concerning both model predictions and measurements of particle dry deposition velocities (e.g., Marggrander and Flothmann 1982), we follow Axenfeld et al. (1984) for an approximation of the v_d values most influential for smaller particles:

$$v_d = \begin{cases} v_g + 0.001u_* : d < 4\mu m \\ v_g + 0.03u_* \;\;: d \geqslant 4\mu m \end{cases} \tag{20}$$

Consistent with the experimental findings of Chamberlain (1966), larger flow velocities cause particle inertia to rise which, according to Eq. (20), favors higher deposition velocities.

Any particle reaching the lower boundary immediately loses some fraction of its probability of existence (p) that was originally set to one at release time. Equivalently, a particle crossing the canopy layer loses a fraction of its mass, i.e., it is irreversibly deposited only to a certain extent, which is determined by w_d. At any deposition occurrence, the actual position of a particle must be related to the corresponding receptor area of the counting grid, and the signs of all vertical velocity components of the particle must be reversed in order to achieve backscattering, i.e., resuspension of the particle.

After $(n-1)$ reflections at ground level, a particle still retains the probability of survival

$$p_n = p_{n-1}(1 - w_d) \tag{21}$$

or equivalently

$$p_n = (1 - w_d)^n \tag{22}$$

Hence, one finally obtains for the cumulative portion of the deposited mass $m_{ij,v}$ after $v \leqslant n$ reflections of a particle in a specific receptor cell (i, j)

$$m_{ij,v} = m_{ij,v-1} + \alpha_c p_{n-1} w_d \tag{23}$$

$$\alpha_c = \frac{Q\Delta\tau}{N_p \Delta x \Delta y}, \tag{24}$$

where α_c is a conversion factor to $kg\ m^{-2}$. Q denotes the emission strength in $kg\ s^{-1}$ per particle and $\Delta\tau$ is the time interval of constant emissivity corresponding to the time resolution of an hour of the available input data for wind, temperature, precipitation and truck frequency. Precipitation and its frequency aids in dust suppression. Hence, we assumed that no dust release takes place during any period of precipitation. The truck frequency is a function of the time of day and weekday. This information was kindly supplied to us by the Alaskan Department of Transportation, which maintains several truck counters along the Dalton Highway. Finally, N_p denotes the amount of particles released, and $\Delta x\ \Delta y$ corresponds to the area of resolution of the receptor grid, i.e., $40\,m \times 40\,m$ for this study.

Dust emission due to road traffic is simulated in the model as a series of subsequent releases of particles with different sizes along the whole length of the considered road section. The size, and thus mass assigned to each emitted particle, was generated by a specific transformation of uniformly distributed random numbers, according to Press et al. (1989), to yield the chosen density function of the emitted particle sizes. Diameters of particles, e.g., smaller than 14 μm, are taken from uniform deviates which, logarithmically displayed and normalized, yield the left slope of the distribution curve shown in curve 3 in Fig. 15.3. The formalism for the determination of the particle imission size spectra is analogous to the one for the computation of the deposition portion in a specific receptor mesh according to Eq. (23).

The calculated amount of dust entrained into the vegetation during a period of 65 days was obtained from an hourly repetition and integration of the above processes with the consideration of the many different conditions of particle transport, diffusion and sedimentation through the turbulent surface layer of the atmosphere. In order to obtain a wide range of statistically significant deposition estimates, more than 10^5 trajectories were computed.

15.7 Numerical Results

An accurate computation of the amount of airborne dust subjected to dry deposition due to gravitational settling requires the knowledge of the particle size distribution at the source site. Because measurements were carried out only at collector sites apart from the road, the particle size distribution at the emission site had to be approximately inferred from these measurements, as well as from the deposition theory explained in the previous section, and complemented by an iterative optimization procedure.

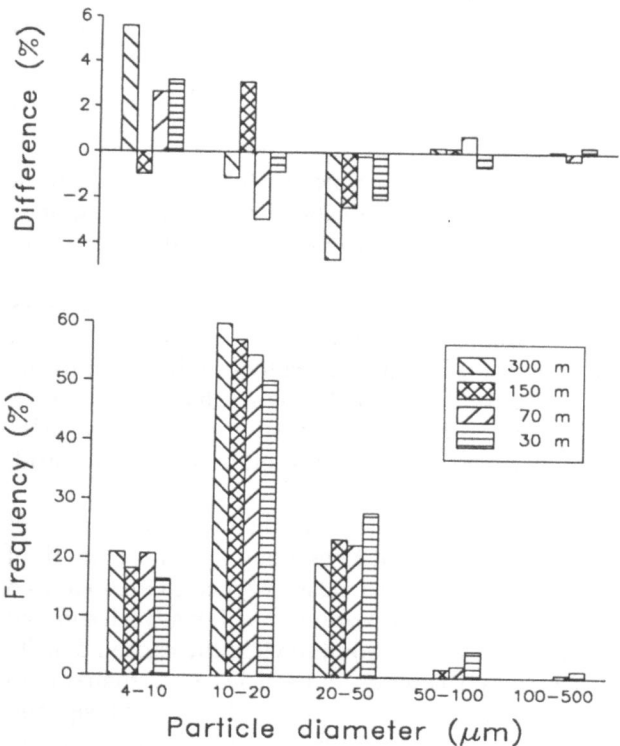

Fig. 15.6. Predicted particle size spectra at different distances from the road (*bottom*) and differences between measured and calculated spectra (*top*)

At the beginning of this study we assumed identical particle size spectra at both the emission and collector locations, i.e., all emitted particles of any size class settled out entirely within the area where dust samples were collected. However, smaller particles that constitute the bulk of suspended dust are known to travel long distances (Patterson et al. 1976), and particles <2 μm may even become part of the stratospheric dust load (Prospero and Ness 1977). The discrepancy we found between measurements and calculations for intermediate-sized and small particles has been used to obtain the desired particle size spectrum at the emission site as displayed in curve 3 in Fig. 15.3. Coarse gravel particles, which are largely suspended by saltation and settle out within a few metres from the road, were neglected. Curve 2 in Fig. 15.3 shows the particle size distribution at the collector sites obtained from the emission spectrum.

A comparison between calculated and measured particle size distributions as a function of different distances from the road is feasible, because only the measured particle size spectrum, which had been evaluated independently of distance (curve 1, Fig. 15.3), was used to determine the particle size distribu-

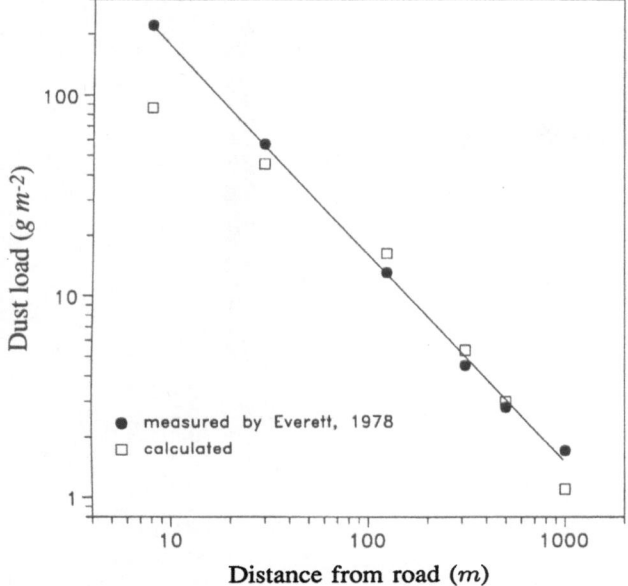

Fig. 15.7. Summer dust loads as measured by Everett (1980) over 96 collection days in comparison with predicted values

tion at the emission site. The computed particle size spectra, as a function of several distances from the road, is shown in Fig. 15.6. The progressive change in particle size, mainly for the sand- and silt-sized material, for the first 300 m from the road, is very obvious. Additionally, the difference spectrum as obtained from the measured minus calculated values is depicted. In view of the various uncertainties concerning both predicted and measured data, the difference spectrum shows acceptable deviations.

A calculation of the total dust load entrained into the vegetation requires knowledge of the average amount of dust thrown into suspension by truck movement. This data was obtained by simulating the dust deposition experiment performed by Everett in 1978 near Toolik (Everett 1980; Walker and Everett 1987). All information, except atmospheric stability, was provided by K. Everett. Hence, the atmosphere was assumed to be moderately stable at night and slightly unstable during daytime. The results are shown in Fig. 15.7. The emission strength due to truck movement was derived by parallel shifting of the computed dust distribution curve until optimal consistency with the measured data was achieved. The model underestimates the amount of dust deposited within the first few metres from the edge of the road (Fig. 15.7). This is partly due to the insufficient resolution (20 m × 20 m for this calibration purpose) of the model receptor grid for calculations in the nearest-field region of the road where deposition loads decrease exponentially with distance. Additionally, the coarsest particles in the emission size spectrum, which largely

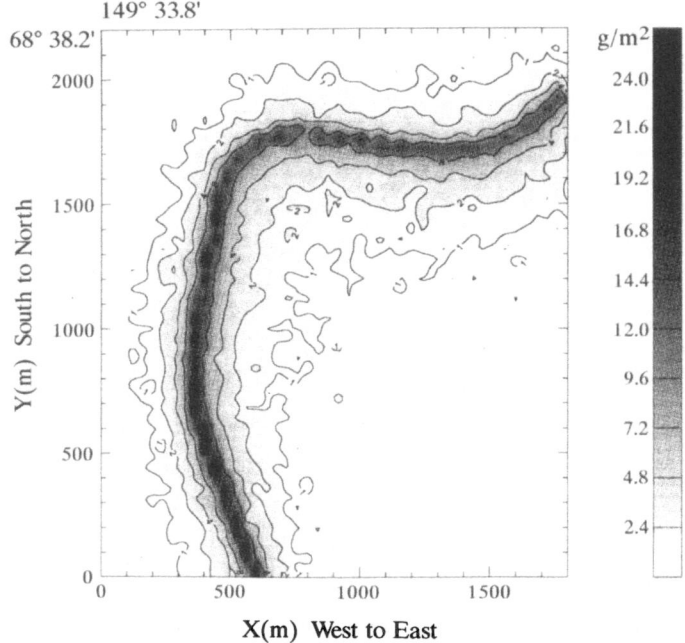

Fig. 15.8. Dust distribution pattern along the Dalton Highway near Toolik Lake. Dust loads are denoted in $g\,m^{-2}$

contribute to the amount of dust deposited within the first 10 m of the road were not considered.

Figure 15.8 shows the calculated dust deposition pattern as obtained using the wind statistics depicted in Fig. 15.2. The higher dust loads in greater distances south of the road are due to comparatively higher wind velocities and wind frequencies from the corresponding wind directions during daytime. Despite the relatively rare occurrences of easterly and westerly winds, the dust loads on either side, east and west, of the road are still remarkably high.

A more detailed analysis of the numerical results revealed that 20–45% of the dust had settled out within the first 40 m of the road; by 200 m, 65–95% had dropped; and by 400 m, 75–98% had been deposited. The higher percentage values were obtained by analyzing the deposition loads along a transect pointing from east to west, taken at a constant y-value of 600 m. Alternatively, the smaller percentage values can be assigned to a vertical transect leading from north to south at a constant x-value of 1500 m.

Walker and Everett (1987) reported that 97% of the total dust load had dropped within the first 125 m from the road, regardless of the specific site. The discrepancy between their observations and the model results we obtained, mainly for the vertical transect, might be due to the following reasons. Firstly, the pronounced dipolar wind regime selected for the computations, which additionally combines enhanced wind velocities right along approximately the same direction from north to south, may carry the suspended dust longer distances away from the road than usually observed at any other location. Secondly, because the coarsest particles have been omitted in the emission size spectrum, the relative portion of dust that has dropped in the near-field region of the road has correspondingly diminished.

15.8 Conclusions

We present a technique for modeling heavy particle dispersion by a random walk model that incorporates a probabilistic approach for calculating dry deposition and resuspension phenomena. Despite current uncertainties in predicting dry deposition, our results show that the processes governing particulate transfer into plant canopies can, in fact, be modeled. Our method could potentially be applied to model dispersal and dry deposition of any sort of particles, i.e., spores, droplets, heavy metal particles, or radioactive compounds. Compared with Eulerian grid models for mesoscale air pollution applications, the Lagrangian approach is physically more convincing and easier to handle. This is particularly true for transport and deposition simulations of heavy particles, which have additional size dependent settling velocities and resuspension capabilities. It is thus likely that random walk theories will continue to be important in understanding and simulating particulate transfer into canopies.

This model has practical applications for ecological studies of disturbance effects in arctic Alaska. Based on our simulations, we conclude that at any location along the Dalton Highway, road dust emitted by truck movement should settle out to 98% within an area of less than 500 m of either side of the road. Such information is essential in order to develop a predictive ability to evaluate the effects of road construction on plants and ecosystems as attempted by Leadley et al. (Chap. 18, this Vol.).

Acknowledgments. The authors are grateful to colleagues from Paul Scherrer Institute who supported the XRD measurements or provided plot software. K. Everett kindly provided data from his dust experiments. Moreover, we would like to thank the editors of this volume who gave helpful comments on a draft of this chapter, and for the inspiration to carry out this work.

References

Auerbach NA (1992) Effects of road and dust disturbance in minerotrophic and acidic tundra ecosystems, northern Alaska. MBS Thesis, Univ Colorado, Boulder
Axenfeld F, Janicke L, Münch J (1984) Entwicklung eines Modells zur Berechnung des Staubniederschlages. Forschungsber 104 02 562, Dtsch Umweltbundesamtes, Berlin

Businger JA, Wyngaard JC, Izumi Y, Bradley EF (1971) Flux-profile relationships in the atmospheric surface layer. J Atmos Sci 28: 181–189

Chepil WS (1957) Sedimentary characteristics of dust storms. III. Composition of suspended dust. Am J Sci 255: 206–213

Csanady GT (1963) Turbulent diffusion of heavy particles in the atmosphere. J Atmos Sci 20: 201–208

Chamberlain AC (1966) Transport of *Lycopodium* spores and other small particles to rough surfaces. Proc R Soc Ser A 296: 45–70

Chamberlain AC, Little P (1981) Transport and capture of particles by vegetation. In: Grace J, Ford ED, Jarvis PG (eds) Plants and their atmospheric environment. Blackwell, London

Davidson CI, Yee-Lin W (1990) Dry deposition of particles and vapors. In: Lindberg SE, Page AL, Norton SA (eds) Acid precipitation, vol 3. Adv Environ Sci, Springer, Berlin Heidelberg New York, pp 103–216

De Baas AF (1988) Some properties of the Langevin model for dispersion. Thesis, Risø Natl Lab, DK-4000 Roskilde, Denmark

Dyer AJ (1974) A review of flux-profile relations. Boundary-Layer Meteorol 1: 363–372

Everett KR (1980) Distribution and properties of road dust along the northern portion of the Haul Road, US Army Cold Region Res Eng Lab, Rep 80-19, pp 101–128

Fuchs NA (1964) The mechanics of aerosols. MacMillan, New York

Garratt JR (1980) Surface influence upon vertical profiles in the atmospheric near-surface layer. Quart J R Met Soc 106: 803–819

Hanna SR (1982) Application in air pollution modeling. In: Nieuwstadt FTM, van Dop H (eds) Atmospheric turbulence and air pollution modeling. Reidel, Dordrecht

Hashem A, Parkin CS (1991) A simplified heavy particle random-walk model for the prediction of drift from agricultural sprays. Atmos Environ 25A(8): 1609–1614

Legg BJ, Raupach MR (1982) Markov-chain simulation of particle dispersion in inhomogeneous flows: the mean drift velocity induced by a gradient in Eulerian velocity variance. Boundary-Layer Meteorol 24: 3–13

Marggrander E, Flothmann D (1982) Dry deposition of particles: comparison of published experimental results with model predictions. In: Georgii HW, Pankrath J (eds) Deposition of atmospheric pollutants. Reidel, Dordrecht

Nicholson KW (1988) A review of particle resuspension. Atmos Environ 22: 2639–51

Noll KE, Fang KYP (1989) Development of a dry deposition model for atmospheric coarse particles. Atmos Environ 23(3): 585–594

Panofsky HA, Dutton JA (1983) Atmospheric turbulence. Models and methods for engineering applications. Wiley, New York

Patterson EM, Gillette DA, Grams GW (1976) The relation between visibility and the size–number distribution of airborne soil particles. J Appl Meteorol 15: 470–478

Press WH, Flannery BP, Teukolsky SA, Vetterling WT (1989) Numerical recipes. The art of scientific computing. Cambridge Univ Press, Cambridge

Prospero JM, Ness RT (1977) Dust concentration in the atmosphere of the equatorial North Atlantic: possible relationships to the Sahelian drought. Science 196: 1196–1198

Raupach MR (1988) Canopy transport processes. In: Steffen WL, Denmead OT (eds) Flow transport in the natural environment. Advances and applications. Springer, Berlin Heidelberg, New York

Santelmann MV, Gorham E (1988) The influence of airborne road dust on the chemistry of *Sphagnum* mosses. J Ecol 76: 1219–1231

Sawford BL (1985) Lagrangian statistical simulation of concentration mean and fluctuation fields. J Climate Appl Meteorol 24: 1152–1166

Sehmel GA (1980) Particle and gas dry deposition: a review. Atmos Environ 14: 983–1011

Sehmel GA, Hodgson WH (1978) A model for predicting dry deposition of particles and gases to environmental surfaces. PNL-SA-6721 Battelle, Pacific North-West Lab, Richland, WA

Snyder HW, Lumeley JL (1971) Some measurements of particle velocity autocorrelation functions in a turbulent flow. J Fluid Mech 48: 41–71

Underwood BY (1991) Deposition velocity and the collision model of atmospheric dispersion. Atmos Environ 25A(12): 2749–2759

VDI Richtlinie 2119 (1991) Grössendifferenzierende Bestimmung der Partikeldepositionsrate mittels der Haftfolienmethode. Verein Deutscher Ingenieure, VDI Verlag GmbH, Düsseldorf

Walker DA, Everett KR (1987) Road dust and its environmental impact on Alaskan Taiga and Tundra. Arct Alp Res 19(4): 479–489

Yudine MI (1959) Physical considerations on heavy particle diffusion. Proc Int Symp Diffusion and air pollution. Adv Geophys 6: 185–191

Zilitinkevich SS (1972) On the determination of the height of the Ekman boundary layer. Boundary-Layer Meteorol 3: 141–145

Zannetti P, Al-Madani N (1983) Simulation of transformation, buoyancy and removal processes by Lagrangian particle methods. Proc 14th Int Tech Meet on Air pollution modeling and its application, Copenhagen, Denmark. Plenum Press, New York, pp 733–744

16 Modeling Decomposition in Arctic Ecosystems

D. L. MOORHEAD and J. F. REYNOLDS

16.1 Introduction

Shifting climatic regimes (Maxwell 1992) and localized impacts of human activities (Chap. 4, this Vol.) are likely to have substantial effects on Arctic ecosystems (Chapin et al. 1992; Chapin and Körner 1995; Oechel, in prep.). Numerous speculations and predictions have been made of the potential response of arctic plants and ecosystems to changes in temperature, season length, moisture availability, and CO_2 concentration (reviewed in Chapin et al. 1992; Chapin and Körner 1995). Nevertheless, plant responses in these ecosystems are generally constrained by low nutrient availability (Shaver and Chapin 1986; Chap. 11, this Vol.), and environmental changes may have little impact on plant productivity unless average nutrient availability also changes (Leadley and Reynolds 1992; Reynolds and Leadley 1992). Thus, it is essential to predict how environmental changes alter decomposition and nutrient availability. In this chapter, we review simulation models for describing decomposition in arctic soils and consider potential effects of climate change, elevated CO_2, and road dust deposition on decomposition and nutrient dynamics in tundra ecosystems.

16.2 Controls on Decomposition

Nadelhoffer et al. (1992), Schimel et al. (Chap. 10, this Vol.), and Oberbauer et al. (Chap. 11, this Vol.) review many of the factors affecting decomposition and nutrient cycling in arctic soils. Arctic ecosystems are nutrient-poor and external input is low, so nutrient release from decomposing organic matter is important in seasonal dynamics (Chapin et al. 1980). However, decay rates of organic matter are typically very low (Giblin et al. 1991; Nadelhoffer et al. 1991, 1992), resulting in the accumulation of peat. Tundra in the Imnavait Creek watershed typically has a 10 to 40 cm layer of organic matter overlying the mineral soil (Walker and Lederer 1987; Chaps. 4 and 5, this Vol.). Physical attributes of this organic mat, i.e., high moisture holding capacity and low thermal conductivity, result in cold, wet conditions that inhibit decomposition.

J.F. Reynolds and J.D. Tenhunen (Eds.)
Ecological Studies, Vol. 120
© Springer-Verlag Berlin Heidelberg 1996

Table 16.1. General characteristics of 28 arctic decomposition models

Type of model	Site description and location	Focus and approach	Reference
Conceptual	Alpine tundra, Hardangervidda, Norway	Mass flows	Sandhaug et al. (1975)
	Alpine tundra, Hardangervidda, Norway	A conceptual model of C and N flows including effects of carbon chemistry	Wielgolaski (1975)
	High arctic tundra, Devon Island, Canada	N and P budgets	Babb and Whitfield (1977)
	High arctic tundra, Devon Island, Canada	A conceptual model of mass and energy flows	Whitfield (1977)
	Coastal wet meadow tundra, Point Barrow, Alaska, USA	C, N, and P budgets	Chapin et al. (1980)
	Coastal wet meadow tundra, Point Barrow, Alaska, USA	Mass and energy flows	MacLean (1980)
	Subarctic mire peatland, Abisko, Sweden	Mass and energy budgets	Svensson and Rosswall (1980)
	Coastal tundra, Spitsbergen, Svalbard	Mass and energy flows	Klekowski and Opalinski (1986)
Budget	High arctic tundra, Devon Island, Canada	Energy flows, including explicit measures of heterotroph biomass and dissolved organic matter concentrations	O'Neill (1976)
	Moss carpet, Signy Island, Antarctica	Mass flows	Davis (1981)
	Tundra sites, International (IBP)	Mass flows	Jones and Gore (1981)
Exponential	*Eriophorum-Calluna* bog, Moor House, UK	Litter mass loss	Gore and Olson (1967)
	Alpine tundra, Hardangervidda, Norway	Litter mass loss, implicitly including litter chemistry, microbial biomass, temperature, and water effects	Goksøyr (1975)
	Sphagnum carpet, Moor House, UK	Litter mass loss, including anaerobiosis	Clymo (1978)

	Location	Description	Reference
	Blanket bog, Moor House, UK	Litter mass loss, including effects of anaerobiosis, plant species, and litter type	Jones and Gore (1978)
	Moss carpet, Signy Island, Antarctica	Litter mass loss; variation with depth	Davis (1980)
	Moss carpet, Signy Island, Antarctica	Litter mass loss	Fenton (1980)
	Subarctic mire peatland, Abisko, Sweden	Litter mass loss focusing on N flows	Rosswall and Granhall (1980)
	Interior tundra sites, North America	Litter mass loss (C), including temperature effects	Miller et al. (1983)
Mechanistic	Tundra sites, International (IBP)	Compartment model of litter mass loss (C); flux based on microbial respiration, including limitations by substrate type, temperature, and moisture (oxygen)	Bunnell and Dowding (1974); Bunnell and Scoullar (1975, 1981); Flanagan and Bunnell (1980)
	Tundra sites, International (IBP); Point Barrow, Prudhoe Bay, and Eagle Summit, Alaska, USA	Compartment model of litter mass loss (C); flux based on microbial respiration including limitations by substrate type, temperature, and moisture (oxygen)	Bunnell et al. (1977a,b)
	Coastal wet meadow tundra, Point Barrow, Alaska, USA	Litter mass loss (C, N, P, and Ca), including nutrient limitations, moisture (oxygen), temperature, and microbial attributes	Barkley et al. (1978)
	Interior tundra sites, North America	Litter mass loss (C, N, and P), based on exoenzyme activity, including effects of litter chemistry, temperature, and moisture	Miller et al. (1984)
	Upland interior tundra, Toolik Lake, Alaska, USA	Compartmental model balancing C and N flows, including effects of litter chemistry, nutrient limitations, soil moisture, and temperature	Rastetter et al. (1992)
	Arctic foothills tundra, Imnavait Creek, Alaska, USA	Litter mass loss (C and N), including effects of litter chemistry, nutrient limitations, soil moisture, and temperature	Moorhead and Reynolds (1993)

Low temperature is the most obvious condition limiting decay, because soils are frozen during most of the year and subsurface soil temperatures seldom exceed 10 °C (Figs. 11.6–11.8, this Vol., Chapin et al. 1979; Kane et al. 1989). Soil moisture content is also important because microbial activity, soil respiration, and decomposition rates decline under water-saturated conditions (Bunnell et al. 1977a; Clymo 1978; Wynn-Williams 1980; Billings et al. 1982, 1983; Peterson et al. 1984; Oberbauer et al. 1991; Chap. 12, this Vol.). This probably explains the large accumulations of organic matter in wetter sites (Walker and Lederer 1987; Walker et al. 1989).

The chemical composition of litter and nutrient availability also affect decomposition (Minderman 1968; Melillo et al. 1982). Berg et al. (1975) noted that pure cellulose decomposed much more rapidly when small amounts of nitrogen fertilizer were applied to tundra soils. Thus, competition between soil microbiota and plants for nutrients is another important factor in tundra systems (Chap. 10, this Vol.; Cheng et al. 1995). In studies of surface-applied ^{15}N-labeled urea, Marion et al. (1982, 1987) initially recovered three to five times as much tracer from soil biomass as from vascular plants.

16.3 Arctic Decomposition Models

Published decomposition models for the arctic utilize many different formulations and approaches (Table 16.1). These models represent a range of complexity including simple conceptual/word models; budget models that assume a dynamic equilibrium and estimate flows of carbon, energy, mass, or nutrients necessary to maintain standing pools; detailed mechanistic formulations based on activities of specific suites of extracellular degradative enzymes; and exponential decay models that describe mass loss of a pool over time using constant decay rate coefficients. The exponential decay models have often been expanded to identify coefficients for different litter types or constituents, and environmental conditions (e.g., Clymo 1978; Jones and Gore 1978; Davis 1980). The progressive development of the ABISKO model by Bunnell and coworkers illustrates this point (Table 16.1).

We present a brief analysis of four specific models to elucidate similarities and differences that may be useful in developing a generic decomposition model for the Imnavait Creek watershed (see Chaps. 1 and 14, this Vol.). Four models – ABISKO (Bunnell and Scoullar 1975, 1981), BARK (Barkley et al. 1978), ARTUS (Miller et al. 1984), and GENDEC (Moorhead and Reynolds 1993) – were chosen for this analysis based on the following criteria:

1. The model was adequately described in publications to generate computer code.
2. The model contained sufficient detail to simulate decay patterns of different litter types under different environmental regimes.
3. The model was representative of the range of formulations used in arctic decomposition models meeting criteria 1 and 2.

Key features of the four models are presented in Table 16.2.

16.3.1 ABISKO

In ABISKO, mass loss of nine substrates (Table 16.3) is a function of microbial respiration and soil water content, oxygenation, and temperature (Table 16.2).

Table 16.2. Core equation for three arctic decomposition models

Model	Core equation	Reference
ABISKO	$R = W/(a_1 + W) \cdot a_2/(a_2 + W) \cdot a_3 \cdot a_4^{([T - 10]/10)}$ R: $\mu l \ O_2 \ g^{-1} \ h^{-1}$ [Respiration rate] W: g $H_2O \ g^{-1}$ [Soil moisture content] T: °C [Soil temperature] a_1: g $H_2O \ g^{-1}$ [Half-saturated moisture content] a_2: g $H_2O \ g^{-1}$ [Half-filled soil air pores] a_3: $\mu l \ O_2 \ g^{-1} \ h^{-1}$ [Maximum rate at 10 °C] a_4: °C^{-1} [Q_{10} coefficient]	Bunnell and Scoullar (1975)
ARTUS	$dC_s/dt = -k_s \cdot C_s \cdot M \cdot S_T$ C_s: g [Substrate mass] s: 1–7 [Substrate pool] $M = 1 - (W - 0.25) / (7.5 - 0.25)$ [Moisture effect] W: g $H_2O \ g^{-1}$ [Soil moisture content] $S_T = 0.25 + 0.75 \cdot (T/20)$ [Temperature effect] T: °C [Soil temperature] k_s: g g^{-1} day^{-1} [Decay rate coefficient]	Miller et al. (1984)
BARK	$R_{s,i} = k_{s,i} \cdot M_i \cdot O_i \cdot N_i \cdot Q_{s,i}^{([T - 10]/10)}$ R: $\mu l \ O_2 \ g^{-1} \ h^{-1}$ [Respiration rate] s: 1–3 [Substrate pool] i: 1–2 [Microbial pool] $k_{s,i}$: g g^{-1} day^{-1} [Decay rate coefficient] $M_i = a_i \cdot W - b_i$ [Moisture effect] W: Percent dry soil weight [soil moisture content] $O_i := c_i \cdot (U \cdot P/2) - d_i$ [Oxygen effect] U: [Unfilled soil pore space] P: [Partial pressure of oxygen] $N_i = N_r/N_a$ [Nitrogen effect] $N_r := $ g [Nitrogen required] N_a: g [Nitrogen available] $Q_{s,i}$: [Q_{10} Coefficient]	Barkley et al. (1978)
GENDEC	$dC_i/dt = -k_i \cdot C_i \cdot S_M \cdot S_T \cdot S_{N(i)}$ C_i: g [Substrate mass] i: 1–3 [Substrate pool] $S_M = a - m \cdot \log(-y)$ [Moisture effect] y: MPa [Soil water potential] $\log_{10}(S_T) = [(T - 25)/10] \cdot \log_{10}(Q)$ [Temperature effect] T: °C [Soil temperature] k_i: g g^{-1} day^{-1} [Decay rate coefficient] $S_{N(i)}$: Dimensionless [nitrogen limitation]	Moorhead and Reynolds (1993)

Respiration is converted to mass loss by assuming a respiratory coefficient of 1.0 (O_2:CO_2) and that litter is 45% carbon by dry weight. Temperature effects are described with a Q_{10} function, but the interacting effects of soil water and oxygen content are combined in a more complex relationship. Essentially, oxygen declines as soil water exceeds a site-specific optimum.

Certain features of ABISKO limit its flexibility. The initial formulation (Bunnell and Scoullar 1975) required site-specific parameters to simulate decay of a multitude of litter types (Table 16.3). Although later work separates ethanol soluble and insoluble fractions of litter (Bunnell et al. 1977b), this division of constituents is not consistent with similar work in other ecosystems, nor are these data readily available for litter at Imnavait Creek. Furthermore, ABISKO does not include the effects of nutrient limitations on decay.

16.3.2 ARTUS

Litter mass loss in ARTUS is described as the sum of losses from several chemically defined litter constituents, each of which has a specific decay rate (Table 16.4). Lignins do not decay in this formulation. Carbon losses are estimated from mass losses by assuming a litter carbon content of 58% dry weight. Carbon evolved as CO_2 is estimated as 40% of substrate carbon loss. The effects of soil temperature and moisture content are approximated by simple empirical relationships based on laboratory incubations (Table 16.2). Decay rates increase as a linear function of temperature and decrease as soil moisture exceeds 25% of the saturation level.

ARTUS also has some limitations that restrict its usefulness. Miller et al. (1984) simulated decay of several chemical constituents that are rarely examined or reported elsewhere, including Imnavait Creek. It also seems improbable that so many substrates decay at the same rate (Table 16.4), and that lignin does not decay. Furthermore, the soil moisture and temperature scalars are very site-specific, and there are no nutrient limitations.

Table 16.3. Parameter values used to describe microbial respiration in ABISKO. (Bunnell and Scoullar 1975)

Material	a_1	a_2	a_3	a_4
Green litter	4.16	4.16	777	8.79
Old standing dead	0.19	30 000	107	2.79
Litter	1.16	28.2	232	3.74
Soil carbon	0.90	25	40	4.00
Dead roots	0.75	20	80	6.00
Soil humus	0.90	2.5	5	1.80
Feces	4.16	4.16	777	8.00
New standing dead	0.74	127	184	2.56
Soluble carbohydrates	0	30 000	1000	2.00

Table 16.4. Decay rate coefficients for chemical constituents in ARTUS. (Miller et al. 1984)

Compound	Decay rate[a]
Cellulose	9.6×10^{-4}
Hemicellulose	9.6×10^{-4}
Pectin	1.44×10^{-5}
Chitin	9.6×10^{-4}
Protein	9.6×10^{-4}
Soluble carbohydrates	5.0×10^{-2}
Lignins	0

[a] $g\ g^{-1}\ day^{-1}$.

Table 16.5. Respiration of decomposer bacteria and fungi on various substrates in BARK. (Barkley et al. 1978)

Decomposer	Parameter/function	Organic matter		
		Soluble	Accessible	Resistant
Fungi	$k\ (g\ g^{-1}\ day^{-1})$	0	4.36×10^{-6}	4.36×10^{-6}
	Q_{10}	–	3	1.5
Bacteria	$k\ (g\ g^{-1}\ day^{-1})$	1.3×10^{-3}	1.3×10^{-3}	1.0×10^{-8}
	Q_{10}	1.2	1.8	5

16.3.3 BARK

The model built by Barkley et al. (1978) is the most comprehensive of those examined. It considers the effects of soil temperature, moisture, oxygenation, mineral nutrient availability and litter quality (carbon and nutrient chemistry) on decomposition (Table 16.2). Litter mass loss is driven by microbial respiration, but unlike ABISKO, the contributions of two distinct microbial pools (bacteria and fungi) are separately evaluated. Litter is divided into three constituents: soluble, accessible (but insoluble), and resistant organic matter. Respiration rates are specific to litter component and microbial pool (Table 16.5) with losses from dead organic matter pools estimated from respiration (Table 16.2), given assimilation efficiency of the decomposer microorganisms on various substrates. The effects of soil moisture and oxygen content are linear relationships of soil water content, whereas the effect of temperature is expressed as a Q_{10} function (Table 16.2). Rate modifiers are specific to both substrate type and microbial pool. Furthermore, assimilation efficiencies are decreased under anaerobic conditions.

Nutrients are associated with all organic matter pools and in soluble and insoluble inorganic forms. The total quantity that is available to microbiota consists of the sum of nutrients released from decomposing substrates (in direct proportion to carbon loss) and dissolved inorganic forms. The nutrient requirement for maximum decomposition is based on the potential amount of carbon utilized by the microbial pool, although details of this relationship are not given in Barkley et al. (1978). Nutrient limitations are calculated as the deficit between need and availability. Microbial biomass is determined by growth and turnover, growth is based on respiration and mortality is a constant fraction of the standing biomass. The dead biomass is partitioned into the substrate pools.

The greatest uncertainty in BARK is the lack of detailed information concerning the soil environment and microbiota at Imnavait Creek. BARK specifies microbial responses for each substrate type and environmental circumstance. At present, we are uncertain as to whether the microbiota or their behavior in the tussock tundra site are similar to the wet coastal site examined by Barkley et al. (1978).

16.3.4 GENDEC

16.3.4.1 General Description

Moorhead and Reynolds (1993) used GENDEC to explore potential impacts of climate change on litter decay and associated aspects of nutrient dynamics in tussock tundra (see Chap. 14, this Vol.). GENDEC is similar to BARK in many regards, but relies on general theoretical considerations of microbial nutrient dynamics and decomposition patterns (e.g., Parnas 1975), instead of explicitly evaluating microbial behavior.

GENDEC uses constant decay rate coefficients to estimate carbon flux from several dead organic matter pools: (1) labile plant compounds, (2) holocellulose (cellulose + hemicellulose), (3) resistant plant compounds (e.g., lignins), and (4) dead microbial biomass (Table 16.2). The loss of carbon from each pool is also controlled by moisture availability, temperature, and nitrogen availability. The effect of soil moisture on decay is estimated as a linear function of soil water potential (Paustian and Schnürer 1987); no effects of waterlogging are included. A simple Q_{10} function is used to estimate the effects of temperature.

Nitrogen flows are assumed to balance carbon flows, and the effect of nitrogen limitation is determined by comparing the available nitrogen to the amount needed to realize the maximum potential decomposition of a carbon pool based on temperature, moisture, and decay rate coefficient. Actual decomposition of nitrogen-limited substrates is reduced in proportion to the existing deficit (if any). Conversely, net nitrogen mineralization occurs when the quantities of nitrogen released from decomposition exceed microbial needs.

The total quantities of carbon and nitrogen available for microbial use consist of the sum of all losses from the dead organic matter pools; available nitrogen includes mineral forms. Microbial growth and respiration are driven by carbon availability, i.e., total losses from the carbon substrates given a fixed microbial carbon assimilation efficiency. Microbial turnover includes: (1) a fraction of the standing biomass, (2) a portion of incremental growth (cf. Parnas 1975), and (3) a fraction of the standing biomass killed by wetting–drying events (cf. Kieft et al. 1987).

16.3.4.2 Validation

Although GENDEC is a simple model, there are few arctic studies that provide sufficient data to estimate parameter values. In this section, we describe our efforts to validate the general structure of GENDEC based on studies in Antarctica and Toolik Lake, Alaska.

Fenton (1980) studied litter decay in moss communities on Signy Island, Antarctica, by excavating intact moss mats, examining structural changes in buried plants, and estimating growth rates. The ages of buried portions of the physically continuous moss were estimated by assuming constant growth rates for the past 100–200 years. We used the chemical characteristics of moss litter reported by Davis (1986) and others (Christie 1987a,b) to parameterize GENDEC, and then compared simulated decomposition as a function of temperature, substrate quality, nutrient availability, and soil moisture to the long-term patterns reported by Fenton (1980; Table 16.6). Temperature was simulated with an amplitude of 10 °C over a 90-day season. Annual amplitude was diminished 0.1 °C year^{-1} over a 100-year period, approximating the point in time when progressively buried materials becomes permanently frozen (cf. Fenton 1980). Soil moisture was assumed to remain optimal (cf. Davis 1986). GENDEC greatly overestimated Fenton's (1980) mass losses (results not shown). However, when we reduced decay rates to 1–1.5% of their original

Table 16.6. Organic soil profile characteristics in an antarctic moss bank (Davis 1986) and arctic tussock tundra. (Miller et al. 1984)

| Material | k[a] | \multicolumn{2}{c}{Antarctic moss bank} | | \multicolumn{2}{c}{Arctic tussock tundra} |
		C:N[b]	Amount[c] (%)	C:N[d]	Amount[e] (%)
Labiles	0.20	10:1	14.5	6.25:1	7
Cellulose and hemicellulose	0.08	∞	58.2	∞	42
Recalcitrants (lignins)	0.01	∞	27.3	∞	28
Dead microbiota	0.30	10:1	0	10:1	0
Live microbiota	NA[f]	10:1	0	10:1	0

[a] Fraction day^{-1} (Van Veen and Paul 1981).
[b] Overall plant material C:N = 69:1.
[c] Davis (1986).
[d] Overall organic material C:N = 38:1.
[e] Miller et al. (1984).
[f] Not applicable.

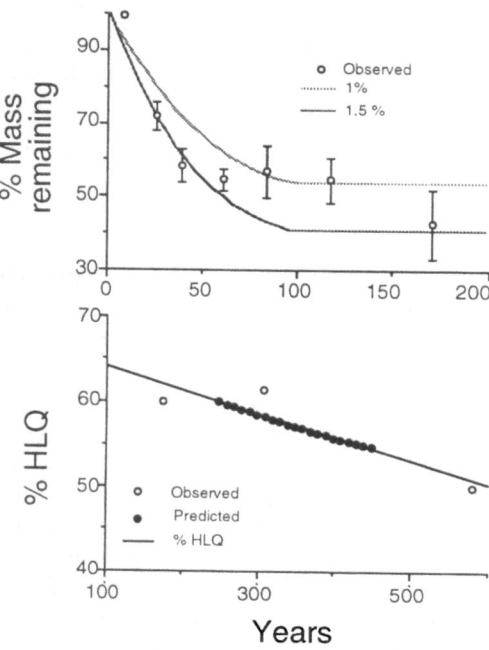

Fig. 16.1. Time course changes in mass of Antarctic peat (*top*) and % HLQ (holocellulose/[holocellulose + lignin]) content of organic soil horizons in Arctic tussock (*bottom*). Relationship used: % HLQ = 66.81−0.27 × years

values, GENDEC's predictions closely approximated observations (Fig. 16.1). Obviously, factors other than those included in the model significantly reduce decomposition. Initially, we considered two possibilities, anaerobiosis and pH, but anaerobiosis was not common at this site and pH effects are likely insufficient to account for this discrepancy (Davis 1986).

At Toolik Lake, Miller et al. (1984) estimated holocellulose and lignin fractions of soil fibric, hemic, and sapric horizons (Table 16.6). Everett (cited in Miller et al. 1984) estimated the ^{14}C ages of these materials. Although there were only three observations (Fig. 16.1), we used these data to compute the change in holocellulose-lignin quotient (HLQ) over time (shown as the regression line in Fig. 16.1). We parameterized GENDEC based on Miller et al.'s (1984) chemical data and used the regression as a long-term record of decomposition patterns for model "validation." Temperature was approximated as a function of time with an amplitude of 10 °C and a period of 90 days. The amplitude was not diminished over time, because thaw depth reaches underlying mineral soils. Decay rates were computed as a function of temperature, substrate quality, soil moisture (held constant at optimum), and nutrient availability. Again, we found that GENDEC overestimated mass losses (results not shown) unless we reduced the decay rates to about 1.5% of their estimated values (Fig. 16.1). As with the Signy Island site, anaerobic conditions are not an obvious feature of the tussock-tundra soil profile at Toolik Lake, nor are they common in tussock tundra at Imnavait Creek. Potential reasons for these overestimations are discussed below.

16.4 Model Comparisons

Because decomposition provides a significant part of soil respiration, Moorhead and Reynolds (1993) used soil efflux data from Imnavait Creek (Chap. 11, this Vol.) to compare predictions of ABISKO, ARTUS, BARK, and GENDEC. The soil profile at nearby Toolik Lake (Miller et al. 1984) is similar to one monitored for respiration at Imnavait Creek (Oberbauer et al. 1991). The soil description provided by Miller et al. (1984) contains all the information required by ARTUS and enough detail to develop initial values of state variables for the other models (Table 16.7). For comparisons, simulations with all models were conducted for a 10-day period at optimum soil moisture content, aerobic conditions, and no nutrient limitations.

Model estimates of CO_2 production due to decomposition were generally less than observed soil respiration (Table 16.8), which was encouraging because total carbon dioxide efflux from soil includes respiration of fauna and plant roots not addressed by these models. In all cases, the ratio of estimated: observed CO_2 production decreased with increasing temperature. ABISKO consistently produced the lowest values (38–44% of observed levels), and BARK predicted both the highest values of CO_2 output and had the greatest

Table 16.7. Soil profile characteristics and state variables used in model simulations

Model	Material allocation	Quantity (g C m^{-2})
ABISKO	Dead roots	700
	Humus	7045
	Litter	17
	Soil carbon	777
	Total	8539[a]
BARK	Accessible	1745
	Resistant	6794
	Soluble	0
	Total	8539
ARTUS	Intermediate	5585
	Resistant	2954
	Soluble	0
	Total	8539[b]
GENDEC	Labile	1745
	Holocellulose	3840
	Recalcitrant	2954
	Total	8539

[a] Assume 45% carbon content of litter mass (Bunnell and Scoullar 1975).
[b] Assume 58% carbon content of litter mass (Miller et al. 1984).

Table 16.8. Predicted CO_2 efflux from an arctic soil profile

Soil moisture (%)	Soil temperature (°C)	Observed[a]	Ratio of predicted:Observed[a]			
			BARK	GENDEC	ABISKO	ARTUS
55	2.5	11.35	1.01	0.70	0.44	0.59
55	5.0	15.62	0.85	0.72	0.44	0.54
60	7.5	23.07	0.73	0.72	0.43	0.44
60	10	36.69	0.53	0.62	0.38	0.33

[a] Observed is carbon evolved as CO_2 (g m^2) over a 10-day period based on regression derived from field observations of hourly CO_2 efflux rates by Oberbauer et al. (1991).

Table 16.9. Intrinsic decay rates used in arctic decomposition models

Model	Intrinsic decay rates (g g^{-1} day^{-1})	
	Intermediate substrates	Resistant substrates
ARTUS	9.6×10^{-4}	1.4×10^{-5}
BARK[a]	1.3×10^{-3}, 4.36×10^{-6}	1.0×10^{-8}, 4.36×10^{-6}
ABISKO	6.6×10^{-3} (General)	1.4×10^{-4} (Humus)
GENDEC	8.0×10^{-2}	1.0×10^{-2}

[a] Bacterial and fungal rates, respectively.

range (53–101% of observations). The predictions of GENDEC and ARTUS were intermediate to the other models, with GENDEC predictions within 62–72% of observations and ARTUS 33–59%. Given the differences in the structure of these models and the uncertainty associated with selecting initial values, we anticipated even larger disparity in these predictions.

Substantial differences in decay rate coefficients exist among decomposition models (Table 16.9). In general, the rates reported by Paul and coworkers (e.g., Paul and Juma 1981), which we used in constructing GENDEC, are 1–3 orders of magnitude greater than the rates used in the other arctic models. Although it is true that decomposition is very slow on cool, wet, acidic sites (see Clymo 1965, 1978), it seems unlikely that maximum intrinsic decay rates for particular substances vary independently of environmental factors. Additional controls on decomposition in arctic systems are implicitly included in the low decay rate coefficients used by other models, although the reasons for these lower rates are not explicitly identified. Hence, it appears that decay rates in arctic systems are much slower than would be expected, based on litter quality, nutrient availability, soil moisture content, and temperature. Although acidity and anaerobic conditions are attributes of moist arctic soils, these factors appear to be insufficient to account for the discrepancies. Reducing the decay rate coefficients in GENDEC to levels comparable to the other models provided good agreement between simulated and observed changes in litter quality, but did not identify the underlying control mechanisms.

16.5 Effects of Environmental Changes

16.5.1 Climate Change

Within the next 50 years, summer precipitation may increase 20–30%, and summer temperatures may increase by 3–6 °C in northern Alaska (Schlesinger and Mitchell 1985; Maxwell 1992). One of the major impacts of these changes on tundra ecosystems will be modifications in litter decay and nutrient cycling processes (Nadelhoffer et al. 1992), which are likely to affect plant production (Chapin et al. 1992). For example, modeling experiments suggest that increasing temperature, season length, and moisture availability will have little impact on growth responses of *Eriophorum vaginatum*, a dominant plant of tussock tundra, unless nutrient availability also changes (Leadley and Reynolds 1992; Reynolds and Leadley 1992). This hypothesis is consistent with experiments showing nitrogen availability to be one of the most important factors controlling plant production (Shaver and Chapin 1980, 1986; Billings et al. 1984; Marion et al. 1987; Chap. 10, this Vol.).

Moorhead and Reynolds (1993) used GENDEC to examine the potential effects of changing climate on decomposition of dead organic matter in tussock tundra. Under current climatic conditions (700% average soil water content, 4.5 °C mean soil temperature and 93-day season length), 5.28 g N m⁻² year⁻¹ was released from decaying organic matter, consistent with observations of decomposition, nitrogen mineralization, and plant growth in this system (Shaver et al. 1990; Giblin et al. 1991; Nadelhoffer et al. 1991; Leadley and Reynolds 1992). Under a predicted range of climate changes, nitrogen release increased as a linear function of both season length and temperature, and as a parabolic function of soil moisture content. The greatest increase in net release (199% of current value) occurred at an optimum 600% soil water content, a

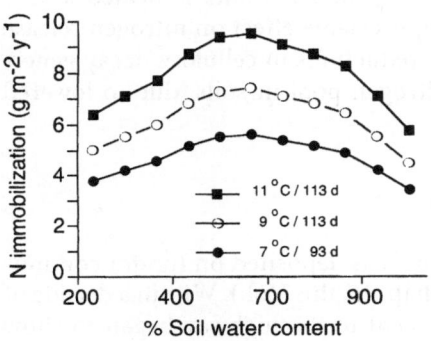

Fig. 16.2. Simulation of cumulative N-immobilization (g m⁻² year⁻¹) under varying climate regimes. (Moorhead and Reynolds 1993)

season length of 140 days, and a mean summer temperature of 8.5 °C (Fig. 16.2). The least amount of nitrogen was released (46% of current value) at 4.5°C, a 70-day season length, and a soil moisture content of 1000%. Season length had little effect on the size of the mineral nitrogen pool, ranging 12–25 mg N m^{-2}, although pool size responded slightly to differences in temperature and soil moisture content.

16.5.2 Effects of Elevated CO_2

Increasing concentrations of atmospheric CO_2 are expected to affect nutrient cycles and soil organic matter dynamics of terrestrial ecosystems in a number of ways (Graham et al. 1990; Comins and McMurtrie 1993; van de Geijn and van Veen 1993). For example, decomposition rates are sensitive to changes in litter quality (e.g., Meetemeyer 1978; Melillo et al. 1982), but plants also can respond to CO_2 fertilization in other ways that affect nutrient cycling (see Rogers et al. 1992; Baker and Allen 1994). For example, greater amounts of labile carbon in plant rhizospheres associated with elevated CO_2 (Zak et al. 1993) could explain changes in rhizosphere bacterial populations, greater soil respiration rates, and higher enzymatic activities in root regions (O'Neill et al. 1987; Körner and Arnone 1992). Also, enhanced CO_2 may modify mycorrhizal activities, stimulating phosphorus uptake (Norby et al. 1986; Conroy et al. 1990).

Following 3-year in situ exposure to 680 ppm CO_2 (Grulke et al. 1990), the enzymatic characteristics of roots, associated mycorrhizae, and surrounding soils were examined in tussock tundra communities of Northern Alaska (A. E. Linkins, unpubl. data). Endocellulase and exocellulase activities were higher in mycorrhizal rhizomorphs, and lower in Oe and Oi horizons, at greater CO_2, suggesting that an increase in carbon exudation from plant roots could be inhibiting cellulase activities in these soils. These data were used to modify GENDEC, reducing the intrinsic decay rate coefficient for cellulose by 45%. This assumes that the observed reduction in endocellulose activity would produce a proportional reduction in the maximum rate of cellulose decay. Under current climatic conditions (average 700% soil water content, mean 4.5 °C soil temperature, and 90-day season), model results indicated a 45% reduction in cellulose turnover with no appreciable effect on nitrogen release from decaying organic matter. However, reductions in cellulose decay generally increased the size of the mineral nitrogen pool in soils, due to lowered microbial immobilization.

16.5.3 Impacts of Road Dust Deposition

Dust raised by traffic on the Dalton Highway is deposited on tundra communities along both sides of the route (see Chap. 15, this Vol.). Within a decade of its construction in 1974, ecosystems adjacent to the highway began to show significant changes in soil chemistry, vegetation, and microclimate, including higher soil pH and calcium concentrations, lower soil moisture, greater thaw depth, and reduction in *Sphagnum* moss occurrence (Spatt and Miller 1981; Walker and Everett 1987; Meininger and Spatt 1988). Although we are aware of no experimental studies of the impacts of road dust on litter decay or nutrient cycling processes, soil enzyme activities in surface organic materials were found to be affected by dust loading (A. E. Linkins, unpubl. data). Activities of endocellulase-, exocellulase-, and phosphatase-associated soil organic materials (0–3 cm depth) increased rapidly with increasing distance from the road (Fig. 16.3).

Previous simulations have shown that changes in the rate of cellulose decay estimated by GENDEC is proportional to modifications in the maximum decay rate coefficient, and that whereas such changes have little effect on nitrogen release from dead organic matter, a reduction in cellulose decay could increase mineral nitrogen pool size in the soil (see above). Therefore, we might expect a decrease in cellulose degradation and concomitant increase in nitrogen availability with dust input. Higher nitrogen availability generally would be expected to enhance plant production in nutrient-poor tundra communities, especially for species that respond to changes in nutrient content of surface soil horizons (e.g., shrubs).

16.5.4 Tussock Phosphorus Dynamics

Although not a decomposition model per se, Moorhead et al. (1993) developed a model of phosphorus dynamics based on phosphatase activities associated with *E. vaginatum* roots, dead organic matter, and mineral soils in *Eriophorum* tussocks. Given a 90-day summer season and organic phosphorus concentrations of 30 µM in the first and last 10-day intervals (15 µM at other times), the model projected an annual release of 155 mg P tussock^{-1}. Enzyme activity associated with living roots of *E. vaginatum* released about 6 mg P, which is almost twice the annual plant demand. These results suggest that *Eriophorum* may obtain much of its phosphorus requirement from the activities of root surface phosphatases.

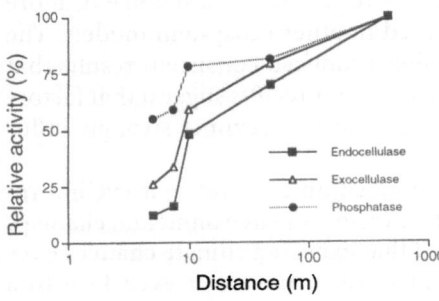

Fig. 16.3. Effects of road dust loading on extracellular enzyme activity in surface organic matter in tussock tundra, northern Alaska. (Linkins, unpubl. data)

Phosphatase activities of tussocks were evaluated following the CO_2 fumigation experiments described by Grulke et al. (1990). Activities were higher on *E. vaginatum* root surfaces, mycorrhizal rhizomorphs, and mantles of the shrub, *Betula nana*, and in surrounding Oe and Oi soil horizons at elevated CO_2 (A.E. Linkins, unpubl. data). These data were used to modify the tussock phosphatase model of Moorhead et al. (1993), assuming that relative change in observed phosphatase activities resulting from elevated CO_2 would result in a proportional shift in the maximum rates of enzyme activities. Therefore, changes were made in model parameters that represented the maximum rates (V_{max}) of phosphatase activity associated with live roots of *Eriophorum*, soil organic matter, and mineral soil. V_{max} was increased by 154% for *Eriophorum* roots and 140% for soil materials, over the original values used in this model, to represent the effects of CO_2 fertilization. Simulations indicated that these higher activities could increase total annual phosphorus release within tussocks by more than 40% including a nearly threefold increase in phosphorus release on root surfaces of *Eriophorum*.

Additional simulations were not conducted to quantify the impacts of dust deposition on phosphorus dynamics in roadside tussock communities. Although substantial reductions in phosphatase activities of surface soils are apparent (Fig. 16.3), the overall impact of dust loading on entire tussock soil profiles is unknown. *Eriophorum* is a deep-rooted species, so that changes in enzymatic characteristics of surface soils may not have much effect on the overall phosphorus dynamics of a tussock or plant nutrient status. In contrast, shallow-rooted species, such as dwarf birch (*Betula nana*), might respond to an increase in near-surface soil phosphorus availability.

16.6 Conclusions

The modeling approaches presented here represent different perspectives of decomposition processes in arctic ecosystems. For example, ABISCO individually describes the decomposition of many litter types under given moisture and temperature regimes, whereas BARK uses a more generalized approach to examine the dynamics of a few common litter constituents. GENDEC shares certain elements of these other models, and yet emphasizes the use of more readily obtained data and formulations used in other ecosystem models. The use of a common data set to drive all models produced consistent results that were comparable to observations. Furthermore, our results suggest that factors other than direct moisture, temperature, and nutrient regimes strongly influence decomposition in arctic systems.

Model simulations suggest that decomposition and nutrient cycling regimes of arctic ecosystems may respond to a variety of environmental changes. Moorhead and Reynolds (1993) concluded that expected climate changes over the next 50 years could nearly double nitrogen turnover, or even lead to a decrease, over current levels. Unfortunately, more precise climate information is needed to predict changes in decomposition and nutrient dynamics. Preliminary observations of changes in belowground exoenzyme activity patterns suggest that tussock-tundra ecosystems also respond to CO_2 fertilization and dust loading, although whole-system responses to the former perturbation are equivocal (e.g., Grulke et al. 1990) and have not been examined with regard to the latter.

In a larger context, tundra soils contain enormous quantities of organically bound carbon and changes in belowground storage may have significant impacts on the global carbon budget (Dixon and Turner 1991). It is not certain whether tussock tundra presently represents a net source or sink for carbon (see Chaps. 11 and 17, this Vol.), but increasing temperatures are likely to increase losses (Oechel and Billings 1992). However, based on simulation (Moorhead and Reynolds 1993) and microcosm studies (Funk et al. 1994), changes in soil moisture regimes may overshadow any direct effects of temperature and promote carbon storage. Responses to increased CO_2 may also modify the net carbon balance of tundra ecosystems through changes in the relative levels of carbon- and nutrient-acquisition activities. Higher nutrient-acquiring activities, lower organic carbon degradation rates, and lower microbial immobilization of mineral nutrients could result in both greater storage of carbon in complex organic molecules (e.g., cellulose and lignins) and greater respiration rates of microbiota metabolizing simpler carbohydrates produced by root exudation.

Acknowledgments. A. E. Linkins generously provided unpublished data. This work was supported by the US Department of Energy OHER/ERD as part of the R4D program (grant no. DE- and DE-F605-92ER61-455).

References

Babb TA, Whitfield DWA (1977) Mineral nutrient cycling and limitation of plant growth in the Truelove Lowland ecosystem. In: Bliss LC (ed) Truelove Lowland, Devon Island, Canada: a high arctic ecosystem. Univ Alberta Press, Edmonton, Canada, pp 589–606

Baker JT, Allen LH Jr (1994) Assessment of the impact of rising carbon dioxide and other potential climate changes on vegetation. Environ Pollut 83: 223–235

Barkley SA, Barèl D, Stoner WA, Miller PC (1978) Controls on decomposition and mineral release in wet meadow tundra – a simulation approach. In: Adriano DC, Brisbin IL Jr (eds) Environmental chemistry and cycling processes. DOE Symp Ser, DOE Tech Inf Serv, Oak Ridge, Tenn, pp 754–778

Berg BL, Karenlampi L, Veum AK (1975) Comparisons of decomposition rates measured by means of cellulose. In: Wielgolaski FE (ed) Fennoscandian tundra ecosystems: plants and microorganisms. Springer, Berlin Heidelberg New York, pp 261–267

Billings WD, Luken JO, Mortenson DA, Peterson KM (1982) Arctic tundra: a source or sink for atmospheric carbon dioxide in a changing environment? Oecologia 53: 7–11

Billings WD, Luken JO, Mortenson DA, Peterson KM (1983) Increasing atmospheric carbon dioxide: possible effects on arctic tundra. Oecologia 58: 286–289

Billings WD, Peterson KM, Luken JO, Mortenson DA (1984) Interaction of increasing atmos-
    pheric carbon dioxide and soil nitrogen on the carbon balance of tundra microcosms.
    Oecologia 65: 26–29
Bunnell FL, Dowding P (1974) ABISKO – a generalized decomposition model for comparisons
    between tundra sites. In: Holding AJ, Heal OW, MacLean WF Jr, Flanagan PW (eds) Soil
    organisms and decomposition in tundra. IBP Tundra Biome Steering Comm, Stockholm, pp
    227–247
Bunnell FL, Scoullar KA (1975) ABISKO II – a computer simulation model of carbon flux in
    tundra ecosystems. In: Rosswall T, Heal OW (eds) Structure and function of tundra ecosys-
    tems. Ecol Bull (Stockholm) 20: 425–448
Bunnell FL, Scoullar KA (1981) Between-site comparisons of carbon flux in tundra using simula-
    tion models. In: Bliss LC, Heal OW, Moore JJ (eds) Tundra ecosystems: a comparative
    analysis. Cambridge Univ Press, Cambridge, pp 685–714
Bunnell FL, Tait DEN, Flanagan PW, Van Cleve K (1977a) Microbial respiration and substrate
    weight loss. I. A general model of the influences of abiotic variables. Soil Biol Biochem 9:
    33–40
Bunnell FL, Tait DEN, Flanagan PW, Van Cleve K (1977b) Microbial respiration and substrate
    weight loss. II. A model of the influences of chemical composition. Soil Biol Biochem
    9: 41–47
Chapin FS III, Körner C (eds) (1995) Arctic and alpine biodiversity: causes and ecosystem
    consequences. Ecological Studies 113, Springer, Berlin Heidelberg New York
Chapin FS III, Van Cleve K, Chapin MC (1979) Soil temperature and nutrient cycling in the
    tussock growth form of Eriophorum vaginatum. J Ecol 67: 169–189
Chapin FS III, Miller PC, Billings WD, Coyne PI (1980) Carbon and nutrient budgets and their
    control in coastal tundra. In: Brown J, Miller PC, Tieszen LL, Bunnell FL (eds) An arctic
    ecosystem: the coastal tundra at Barrow, Alaska. Dowden, Hutchinson and Ross,
    Stroudsburg, pp 458–482
Chapin FS III, Jefferies RL, Reynolds JF, Shaver GR, Svoboda J (eds) (1992) Arctic ecosystems in
    a changing climate: an ecophysiological perspective. Academic Press, San Diego
Cheng W, Virginia RA, Oberbauer SF, Tenhunen JD, Gillespie CT, Reynolds JF (1995) Spatial and
    temporal variation in soil nitrogen, microbial biomass, and respiration in an arctic catena
    (in preparation)
Christie P (1987a) Nitrogen in two contrasting Antarctic bryophyte communities. J Ecol
    75: 73–93
Christie P (1987b) C:N ratios in two contrasting antarctic peat profiles. Soil Biol Biochem 19:
    777–778
Clymo RS (1965) Experiments on breakdown of Sphagnum in two bogs. J Ecol 53: 747–758
Clymo RS (1978) A model of peat bog growth. In: Heal OW, Perkins DF (eds) Production ecology
    of British moors and montane grasslands. Ecological Studies 27. Springer, Berlin Heidelberg
    New York, pp 213–223
Comins HN, McMurtrie RE (1993) Long-term response of nutrient-limited forests to CO_2 enrich-
    ment: equilibrium behavior of plant-soil models. Ecol Appl 3: 666–681
Conroy JP, Milnam PH, Reed ML, Barlow EW (1990) Increases in phosphorus requirements for
    CO_2-enriched pine species. Plant Phys 92: 977–982
Davis RC (1980) Peat respiration and decomposition in Antarctic terrestrial moss communities.
    Biol J Linn Soc 14: 39–49
Davis RC (1981) Structure and function of two Antarctic terrestrial moss communities. Ecol
    Monogr 51: 125–143
Davis RC (1986) Environmental factors influencing decomposition rates in two antarctic moss
    communities. Polar Biol 5: 95–103
Dixon RK, Turner DP (1991) The global carbon cycle and climate change: responses and
    feedbacks from below ground systems. Environ Pollut 73: 245–262
Fenton JHC (1980) The rate of peat accumulation in antarctic moss banks. J Ecol 68: 211–228

Flanagan PW, Bunnell FL (1980) Microflora activities and decomposition. In: Brown J, Miller PC, Tieszen LL, Bunnell FL (eds) An arctic ecosystem: the coastal tundra at Barrow, Alaska. Dowden, Hutchinson and Ross, Stroudsburg, pp 291–334

Funk DW, Pullman ER, Peterson KM, Crill PM, Billings WD (1994) Influence of water table on carbon dioxide, carbon monoxide, and methane fluxes from taiga bog microcosms. Global Biogeochem Cycles 8: 271–278

Giblin AE, Nadelhoffer KJ, Shaver GR, Laundre JA, McKerrow AJ (1991) Biogeochemical diversity along a riverside toposequence in arctic Alaska. Ecol Monogr 61: 425–435

Goksøyr J (1975) Decomposition, microbiology, and ecosystem analysis. In: Wielgolaski FE (ed) Fennoscandian tundra ecosystems: plants and microorganisms. Springer, Berlin Heidelberg New York, pp 230–238

Gore AJP, Olson JS (1967) Preliminary models for accumulation of organic matter in an Eriophorum/Calluna ecosystem. Aquilo Ser Bot 6: 297–313

Graham RL, Turner MG, Dale VH (1990) How increasing CO_2 and climate change affect forests. BioScience 40: 575–587

Grulke NE, Riechers GH, Oechel WC, Hjelm U, Jaeger C (1990) Carbon balance in tussock tundra under ambient and elevated atmospheric CO_2. Oecologia 83: 485–494

Jones HE, Gore AJP (1978) A simulation of production and decay in blanket bog. In: Heal OW, Perkins DF (eds) Production ecology of British moors and montane grasslands. Ecological Studies, vol 27. Springer, Berlin Heidelberg New York, pp 160–186

Jones HE, Gore AJP (1981) A simulation approach to primary production. In: Bliss LC, Heal OW, Moore JJ (eds) Tundra ecosystems: a comparative analysis. Cambridge Univ Press, Cambridge, pp 239–256

Kane DL, Hinzman LD, Benson CS, Everett KR (1989) Hydrology of Imnavait Creek, an arctic watershed. Holarct Ecol 12: 262–269

Kieft TL, Soroker E, Firestone MK (1987) Microbial biomass response to a rapid increase in water potential when dry soil is wetted. Soil Biol Biochem 19: 119–126

Klekowski RZ, Opalinski KW (1986) Matter and energy flow in Spitsbergen ornithogenic tundra. Polar Res 4: 187–197

Körner C, Arnone JA (1992) Responses to elevated carbon dioxide in artificial tropical ecosystems. Science 257: 1672–1675

Leadley PW, Reynolds JF (1992) Long-term response of an arctic sedge to climate change: a simulation study. Ecol Appl 2: 323–340

MacLean WF (1980) The detritus-based trophic system. In: Brown J, Miller PC, Tieszen LL, Bunnell FL (eds) An arctic ecosystem: the coastal tundra at Barrow, Alaska. Dowden, Hutchinson and Ross, Stroudsburg, pp 411–457

Marion GM, Miller PC, Kummerow J, Oechel WC (1982) Competition for nitrogen in a tussock tundra ecosystem. Plant Soil 66: 317–327

Marion GM, Miller PC, Black CH (1987) Competition for tracer ^{15}N in tussock tundra ecosystems. Holarct Ecol 10: 230–234

Maxwell B (1992) Arctic climate: potential for a change under global warming. In: Chapin FS III, Jefferies RL, Reynolds JF, Shaver GR, Svoboda J (eds) Arctic ecosystems in a changing climate: an ecophysiological perspective. Academic Press, San Diego, pp 11–34

Meetemeyer V (1978) Macroclimate and lignin control of litter decomposition rates. Ecology 59: 465–472

Meininger CA, Spatt PD (1988) Variations of tardigrade assemblages in dust-impacted Arctic mosses. Arct Alp Res 20: 24–30

Melillo JM, Aber JD, Muratore JF (1982) Nitrogen and lignin control of hardwood leaf litter decomposition dynamics. Ecology 63: 621–626

Miller PC, Kendall R, Oechel WC (1983) Simulating carbon accumulation in northern ecosystems. Simulation 40: 119–131

Miller PC, Miller PM, Blake-Jacobson M, Chapin FS III, Everett KR, Hilbert DW, Kummerow J, Linkins AE, Marion GM, Oechel WC, Roberts SW, Stuart L (1984) Plant-soil processes in

Eriophorum vaginatum tussock tundra in Alaska: a systems modeling approach. Ecol Monogr 54: 361–405

Minderman G (1968) Addition, decomposition and accumulation of organic matter in forests. J Ecol 56: 355–362

Moorhead DL, Reynolds JF (1993) Effects of climate change on decomposition in arctic tussock tundra: a modeling synthesis. Arct Alp Res 25: 403–412

Moorhead DL, Kroehler CA, Linkins AE, Reynolds JF (1993) Dynamics of extracellular phosphatase activities in *Eriophorum vaginatum* tussocks. Arct Alp Res 25: 50–55

Nadelhoffer KJ, Giblin AE, Shaver GR, Laundre JA (1991) Effects of temperature and substrate quality on element mineralization in six arctic soils. Ecology 72: 242–253

Nadelhoffer KJ, Giblin AE, Shaver GR, Linkins AE (1992) Microbial processes and plant nutrient availability in arctic soils. In: Chapin FS III, Jefferies RL, Reynolds JF, Shaver GR, Svoboda J (eds) Arctic ecosystems in a changing climate: an ecophysiological perspective. Academic Press, San Diego, pp 281–300

Norby RJ, O'Neill EG, Luxmoore RJ (1986) Effects of atmospheric CO_2 enrichment on the growth and mineral nutrition of *Quercus alba* seedlings in nutrient poor soil. Plant Physiol 82: 83–89

Oberbauer SF, Tenhunen JD, Reynolds JF (1991) Environmental effects on CO_2 efflux from water track and tussock tundra in arctic Alaska, USA. Arct Alp Res 23: 162–169

Oechel WC (ed) Global change and arctic terrestrial ecosystems. Ecological Studies, Springer, Berlin Heidelberg New York (in preparation)

Oechel WC, Billings WD (1992) Effects of global change on the carbon balance of arctic plants and ecosystems. In: Chapin FS III, Jefferies RL, Reynolds JF, Shaver GR, Svoboda J (eds) Arctic ecosystems in a changing climate: an ecophysiological perspective. Academic Press, San Diego, pp 139–168

O'Neill EG, Luxmoore RJ, Norby RJ (1987) Elevated atmospheric CO_2 effects on seedling growth, nutrient uptake, and rhizosphere bacterial populations of *Liriodendron tulipifera* L. Plant Soil 104: 3–11

O'Neill RV (1976) Ecosystem persistence and heterotrophic regulation. Ecology 57: 1244–1253

Parnas H (1975) Model for decomposition of organic material by microorganisms. Soil Biol Biochem 7: 161–169

Paul EA, Juma NG (1981) Mineralization and immobilization of soil nitrogen by microorganisms. In: Clark FE, Rosswall T (eds) Terrestrial nitrogen cycles. Ecol Bull (Stockholm) 33: 179–199

Paustian K, Schnürer J (1987) Fungal growth response to carbon and nitrogen limitation: application of a model to laboratory and field data. Soil Biol Biochem 19: 621–629

Peterson KM, Billings WD, Reynolds DN (1984) Influence of water table and atmospheric CO_2 concentration on the carbon balance of arctic tundra. Arct Alp Res 16: 331–335

Rastetter EB, McKane RB, Shaver GR, Melillo JM (1992) Changes in C storage by terrestrial ecosystems: how C–N interactions restrict responses to CO_2 and temperature. Water Air Soil Pollut 64: 327–344

Reynolds JF, Leadley PW (1992) Modeling the response of arctic plants to climate change. In: Chapin FS III, Jefferies RL, Reynolds JF, Shaver GR, Svoboda J (eds) Arctic ecosystems in a changing climate: an ecophysiological perspective. Academic Press, San Diego, pp 413–438

Rogers HH, Prior SA, O'Neill EG (1992) Cotton root and rhizosphere responses to free-air CO_2 enrichment. Crit Rev Plant Sci 11: 251–263

Rosswall T, Granhall U (1980) Nitrogen cycling in a subarctic ombrotrophic mire. In: Sonesson M (ed) Ecology of a subarctic mire. Ecol Bull (Stockholm) 30: 209–234

Sandhaug A, Kjelvik S, Wielgolaski FW (1975) A mathematical simulation model for terrestrial tundra ecosystems. In: Wielgolaski FE (ed) Fennoscandian tundra ecosystems. Part 2. Animals and systems analysis. Springer, Berlin Heidelberg New York, pp 251–266

Schlesinger ME, Mitchell JFB (1985) Model projections of the equilibrium climatic response to increased carbon dioxide. In: MacCracken MC, Luther FM (eds) Projecting the climatic effects of increasing carbon dioxide. US Dep Energy Rep DOE/ER-0237, Washington DC, pp 80–147

Shaver GR, Chapin FS III (1980) Response to fertilization by various plant growth forms in an Alaskan tundra: nutrient accumulation and growth. Ecology 61: 662–675

Shaver GR, Chapin FS III (1986) Effect of fertilizer on production and biomass of tussock tundra Alaska, USA. Arct Alp Res 18: 261–268

Shaver GR, Nadelhoffer KH, Giblin AE (1990) Biogeochemical diversity and element transport in a heterogeneous landscape, the north slope of Alaska. In: Turner MG, Gardner RH (eds) Quantitative methods in landscape ecology. Ecological Studies 82. Springer, Berlin Heidelberg New York, pp 105–125

Spatt PD, Miller PC (1981) Growth conditions and vitality of Sphagnum in a tundra community along the Alaska Pipeline Haul Road. Arctic 34: 48–54

Svensson BH, Rosswall T (1980) Energy flow through the subarctic mire at Stordalen. In: Sonesson M (ed) Ecology of a subarctic mire. Ecol Bull (Stockholm) 30: 283–301

Van de Geijn SC, Van Veen JA (1993) Implications of increased carbon dioxide levels for carbon input and turnover in soils. Vegetatio 104: 283–292

Van Veen JA, Paul EA (1981) Organic carbon dynamics in grassland soils. 1. Background information and computer simulation. Can J Soil Sci 61: 185–201

Walker DA, Everett KR (1987) Road dust and its environmental impact on Alaskan taiga and tundra. Arct Alp Res 19: 479–489

Walker DA, Lederer ND (1987) A toposequence study: site factors, soil physical and chemical properties and plant species cover. DOE-R4D Progr Data Rep, Inst Arct Apine Res, Boulder, 29 pp

Walker DA, Binnian E, Evans BM, Lederer ND, Nordstrand E, Webber PJ (1989) Terrain, vegetation and landscape evolution of the R4D research site, Brooks Range foothills, Alaska. Holarct Ecol 12: 238–261

Whitfield DWA (1977) Energy budgets and ecological efficiencies on Truelove Lowland. In: Bliss LC (ed) Truelove Lowland, Devon Island, Canada: a high arctic ecosystem. Univ Alberta Press, Edmonton, Canada, pp 607–620

Wielgolaski FE (1975) Principles in the use of wide-scale models on tundra data. In: Wielgolaski FE (ed) Fennoscandian tundra ecosystems. Part 2. Animals and systems analysis. Springer, Berlin Heidelberg New York, pp 245–250

Wynn-Williams DD (1980) Seasonal fluctuations in microbial activity in antarctic moss peat. Biol J Linn Soc 12: 11–28

Zak DR, Pregitzer KS, Curtis PS, Teeri JA, Fogel R, Randlett DL (1993) Elevated atmospheric CO_2 and feedback between carbon and nitrogen cycles. Plant Soil 151: 105–117

17 Hydrological Controls on Ecosystem Gas Exchange in an Arctic Landscape

B. Ostendorf, P. Quinn, K. Beven, and J. D. Tenhunen

17.1 Introduction

Excluding those areas disturbed by development, the distribution of plant communities over large expanses of Alaskan tundra has been determined by an integrated, long-term response to natural environmental gradients. Both in relatively flat coastal tundra (Webber 1978) and in the Foothills Province of the Alaskan North Slope (Ostendorf and Reynolds 1993; Chaps. 4 and 5, this Vol.), gradients in water availability caused by topography or microrelief play a particularly important role in determining landscape vegetation patterns. As a result, some aspects of these general patterns are predictable with relatively simple models based solely on topography. Ostendorf and Reynolds (1996; Chap. 14, this Vol.) developed topographically derived models for predicting landscape distribution of communities based on slope and local water discharge. Leadley et al. (Chap. 18, this Vol.) linked patterns of water discharge with a simple ecosystem model to predict landscape patterns of nitrogen availability, vegetation type, plant biomass, and net primary productivity within the Imnavait Creek watershed.

The success of these topographically derived models is attributable to the recognizable long-term equilibration of tundra ecosystems with physical climate factors especially hydrology and hydrological transport (see Fig. 14.9, this Vol.). To further examine the utility of these models for modeling ecological response to anthropogenic influences, we must also consider short-term dynamics and feedbacks for non-equilibrium situations. An extremely important and naturally occurring aspect of tundra ecosystem dynamics is the control by hydrological processes of net ecosystem carbon exchange (cf. Chap. 11, this Vol.). We have begun to model the coupling between carbon balance and spatial variation in topography, subsurface flow, and water table. These results are not only of interest with regard to global climate change and storage of carbon in tundra soils, but also with regard to subsequent short-term modeling of disturbances because of the implications for nutrient redistribution, productivity, and stability of communities.

In our mechanistic GAS-FLUX simulator (Chap. 14, this Vol.; Tenhunen et al. 1994), the coupling at the patch scale occurs via water table fluctuations, which directly influence soil aeration, soil respiration, and, thus, carbon balance. Whereas seasonal changes in phenology and leaf area development are considered in GAS-FLUX, seasonal variations in water table that alter these

J.F. Reynolds and J.D. Tenhunen (Eds.)
Ecological Studies, Vol. 120
© Springer-Verlag Berlin Heidelberg 1996

parameters and, ultimately, the magnitude of carbon dioxide flux, are not. In this chapter we describe an exploratory study to link GAS-FLUX with TOPMODEL, a quasidistributed hydrological model (Beven and Kirby 1979; Beven 1986a,b; Beven et al. 1988), for the Imnavait Creek watershed. The analysis focuses on the summer of 1986 from snowmelt (the first hydrograph recession is during the snow-free period of the snowmelt event) until the first freeze in August, a year during which required data are relatively complete and extensive drying of the watershed occurred during June and early July. We described the extent to which process information collected on very small time and spatial scales can be applied to landscape processes on longer time and larger spatial scales (Acock and Reynolds 1990).

17.2 Description of Models

17.2.1 Community Gas Exchange

GAS-FLUX simulates short-term (hourly) dynamics of canopy water and CO_2 exchange at the patch scale. Gas exchange is integrated over a vertically

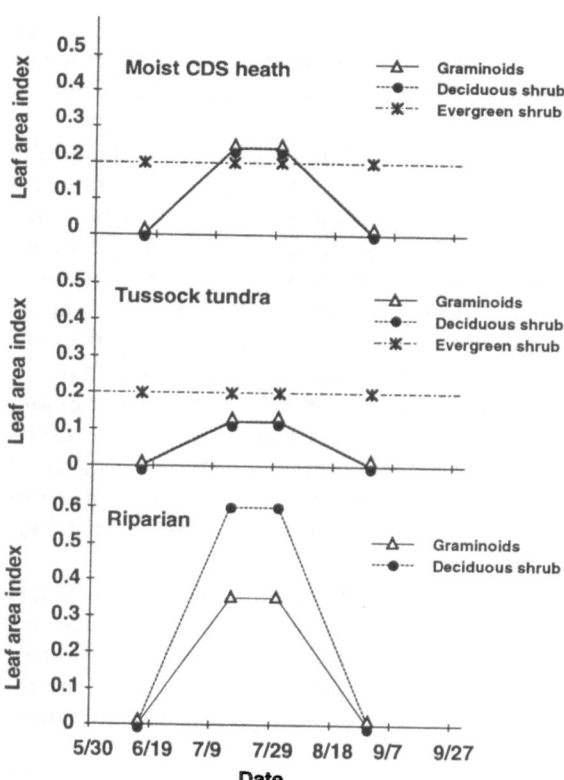

Fig. 17.1. Assumed seasonal changes in leaf area index (LAI) of graminoid species and deciduous shrubs in three plant communities used to represent vegetation development within the Imnavait Creek catchment during 1986. Evergreen shrubs are present only in tussock-tundra and heath communities, and LAI remains constant. Harvest data related to maximum LAI at mid-season are discussed in detail by Tenhunen et al. (1994). *CDS Cassiope*-dwarf shrub

stratified canopy (Chap. 14). Photosynthesis and transpiration of vascular plants are computed as a function of incident radiation, leaf temperatures, atmospheric CO_2 concentration, and vapor pressure deficit. Canopy humidity and wind speed profiles are estimated from above-canopy measurements based on field observations. Phenological development is included in the seasonal simulations via variation in leaf area index (LAI; Fig. 17.1). Additionally, the leaf scaling factors for carboxylation capacity of photosynthesis or electron transport capacity are varied seasonally to represent leaf maturation and senescence (cf. Chap. 11, this Vol.; Tenhunen et al. 1994) using linear functions similar to those in Fig. 17.1. Empirical equations describing soil respiration as a function of soil temperature and depth to water table were derived from field cuvette observations (cf. Chap. 11, this Vol.; Oberbauer et al. 1991, 1992). The model and its parameterization for three communities in the watershed – riparian, tussock tundra, and moist *Cassiope*-dwarf shrub heath (Chap. 5, this Vol.) – is described in Chapter 14 (this Vol.) and by Tenhunen et al. (1994).

17.2.2 Spatial Variation in Water Availability

Watershed models, such as T-HYDRO (Chap. 14, this Vol.; Ostendorf and Reynolds 1993) and HBV (Chap. 6, this Vol.; Bergström 1976), can predict the hydrological response of the Imnavait Creek watershed, but they do not consider spatial variability in water storage and saturation profiles with depth. Furthermore, their resolution is too coarse to represent patterns relevant to many biological processes. Theoretically, hillslope models that are based on a numeric solutions of the Darcy equation (e.g., the SHE model; Bathurst 1986) might be applied with high spatial and temporal resolution. Such models require extensive input data, e.g., hydrological conductivity, which varies greatly in both horizontal and vertical dimensions, and realistically such data are impossible to obtain (Beven 1993).

TOPMODEL is an alternative modeling approach that has been applied in numerous watersheds and has been successfully linked with an ecosystem model (Band et al. 1993). The theory underlying TOPMODEL and the derivation of the main dependencies are discussed in Beven (1986a,b), whereas the current application (with exception of the modifications discussed below) is summarized by Quinn et al. (1995). The model is not fully distributed, because lateral water flows are only implicitly computed.

Precipitation is routed vertically through a set of soil stores, first entering the root store (Fig. 17.2). In the case of tundra vegetation, water loss from the surface includes the evaporative flux from the moss surface ($T_{M,i}$) and transpiration from vascular plants ($T_{P,i}$). These fluxes influence the movement of water to the unsaturated zone from where it moves over time to the saturated zone ($Q_{V,i}$; Fig. 17.2). In the original version of TOPMODEL, evapotranspiration rate is linearly related to storage in the root zone resulting in a decrease in water loss as the soil dries, due to stomatal closure. However,

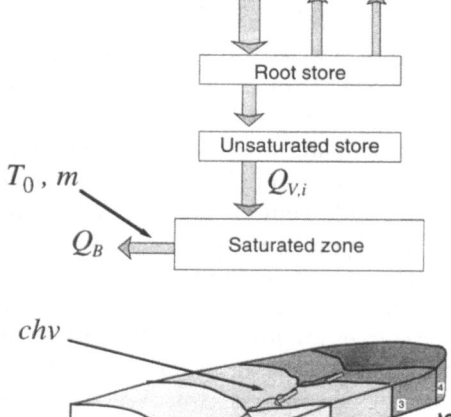

PPT $T_{M,i}$ $T_{P,i}$

Root store

Unsaturated store

T_0, m

$Q_{V,i}$

Q_B

Saturated zone

chv

Isochrone bands

Q_B

Fig. 17.2. Representation of TOPMODEL as related to the Imnavait Creek catchment. *PPT*, precipitation input; $T_{M,i}$, local moss evaporation; $T_{P,i}$, local vascular plant transpiration; $O_{V,i}$, local flux to the saturated zone; Q_B, watershed discharge. T_0, m, and *chv* are TOPMODEL parameters as described in text. Orientation of isochrone bands is shown

in tundra ecosystems field observations have consistently shown that the water supply to vascular plants is not restricted – predawn water potential and maximum stomatal conductance remain high, even during long dry periods (Stoner and Miller 1975; Oberbauer and Miller 1979, 1982; Chapin and Shaver 1985a; Matthes-Sears et al. 1988). Capillary rise of water is associated with the dense growth of *Sphagnum* mosses (Clymo and Hayward 1982), which tends to maintain hydration of the tundra surface. Although probably an oversimplification, we modified TOPMODEL to allow unrestricted moss and vascular plant water use from the saturated zone.

The water routing scheme and the distribution of soil water deficit are based on topography; the term "quasidistributed" characterizes this phenomenological description of lateral water movement. Drainage area (A) upslope from a particular landscape location and slope angle (β) at that location are combined into a topographic index in ($A/tan\,\beta$) (hereafter referred to as ATB). Whereas hydrograph dynamics are related to average catchment conditions, the ATB index allows estimation of local (individual map pixel) soil moisture. This is illustrated for the Imnavait Creek watershed in the top panel of Fig. 17.3, where each pixel was placed into one of 17 categories of hydrological similarity (see Sect. 14.2.1, this Vol.) based on the frequency distribution of the ATB index.

In applying TOPMODEL to the Imnavait Creek catchment, the landscape was divided into four isochrone bands with connected channel segments (Fig. 17.2). Discharge is estimated separately for each isochrone band, and there are longer delays in discharge for upstream bands, which influences the catchment hydrograph (Fig. 17.2). An accounting is separately maintained for stores in all

ATB categories (Fig. 17.3) and in all isochrone bands. Two parameters, T_o (lateral transmissivity at the soil surface) and m (rate of change in transmissivity with increasing soil moisture deficit), quantify the decline of discharge as a function of water storage in the saturated zone, whereas another, *chv*, characterizes the velocity of water flowing as overland flow and through the channel (Fig. 17.2; see Quinn et al. 1995 for methods related to estimating T_o, m, and *chv*). The hydrograph recession is influenced primarily by time-dependent changes in discharge from the hillslopes and secondarily by channel flow.

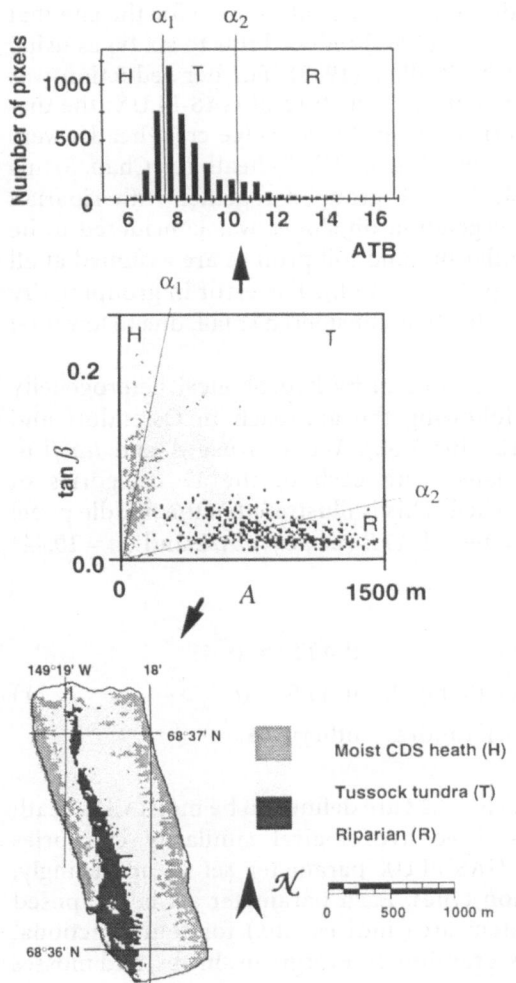

Predicted vegetation

Fig. 17.3. The distribution of three communities within the Imnavait Creek catchment as dependent on the TOPMODEL topographic index and determined by the threshold parameters (α_1 and α_2) as described in text

17.3 Coupling of Hydrology and Ecosystem Gas Exchange

17.3.1 Vegetation Distribution

A digital elevation model (DEM) for the watershed was generated from USGS topographic survey maps at the scale of 1:5000. Isolines (5-m contour) were digitized using ARC/INFO and interpolated with a method based on the steepest decent. The DEM obtained without flat areas and local water sinks allows water from any location to be routed to the outlet. Topographic analysis (identification of isochrone bands and the generation of the ATB map) was conducted according to Quinn et al. (1991, 1995).

Walker et al. (1989; Chap. 4) developed a vegetation map for the site that included 34 vegetation types. Stow et al. (1989) reduced this to six types using a classification scheme developed by Walker (1983). Further reduction was required to be compatible with the current structure of GAS-FLUX: the disturbed site in the area (gravel quarry) was excluded; ridge crest heaths were grouped together as moist *Cassiope*-dwarf shrub (CDS) heath (cf. Chap. 5, this Vol.); water tracks low on the slope were grouped together with riparian communities; and the remaining vegetation on slopes was considered to be homogeneous tussock tundra. Similar organic soil profiles are assumed at all locations in the catchment (cf. Chap. 11, this Vol.). The error in grouping dry heath types together with moist CDS heath is considered small, due to low total surface cover (<5% of catchment).

Whereas the ATB index is used to quantify hydrological heterogeneity in the watershed (Sect. 17.2.2), following the approach in Ostendorf and Reynolds (1996; Chaps. 14 and 18, this Vol.). We also use A and $\tan\beta$ to define the vegetation type associated with each of the 17 categories of hydrological similarity in the watershed. This is illustrated in the middle panel of Fig. 17.3 where two threshold values of ATB (i.e., $\alpha_1 = 7.22$ and $\alpha_2 = 10.44$) were used:

$$\text{Predicted vegetation} = \begin{cases} \text{Riparian} & \text{if } ATB > \alpha_2 \\ \text{Moist CDS heath} & \text{if } ATB < \alpha_1 \\ \text{Tussock tundra} & \text{otherwise} \end{cases} \quad (1)$$

For example, all map pixels with $ATB < 7.22$ are defined to be moist CDS heath vegetation and fall into one of three hydrological similarity categories (Fig. 17.3, top panel). A specific GAS-FLUX parameter set is, accordingly, associated with the three vegetation types. Each parameter set is composed of specific structural data (LAI, stem area indices, etc.) for four functional groups of plants: deciduous shrubs, graminoids, evergreen shrubs, and mosses (see Table 14.2, this Vol.; Tenhunen et al. 1994). Equation (1) is a quick, simple approach for linking GAS-FLUX and TOPMODEL (see Fig. 17.3) and, on pixel by pixel comparison of predicted vegetation with the original vegetation map of Walker et al. (1889), there is correspondence of 73% (Fig. 17.3, lower panel). Equation (1) results in only slightly different predictions than those using slope and discharge from T-HYDRO [see Eq. (3) in Chap. 18, this Vol.].

17.3.2 Spatial Variation in Water Table

Overall ecosystem carbon flux is positive or negative depending on the balance between aboveground CO_2 fixation and soil respiration. The largest variations in soil respiration we have observed in tundra occur in response to changes in water table and soil aeration (Chap. 11, this Vol.). The coupling of GAS-FLUX to TOPMODEL allows us to consider these dynamics, because TOPMODEL estimates water table depth (w) from the saturation deficit S_i at location i. $\bar{S}$ is the integral of the available pore space (p) over soil depth (s) between the soil surface and the water table (w):

$$\bar{S} = \int_0^w P(s)ds. \tag{2}$$

$\bar{S}$ refers to the part of the pore space that rapidly drains under field conditions, in contrast to porosity values derived from oven-dried soil cores.

The function assumed to describe Eq. (2) for soils of the Imnavait Creek watershed is illustrated in Fig. 17.4. The integral of pore space is evaluated below a reference level of −5 cm at the base of the green moss layer that has been used as $s = 0$ in the soil respiration model. Seasonal changes in soil moisture decrease as bulk density increases with depth (shaded area; Fig. 17.4). At ca. 20 cm, a discontinuity occurs due to the transition from organic to

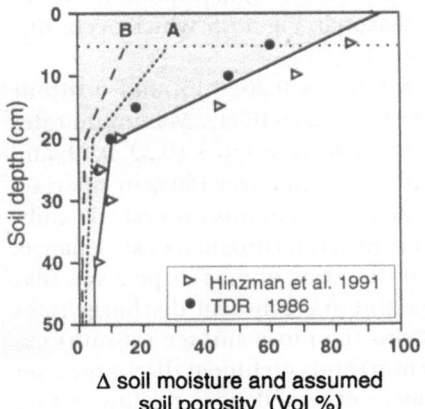

Fig. 17.4. Assumed relationship between vertical profiles in the maximal changes in soil moisture (*symbols*) and soil porosity. *Broken lines A* and *B* indicate porosity profiles for two versions of the hydrological model with and without interception and surface retention store as discussed in text. *TDR* Time domain reflectometry

mineral soil. Estimating the actual pore space filled and emptied along with the measured changes in soil moisture is extremely difficult; pore space measurements in these arctic soils are scarce and complications arise due to strong swelling and shrinking of substrates with changes in hydration. To describe the vertical profile of porosity with depth, we used a piecewise linear function calibrated with depth according to maximal changes in midslope tussock-tundra time domain reflectometry (TDR) soil moisture (Hinzman et al. 1991, pers. comm.). Below 40 cm a constant soil porosity is assumed. As indicated by lines A and B in Fig. 17.4, two versions of the hydrological model (with and without interception and surface retention store) were used. In cases A and B coefficients of the linear model were adjusted to reproduce observed water table depths as a function of S_i in specific study plots after a long period of watershed drying (see Fig. 11.4, this Vol.; see also Sect. 17.5.1). Additional field studies are required to better define the manner in which water table depth and water content profiles simultaneously vary in time and space.

17.4 Water Balance and Seasonal Changes in Water Fluxes

17.4.1 Evapotranspiration

Because stomatal behavior of vascular plants and evapotranspiration from the moss is considered unaffected by soil water availability (Sect. 17.2.2), the seasonal course of both evapotranspiration and CO_2 uptake of vascular plants and mosses may be computed independently and uncoupled from TOPMODEL for each vegetation zone. Radiation, air temperature, humidity, and wind speed inputs were assumed homogeneous over the catchment. The predicted transpiration rates and evaporation from the moss surface (illustrated in Fig. 17.5 for the tussock-tundra zone) reflect the cyclic changes between fair weather and periods with overcast, low temperature, and potential precipitation (Fig. 17.6, lower panel). Reductions in transpiration of vascular plants are apparent early and late in the season in Fig. 17.5, which occur due to changes in LAI (expansion and senescence).

Maximum vascular plant transpiration rates of 0.20, 0.13, and 0.18 mm h^{-1} occur in riparian, tussock tundra, and heath, respectively. Maximum rates of evaporation from mosses in each of these vegetation types (0.22, 0.19, and 0.15 mm h^{-1}, respectively) varies, due to differences in cover (60% in riparian, 50% in tussock tundra, and 40% in heath). Vegetation in this watershed exhibits a major control on the long-term water balance (cf. Hinzman et al., Chap. 6, this Vol.). Diurnal maxima in evapotranspiration rate of mosses plus vascular plants are comparable in magnitude to maximum catchment discharge rates during heavy storms. Total evaporation from the moss surface (56 mm) exceeds both vascular plant transpiration (39 mm) and catchment discharge over the summer study period (37 mm). This may be expected given the low stature

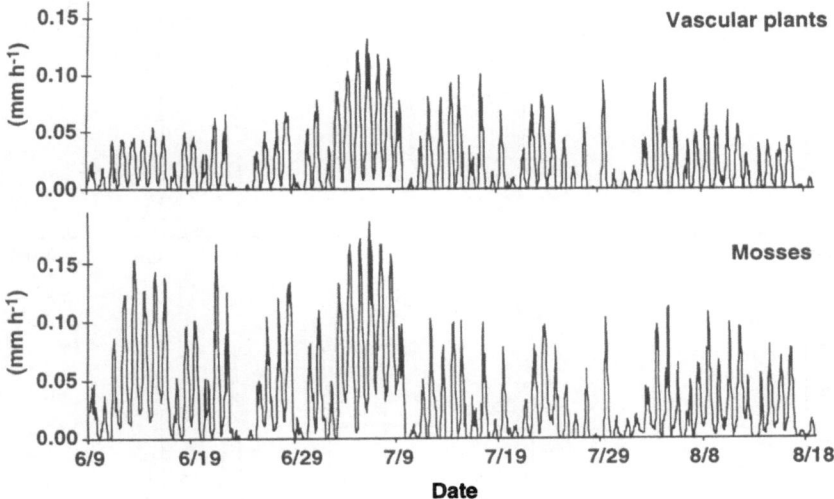

Fig. 17.5. Seasonal course of predicted vascular plant transpiration and evaporation from the moss surface in tussock-tundra vegetation zone for summer of 1986. Rates were very similar in heath and riparian zones, but proportionally shifted according to differences in LAI and percent moss cover

of the vegetation, extensive moss cover, and the prevalence of windy conditions. Very few experimental studies have directly examined evaporation from mosses. As discussed in Chapter 14 (this Vol.), a constant transfer resistance for water vapor is assumed at the moss surface (based on Alpert and Oechel 1984), whereby actual time-dependent changes in this transfer resistance could be very important to understanding the tundra water balance.

17.4.2 Discharge

Best-fit solutions were obtained with a very small value for m (0.003 m) in conjunction with an apparently very high transmissivity for the watershed at saturation ($T_o = 6\,m^2\,h^{-1}$). The model does not react very sensitively to changes in the parameter chv, and it was set in all simulations at $100\,m\,h^{-1}$. These settings define a very fast decline of the hydrograph (Fig. 17.6, upper two panels) during the late snow melt period and after major storm events in summer (with twice the thaw depth) equally well. As discussed in Chapters 6 and 9 (this Vol.), hydraulic conductivity decreases rapidly with depth, and discharge dynamics depend on lateral transfer processes in upper soil layers (see also Everett and Ostendorf 1988). Using these values for m and T_o, an essentially immediate transfer of water occurs from the unsaturated zone to the saturated zone, which eliminates the need for an unsaturated store (cf. Figs.

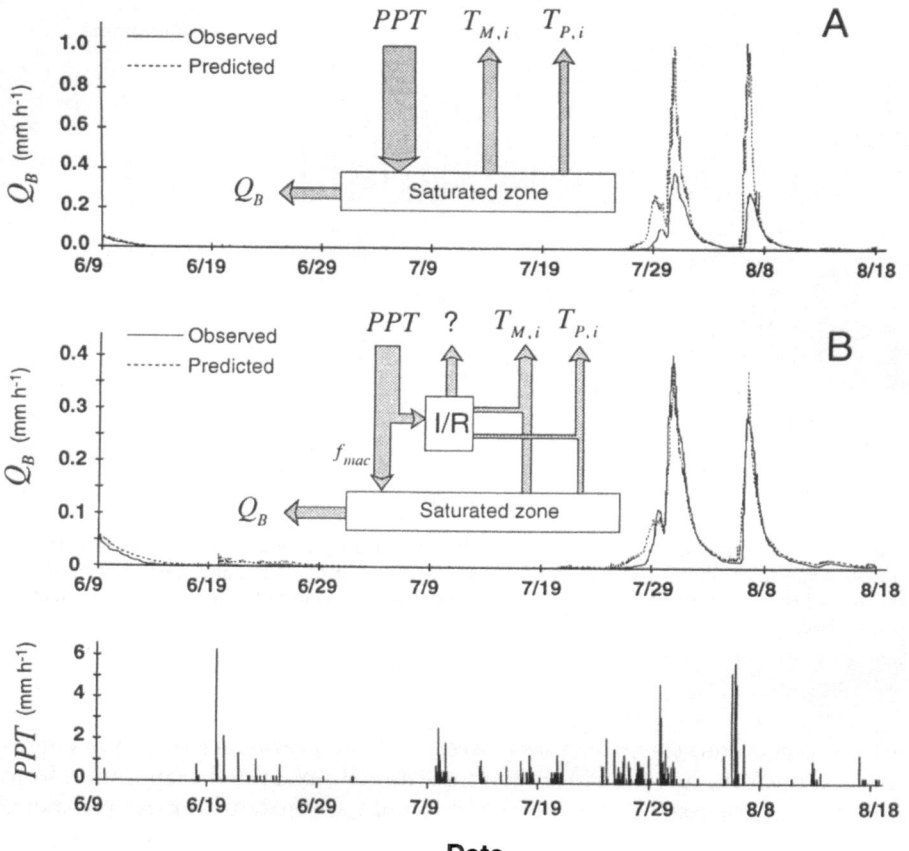

Fig. 17.6. *Upper panel:* diagram of water routing in TOPMODEL version A together with observed and predicted discharge (Q_B) for the Imnavait Creek catchment during summer 1986. *Center panel* Diagram of water routing in TOPMODEL version B and discharge as in upper panel. I/R indicates interception/retention store at the tundra surface. Discharge of $1\,\mathrm{mm\,h^{-1}} = 0.6\,\mathrm{m^3}$ $\mathrm{s^{-1}}$ for the $2.18\,\mathrm{km^2}$ catchment. *Lower panel* Precipitation input measured on the northeast ridge during summer 1986 (Fig. 9.2 in Chap. 9, this Vol.)

17.2 and 17.6, upper two panels). It is noteworthy that the shape of the hydrograph is correctly predicted with only three parameters.

Although the general time course of discharge is predicted well, the magnitudes of peak discharge are overestimated. Thus, version A of the model failed to represent important dynamics during the storm periods. The predicted seasonal sum of evapotranspiration from vascular plants and mosses from GAS-FLUX is 95 mm. The estimated evapotranspiration during this period based on measured precipitation (155 mm) and discharge (37 mm) was 118 mm. The difference of 23 mm may result from a variety of sources including:

1. Inaccurate spatial estimates of precipitation;
2. Inaccurate discharge estimates resulting from permafrost melting around the weir;
3. An undefined catchment boundary related to the flat topography at the southern extreme of the experimental area;
4. Improper LAI estimates in GAS-FLUX;
5. Lack of an interception store;
6. Constant transfer resistance for water vapor assumed at the moss surface.

17.4.3 Interception and Surface Water Retention

To further examine seasonal water dynamics, an interception and surface water retention store was included in the model structure (Fig. 17.6, center panel). In model version B, only a portion of precipitation enters the saturated zone immediately as macropore flow (f_{mac}). The remainder of the precipitation is withheld in an interception store that includes rehydration and expansion of moss tissues (*Sphagnum* makes up ca. 50% of living biomass in the system). To fit the hydrograph with model B we assumed that 50% of total precipitation was retained by interception and surface absorption (I/R store, Fig. 17.6), and that 55% of transpiration is subsequently derived from this store (the remaining 45% comes from the saturated zone). The three other TOPMODEL parameters were the same as indicated previously for model A. The hydrograph fit has an efficiency of $r^2 = 0.95$. Whereas errors of the type mentioned in Section 4.2, especially short-term interception storage, must be more carefully considered in future studies of tundra water balance, the results suggest that dynamics of water storage in mosses is probably an extremely important component that determines system response.

17.5 Carbon Balance and Seasonal Changes in Carbon Fluxes

17.5.1 Predicted Water Table and Soil Respiration

The differences in versions A and B of TOPMODEL affect the time course of saturation deficit (Fig. 17.7, upper two panels) and, consequently, water table depth predictions (Fig. 17.7, center two panels). Nevertheless, both versions are in reasonable agreement with TDR point estimates of soil moisture in tussock tundra. The correspondence obtained with version B is striking. A strong correlation exists between surface moss water content and water table depth (Clymo and Hayward 1982). Thus, both versions of the model may be brought into agreement with local changes in water table in the three vegetation zones after a long "drought" period by adjusting the vertical profile assumed for porosity as shown in Fig. 17.4. Seasonal maximum water table depths observed during 1990 in heath, tussock tundra, and riparian areas were 25, 23, and 16 cm, respectively (indicated as dashed lines in Fig. 17.7, two center panels). With additional field data this adjustment of the porosity function can be improved.

The effects of seasonal variation in water table depth, as well as short-term diurnal temperature changes, on average soil respiration rate are shown in Fig. 17.7 (lower two panels). Whereas soil respiration in heath and tussock-tundra areas is similar, the inhibitory effects of near-surface water tables on soil respiration in the riparian zone are evident throughout the season, especially after August storm events. Although uncertainty remains with regard to the best formulation of TOPMODEL, both versions applied here permit assessment of spatial variation in water table and consequent influences of water table variation on soil respiration and net CO_2 exchange. The differences shown in soil respiration as derived from either model A or B are not large.

The dynamics of soil respiration over time and along a ridge-to-ridge transect across the catchment are illustrated in Fig. 17.8. The banded pattern indicates diurnal effects of soil temperature. Maximum respiration occurs during a 7-day period with maximum water table depth just previous to July 10. Respiration is predicted to continue at substantial levels even during periods with below-zero air temperature (June 17, July 11, August 19). The lowest rates during these cold periods are ca. -4.2 mmol $m^{-2} h^{-1}$ in heath and tussock tundra, and -3.5 mmol $m^{-2} h^{-1}$ in riparian areas. Variations in water table during the storm events in August strongly reduce soil respiration – not only adjacent to Imnavait Creek, but extensively up the lateral slopes. Zero soil respiration (the white patterning in Fig. 17.8) indicates saturated conditions and overland flow. Thus, from simulations during the single summer of 1986, a series of pictures are obtained that demonstrate the spectrum of response potentials. The signature shown in Fig. 17.8 will change annually in response to long-term climate fluctuations and trends.

17.5.2 Predicted Watershed Level Net CO_2 Balance

Maps of seasonal net CO_2 exchange in the watershed for both versions of the coupled GAS-FLUX and TOPMODEL simulator are shown in Fig. 17.9. The cover extent of each vegetation type is easily recognized by abrupt changes in the magnitude of net CO_2 exchange. At these boundaries LAI changes as a step function, and it is clear that future efforts must consider more gradual shifts in community characteristics perhaps by utilizing remotely sensed vegetation index in addition to topographic information. In contrast, smooth gradients in community response are apparent within the band of tussock tundra, with pixels further downslope exhibiting more positive CO_2 fixation. Riparian areas have the greatest leaf area and highest water table, and accordingly are a strong sink for CO_2. In version A, the entire landscape is a net sink for CO_2, whereas

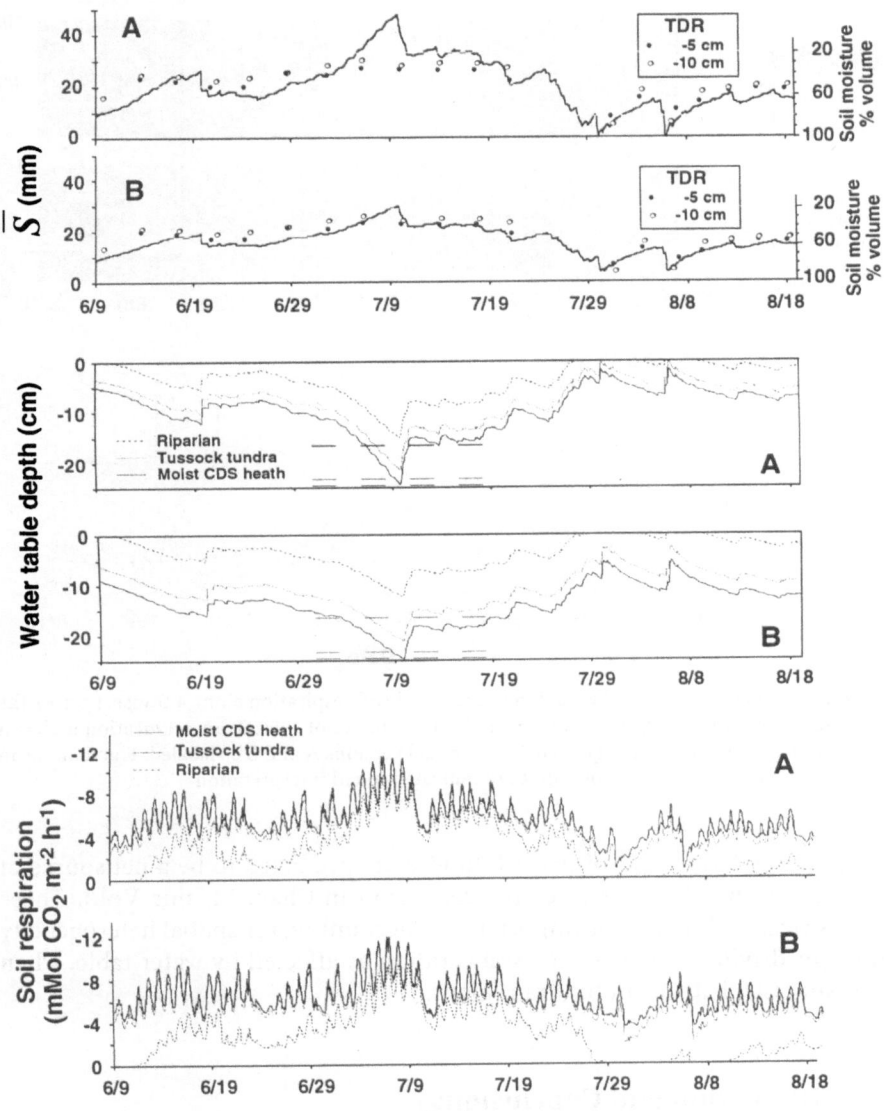

Fig. 17.7 Seasonal course during summer 1986 for predicted average catchment saturation deficit (*upper two panels*), average water table depth in each community (*center two panels*), and average soil respiration in each community (*lower two panels*) as obtained with TOPMODEL versions A and B (as labeled) linked with the soil respiration submodel of GAS-FLUX. *Symbols* in upper two panels indicate measured TDR soil moisture at tussock-tundra runoff plots (Fig. 9.2 in Chap. 9, this Vol.). *Dashed lines* in center two panels indicate maximum depth of water table measured in the three communities after an extensive period of drying in 1990

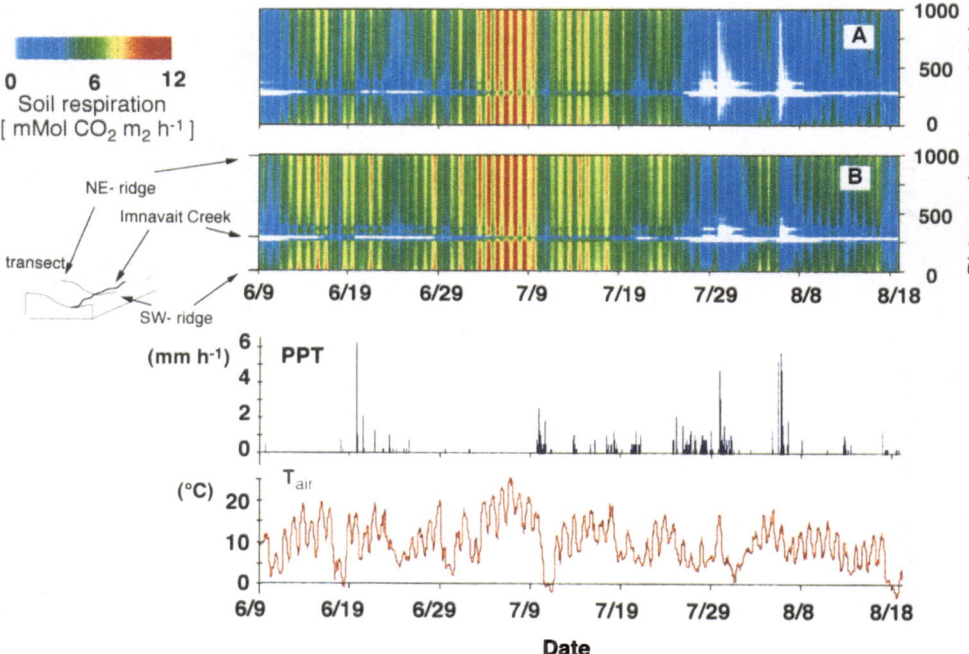

Fig. 17.8. Spatial and temporal variation in predicted soil respiration along a transect across the catchment perpendicular to Imnavait Creek during summer of 1986. Transect location is shown in map of Fig. 17.9. Results are given for both model versions A and B as labeled. Fluctuations in air temperature and precipitation input are indicated to aid interpretation

in version B, large areas of tussock tundra are predicted to be a net source of CO_2 during this dry year (see also discussion in Chap. 11, this Vol.). These results clearly illustrate the importance of accounting for spatial heterogeneity in major driving variables, e.g., soil aeration as affected by water table, when assessing CO_2 balance in tundra regions.

17.6 Discussion and Conclusions

As an initial step in linking transport and ecosystem process models, this study must be viewed as relatively successful. New insight has been gained both with regard to tundra ecosystem hydrology and gas exchange. Future research efforts must be directed at adjusting and optimizing the complexity of the model descriptions in order to obtain computational efficiency. Simplifications of GAS-FLUX are visualized that maintain ecosystem sensitivity to relevant environmental factors and ecosystem physiology. For general applications in tundra regions, a capability must still be developed for describing

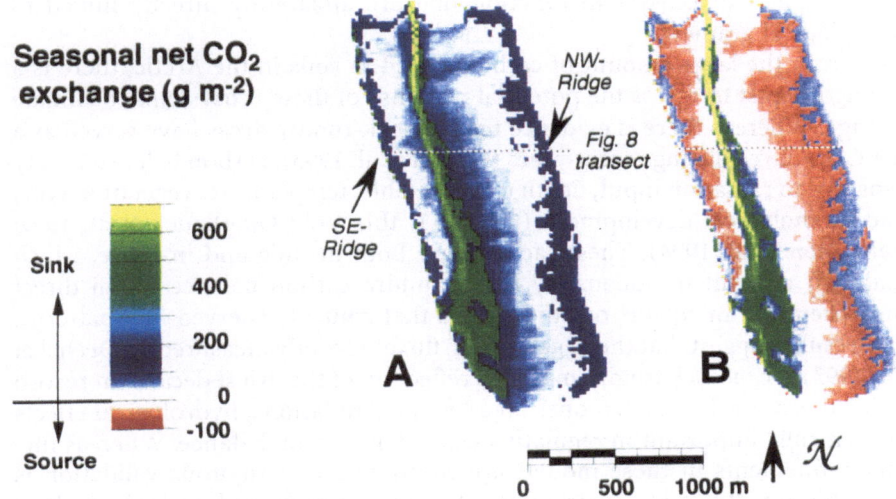

Seasonal net CO$_2$ exchange (g m^{-2})

Sink

600
400
200
0
-100

Source

NW-Ridge
Fig. 8 transect
SE-Ridge

A

B

0   500   1000 m

N

Fig. 17.9 Predicted spatial variation within the Imnavait Creek catchment in seasonal net CO$_2$ exchange for the summer season of 1986 as described in text. Results are given for both model versions A and B as labeled

carbon balances in dry heath habitats with shallow organic soil horizons and high lichen biomass (Chap. 5; see also Chap. 18, this Vol.). The most appropriate formulation of water flows and storage in TOPMODEL must be reconsidered, and further efforts must be made to include spatial detail. The clear spatial patterning of soil types along the hillslope, and the observation that vegetation pattern is predictable from the topographic index, suggests that it is possible to link more complex assumptions about soil and vegetation to the topographic index.

Topographically related variables, such as drainage, wind exposure, and snow distribution, have a broad influence on general aspects of ecosystem structure and function in arctic landscapes (Bliss 1988). Water limitation is viewed as a major determinant of vegetation patterning (Bliss 1962; Jonasson 1982; Billings 1987; Walker et al. 1989); Peterson and Billings (1980) used soil moisture to classify vegetation sequences along the margins of the Meade River; and the first axis of species ordination studies of tundra systems is generally correlated with moisture gradients (Webber 1977; Jasienuik and Johnson 1982). Bliss et al. (1984) have suggested that species diversity and plant cover are moisture controlled. *Eriophorum vaginatum* located in water-track drainages exhibits a tenfold increase in productivity (Chapin et al. 1988). Matthes-Sears et al. (1988) reported higher aboveground biomass, tissue nitrogen concentrations, and tissue phosphorus concentrations of *Betula nana* and *Salix pulchra* growing in water tracks. Spatial modeling of the type explored in this chapter offers new potentials for analyzing the many aspects of ecosystem function that are apparently directly linked to hydrological gradients.

Given the large amount of carbon stored in soils in the Arctic, there is a strong interest to assess the potential response of these ecosystems to climate change. Whereas there is evidence that tussock-tundra areas have served as a net CO_2 source during recent years (Oechel et al. 1993), carbon balance is very sensitive to radiation input, depth to water table, temperature, vegetation type, and phenological development (Chap. 11, this Vol.; Oberbauer et al. 1992; Tenhunen et al. 1994). These factors vary both in time and in space, which makes it difficult to adequately assess tundra carbon balances from direct measurements or to sort out the factors that control observed flux patterns. Our results suggest that the negative CO_2 fluxes recently measured by Oechel et al. (1993) for tussock tundra may be a reflection of the driest decade on record in northern Alaska. As demonstrated by our simulations, hydrological effects are critically important in regulating ecosystem carbon balance. Whereas further refinements in these models are continuing and rigorous validation is desirable, our study demonstrates that linked ecosystem and spatial hydrology models are an essential tool for examining the effects of disturbances or climate change in tundra ecosystems.

Acknowledgments. This work was supported by the U.S. Department of Energy (grant no. DEFG03-84ER60250) and German Federal Ministry for Science and Technology (grant no. BE051-0339476A).

References

Acock B, Reynolds JF (1990) Model structure and data base management. In: Dixon RK, Meldahl RS, Ruark GA, Warren WG (eds) Process modeling of forest growth responses to environmental stress. Timber Press, Portland, pp 169–179

Alpert P, Oechel WC (1984) Microdistribution and water loss resistances of selected bryophytes in an Alaskan *Eriophorum* tussock tundra. Holarct Ecol 7: 111–118

Band LE, Patterson P, Nemani R, Running SW (1993) Forest ecosystem processes at the watershed scale: incorporating hillslope hydrology. Agric For Meteorol 63: 93–126

Bathurst JC (1986) Physically based distributed modelling of an upland catchment using the System Hydrologique Europeen. J Hydrol 87: 97–102

Bergström S (1976) Development and application of a conceptual runoff model for Scandinavian catchments. Swed Meteorol Hydrol Inst, Norrköping, Sweden, Rep RH07, 118 pp

Beven KJ (1986a) Runoff production and flood frequency in catchments of order *n*: an alternative approach. In: Gupta VK, Rodriguez-Irtube I, Wood EF (eds) Scale problems in hydrology. Reidel, Dordrecht, pp 191–219

Beven KJ (1986b) Hillslope runoff processes and flood frequency characteristics. In: Abrahams AD (ed) Hillslope processes. Allen and Unwin, Boston, pp 187–202

Beven K (1993) Prophecy, reality and uncertainty in distributed hydrological modeling. Adv Water Res 16: 41–51

Beven KJ, Kirkby MJ (1979) A physically based, variable contributing area model of basin hydrology. Hydrol Sci Bull 24: 43–69

Beven KJ, Wood EF, Sivapalan M (1988) On hydrological heterogeneity – catchment morphology and catchment response. J Hydrol 100: 353–375

Billings DW (1987) Constraints to plant growth, reproduction, and establishment in arctic environments. Arct Alp Res 19: 357–365

Bliss LC (1962) Adaptations of arctic and alpine plants to environmental conditions. Arctic 15: 117–144

Bliss LC (1988) Arctic tundra and polar desert biome. In: Barbour MG, Billings WD (eds) North American terrestrial vegetation. Cambridge Univ Press, Cambridge, pp 1–32

Bliss LC, Svoboda J, Bliss DI (1984) Polar deserts, their plant cover and plant production in the Canadian High Arctic. Holarct Ecol 7: 305–324

Chapin FS III, Shaver GR (1985a) Arctic. In: Chabot BF, Mooney HA (eds) Physiological ecology of North American plant communities. Chapman and Hall, New York pp 16–40

Chapin FS III, Fetcher N, Kielland K, Everett KR, Linkins AE (1988) Productivity and nutrient cycling of Alaskan tundra: enhancement by flowing soil water. Ecology 69: 693–702

Clymo RS, Hayward PM (1982) The ecology of Sphagnum. In: Smith AJE (ed) Bryophyte ecology. Chapman and Hall, New York, pp 229–289

Everett KR, Ostendorf B (1988) Hydrology and geochemistry of a small arctic drainage basin in upland tundra, northern Alaska. In: Senneset K (ed) Proc 5th Int Permafrost Conf, vol 1. Tapir, Trondheim, Norway, pp 574–579   –

Hinzman LD, Kane DL, Gieck RE, Everett KR (1991) Hydrological and thermal properties of the active layer in the Alaskan Arctic. Cold Reg Sci Tech 19: 95–110

Jasienuik MA, Johnson EA (1982) Peatland vegetation organization and dynamics in the western subarctic, Northwest Territories, Canada. Can J Bot 60: 2581–2593

Jonasson S (1982) Organic matter and phytomass on three north Swedish tundra sites, and some connections with adjacent tundra areas. Holarct Ecol 5: 367–375

Matthes-Sears U, Matthes-Sears WC, Hastings SJ, Oechel WC (1988) The effects of topography and nutrient status on the biomass, vegetative characteristics, and gas exchange of two deciduous shrubs on an arctic tundra slope. Arct Alp Res 20: 342–351

Oberbauer S, Miller PC (1979) Plant water relations in montane and tussock tundra vegetation types in Alaska. Arct Alp Res 11: 69–81

Oberbauer S, Miller PC (1982) Growth of Alaskan tundra plants in relation to water potential. Holarct Ecol 5: 194–199

Oberbauer SF, Tenhunen JD, Reynolds JF (1991) Environmental effects on CO_2 efflux from water track and tussock tundra in Arctic Alaska, USA. Arct Alp Res 23: 162–169

Oberbauer SF, Gillespie CT, Cheng W, Gebauer R, Sala Serra A, Tenhunen JD (1992) Environmental effects on CO_2 efflux from riparian tundra in the northern foothills of the Brooks Range, Alaska, USA. Oecologia 92: 568–577

Oechel WC, Hastings SJ, Vourlites G, Jenkins M, Riechers G, Grulke N (1993) Recent change of arctic tundra ecosystems from a net carbon dioxide sink to a source. Nature 361: 520–532

Ostendorf B, Reynolds JF (1993) Relationships between a terrain-based hydrologic model and patch-scale vegetation patterns in an arctic tundra landscape. Landscape Ecol 8: 229–237

Ostendorf B, Reynolds JF (1996) A model of arctic tundra vegetation derived from topographic gradients. (submitted)

Peterson KM, Billings WD (1980) Tundra vegetational patterns and succession in relation to microtopography near Atkasook, Alaska. Arct Alp Res 12: 473–482

Quinn p, Beven K, Chevallier P, Planchon O (1991) The prediction of hillslope flow paths for distributed hydrological modelling using digital terrain models. Hydrol Proc 5: 59–79

Quinn P, Ostendorf B, Beven KJ, Tenhunen JD (1995) Spatial and temporal soil moisture predictions for an Alaskan catchment using TOPMODEL. (in preparation)

Stoner WA, Miller PC (1975) Water relations of plant species in the wet coastal tundra at Barrow, Alaska. Arct Alp Res 7: 109–124

Stow D, Burns B, Hope A (1989) Mapping arctic tundra vegetation types using digital SPOT/HRV-XS data. A preliminary assessment. Int J Rem Sens 10: 1451–1457

Tenhunen JD, Siegwolf RA, Oberbauer SF (1994) Effects of phenology, physiology, and gradients in community composition, structure, and microclimate on tundra ecosystem CO_2 exchange. In: Schulze E-D, Caldwell MM (eds) Ecophysiology of photosynthesis. Springer, Berlin Heidelberg New York, pp 431–460

Walker DA (1983) A hierarchical tundra vegetation classification especially designed for mapping in northern Alaska. Proc 4th Int Permafrost Conf. Natl Academic Press, Washington DC, pp 1332–1337

Walker DA, Binnian E, Evans BM, Lederer ND, Nordstrand E, Weber PJ (1989) Terrain, vegetation, and landscape evolution of the R4D research site, Brooks Range Foothills, Alaska. Holarct Ecol 12: 238–261

Webber PJ (1977) Tundra primary productivity. In: Ives JD, Barry RG (eds) Arctic and alpine environments. Methuen, London, pp 445–473

Webber PJ (1978) Spatial and temporal variation of the vegetation and its production, Barrow, Alaska. In: Tieszen L (ed) Vegetation and production ecology of an Alaskan Arctic tundra. Ecological Studies 29. Springer, Berlin Heidelberg New York, pp 37–112

18 Road-Related Disturbances in an Arctic Watershed: Analyses by a Spatially Explicit Model of Vegetation and Ecosystem Processes

P. W. Leadley, H. Li, B. Ostendorf, and J. F. Reynolds

18.1 Introduction

Landscape models have proven very useful in assessing historical change in vegetation patterns, for predicting the impacts of human disturbance on ecosystems, and for developing strategies to manage natural resources (e.g., Shugart 1984; Turner 1987; Costanza et al. 1990; Turner and Dale 1991; Wu and Levin 1994). Landscape models may implicitly or explicitly consider the *spatial heterogeneity* of system properties such as plant biomass, soil nutrient concentration, and topography – defined by either qualitative indices (e.g., patchiness, diversity, contagion) and/or quantitative indices (e.g., autocorrelation, variance, trend) (Li and Reynolds 1995). Accounting for such spatial heterogeneity has been shown to be essential for modeling ecosystem response to disturbance (Turner 1989; Costanza et al. 1990; Turner and Dale 1991; DeAngelis and White 1994). Although the use of simplified, aggregate models to represent vegetation and ecosystem processes (particularly at the scale of a landscape) has inherent dangers (Bonan 1993), the questions posed by resource managers require the development of models that summarize our "state-of-the-art" ecological knowledge and realistically represent the dynamic function of ecosystems in time and space.

Most existing ecosystem models for the Arctic emphasize properties at the level of the plant or plant community and do not incorporate landscape level properties (see review in Reynolds and Leadley 1992). Reich et al. (1991) and McGuire et al. (1992) parameterized a spatially explicit model for arctic ecosystems, but it did not include transport processes and operates at a grid size (0.5° latitude × 0.5° longitude) that is too large to assess the impacts of common disturbances (Fig. 3.2 in Chap. 3, this Vol.). In this chapter we present T-MAP (Terrain Models for Arctic Processes), a set of spatially explicit models for examining the effects of road construction on tundra ecosystems. T-MAP predicts landscape patterns of water discharge, nitrogen (N) availability, vegetation types, plant biomass, and net primary productivity within the Imnavait Creek watershed in the foothills of the Brooks Range in Alaska. The development of T-MAP was motivated by the R4D goal of predicting the potential effects of energy-related development on ecosystems of the North Slope of Alaska (Chaps. 1 and 14, this Vol.).

We describe T-MAP and present comparisons of simulated ecosystem properties with a vegetation index derived from satellite measurements of

J.F. Reynolds and J.D. Tenhunen (Eds.)
Ecological Studies, Vol. 120
© Springer-Verlag Berlin Heidelberg 1996

spectral reflectance. We used T-MAP to explore the ecosystem consequences of several scenarios for hypothetical road construction through the Imnavait Creek watershed including: (1) the obstruction of natural water drainage, and (2) the deposition of dust originating from high-speed vehicle traffic. We focused on road-related disturbances, because the Dalton Highway that serves the trans-Alaskan pipeline is one of the most prominent energy-related ecosystem disturbances in the foothills of the Brooks Range (Chap. 3, this Vol.). Gravel roads built to support energy exploration in the Arctic alter ecosystem properties mainly by (1) disrupting natural water drainage and (2) via dust deposition that accompanies the high-speed vehicle traffic (Chap. 15, this Vol.). The disruption of water drainage can cause impoundments and result in severe thermal erosion (i.e., thermokarst) particularly in areas with massive ground ice (Sect. 3.3.4.7, this Vol.). Dust damages or eliminates most of the acidophyllic plant species, e.g., *Sphagnum* mosses and lichens (Walker and Everett 1987; Meininger and Spatt 1988; Sect. 3.3.4.6, this Vol.). Our simulations with T-MAP, although preliminary, provide information that may be useful for considering options for road placement across watersheds, and suggest that this modeling approach has high potential for development as a management tool in arctic landscapes.

18.2 Environmental Gradients and Vegetation Distribution

18.2.1 Vegetation and Topography

Walker et al. (1989) grouped the vegetation of the Brooks Range Foothills into five main classes:

1. Dry types dominated by ericaceous dwarf shrubs and lichens;
2. Moist types dominated by mesic dwarf shrubs, sedges, and mosses;
3. Wet types dominated by sedges, dwarf shrubs, and mosses;
4. Aquatic types dominated by sedges, grasses, or forbs;
5. Barren or partially vegetated areas.

In developing T-MAP we focused on the first three types following the definitions used by Oberbauer et al. (Chap. 11, this Vol.) and Ostendorf et al. (Chap. 17, this Vol.), i.e., we refer to the dry types as *heath*, the moist types as *tussock tundra*, and the wet types as *riparian*. These vegetation types account for more than 95% of the vegetation in the Imnavait Creek and adjacent watersheds (Walker et al. 1989; Chap. 4, this Vol.).

The Imnavait Creek watershed is elongated in the N–S direction with the creek running northward within the basin. Both ridges of the watershed are dominated by dry heath vegetation. The southwest facing slope is dominated by moist tussock tundra and is characterized by the presence of well-developed water tracks. The northeast facing slope of the watershed is steeper with a relatively narrow band of tussock tundra vegetation separating heath from riparian vegetation. The watershed basin – characterized by very low slopes – is dominated by wet riparian vegetation.

18.2.2 Role of Water and Light

Soil moisture is one of the most important determinants of the distribution, biomass, and productivity of poikilohydric plants in the Arctic (Fig. 14.9, this Vol.). Along a moisture gradient in the Imnavait Creek watershed, Tenhunen et al. (1992) reported a transition from a lichen-dominated system (lowest water availability) to one dominated by *Sphagnum* moss, and then to one where poikilohydric plants were eliminated (highest water availability). Lichens are restricted to relatively dry sites, because mosses outcompete the slower growing lichens as surface moisture increases. Mosses grow in abundance in most tundra ecosystems with the exception of the driest and wettest habitats. Moss biomass and productivity generally increase with increasing moisture (Hastings et al. 1989; Murray et al. 1989), although biomass tends to decrease in the wettest sites. Light is also an important determinant of moss productivity. Shading by vascular plants may reduce moss biomass (Hastings et al. 1989), but may increase the productivity: biomass ratio (Harley et al. 1989; Murray et al. 1989).

18.2.3 Role of Nutrients

Although numerous process studies in the Arctic have established a strong relationship between water and plant structure and function, water per se does not strongly affect vascular plant productivity in most low arctic ecosystems. Much evidence suggests that N availability is the primary constraint on vascular plant productivity (e.g., Shaver and Chapin 1980; Shaver et al. 1986). Simulations of N uptake and movement in tussock tundra soils (Leadley et al. 1995) and simulations of the growth of *Eriophorum vaginatum* (Leadley and Reynolds 1992) suggest that (1) N uptake by tussock-tundra plants is primarily limited by the N availability in the soils, and not by plant uptake characteristics or root biomass, and (2) nitrogen – not CO_2 concentration, temperature, or radiation – is the primary determinant of plant growth.

There appears to be a strong relationship between water and nutrient availability in arctic tundra ecosystems. In field studies, the observed changes in vascular plant productivity, due to increased moisture, have also been attributed to increased nutrient availability (Chapin et al. 1988; Kielland and Chapin 1992; Oberbauer and Dawson 1992; Chap. 10, this Vol.). Chapin et al. (1988) found that areas of moist tundra where water is channeled (water tracks) have higher vascular productivity and N availability than areas that do not. Increased N availability in water tracks has been attributed to movement of N in water, increased thaw depth, warmer soil temperatures, and increased

N diffusion in the soil (Chapin et al. 1988; Oberbauer and Dawson 1992). Without water movement wet soils are associated with relatively low N availability, due to anaerobic conditions that inhibit microbial mineralization. In this case, decreases in soil moisture improve soil aeration and tend to enhance N availability (Kielland and Chapin 1992; Nadelhoffer et al. 1992).

18.3 Description of Model

18.3.1 Overview

T-MAP is based on the following assumptions:

1. "Average" soil moisture during the growing season, computed as the variable *Soil Mois*, can be derived from local water discharge and slope.
2. The pattern of vegetation types in arctic landscapes can be predicted from local water discharge and slope.
3. The productivity and biomass of poikilohydric plants can be directly related to soil moisture.
4. N availability can be related to soil moisture.
5. The long-term productivity of vascular plants in tundra ecosystems can be reasonably estimated from the amount of N available and the efficiency of N use in production.

These assumptions are based on the experimental and modeling results discussed in Section 18.2 and in Chapters 5 and 14 (this Vol.).

T-MAP is composed of four models (T-HYDRO, T-VEG, T-NUT, and T-PLT) that describe spatially explicit relationships between elevation, slope, soil moisture, vegetation types, N availability, and plant biomass and production for each of ca. 5200 pixels in the Imnavait Creek watershed map (see Fig. 18.1). The only inputs to T-MAP are the digital elevation model (DEM, ca. 20 m × 20 m pixel size) and annual precipitation. All variables and parameters are defined in Table 18.1.

18.3.1.1 T-HYDRO

T-HYDRO computes total *Discharge* for each pixel in the watershed as:

$$Discharge = Runon + Ppt - Evapo, \tag{1}$$

where *Discharge* is total volume of water (m^3 pixel^{-1} or m^3 20 m^{-2}), *Runon* is total runon, *Ppt* is precipitation and *Evapo* is evapotranspiration. Details and assumptions of the water-routing algorithm used in T-HYDRO are given in Ostendorf and Reynolds (1993), and a brief overview is provided in Chap. 14 (this Vol.). We assume no net change in water storage in a pixel during the time interval for which the budget is calculated (i.e., between years). Conceptually, all pixels are linked by a network of nodes (located at the center of a pixel) that

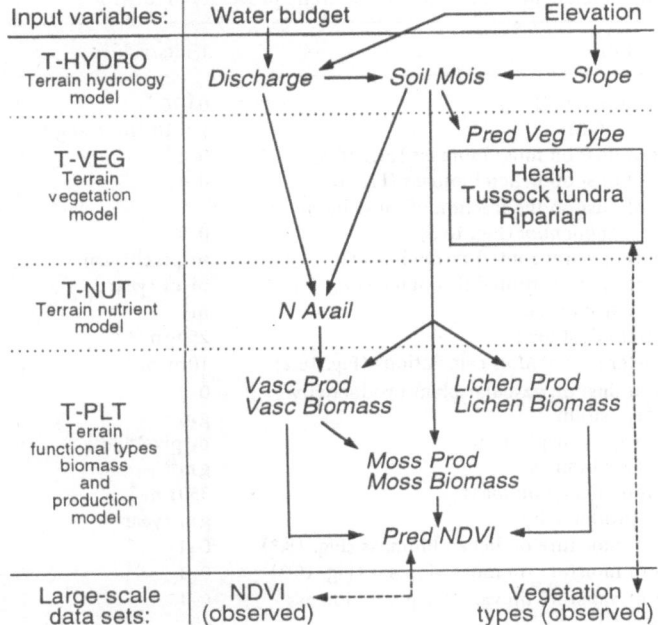

Fig.18.1 Overview of models contained in terrain models for artic processes (T-MAP)

are connected by pipes of equal diameter filled with soil. The elevation difference between pixels is the gravitational potential between nodes. Darcy's law is used to estimate *Discharge* through these connections and, ultimately, its routing through the pixels of the landscape (Fig. 14.10 in Chap. 14, this Vol.). Model output is a two-dimensional "drainage area" map that gives total upslope discharge (m³ water) from each pixel. In order to obtain units independent of pixel size, *Discharge* was scaled to units of m³ m⁻¹ (hereafter noted as *Discharge**) by multiplying by the length of a pixel (20 m; Quinn et al. 1991). The drainage map was scaled assuming (*Ppt* − *Evapo*) = 0.098 m³ m⁻² for each pixel based on the total 1986 discharge of 209 000 m³ in Imnavait Creek (Chap. 6, this Vol.).

Following Burt and Butcher (1985) a simple index of average soil moisture, *Soil Mois*, is derived from Darcy's law as a function of *Discharge** and *Slope*. *Slope* determines the gravitational gradient that influences flow velocity. *Slope* was obtained from the Imnavait Creek DEM grid using the scheme described in Ostendorf and Reynolds (1995; see also Chap. 17, this Vol.). The index was modified because it appeared to overestimate soil moisture at low slopes and discharges in the Imnavait Creek catchment:

$$Soil\ Mois = \begin{cases} Discharge*/S_{min} & \text{if } Discharge* < D_{min} \text{ and } Slope < S_{min} \\ Dishcarge*/Slope & \text{otherwise,} \end{cases} \qquad (2)$$

Table 18.1. Variables and parameters in terrain models for arctic processes (T-MAP)

Name	Description	Units/values
a	Intercept, Eq. (20)	0.026
b	Slope, Eq. (20)	$7 \times 10^{-4} \, m^2 \, year \, g^{-1}$
$D_{Moss}^{Biomass}$	Effect of dust on moss biomass (Fig. 18.3)	0–1
$D_{Lichen}^{Biomass}$	Effect of dust on lichen biomass (Fig. 18.3)	0–1
D_{Sphag}^{Frac}	Effect of dust on the fraction of moss biomass that is *Sphagnum* (Fig. 18.3)	0–1
Discharge	Amount of water routed through a site	$m^3 \, pixel^{-1} \, year^{-1}$
Discharge*	Amount of water routed through a site	$m^3 \, m^{-1} \, year^{-1}$
Dist	Distance from road	m
D_{max}	Maximum dust load	$250 \, g \, m^{-2}$
D_{min}	Parameter in *Soil Mois* calculations (Fig. 18.2)	$10 \, m^3 \, m^{-1}$
D_{Vasc}^{Prod}	Effect of dust on vascular plant productivity	0–1
Dust Load	Dust deposition	$g \, m^{-2}$
Evapo	Total evapotranspiration	$m^3 \, pixel^{-1} year^{-1}$
Lichen Biom	Live lichen biomass	$g \, m^{-2}$
Lichen Biom Max	Maximum lichen biomass	$350 \, g \, m^{-2}$
Lichen Prod	Lichen productivity	$g \, m^{-2} year^{-1}$
$M_{Lichen}^{Biomass}$	Effect of moisture on lichen biomass (Fig. 18.3)	0–1
$M_{Moss}^{Biomass}$	Effect of moisture on moss biomass (Fig. 18.3)	0–1
$M_{Vasc}^{Biomass}$	Effect of moisture on vascular plant biomass (Fig. 18.3)	0–1
M_{Sphag}^{Frac}	Effect of moisture on the fraction of moss biomass that is *Sphagnum* (Fig. 18.3)	0–1
M_N^{min}	Minimum value of M_N (Fig. 18.3)	0.55
M_N	Effect of *Soil Mois* on *N Avail* (Fig. 8.3)	0–1
Moss Biom	Live (i.e., green) moss biomass	$g \, m^{-2}$
Moss Biom Max	Maximum moss biomass	$400 \, g \, m^{-2}$
Moss Biom Sphag	Biomass of the moss, *Sphagnum*	$g \, m^{-2}$
Moss Prod	Moss productivity	$g \, m^{-2} year^{-1}$
N Avail	N availability	$g \, m^{-2} year^{-1}$
N_{max}	Maximum nitrogen availability	$1.6 \, g \, m^{-2} year^{-1}$
NPP	Net primary productivity	$g \, m^{-2} year^{-1}$
NUE	Nitrogen-use efficiency for vascular plants	$140 \, g$ dry weight $g^{-1} \, N$
Obs NDVI	Observed normalized difference vegetation index derived from a SPOT/HRV-XS scheme June 22, 1987 (Stow et al. 1986)	–
Obs Veg Type	Observed vegetation type from Walker et al. (1989): heath, tussock tundra, or riparian	–
PB_{Lichen}	Lichen productivity: biomass ratio	$0.025 \, year^{-1}$
PB_{Moss}	Moss productivity: biomass ratio	$0.25 \, year^{-1}$
PB_{Vasc}	Vascular plant productivity: biomass ratio	$0.25 \, year^{-1}$
Ppt	Total precipitation	$m^3 \, pixel^{-1} \, year^{-1}$
Pred NDVI	Predicted normalized difference vegetation index	–
Pred Veg Type	Predicted vegetation type: heath, tussock tundra, riparian	–
Runon	Total water run-on to a pixel	$m^3 \, pixel^{-1} year^{-1}$
$S_{Moss}^{Biomass}$	Effect of shading on moss biomass (Fig. 18.3)	0–1
Slope	Slope	$m \, m^{-1}$
S_{min}	Parameter in *Soil Mois* calculations (Fig. 18.2)	2 degrees

Table 18.1. (*continued*)

Name	Description	Units/values
Soil Mois	Soil moisture index	$m^3\,m^{-1}\,degree^{-1}$
S^{Prod}_{Moss}	Effect of shading on moss productivity (Fig. 18.3)	0–1
TAGB	Total aboveground plant biomass	$g\,m^{-2}$
T_{dry}	Value of *Soil Mois* index, below which *Pred Veg Type* is heath	2.5
T_{wet}	Value of *Soil Mois* index, above which *Pred Veg Type* is riparian	65.0
Vasc AGB	Vascular plants aboveground biomass	$g\,m^{-2}$
Vasc Biom	Above- and belowground biomass of vascular plants	$g\,m^{-2}$
Vasc Cover	Vascular plant cover	–
Vasc Prod	Vascular plant productivity	$g\,m^{-2}\,year^{-1}$
W_N	Effect of *Soil Mois* and *Discharge* on *N Avail* (Fig. 18.3)	0–1

where S_{min} and D_{min} are parameters as shown in Fig. 18.2. Thus, soil moisture is high if discharge is high or if the slope is low; soil moisture is low if the amount of water routed through a site is low or the slope is steep.

18.3.1.2 T-VEG

T-VEG is similar to the model of Ostendorf and Reynolds (1996; Sect. 14.4.2; Sect. 17.3.1, this Vol.) and predicts vegetation types based on thresholds of *Soil Mois*:

$$\text{Pred Veg Type} = \begin{cases} \text{Heath} & \text{if Soil Mois} < T_{dry} \\ \text{Riparian} & \text{if Soil Mois} > T_{wet} \\ \text{Tussock tundra} & \text{otherwise,} \end{cases} \tag{3}$$

where T_{dry} and T_{wet} are thresholds such that the heath vegetation type is assumed to occur in sites with low soil moisture, the riparian vegetation type on sites with high moisture, and tussock tundra with intermediate moisture.

18.3.1.3 T-NUT

T-NUT computes N availability (*N Avail*), which is assumed to increase with increasing soil moisture, up to a maximum level. Beyond this maximum additional soil moisture is assumed to lead to saturated, anaerobic soils with a subsequent reduction in *N Avail*. This reduction is ameliorated by high discharge, i.e., increased soil aeration when large amounts of water are routed

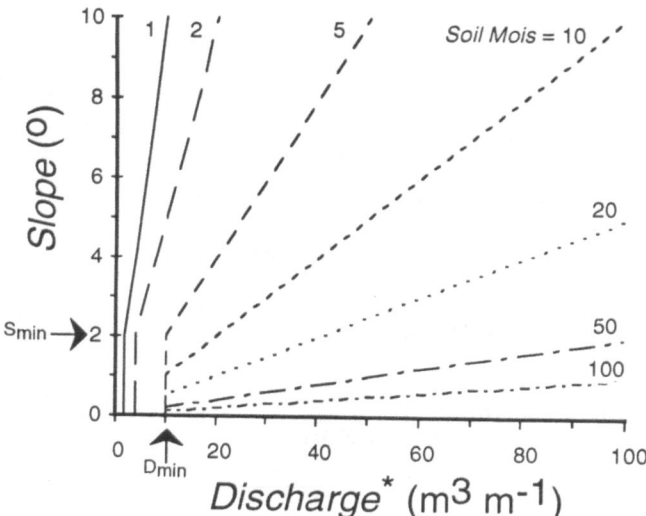

Fig.18.2 *Soil Mois* as a function of *Slope* and *Discharge*

through a site. We compute two scalars, M_N and W_N, that reduce *N Avail* from a maximum based on the relative values of *Soil Mois* and *Discharge**, respectively:

$$N\,Avail = N_{max} \cdot (M_N + W_N). \tag{4}$$

M_N and W_N are illustrated in Fig. 18.3, where M_N^{min} is the minimum allowable value of M_N.

18.3.1.4 T-PLT

T-PLT computes productivity and biomass for the three functional groups of plants, i.e., lichens, mosses, and vascular plants. In contrast to T-VEG, witch predicts a single vegetation type per pixel Eq. (3), T-PLT computes productivity and biomass for each functional group that may occur in a given pixel. Scalars, with values that range from 0 to 1, are used to represent the effect of soil moisture (M_k^i), vascular plant shading (S_k^i), and dust (D_k^i) on process i in functional group k.

The productivity equations are given by:

$$Vasc\,Prod \quad = \quad NUE \cdot N\,Avail \tag{5}$$

$$Moss\,Prod \quad = Moss\,Biom \cdot PB_{Moss} \cdot S_{Moss}^{Prod} \tag{6}$$

$$Lichen\,Prod = Lichen\,Biom \cdot PB_{Lichen} \tag{7}$$

Vascular plant productivity is assumed to be directly proportional to *N Avail* via a nitrogen use efficiency (NUE). Moss and lichen productivities are

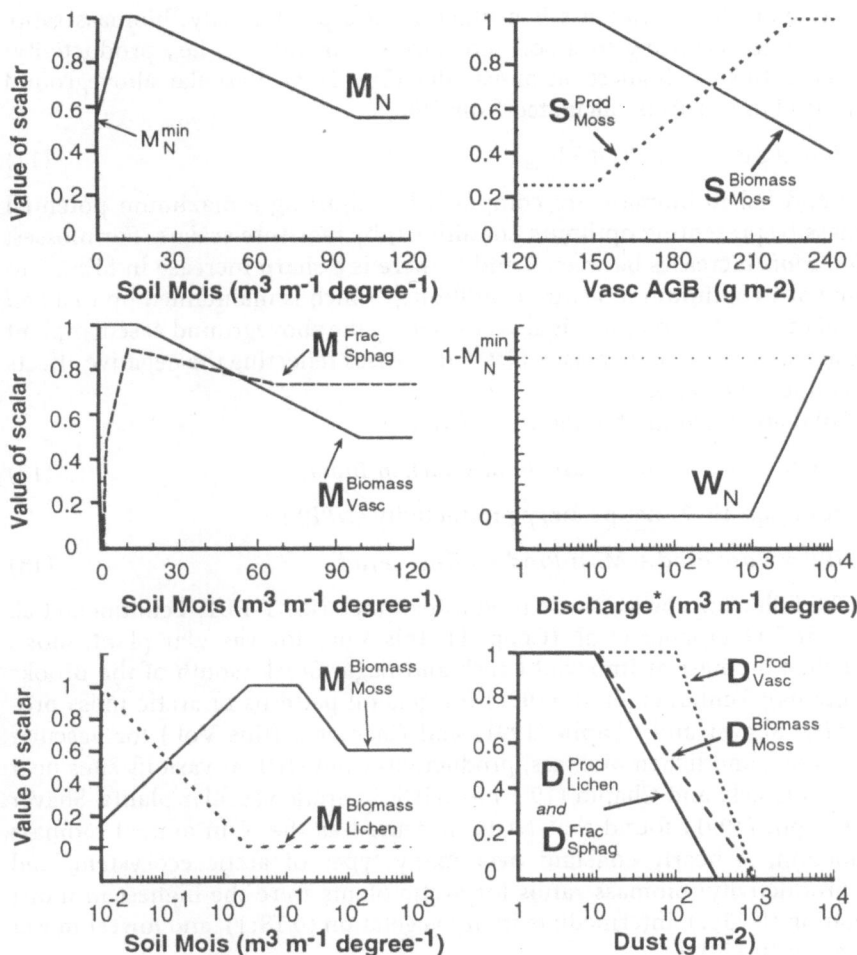

Fig.18.3 Scalars used in T-MAP. (See Table 18.1 for definitions)

computed as functions of total biomass (see below) and productivity: biomass ratios (PB_k). The moss productivity: biomass ratio varies as a function of vascular aboveground biomass via the shading scalar, S_{Moss}^{Prod} (Fig. 18.3), which represents increased moss productivity with increasing shading by vascular plants.

The biomass equations are given by:

$$Vasc\ Biomass = Vasc\ Prod\ /\ (PB_{Vasc} \cdot M_{Vasc}^{Biomass}) \tag{8}$$

$$Moss\ Biom\quad = Moss\ Biom\ Max \cdot M_{Moss}^{Biomass} \cdot S_{Moss}^{Biomass} \tag{9}$$

$$Lichen\ Biom = Lichen\ Biom\ Max \cdot M_{Lichen}^{Biomass} \tag{10}$$

For vascular plants, biomass is a function of a productivity: biomass ratio, which varies according to a soil moisture scalar, $M_{Vasc}^{Biomass}$, i.e., productivity: biomass ratios are highest on moist sites (Fig. 18.3). Vascular aboveground biomass (*Vasc AGB*) is calculated from PB_{Vasc}:

$$Vasc\ AGB = Vasc\ Prod/PB_{Vasc}. \tag{11}$$

Moss and lichen biomass are computed by adjusting a maximum potential biomass (representing optimum conditions) by moisture scalars. For mosses, as *Soil Mois* increases between 0 and 5, there is a sharp increase in $M_{Moss}^{Biomass}$ to a value of 1 (optimum moisture conditions), which is maintained up to a *Soil Mois* of ca. 15. Moss biomass is also affected by the aboveground vascular plant biomass: as *Vasc AGB* increases, $S_{Moss}^{Biomass}$ decreases reflecting the negative effects of shading (Fig. 18.3).

Total aboveground biomass (*TAGB*) is:

$$TAGB = Vasc\ AGB + Moss\ Biom + Lichen\ Biom, \tag{12}$$

and from Eqs. (5–7) net primary productivity (*NPP*) is:

$$NPP = Vasc\ Prod + Moss\ Prod + Lichen\ Prod. \tag{13}$$

The following sources were used to parameterize T-PLT: Tenhunen et al. (1992) and Oberbauer et al. (Chap. 11, this Vol.) for vascular plant, moss, and lichen biomass at Imnavait Creek and Eagle Creek (south of the Brooks Mountains); Tenhunen et al. (1992) for general patterns of arctic moss productivity; Shaver and Chapin (1991) and Chapter 5 (this Vol.) for vascular plant, moss, and lichen biomass, productivity, and NUE at various sites near Imnavait Creek; and Chapin (1989) for NUE in arctic vascular plants. Shaver and Chapin (1991) found that NUE, measured as the N in annual biomass production, is nearly constant over many types of arctic ecosystem, and that productivity: biomass ratios for arctic plants were the highest in moist vegetation (0.25:1), intermediate in dry vegetation (0.18:1), and lowest in wet vegetation (0.12:1).

18.3.2 Disturbance Scenarios

The effects of two categories of road-related disturbances are considered: (1) changes in patterns of water discharge, which modify vegetation and potential for thermokarst erosion, and (2) dust deposition, which has a direct impact on plant biomass and productivity. Simulation results are presented for four differing placements of roads through the watershed as described in Table 18.2.

18.3.2.1 Effects of Altering Discharge

The construction of a road may alter natural patterns of water drainage, effectively acting as a dam. The modification of soil moisture may severely impact

Table 18.2. Characteristics of four simulated roads through the Imnavait Creek watershed

Road no.	Relative to Imnavait Creek	Length (number of pixels)	Number of culverts	Communities impacted[a]	Compass direction
1	Parallel	145	5[b]	Footslope only	S to N, E side of creek
2	Parallel	133	0	Crest only	S to N, E side of creek
3	Perpendicular	56	1	All	SW to NE; crosses creek near N end of watershed
4	Perpendicular	52	1	All	SW to NE; crosses creek near S end of watershed

[a] See toposequence of Walker (Fig. 4.8 in Chap. 4, this Vol.).
[b] One at each end of the road and three in areas of high discharge. The elevation map was altered to allow water to flow freely along the road to the culverts.

vegetation at both the patch and landscape scales and over the short and long term. T-MAP allows us to compare predicted vegetation distribution, biomass, and production before and after road building, assuming long-term equilibration has occurred.

18.3.2.2 Effects of Road Dust

Dust loads in the vicinity of the road were simulated based on the observations made by Walker and Everett (1987) near Imnavait Creek during 1978, and are similar to the model output from Lamprecht and Graber (Chap. 15, this Vol.). The following approximation to the dust model results was used:

$$Dust\ Load = \begin{cases} D_{max} & \text{if } Dist < 8 \\ D_{max}/(Dist - 7) + \text{Random} & \text{otherwise,} \end{cases} \tag{14}$$

where $Dist$ is distance from the road, D_{max} is the maximum dust load, and Random is random variance.

The most detrimental effects of road dust are on lichens and *Sphagnum* species (Walker and Everett 1987; Chap. 4, this Vol.). Some non-*Sphagnum* mosses and some lichens may actually benefit from dust damage to other nonvascular species. Four scalars were used to represent the effect of dust on vascular productivity (D_{Vasc}^{Prod}), moss biomass ($D_{Moss}^{Biomass}$), fraction *Sphagnum* biomass (D_{Sphag}^{Frac}), and lichen biomass ($D_{Lichen}^{Biomass}$). These dust scalars are illustrated in Fig. 18.3 and are based on the work of Walker and Everett (1987), Meininger and Spatt (1988), and Santelmann and Gorham (1988).

The effects of dust on the productivity and biomass of the functional groups are given by:

$$Vasc\ Prod = \text{NUE} \cdot N\ Avail \cdot \mathbf{D}_{\text{Vasc}}^{\text{Prod}} \tag{15}$$

$$Moss\ Biom = Moss\ Biom\ Max \cdot \mathbf{M}_{\text{Moss}}^{\text{Biomass}} \cdot \mathbf{S}_{\text{Moss}}^{\text{Biomass}} \cdot \mathbf{D}_{\text{Moss}}^{\text{Biomass}} \tag{16}$$

$$Moss\ Biom\ Sphag = Moss\ Biom \cdot \mathbf{M}_{\text{Sphag}}^{\text{Frac}} \cdot \mathbf{D}_{\text{Sphag}}^{\text{Frac}} \tag{17}$$

$$Lichen\ Biom = Lichen\ Biom\ Max \cdot \mathbf{M}_{\text{Lichen}}^{\text{Biomass}} \cdot \mathbf{D}_{\text{Lichen}}^{\text{Biomass}} \tag{18}$$

In Eq. (17) we assume that the fraction of moss biomass and productivity attributable to *Sphagnum* moss species (i.e., $\mathbf{M}_{\text{Sphag}}^{\text{Frac}}$) peaks at moderate moisture and declines with increasing moisture.

18.3.3 Model Validation and Limitations

It is not possible to validate T-MAP directly at the present time. However, we compared output from T-MAP to a satellite-derived map of normalized difference vegetation index (NDVI, derived from a SPOT/HRV-XS scene, June 22, 1987; Stow et al. 1989) and to a map of observed vegetation types (Walker et al. 1989) for the lmnavait Creek watershed. These observed data sets are referred to as *Obs NDVI* and *Obs Veg Type*, respectively. The remotely sensed NDVI is linearly related to the amount of photosynthetically active radiation (PAR) intercepted by vegetation in other systems (Hatfield et al. 1984; Seller 1987). NDVI may provide a measure of net primary productivity, because there is a close relationship between integrated intercepted PAR and NPP in many types of vegetation (Monteith 1981).

18.4 Model Predictions for Undisturbed Watershed

The dominant features of the lmnavait Creek watershed are apparent in the maps of *Obs Veg Type* and *Obs NDVI* (Fig. 18.4). Both the dry heath vegetation on the ridges and the wet riparian vegetation in the basin of the watershed are characterized by a low NDVI. Higher values of NDVI delimit the slopes with tussock tundra. The extent of the southwest facing (eastern) slope contrasts strongly with that of the steep northeast facing (western) slope. The eastern slope has numerous well-developed water tracks (Fig. 18.4), which are associated with high *Obs NDVI* (Stow et al. 1989). Within a class of *Obs Veg Type*, we suspect that leaf area index is probably directly proportional *Obs NDVI*, but this has not been validated (cf. Chap. 7, this Vol.).

18.4.1 Vegetation

The spatial agreement between *Pred Veg Type* determined with T-VEG and *Obs Veg Type* in the watershed was ca. 76% (Table 18.3; Fig. 18.4). T-VEG is most

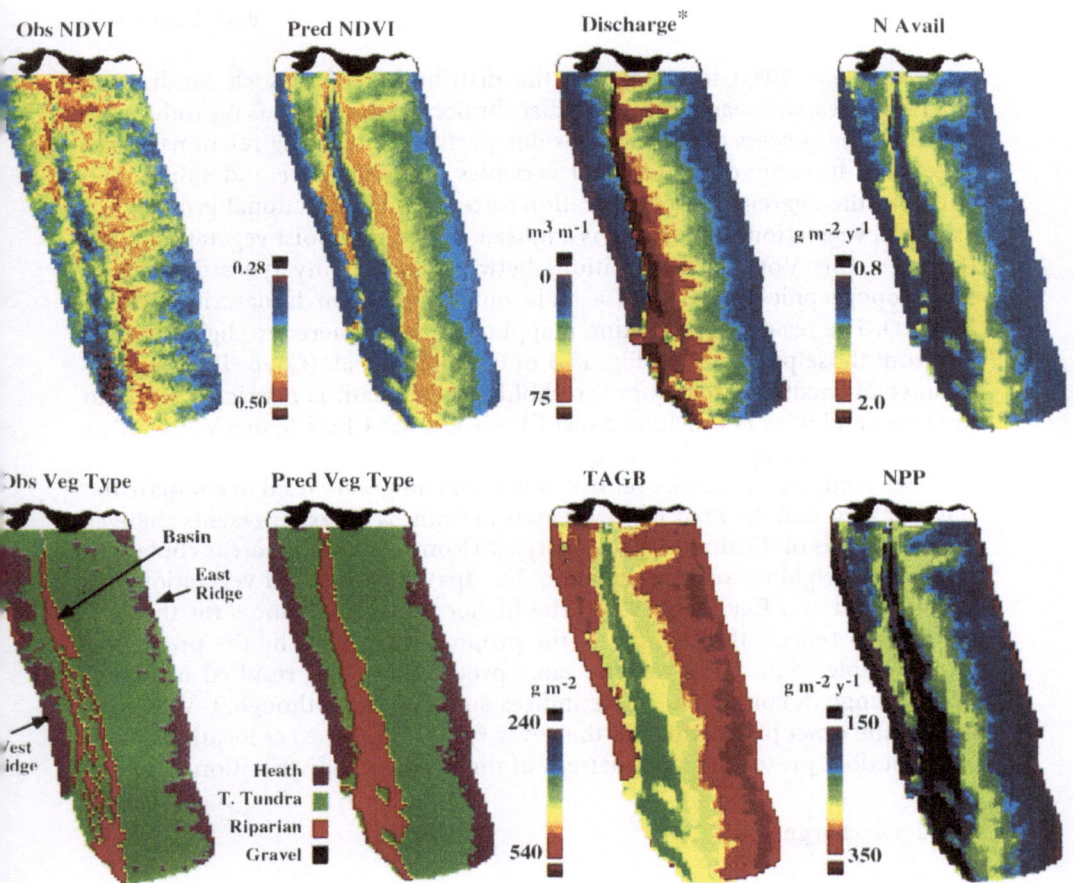

Fig.18.4 Observed NDVI and vegetation in Imnavait Creek watershed compared with predicted variables from T-MAP

Table 18.3. Comparison of *Obs Veg* and *Pred Veg* maps in the Imnavait Creek watershed

	Number of map pixels		
Vagetation types	Obs Veg[a]	Pred Veg	% Matching
Dry heath	950	1053	68.0
Tussock tundra	3533	3128	77.5
Riparian	549	851	79.0
Total	5032	5032	75.9
Evenness index	0.56	0.71	
Contagion index	0.36	0.37	

[a] Aquatic and barren types (a total of 13 pixels) were assigned to wet and moist types, respectively, because they are not included in T-MAP.

successful (ca. 80%) in predicting the distribution of tussock tundra and riparian types, whereas community distribution on the ridges is reproduced at ca. 68%. The success of the model is due partly to the strong relationships of vegetation to some environmental variables (e.g., moisture and slope), and partly to the aggregation of vegetation into the three functional groups. For example, vegetation in the basin is a mosaic of wet and moist vegetation types (Chap. 4, this Vol.). The transitions between community types depend on microtopographic variation at a scale much finer than is described by T-HYDRO. The results of vegetation mapping presented here are slightly different from those presented in Fig. 17.3 of Ostendorf et al. (Chap. 17, this Vol.) because of small differences in threshold values (heath is restricted solely to dry types and does not include moist CDS heath; see Chap. 5, this Vol.) and as a result of including S_{min} and D_{min}.

Two landscape indices, evenness and contagion, were used to compare the *Obs Veg Type* and the *Pred Veg Type* maps. Evenness index represents changes in proportions of the three vegetation types (Romme 1982), whereas contagion represents neighborhood information. i.e., spatial patterns of vegetation distribution (Li and Reynolds 1993). The higher value of evenness for the predicted map reflects the changes of the proportions caused by the prediction errors (Table 18.3). However, the same prediction errors resulted in only a small change in contagion. These indices suggest that, although T-VEG may not get the exact proportions of the cover types and the exact locations of all pixels, it does preserve spatial pattern of the vegetation distribution.

18.4.2 Discharge

Annual *Discharge** is strongly correlated with many of the dominant features of the watershed as evident in the map produced by T-HYDRO (Fig. 18.4). The ridge tops are characterized by low discharge and the basin by very high discharge. Water track channels in the Imnavait Creek catchment are ca. 5–20 m wide and are depressed ca. 0.25–1 m in relation to adjacent tussock tundra (Hastings et al. 1989; Chap. 4, this Vol.). The most prominent water track in the watershed (water track 3, Fig. 9.2, this Vol.) is on the northern end of the eastern slope, which shows up as an area of high *Discharge** (Fig. 18.4). Other areas of high *Discharge** on the eastern slope, i.e., the patterning with a series of stripes with high discharge, also correspond to areas with well-developed water tracks. *Soil Mois* and *Discharge** have nearly identical patterns for the ridges and slopes, but due to a rapid decrease in *Slope*, *Soil Mois* increases much more rapidly than *Discharge** at the boundary between the bottom of the slopes and the basin (not shown).

18.4.3 N Availability and NPP

Predicted total aboveground biomass [*TAGB*, Eq. (12)] is highest on the ridges, declines toward the bottoms of the slopes, and is the lowest in the basin (Fig.

18.4). The comparison to actual biomass harvests in the watershed is very good (see Table 5.5, this Vol.). Vascular plant biomass increases from the ridge to the bottom of the slope on both sides of the watershed, lichen biomass decreases to zero within 100 m of the ridges, and moss biomass declines in areas of high discharge, due to high vascular plant biomass (not shown). *N Avail* and *NPP* (Fig. 18.4) are low on the ridges and in the basin, and highest at the bottoms of the slopes and in water tracks. Most of the other variables that are directly related to *NPP*, i.e., vascular plant productivity, moss productivity, and total vascular plant biomass, also follow these patterns.

18.4.4 Model Evaluation

Spatial correlations between various output variables of T-MAP and observed NDVI are given in Table 18.4. The correlation between *Discharge** and *Obs NDVI* is poor, because NDVI is low in both the basin and ridges, whereas *Discharge* is high in the basin and low on the ridges. *Soil Mois* is also poorly correlated to NDVI. On the other hand, *N Avail*, which is a function of both *Soil Mois* and *Discharge**, has a much higher correlation with NDVI. The correlation between *TAGB* and *Obs NDVI* is poor, because T-MAP simulates low *TAGB* in areas with the highest NDVI (e.g., at the foot of the eastern slope). This may be part of the reason why NDVI was poorly correlated with leaf area index, total community biomass, and total green biomass in "ground truth" measurements (Chap. 7, this Vol.). It must also be noted that NDVI may be sensitive to aspects of tundra ecosystem structure that have little to do with plant biomass or productivity, e.g., surface water, exposure of mineral soil in the ridge heath, etc. (Sect. 7.3, this Vol.).

NPP has the highest correlation ($r = 0.66$) with *Obs NDVI*. Both variables are strongly dependent on community structure. We regressed *NPP* on *Obs*

Table 18.4. Correlation coefficients for comparisons of maps of modeled variables with the *Obs NDVI* map. Resample statistics, which are less influenced by spatial autocorrelations, are based on 50 resamples with $n = 30$ for each sample

Model output	r	r²	Resample mean r²	Resample variance
Abiotic variables				
*Discharge**	−0.07	<0.01	0.06	0.003
Soil Mois	−0.13	0.02	0.08	0.005
Slope	0.20	0.04	0.05	0.004
Biotic variables				
N Avail	0.65	0.42	0.43	0.006
Vasc AGB	0.65	0.42	0.43	0.012
Moss Biom	<0.01	<0.01	0.01	0.001
Lichen Biom	−0.1	0.01	0.02	<0.001
TAGB	−0.10	0.01	0.05	0.004
NPP	0.66	0.44	0.45	0.006

NDVI to obtain a "predicted" NDVI, i.e., *Pred NDVI* = a + b * *NPP*. *Pred NDVI* does not add new information, but instead transforms *NPP* into NDVI units, which allows us to make a better visual comparison with the *Obs NDVI* map (Fig. 18.4). Low *Pred NDVI* occurs on the ridges and in the basin. *Obs NDVI* increases toward the foot of both slopes, and some of this variation along the slope is also apparent in the *Pred NDVI* map. The most highly developed water track in the watershed – a region of high NDVI at the north end of the east slope – is obvious in the *Pred NDVI* map. The primary discrepancies between *Pred NDVI* and *Obs NDVI* are (1) less variation in *Pred NDVI* along the slopes and ridges than in the observed NDVI; (2) a wider and longer area of low *Pred NDVI* in the basin than observed; and (3) a band of low *Pred NDVI* along the west ridge that is much more prominent than observed. These discrepancies may be attributed to inherent limitations in the resolution of the DEM map, the assumption of a uniform water budget, a lack of stochasticity in the hydrology and vegetation models, and other factors not included in the model. Despite these limitations, the *Pred NDVI* predicted by the T-MAP model accounts for 45% of the variation in the *Obs NDVI* (Table 18.4).

18.5 Model Predictions for Disturbed Watershed

The predicted changes in vegetation type and biomass presented here must be considered as qualitative measures (relative indicators) of the effects of distur- bance on vegetation, rather than quantitative measures (absolute amounts). The response to disturbances is a relatively short-term adjustment, yet the relationships between soil moisture and vegetation pattern and processes upon which T-MAP is based also depend on long-term system evolution. e.g., soil development, population redistribution, snow accumulation patterns, etc., which are not considered. Other limitations are discussed in Section 18.6.3.

18.5.1 Discharge Disturbance

The four roads (Table 18.2) alter patterns of discharge by impeding downslope movement of water. The resulting changes in the soil moisture index are input into the T-VEG, T-NUT, and T-PLT models (see Fig. 18.1). The predicted changes in ecosystem structure and function in response to disturbance are summarized in Figs. 18.5 and 18.6, and in Table 18.5. While altered spatial patterns are found for all the variables discussed previously, predicted changes in biomass are emphasized.

18.5.1.1 Road #1

Water moves down the eastern slope until it is blocked by the road. It is then channeled along the road to the culverts where it again moves downslope (Fig.

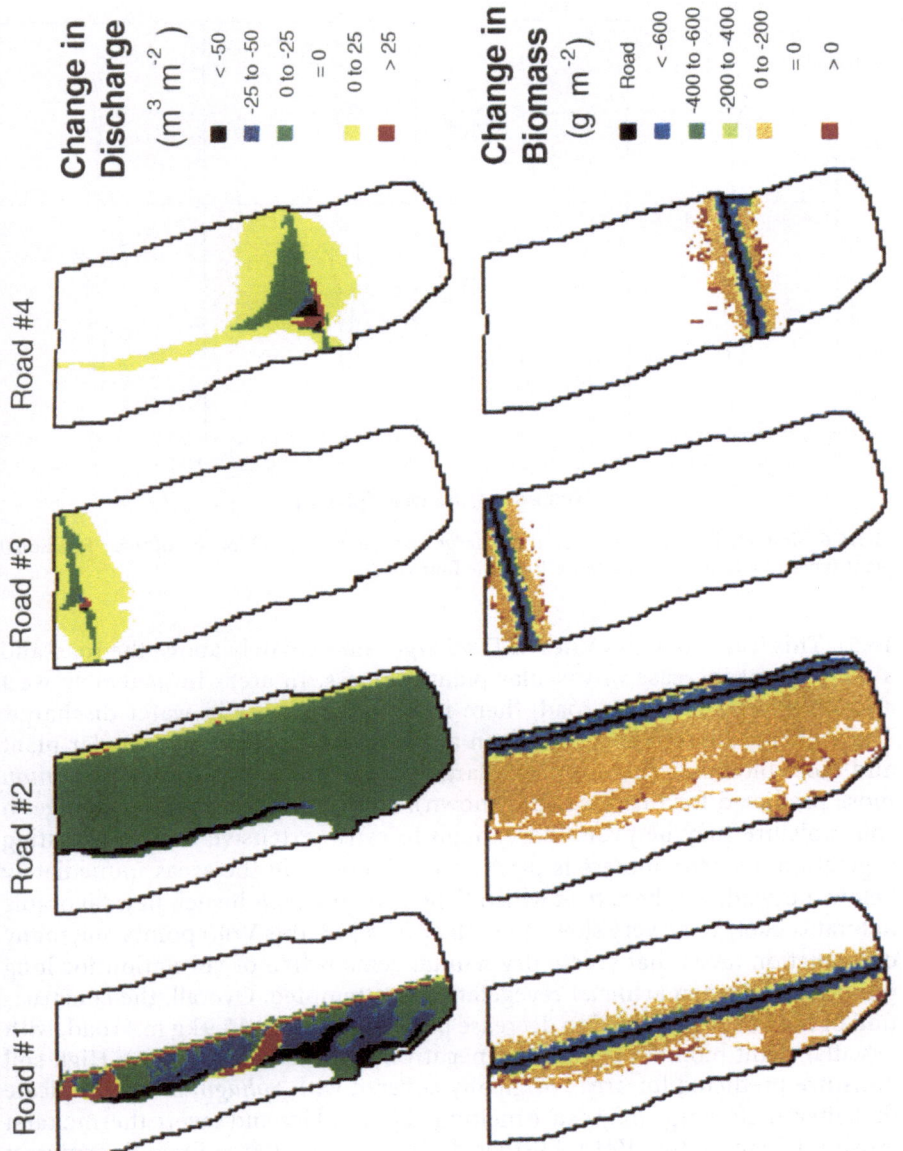

Fig. 18.5 Simulated changes in discharge and total aboveground biomass caused by the construction of four roads through the watershed

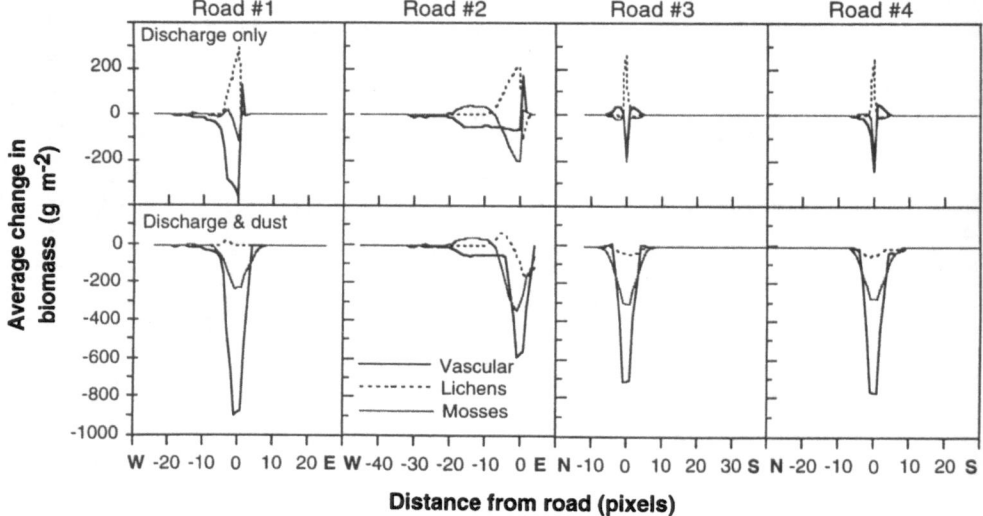

Fig.18.6 Simulated average changes in discharge and aboveground biomass of three functional plant types as a function of distance from the four roads

18.5). This leads to high values of *Discharge** immediately above the road and a consequent increase in vascular plant biomass. In areas immediately west and downslope from the road, there is a large decrease in water discharge. These changes in discharge result in a substantial decrease in vascular plant and moss biomass. There is also a large change in the fraction of *Sphagnum* moss predicted for this area (not shown). Under this scenario the change in soil moisture may be profound enough to cause extensive death to existing vegetation. Lichen biomass is predicted to increase in the areas immediately below the road; yet the rate at which lichens might exploit such new favorable habitat is likely to be very slow. As Walker (Chap. 3, this Vol.) points out, many construction areas that create dry habitat remain free of vegetation for long periods, even when artificial revegetation is attempted. Overall, the construction of road #1 is predicted to decrease plant biomass ca. 15.9 kg m^{-1} road, with vascular plant biomass being most negatively affected (Table 18.5). High soil moisture predicted for areas originally covered with *Sphagnum* may indicate that altered discharge may cause melting of ground ice and severe thermokarst erosion (Chap. 3, this Vol.). Further downslope, discharge from the culverts fans out and the effects on vegetation decrease substantially (Figs. 18.5 and 18.6).

18.5.1.2 Road #2

This road is placed high enough on the east ridge so that relatively little water accumulates on the upslope side of the road. However, the road passes through

Table 18.5. Predicted changes in biomass and percentage of watershed affected for the three plant functional types in the Imnavait Creek watershed resulting from the construction of four roads (see Table 18.2). Statistics are for percentages of watershed affected

	Road no.1 (145 pixels)		Road no.2 (133 pixels)		Road no.3 (56 pixels)		Road no.4 (52 pixels)	
Changes in discharge only								
Vascular Plants								
Mean increase (kg/pixel[a])	34.2	4%[b]	52.9	3%	13.0	7%	9.7	5%
Mean decrease (kg/pixel)	−69.4	24%	−16.1	63%	−41.3	2%	−35.6	5%
Net change (kg biomass/m of road length[c])	−26.3		−16.4		−0.6		−5.8	
Lichens								
Mean increase (kg/pixel)	85.7	9%	53.8	17%	93.0	1%	76.2	2%
Mean decrease (kg/pixel)	0.0	0%	−35.9	3%	−14.0	1%	−8.4	2%
Net change (kg biomass/m of road length)	14.1		15.8		4.8		5.3	
Mosses								
Mean increase (kg/pixel)	12.3	11%	12.7	35%	8.2	2%	7.9	4%
Mean decrease (kg/pixel)	−21.5	16%	−30.5	31%	−22.1	6%	−22.0	5%
Net change (kg biomass/m of road length)	−3.7		−9.5		−5.5		−3.7	
Total net change (kg biomass/m road length)	−15.9		−10.1		−1.3		−4.2	
Vascular Plants								
Mean increase (kg/pixel)	10.0	1%	0.0	0%	6.2	4%	2.5	3%
Mean decrease (kg/pixel)	−157.0	30%	−46.0	67%	−210.4	6%	−172.5	8%
Net change (kg biomass/m of road length)	−80.6		−58.7		−53.1		−63.4	
Lichens								
Mean increase (kg/pixel)	15.6	3%	22.0	10%	0.0	0%	2.4	0%
Mean decrease (kg/pixel)	0.0	0%	−47.1	10%	−52.2	2%	−33.3	4%
Net change (kg biomass/m of road length)	0.8		−4.7		−4.3		−5.7	
Mosses								
Mean increase (kg/pixel)	8.6	5%	11.9	31%	0.4	0%	1.6	1%
Mean decrease (kg/pixel)	−45.2	34%	−53.1	36%	−61.6	12%	−54.1	12%
Net change (kg biomass/m of road length)	−26.2		−28.9		−32.0		−32.4	
Total net change (kg biomass/m road length)	−106.0		−92.2		−89.4		−101.5	

[a] Pixel = 20 m × 20 m.
[b] Based on a total of 5032 pixels in watershed.
[c] Length of road = no. pixels × 20 m.

an area where small changes in soil moisture have relatively large impacts on the lichens and mosses. Overall, road #2 is predicted to decrease plant biomass ca. 10.1 kg m^{-1} road (Table 18.5). Areas immediately downslope from the road become drier resulting in a decline in moss biomass, an increase in lichen biomass, and a slight decline in vascular plant biomass (initially low due to dry conditions). The prediction of short-term increases in lichen biomass seems more likely in this scenario compared with that for road #1, because the affected areas are in (or near) areas with high initial lichen biomass. Changes in biomass propagate much further downslope than for road #1 (see number of pixels affected in Table 18.5).

18.5.1.3 Roads #3 and #4

These roads are perpendicular to the elevation contours with culverts placed at the lowest elevation. Therefore, the downslope movement of water is not impeded. Consequently, only small areas on either side of the roads are affected. Large changes in biomass occur only in the immediate vicinity of the roads, and in pixels where localized damming or drying occurs. Road #4 results in greater biomass changes than road #3 (Table 18.5), because the orientation is not exactly perpendicular to the elevation contours. The most likely negative effect of these roads would be damage due to melting of massive ground ice and thermokarst erosion immediately adjacent to the road in the basin of the watershed where soil moisture and *Sphagnum* moss biomass are high.

18.5.2 Dust and Discharge Disturbance

Dust combined with discharge disturbance is predicted to cause substantially greater changes in vascular plant and moss biomass than discharge disturbance alone (Fig. 18.6). The predicted net loss of biomass per meter of road is roughly the same in all scenarios when dust and discharge effects are combined, ca. 90–110 kg (Table 18.5). There are, however, relatively large differences in how this change is distributed among functional groups.

Road #1 has relatively large effects on vascular plant biomass, because this road passes through communities with high vascular plant biomass. In contrast, there is minimal effect on lichens as they are not predicted to occur in the neighborhood of this road. Road #2 has larger negative and positive effects on lichens (see number of pixels affected, Table 18.5), because this road passes through areas with high lichen biomass. Some damage occurs in an adjacent watershed that is not included in the analysis. Roads #3 and #4 pass through all vegetation types and, therefore, change biomass of all functional groups. In all scenarios, the order in which biomass is negatively affected is: vascular plants > mosses > lichens.

18.5.3 Effect of Disturbance on Spatial patterns

For both the undisturbed and disturbed watershed (the four road scenarios with both discharge and dust) we constructed omnidirectional (i.e., average of all directions) and directional (N15°W and N75°E) semivariograms for all variables (abiotic and biotic). Semivariograms, which relate variation to scales, are useful tools to depict and analyze spatial patterns (Fortin et al. 1989; Rossi et al. 1992). The directional semivariograms coincide with the major environmental gradients in the watershed, i.e., N75°E is perpendicular to the creek and N15°W is parallel to the creek. Only the results for biomass of the three

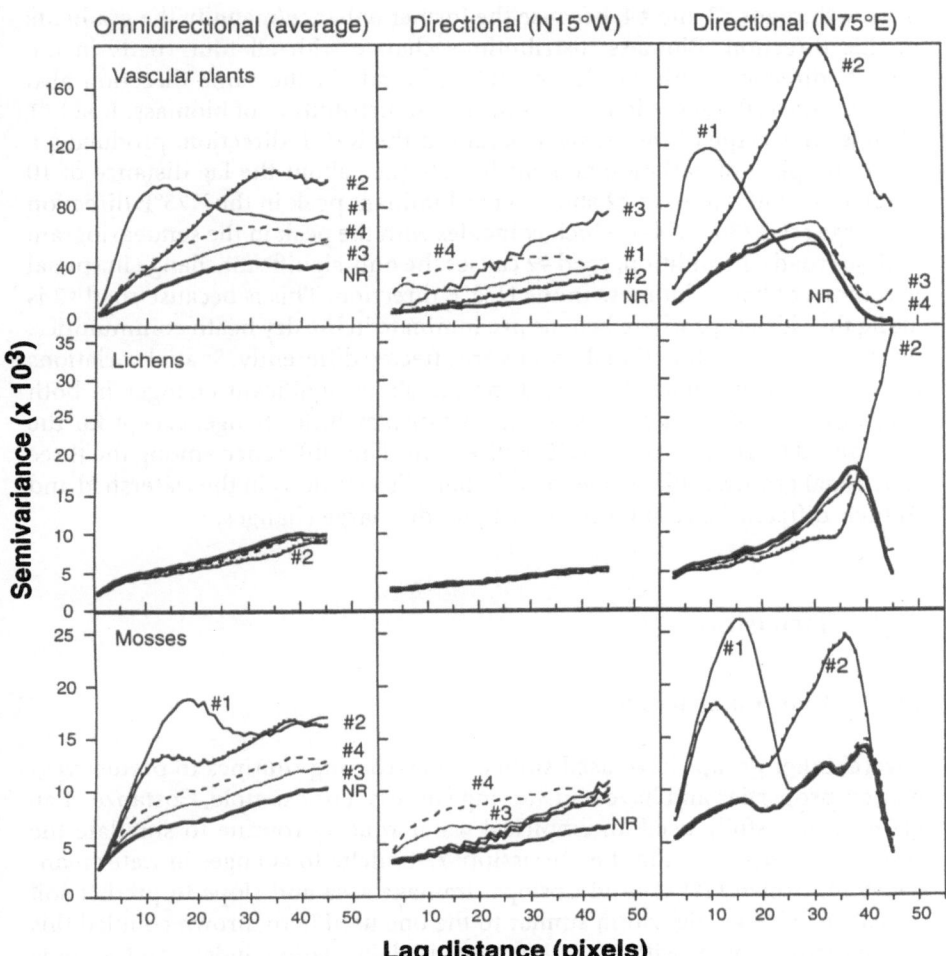

Fig.18.7 Omnidirectional and directional semivariograms of biomass of the three functional groups under four road scenarios (see Table 18.2). NR = no road

functional types of plants are described here (see Chap. 14, this Vol. for other variables). The semivariograms without road disturbance depict the background variation of plant biomass in the watershed. Of course, the general shape of the semivariograms is to a great extent controlled by spatial patterns of the driving variables in T-MAP (e.g., discharge, soil moisture, and nutrient availability).

The semivariogram analysis show that different road scenarios exert distinct effects on biomass distribution of the three functional groups (Fig. 18.7) Firstly, the omnidirectional semivariograms indicate that roads #1 and #2 generally cause more changes in spatial variations of biomass than roads #3 and #4. Secondly, the directions of the roads produced different effects. In the N75°E direction biomass distributions change greatly with roads #1 and #2, but little with roads #3 and #4, because the former enhance/magnify the gradients in this direction. Biomass distributions change with all four roads in the N15°W direction. Thirdly, the location of roads in the same direction also makes large differences in changes in spatial distribution of biomass. Road #1 changes the shape of the semivariogram in the N75°E direction, producing a profound peak of variation at a small scale (i.e., about the lag distance of 10 pixels). In contrast, road #2 shows a predominant peak in the N75°E direction at a large scale (30 pixels), which coincides with the peak in the semivariogram without roads. In addition, road #2 causes the only significant change in spatial variations of lichen biomass in the N75°E direction. This is because road #2 is along the ridge tops where lichens predominate in the dry heath communities. Fourthly, the three functional groups are affected differently. Spatial variations of biomass of vascular plants and mosses show significant changes in both directions examined, but those of lichens display little change, except for the one caused by road #2 in the N75°E direction. This difference among the three functional groups is likely due to both their distributions in the watershed and to their differential responses to dust and discharge changes.

18.6 Discussion

18.6.1 Model Comparisons

Several other groups have used simple water-routing routines to predict vegetation properties and have had varying success. For example, Costanza et al. (1990) successfully used an empirical water routing routine to simulate the response of vegetation in the Mississippi river delta to changes in water management. Brown (1994) used upslope drainage area and slope to predict soil moisture using an algorithm similar to the one used here. Brown coupled this soil moisture index with elevation, potential insolation index, and a snow potential index to predict alpine vegetation types in Glacier National Park. He was able to explain only about 38% of the variation in vegetation type. The success of our approach and that of Costanza et al. probably lies in the importance of water in determining vegetation type in these ecosystems, and in the lack of deep seepage of water.

T-MAP builds on the discharge models developed by Ostendorf and Reynolds (1993 and Chaps. 14 and 17, this Vol.), and integrates hydrological perspectives with an ecosystem model. Whereas Ostendorf and Reynolds correlated NDVI with predicted discharge patterns in the Imnavait Creek watershed, added explanation is achieved with T-MAP, because discharge patterns can be related to nutrient and vegetation patterns, which can then be related to remotely sensed NDVI. The patterns of discharge in our model (with the exception of the basin) are similar to those described by Ostendorf and Reynolds who used a 10- × 10-m pixel elevation map. *Discharge** is uniformly high in the basin in our maps and striations of low and high *Discharge** that they reported for their simulations do not occur. With coarser resolution maps there is also a decrease in the definition of water tracks as compared with Ostendorf and Reynolds.

T-MAP improves on the model of Ostendorf and Reynolds by predicting patterns for the entire watershed, not just the ridges and slopes. Ostendorf and Reynolds excluded values of *Discharge** over $60\,m^3m^{-1}$ (i.e., basin values) in their analysis of the relationship between *Discharge** and observed NDVI. We find a nearly identical relationship when we exclude values of Discharge over $60\ m^3\,m^{-1}$ in T-MAP (Fig. 18.8). Inclusion of high *Discharge** values substantially degrades this relationship, because *Obs NDVI* is low on the ridges and in the basin, whereas *Discharge** is low on the ridges and high in the basin. The relationship of *NPP* to *Obs NDVI* is much better than the relationship between *Discharge** and *Obs NDVI* when the entire basin is included in the analysis (Fig. 18.8).

18.6.2 Patterns of N Availability

Our predicted patterns of N availability, i.e., high in water tracks, intermediate in moist sites, and low in dry and wet sites, contrast with those measured by Giblin et al. (1991) and reported in Nadelhoffer et al. (1992). In addition, their annual rates of N mineralization for dry and wet vegetation (ca. $0.5\,g\ m^{-2}\ year^1$) are considerably lower than our estimates (1.2–$1.4\,g\ m^{-2}\ year^{-1}$), and they report a rate of N mineralization in tussock tundra (ca. $0.1\,g\,m^{-2}\ year^{-1}$) that is more than an order of magnitude lower than our predicted values (1.2–$1.6\,g\ m^{-2}\,year^{-1}$). There are several lines of evidence that suggest the data of Giblin et al. (1991) may not reflect typical patterns or magnitudes of N availability on tussock-tundra slopes (cf. Chap. 10, this Vol.).

Our estimates of N cycling rates are based on (1) measurements of plant productivity and N use efficiency from a wide variety of North Slope tundra sites, and (2) simulation models of N uptake by arctic tundra plants (Leadley and Reynolds 1992; Leadley et al. 1996, see Chap. 14, this Vol). McGuire et al.

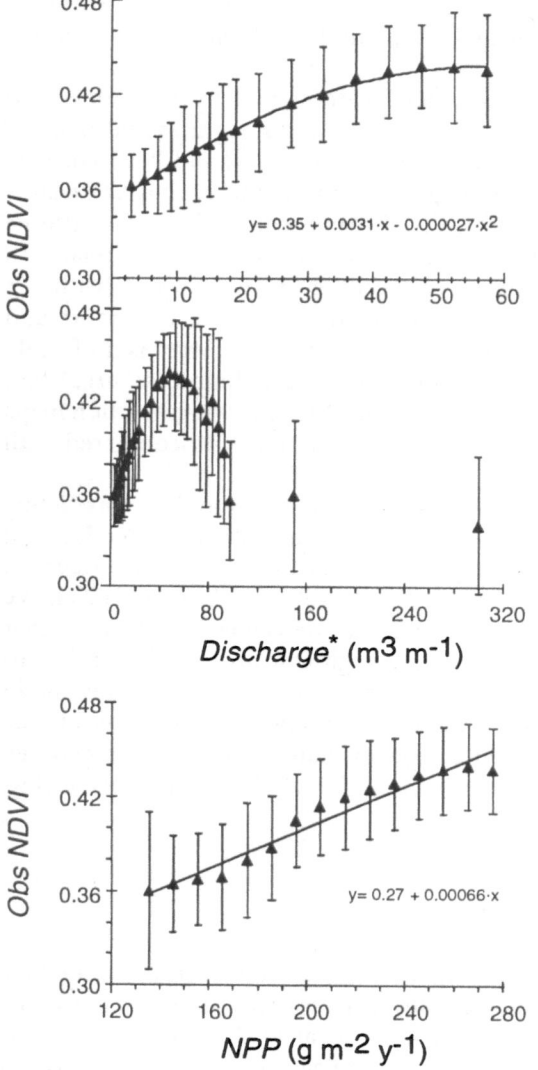

Fig.18.8 Relationship between *Obs NDVI* and *Discharge** and *NPP*. Values represent the means and standard errors of all values in the map that are equidistant from the *Discharge** value. In *top panel*, analysis is limited to values of *Discharge** less than $60\,m^3\,m^{-1}$; in *middle panel*, means were unequally spaced to reduce the number of points on the graph and to roughly equalize the number of pixels represented by each point. Comparisons of variances between unequally spaced points is not appropriate

(1992) found that simulated N mineralization for North American sites overlapped observed N mineralization rates reported in Nadelhoffer et al. (1992) in all ecosystem types except in the Arctic. They found that they needed 0.3–$1.43\,g\,m^{-2}year^{-1}$ of N mineralization to account for plant productivity in wet and moist tundra ecosystems, whereas Nadelhoffer et al. give a range of ca. 0.04–$0.5\,g\,m^{-2}year^{-1}$. Our review of productivity and N-use efficiency by tussock-tundra plants would lead to a value of N availability that is not less than $0.6\,g\,m^{-2}\,year^{-1}$ (compared with $0.1\,g\,m^{-2}\,year^{-1}$ reported by Giblin et al 1991).

Giblin et al. and Nadelhoffer et al. suggest that uptake of organic N (only inorganic N mineralization rates were measured in their study) may account for the discrepancy between observed plant N uptake and their estimates of net N mineralization. There is evidence for at least one important tussock-tundra species, *Eriophorum vaginatum*, that organic sources of N can account for more than 60% of total N uptake in this species (Chapin et al. 1993; Leadley et al. 1996).

The patterns of N mineralization reported by Giblin et al. do not correspond to other patterns that should be related to N uptake by plants. The pattern of *Obs NDVI* indicates an increase in productivity from dry sites to moist sites. It is unlikely that these patterns in NDVI correspond to the order of magnitude decrease in N mineralization reported by Giblin et al. (1991). We believe that Giblin et al. (1991) found such low mineralization rates in moist tundra because they measured N mineralization in intertussock areas. These intertussock areas are often dominated by mosses that provide a very poor substrate for decomposition. Alternatively, the low rates of mineralization may relate to local substrate conditions (see comments in Chap. 10, this Vol.). Measured and modeled estimates of annual N cycling within tussocks (Chapin et al. 1988; Leadley and Reynolds 1992) are much higher than the estimates of Giblin et al. (1991). There may also be problems in relating N mineralization measurements made with buried bags with N uptake by plants (Binkley and Vitousek 1989).

18.6.3 Extrapolation Potential: Some Cautionary Notes

Ostendorf and Reynolds (1996; Chap. 14, this Vol.) show that discharge and slope patterns can be used to explain vegetation type patterns in the Imnavait Creek watershed, and that these relationships also hold outside the watershed. However, they also show that these relationships are poor at tundra sites with glaciation histories that differ from the lmnavait Creek watershed. Thus, the results of our modeling work can be extrapolated beyond the lmnavait Creek watershed, but caution must be used when doing so. As in Ostendorf and Reynolds (1996), T-MAP uses slope and discharge to predict vegetation patterns, but the basis is mechanistic (albeit simple mechanism), rather than correlative per se.

There are several caveats that must be considered before using T-MAP for simulations of ecosystem response in larger areas. Firstly, the magnitude of many parameter values are heavily influenced by data collected at the Imnavait Creek site, one of the most intensively studied tussock-tundra sites in Alaska. Although the structure of the model and the form of the relationships should be applicable to most watersheds in the Brooks Mountain foothills, the parameter settings must be tested before they are applied to other watersheds particularly if the glaciation history is different from that of Imnavait Creek catchment. Secondly, T-MAP does not predict the presence of several important vegetation types that occur in the Brooks Mountain foothills, e.g., aquatic or tall shrub communities. Thirdly, model development is hindered by a lack of information about vegetation structure. Labor-intensive harvesting of biomass has generally not been designed to be linked with or to support spatial modeling studies. For example, the areas of low observed NDVI at the south end of the basin (Fig. 18.4) received little attention from researchers at Imnavait Creek. Due to concentrated investment of resources in the study of processes in tussock tundra, the dominant vegetation in this region, we have much less confidence in our predictions of rates of processes and patterns for the riparian basin. Fourthly, the discharge model works best in areas with significant topographic relief. It is not possible to accurately route water in relatively flat areas because of the resolution limits in the elevation data base. Fifthly, the discharge model is parameterized with an estimate of the water budget based on a single year. Given the nature of the model, it would be better if it were parameterized with a long-term average of the annual water budget. Finally, we parameterized the *Pred NDVI* regression model based on a single NDVI scene. It would be preferable to use several scenes of NDVI within a year to obtain a better estimate of intercepted PAR. However, because the patterns of variation in NDVI are of primary concern, and less so the absolute values of NDVI, other mid-season NDVI scenes should provide similar correlations.

18.7 Conclusions

Roads that disrupt discharge may cause substantial disturbance to arctic eco-systems (Figs. 18.5 and 18.6). Our simulations suggest that disturbance can be minimized by building roads in areas with the smallest effect on discharge patterns. Roads that ascend and descend slopes perpendicular to elevation contours and roads along ridges (i.e., watershed boundaries) are the least likely to cause vegetation disturbance through altered discharge. Dust from roads will have significant impacts regardless of road location.

T-MAP predicts that on a meter-of-road basis, the severity of changes in discharge on vascular plants is: road #1 > road #2 >> road #4 > road #3 (Table 18.5.). Negative effects on moss biomass, due to changes in discharge, are the most severe in road #2 with little difference among the other three scenarios. Negative effects of changing discharge on lichen biomass are restricted to areas where roads pass along the ridges, so the severity of effects is: road #2 > roads #3 and #4 >> road #1. The likelihood of large-scale thermokarst erosion based on *Sphagnum* moss distributions and changes in soil moisture from the most severe to least severe is: road #1 >> road #3 and #4 >> road #2. Dust effects are similar for all roads when all vegetation types are taken into account, but road #2 has greater negative effects on mosses and lichens, and Road #1 has proportionally greater negative effects on vascular plants.

Our conclusions and recommendations based on simulations with T-MAP are:

1. Any road construction has the potential to substantially alter vegetation biomass and community structure over large areas of the watershed. The net change in vegetation biomass and productivity is likely to be large and negative.
2. Roads through the basins of watersheds (e.g., road #1) should be avoided because of the potential for severe discharge-related damage to vascular plant and moss communities, and the potential for widespread thermokarst erosion.
3. Roads along ridges (e.g., road #2) may cause severe damage to lichens and mosses, and the effects of these roads propagate over large areas of the watershed and probably should also be avoided.
4. Roads placed perpendicular to the elevation contours (e.g., roads #3 and #4) are likely to cause the least damage, but these roads disturb small areas of the watershed that are likely to be susceptible to thermokarst erosion.

We show that a model based solely on elevation data and annual water budget can predict the major vegetation types in an arctic watershed (76% success). Furthermore, predictions of net primary productivity can explain ca. 45% of the variation in observed NDVI. Attempts to find direct correlations between ecosystem properties and remotely sensed data often fail. Our approach demonstrates how a simulation model can be used to relate remotely sensed vegetation indices to a wide range of important ecosystem properties, and has potential to become an applied management tool with further development and testing. Such models may serve as valuable tools for generating maps of risk potential of arctic landscapes to various types of disturbances.

Acknowledgements. We thank F. S. Chapin III for helpful comments on the manuscript, and A. Hope, D. Stow, and D. A. Walker for helpful discussions of their data. Financial support was provided by NASA Global Change Fellowship program (to P. Leadley) and the US Dept. of Energy (grant nos. DE-FG03-84R60250 and DE-FG05-92ER61455).

References

Binkley D, Vitousek P (1989) Soil nutrient availability In: Pearcy RW, Ehleringer J, Monney HA, Rundel PW (eds) Plant physiological ecology: field methods and instrumentation. Chapman and Hall, London, pp 75–96

Bonan GB (1993) Do biophysics and physiology matter in ecosystem models? Climate Change 24: 281–285

Brown DG (1994) Predicting vegetation types at treeline using topography and biophysical distubance variables. J Veg Sci 5: 641–656

Burt TP, Butcher DP (1985) Topographic controls of soil moisture distributions. J Soil sci 36: 469–486

Chapin FS III (1989) The cost of tundra plant structures: evaluation of concepts and currencies. Am Nat 133: 1-19

Chapin FS III, Fetcher N, Kielland K, Everett K, Linkins AE (1988) Productivity and nutrient cycling of Alaskan tundra: enhancement by flowing soil water. Ecology 69: 693-702

Chapin FS III, Moilanen L, Kielland K (1993) Preferential use of organic nitrogen for growth by a non-mycorrhizal arctic sedge. Nature 361: 150-153

Costanza R, Sklar FH, White ML (1990) Modeling coastal landscape dynamics. BioScience 40: 91-107

DeAngelis DL, White PS (1994) Ecosystems as products of spatially and temporally varying forces, ecological processes, and landscapes: a theoretical perspective. In: Davis SM, Ogden JC (eds) Everglades: the ecosystem and its restoration. St Lucie Press, Delray Beach, pp 9-27

Fortin M-J, Drapeau P, Legendre P (1989) Spatial autocorrelation and sampling design in plant ecology. Vegetatio 83: 209-222

Giblin, AE, Nadelhoffer K, Shaver GR, Laundre JA, McKerrow AJ (1991) Biogeochemical diversity along a riverside toposequence in arctic Alaska. Ecol Monogr 61: 415-435

Harley PC, Tenhunen JD, Murray KJ, Beyers J (1989) Irradiance and temperature effects on photosynthesis of tussock tundra Sphagnum mosses from the foothills of the Philip Smith Mountains, Alaska. Oecologia 79: 251-259

Hastings SJ, Luchessa SA, Oechel W, Tenhunen JD (1989) Standing biomass and production in water drainages of the foothills of the Philip Smith Mountains, Alaska. Holarct Ecol 12: 304-311

Hatfield JL, Asrar G, Kanemasu ET (1984) Intercepted photosynthetically active radiation estimated by spectral reflectance. Remote Sens Envir on 14: 65-75

Kielland K, Chapin FS III (1992) Nutrient absorption and accumulation in arctic plants. In: Chapin FS III, Jefferies RL, Reynolds JF, Shaver GR, Svoboda J (eds) Arctic ecosystems in a changing climate: an ecophysiological perspective. Academic Press, San Diego, pp 321-336

Leadley PW, Reynolds JF (1992) Long-term response of an arctic sedge to climate change: a simulation study. Ecol Appl 2: 323-340

Leadley PW, Reynolds JF, Chapin FS III (1996) A model of ammonium, nitrate, and glycine uptake by Eriophorum vaginatum roots in the field: ecological implications (submitted)

Li H, Reynolds JF (1993) A new contagion index to quantify spatial patterns of landscapes. Landscape Ecol 8(3): 155-162

Li H, Reynolds JF (1995) On definition and quantification of heterogeneity. Oikos 73(2): 280-284

McGuire AD, Melillo JM, Joyce LA, Kicklighter DW, Grace AL, Moore B III, Vorosmarty CJ (1992) Interactions between carbon and nitrogen dynamics in estimating net primary productivity for potential vegetation in North America. Global Biogeochem Cycles 6: 101-124

Meininger CA, Spatt CD (1988) Variations of tardigrade assemblages in dust-impacted arctic mosses. Arct Alp Res 20: 24-30

Monteith JL (1981) Does light limit crop production? In: Johnson CB (ed) Physiological processes limiting plant productivity, Butterworths, London, pp 23-38

Murray KJ, Tenhunen JD, Kummerow J (1989) Limitations on moss growth and net primary production in tussock tundra areas of the foothills of the Philip Smith Mountains, Alaska. Oecologia 20: 256-262

Nadelhoffer KJ, Giblin AE, Shaver GR, Linkins Ae (1992) Microbial processes and plant nutrient availability in arctic soils. In: Chapin FS III, Jefferies RL, Reynolds JF, Shaver GR, Svoboda J (eds) Arctic ecosystems in a changing climate: an ecophysiological perspective. Academic Press, San Diego, pp 281-300

Oberbauer SF, Dawson TE (1992) Water relations of arctic vascular plants. In: Chapin FS III, Jefferies RL, Reynolds JF, Shaver GR, Svoboda J (eds) Arctic ecosystems in a changing climate: an ecophysiological perspective. Academic Press, San Diego, pp 259-280

Ostendorf B, Reynolds JF (1993) Relationships between a terrain-based hydrologic model and patch-scale vegetation patterns in an arctic tundra landscape. Landscape Ecol 8: 229-237

Ostendorf B, Reynolds JF (1996) A model of arctic tundra vegetation derived from topographic gradients (submitted)

Quinn P, Beven K, Chevallier P, Planchon O (1991) The prediction of hillslope flow paths for distributed hydrological modelling using digital terrain models. Hydrol Proc 5: 59–79

Reich JW, Rastetter EB, Melillo JM, Kicklighter DW, Steudler PA, Peterson BJ, Grace AL, Moore B III, Vörösmarty CJ (1991) Potential net primary productivity in South America: application of a global model. Ecol Appl 14: 399–429

Reynolds JF, Leadley PW (1992) Modeling the response of arctic plants to changing climate. In: Chapin FS III, Jefferies RL, Reynolds JF, Shaver GR, Svoboda J (eds) Arctic ecosystems in a changing climate: an ecophysiological perspective. Academic Press, San Diego pp 413–440

Romme WH (1982) Fire and landscape diversity in subalpine forests of Yellowstone National Park. Ecol Monogr 52: 199–221

Rossi RE, Mulla DJ, Journel AG, Franz EH (1992) Geostatistical tools for modeling and interpreting ecological spatial dependence. Ecol Monogr 62(2): 277–314

Santelmann MV, Gorham E (1988) The influence of airborne road dust on the chemistry of *Sphagnum* mosses. J Ecol 76:1219–1231

Sellers PJ (1987) Canopy reflectance, photosynthesis, transpiration: the role of biophysics in the linearity of their interdependence. Remote Sens Environ 21: 143–183

Shaver GR, Chapin FS III (1980) Response to fertilization by various plant growth forms in an Alaskan tundra: nutrient accumulation and growth. Ecology 61: 662–675

Shaver GR, Chapin FS III (1991) Production: biomass relationships and element cycling in contrasting arctic vegetation types. Ecol Monogr 61: 1–31

Shaver GR, Chapin FS III, Gartner BL (1986) Factors limiting seasonal growth and peak biomass accumulation in *Eriophorum vaginatum* in Alaskan tussock tundra. J Ecol 74: 257–278

Shugart HH (1984) A theory of forest dynamics. Springer, Berlin Heidelberg New York

Stow D, Burns B, Hope A (1989) Mapping arctic tundra vegetation using digital SPOT/HRV-XS data: a preliminary assessment. Int J Remote Sens 10: 1451–1457

Tehnunen JD, Lange OL, Hahn S, Siegwolf R, Oberbauer SF (1992) The ecosystem role of poikilohydric tundra plants In: Chapin FS III, Jefferies RL, Reynolds JF, Shaver GR, Svoboda J (eds) Arctic ecosystems in a changing climate: an ecophysiological perspective. Academic Press, San Diego, pp 213–238

Turner MG (ed) (1987) Landscape heterogeneity and disturbance. Springer, Berlin Heidelberg New York

Turner MG (1989) Landscape ecology: the effect of pattern on process. Annu Rev Ecol Syst 20: 171–197

Turner MG, Dale VH (1991) Modeling landscape disturbance. In: Turner MG, Gardner RH (eds) Quantitative methods in landscape ecology. Springer, Berlin Heidelberg New York, pp 323–35

Walker DA, Everett KR (1987) Road dust and its environmental impact on Alaskan taiga and tundra. Arct Alp Res 19: 479–489

Walker DA, Binnan E, Evans BM, Lederer ND, Nordstrand E, Webber PJ (1989) Terrain, vegetation, landscape evolution of the R4D research site, Brooks Range Foothills, Alaska. Holarct Ecol 12: 238–261

Wu J, Levin SA (1994) A spatial patch dynamic modeling approach to pattern and process in an annual grassland. Ecol Monogr 64: 447–464

V Summary

19 Ecosystem Response, Resistance, Resilience, and Recovery in Arctic Landscapes: Progress and Prospects

J. D. TENHUNEN and J. F. REYNOLDS

19.1 The NRC Tasks and R4D Accomplishments

Research on terrestrial tundra ecosystems of northern Alaska has been directly and indirectly linked to social and economic interests since World War II (Washburn and Weller 1986). The discovery of large petroleum reserves (Energy Resource Map of Alaska 1977) made it clear that compromises would have to be made to satisfy both economic demands and environmental concerns. This led to the specific recommendations made by the National Research Council (NRC) committee on ecological research priorities for the Arctic (NAS 1982) and the subsequent implementation of the R4D program by the Department of Energy (DOE) at the Imnavait Creek watershed near Toolik Lake, Alaska (Fig. 1.1, this Vol.). The series of recommendations – and expected benefits – made by the NRC committee (Chap. 1, this Vol.) are reviewed here in the context of R4D results (italics indicate a direct paraphrasing of the recommendations).

Results of R4D can be generalized to other Arctic sites. This is possible because basic processes of the natural environment were studied. Webber (1978) identified the degree of hydration as the principal environmental variable influencing vegetation distribution in coastal tundra at Barrow, Alaska. Webber's (1978) general conclusions concerning the central role of hydrology and water status in determining both structure and function in tundra communities (Fig. 14.9, this Vol.) has been significantly reinforced by R4D research. Hydrological effects on tussock tundra ecosystem structure and function in the Foothills Region of the Brooks Range may be viewed as falling into two major categories: (1) those related to winter snow accumulation and melt; and (2) those affecting water table, lateral flow, leaching, and soil aeration during the summer season. Soil freezing and thawing and snow accumulation affect ecosystem structure directly via redistribution of soil, severe scouring of basins (Chaps. 6 and 9, this Vol.; Scherer and Parlow 1994), and redistribution of nutrients from the snowpack and immediate surface layer of soil and moss during spring snowmelt. Hence, at a particular location along a toposequence (see Fig. 4.8, this Vol.), there is a characteristic suite of physical factors to which tundra communities are subjected during growth (Chap. 14 and Fig. 4.9, this Vol.; de Molenaar 1987; Shaver et al. 1991).

We used these strong correlations between topography, soil hydration, and community function as the basis for landscape models for assessing tun-

J.F. Reynolds and J.D. Tenhunen (Eds.)
Ecological Studies, Vol. 120
© Springer-Verlag Berlin Heidelberg 1996

dra carbon balance (Chap. 17, this Vol.) and for examining the consequences of road building on tundra community biomass and production (Chap. 18, this Vol.). We recognize that functioning of the actual system is more complex, i.e., that multiple pathways of nutrient transport in the soil may exist that are independently regulated (e.g., Peterjohn and Correll 1984) or that atmospheric transport of materials could be positively or negatively correlated with hydrologically based transport (Chap. 15, this Vol.). Nevertheless, general topographically derived relationships proved to be valuable building blocks for our initial landscape modeling efforts and suggest potential ways to extrapolate results to other tundra locations (Chaps. 14 and 18, this Vol.).

   The whole landscape should be examined so that unanticipated effects at interfaces of landscape units can be discovered. Shifts in tundra vegetation structure (e.g., species composition and biomass) and in process rates (e.g., decomposition, production) were causally linked in R4D to topographical gradients in physical environmental variables, i.e., energy, water, and nutrients [see Lauenroth et al. 1993 for similar comparisons at large scales between Long Term Ecological Research (LTER) sites]. Cold soils during the summer lead to decreased microbial activity and slow process rates for nutrient cycling (Chaps. 10 and 16, this Vol.). Permafrost and topography determine soil aeration (Chaps. 17 and 18, this Vol.), lateral water flux rates, and the water table, which are important in the transport of nutrients and dissolved and particulate organic carbon between communities (Chap. 13, this Vol.). Cold temperatures, topographic variation, the precipitation regime, and incoming radiation interact to determine landscape patterns in depth of thaw and, thus, nutrient supply in the rooting zone. Vegetation develops in positive correlation with nutrient supply, increasing in stature, in productivity, and in capacity for latent heat exchange (Chaps. 5 and 11, this Vol.).

   Tundra landscapes may be studied along toposequences in order to identify relationships among landscape units (see Fig. 4.9 and Table 14.1, this Vol.). In the Imnavait Creek watershed, distinct patterns of vegetation, soils, and hydrology occur along a gradient from the dry ridges upslope (dominated by mosses and lichens), through mid-slope tussock tundra (dominated by *E. vaginatum*), down to the beaded stream at the valley bottom (see Chaps. 4 and 5, this Vol.). The dynamics of these upslope ecosystems ("source areas"; de Molenaar 1987) are closely tied to precipitation events and are highly subject to leaching and exposure to extreme conditions; the mid-slope ecosystems ("transitional"; de Molenaar 1987) are relatively diverse as species respond to intricate spatial patterns and small differences in resource availability; and the downslope ("sink areas"; de Molenaar 1987) are generally water-saturated, experience less fluctuation in environmental conditions, and have high recovery potentials. The strong linkage between hydrological transport of materials and landscape function may be recognized independently in three aspects of investigations described in this volume: (1) the historical development that has occurred in ecosystem studies focusing on nutrient cycling at the landscape level (Chap. 2, this Vol.); (2) the recognition that anthropogenic disturbances in tundra regions, although of very different types, impact arctic ecosystems in a similar manner – via altered energy and water balance and subsequently thermokarst subsidence (Chaps. 3 and 6, this Vol.); Haag and Bliss 1974); and (3) landscape level interpretations of disturbance effects may be achieved via spatial modeling of the coupling between topography, water fluxes, and ecosystem process rates (Chaps. 14, 17 and 18, this Vol.).

Studies in arctic tundra should emphasize the effects of disturbance manipulations on species and on ecosystem processes. With respect to actual disturbance manipulations, a series of plot-scale water additions and fertilization manipulations were conducted in R4D to test their feasibility for long-term studies in a remote location and the potential for larger scale applications (Marion and Everett 1989; Oberbauer et al. 1989). Comparative measurements of process rates were made between the Imnavait Creek site and nearby disturbed areas (Chaps. 7 and 10, this Vol.). Extensive information from historical disturbances at a variety of locations was used in conjunction with Imnavait Creek data and modeling efforts to achieve a new perspective on research priorities (see below and Chaps. 3, 6, 9, 10, 14–18, this Vol.).

Many types of disturbances in tundra regions result in decreased albedo, decreased latent heat flux, greater heat flux to the soil, and greater thaw (Haag and Bliss 1974; Chap. 4, this Vol.). Ives (1970) suggested that the "fragility" of tundra sites – a quality related to the degree of ecosystem stability – is directly proportional to the ice content of the permafrost and inversely proportional to the naturally occurring mean annual soil temperature. Disturbances that lead to soil heating will depend greatly on impacts to the moss layer, which has a large effect on albedo, evapotranspiration, and insulation of the soil (Tenhunen et al. 1992; Chaps. 4 and 6, this Vol.). What is required next is the direct formulation of specific hypotheses about disturbance effects, an experimental design at appropriate time and space scales, and further development of the modeling approaches to develop specific insights for input into ecosystem-based management models.

A major problem that remains in evaluating the effects of disturbance in the Arctic is the lack of information describing the *dynamic response* of ecosystems to altered hydrological regimes (Chap. 18, this Vol.) and the accompanying changes in water quality (Chaps. 10 and 16, this Vol.; de Molenaar 1987; Siegel 1988; Chapin et al. 1988). Webber and Ives (1978) reported two decades ago that even succession in response to natural cycles of disturbance is not well understood in the Arctic because it has not been consistently examined in conjunction with hydrological gradients.

Future research must produce information on long-term effects and recovery. Walker et al. (1987) showed that the cumulative effects of development at Prudhoe Bay progressively affected larger areas as road networks increasingly modified water flows. Impacts that were not immediately obvious after road construction, e.g., thermokarst erosion, developed over time (Walker et al. 1987). These additive, synergistic, and indirect effects of cumulative impacts must be anticipated as we plan future disturbance research (Risser 1988; see

Chap. 1, this Vol.). The fragmentation of landscapes that occurs along with continued intensive development of a region can also modify animal population dynamics and distributions (Robinson et al. 1992). Nevertheless, even in the case of cumulative impacts, modifications of energy balance and hydrology are of foremost concern and deserve the most immediate attention with respect to the complex task of developing new environmental management perspectives (e.g., Siegel 1988; Winter 1988). It has become very clear from R4D that characterization of the hydrology of sites and the initiation of long-term time series data are essential steps in achieving a comprehensive understanding of natural and long-term anthropogenic disturbance effects in tundra ecosystems.

R4D has provided a framework for prioritizing future research and evaluating performance on research already under way. Based on the above considerations, there is much to be gained in future research on the disturbance of tundra landscapes if it is based on studies along toposequences, specifically comparing the response and recovery of crest, slope, and riparian communities to applied disturbances and examining processes in relation to energy balance and hydrology. We suggest that comparative studies be conducted along a north-south gradient at permafrost sites in the coastal, foothills, and Brooks Range physiographic provinces of the North Slope of Alaska as well as in the taiga region, thus allowing for variation in both geomorphology and mean annual temperature (Fig. 19.1). Observations must be made over the long term in order to understand the relative roles played by slow- vs fast-growing species and the complexity of ecosystem interactions. While true replication is impractical at this scale (Hargrove and Pickering 1992), the experimental design should strive to cover a range of variabilities. Information on succession and population dynamics will allow improvements in topographically derived landscape models (Chaps. 14, 17 and 18, this Vol.). In spite of its limitations (Hargrove and Pickering 1992), a space-for-time substitution would increase the relevance of the results for study of climate change effects (Chaps. 6 and 17, this Vol.), especially with a comparison of foothill tundra with taiga treeline. These studies could potentially build upon existing research at the Toolik Lake and Bonanza Creek LTER sites as well as previous research at Barrow, Alaska (Chap. 2, this Vol.).

It may be possible to predict tundra sensitivity to disturbance based only on geomorphological and geobotanical features of the landscape (Webber and Ives 1978) and by relating these features to key ecosystem variables such as active layer thaw, thickness of insulation by peat, degree of ground ice development, standing biomass, and annual production, etc. However, predictions of ecosystem recovery and assessments of landscape status after disturbance, e.g., regrowth, the reoccupation of disturbed areas by individual species, or the potential for nutrient sequestering or nutrient release to stream systems, will require the use of "fully coupled ecosystem models" (Lauenroth et al. 1993; see also Acker 1990). In these descriptions of arctic ecosystem dynamics, we think that it is necessary that (1) explicit plant-soil interactions be consid-

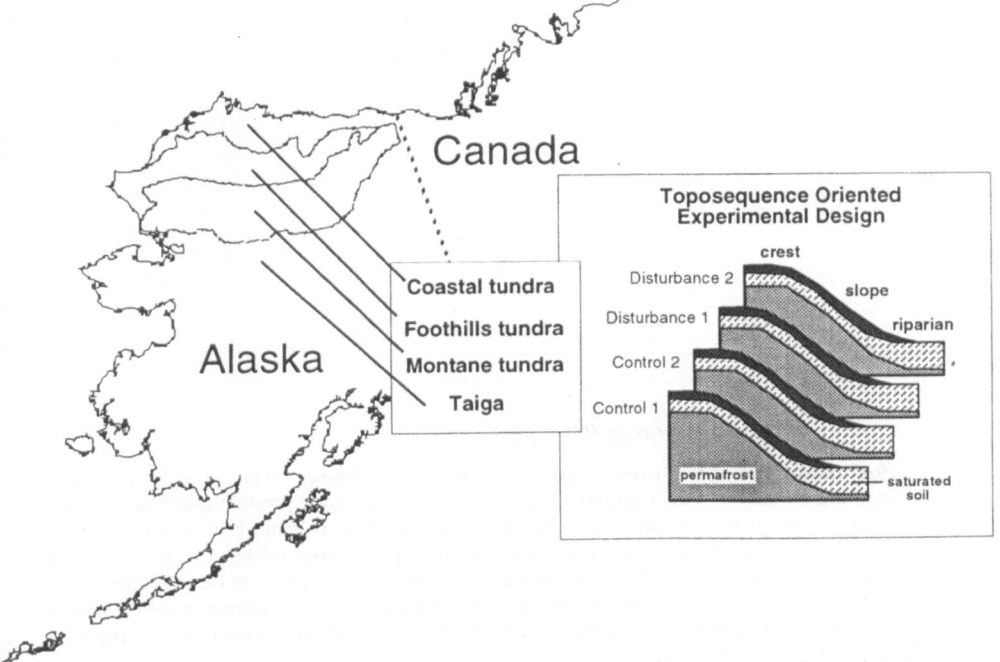

Fig. 19.1. Hierarchical scheme for the study of tundra landscape response to disturbance as described in the text. Emphasis is placed on the examination of response and recovery of crest, slope, and riparian communities along toposequences. Comparative studies are suggested in the coastal, foothills, and Brooks Range physiographic provinces of the North Slope of Alaska as well as in the taiga region

ered; (2) vegetation structure and composition must be influenced by individual species' abilities to obtain and utilize available resources; and (3) lateral material flows affecting soil-plant dynamics must be a function of topography.

We believe that the next integrated research effort on energy-related impacts in the arctic should develop a number of hypotheses that could be tested at both local and regional scales, following the experimental approach outlined in Fig. 19.1 and findings to date. For example, Pomeroy (1970) and Webster et al. (1974) suggested that ecosystem stability is related to nutrient cycling, standing biomass, turnover time, and productivity. Webster et al. (1974) proposed that stability depends on system characteristics that control the balance between structural mass increment and dissipative forces tending to break down structure. Thus, resistance may be viewed as the ability of an ecosystem to maintain its state (reflected in standing biomass) and resilience is defined as the ability of a system to rebuild structure after disturbance (reflected in net primary productivity per unit biomass). Based on this definition, resistance and resilience estimates for tundra communities are inversely related (Fig.

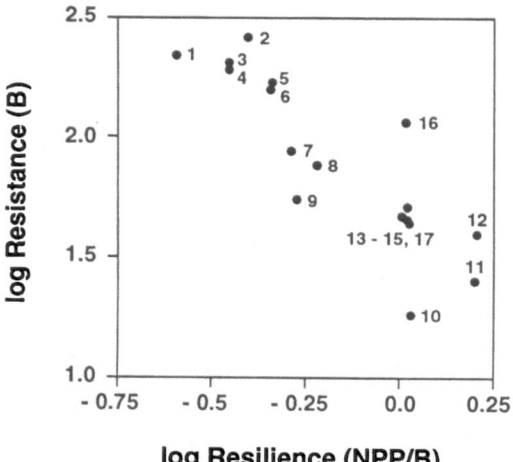

Fig. 19.2. Illustration of the inverse relationship between resistance (*B* is summer mid-season aboveground biomass in g m^{-2}) and resilience (*NPP/B* or net primary production/biomass; turnover in year^{-1}) estimated for tundra communities. *Numbers 1–4* are water track communities at the R4D research site (unpubl.), *5* is tussock tundra at Eagle Summit (Miller 1982), *6* is tussock tundra at Toolik Lake (Shaver, pers. comm.), *7–9* are sedge communities at Devon Island (Muc 1972), *10–17* are communities studied during the US IBP tundra biome project in coastal tundra at Barrow, Alaska, numbered in sequence as described by Webber (1978). Stakhiv (1988) reported a similar pattern for temperate wetlands

19.2). The studies from a number of sites suggest that tussock tundra communities found on mid-slopes of toposequences have higher resistance but lower resilience than riparian or coastal tundra communities. In the context of specific disturbances, we should perform experiments to see whether these are good surrogates of resistance and resilience. Do these patterns vary along a longitudinal gradient (e.g., Fig. 19.1)? Within a toposequence in coastal tundra? In the taiga region?

Continued research along these lines is also relevant with respect to potential climate change in the Arctic. Climate models are sensitive to biological processes and carbon storage is sensitive to hydrology and ecosystem cycling rates (Chap. 17, this Vol.). Comparisons of some of the leading ecosystem models have demonstrated that our knowledge with respect to biotic-abiotic linkages is inadequate (Bonan 1993). Landscape experiments and further model development are required to obtain better estimates of regional vegetation/atmosphere exchange parameters (Risser 1990).

The initial landscape-oriented studies of R4D should be centered at a single site (which is cost-effective for northern research). As a result of the landscape and toposequence orientation of the R4D program, we were able to quantify the shifts in the controls on processes and in process rates along gradients of microclimate, water availability and nutrient availability. These observations stimulated a comparison with results from other sites and a reevaluation of key site characteristics. The interdisciplinary investigation from many different perspectives of the Imnavait Creek watershed has also produced a new overall data structure that supports examination of process linkages and a clear perspective on the relative time scales at which different processes vary. Thus, the recommendation of NRC committee to study a single landscape unit has proven to be very useful.

Prior to the R4D program, extensive terrestrial ecosystem studies had not been undertaken in the Foothills Province of the North Slope of Alaska. Thus, the landscape-oriented studies in the Imnavait Creek watershed required considerable investment (both financial and man-power) in basic site description and quantification of initial conditions. The Imnavait Creek catchment should now be considered a valuable resource. The difficult logistic conditions for conducting field research on the North Slope of Alaska certainly made the decision to locate the R4D at the Imnavait Creek site a practical one. A high priority should be given to see whether it is feasible to consider this site as a surrogate for the Arctic National Wildlife Refuge, where the majority of the area being proposed for drilling is tussock tundra (Walker et al. 1982).

The R4D studies as well as future efforts must enhance the training of Arctic scientists who by virtue of their experience in tundra and taiga environments will be able to predict the impact of environmental disturbance. Future research initiatives in the Arctic must stimulate ecologists to develop new, robust predictive capabilities jointly with those involved in the design of ecosystem-based management protocols (Cairns 1990; Slocombe 1993). The experiences of the R4D project suggest certain types of experimental approaches (see above) relevant to management. The R4D program provided training for many students, both undergraduate and graduate, in both basic research and in the challenge of examining ecological data with respect to applied problems. Individuals with this type of expertise will be required to interact with resource managers (Slocombe 1993) and meet the challenge of developing cost-effective management tools for arctic tundra landscapes.

19.2 Conclusion

While the task of the R4D program was well defined from the onset (Chap. 1, this Vol.), the objectives have only been partly achieved. This is because of the large gap that exists between our basic understanding of the ecology of tundra ecosystems and the information needed to implement ecosystem-based management approaches. While important gains were made in the R4D research effort to develop models to help bridge this gap, the scope of the problem is enormous. Additional research is essential if we are to fully realize the means by which assessments of risks, gains, and losses to ecological, social, and economic systems may be made (NAS 1982; see also Kelly and Harwell 1990).

In Chapter 1 (this Vol.) we remarked that the NRC committee's challenge to arctic ecologists to document and to model the effects of disturbances associated with energy development in tundra regions represented a "real world" problem. That is, the problem is very broad, it involves high spatial and temporal variability, and defies plausible experimentation. Consequently, the R4D program could tackle only a small subset of this "real world" problem. The research at Imnavait Creek watershed was much simpler and narrower in scope than the NRC challenge, but the research was solidly structured and had well-defined experimental components. As a result, the research presented in this volume represents only a preliminary contribution towards the goal of developing a capability for ecosystem-based management in the Arctic. Ecosystem-based management involves a complex of interacting factors that include politics, management frameworks, economics, sociology, and basic ecosystem knowledge (Slocombe 1993). The R4D results contribute to basic ecosystem knowledge for arctic tundra regions. Key relationships were identified, ecosystem processes were expressed at the landscape scale, a better understanding of uncertainties was achieved, and conclusions previously reached in studies of coastal tundra were placed in the context of tussock tundra landscapes. The next step is to start integrating this ecological knowledge within a management framework that considers various economic, sociological, and political concerns of the North Slope of Alaska. While much work remains, we are confident that ecosystem-based management is an achievable goal for these ecosystems in the near future if the resources for supporting this goal are forthcoming.

References

Acker SA (1990) Vegetation as a component of a non-nested hierarchy: a conceptual model. J Veg Sci 1: 683–690

Bonan GB (1993) Do biophysics and physiology matter in ecosystem models? Climatic Change 24: 281–285

Cairns J Jr (1990) Lack of theoretical basis for predicting rate and pathways of recovery. Environ Manage 14: 517–526

Chapin FS III, Fetcher N, Kielland K, Everett K, Linkins AE (1988) Productivity and nutrient cycling of Alaskan tundra: enhancement by flowing soil water. Ecology 69: 693–702

de Molenaar JG (1987) An ecohydrological approach to floral and vegetational patterns in arctic landscape ecology. Arct Alp Res 19: 414–424

Energy Resource Map of Alaska (1977) Alaska Department of Natural Resources, Division of Geological and Geophysical Surveys, College, Alaska 99708

Haag RW, Bliss LC (1974) Energy budget changes following surface disturbance to upland tundra. J App Ecol 11: 355–374

Hargrove WW, Pickering J (1992) Pseudoreplication: a *sine qua non* for regional ecology. Landscape Ecol 6: 251–258

Ives JD (1970) Arctic tundra; how fragile? A geomorphologist's point of view. Trans R Soc Can, 4th Ser, VII: 39–42

Kelley JR, Harwell MA (1990) Indicators of ecosystem recovery. Environ Manage 14: 527–545

Lauenroth WK, Urban DL, Coffin DP, Parton WJ, Shugart HH, Kirchner TB, Smith TM (1993) Modeling vegetation structure-ecosystem process interactions across sites and ecosystems. Ecol Modeling 67: 49–80

Marion GM, Everett KR (1989) The effect of nutrient and water additions on elemental mobility through small tundra watersheds. Holarct Ecol 12: 317–323

Miller PC (1982) Environmental and vegetational variation across a snow accumulation area in montane tundra in central Alaska. Holarct Ecol 5: 85–98

Muc M (1972) Vascular plant production in the sedge meadows of the Truelove Lowland. In: Bliss LC (ed) Devon Island IBP Project. High Arctic Ecosystem, Project Rep 1970 and 1971, University of Alberta, Edmonton, pp 113–145

NAS (1982) Arctic Terrestrial Environmental Research Programs of the Office of Energy Research, Department of Energy: evaluation and recommendations. National Academy Press, Washington, DC, p 63

Oberbauer SF, Hasting SJ, Beyers JL, Oechel WC (1989) Comparative effects of downslope water and nutrient movement on plant nutrition, photosynthesis, and growth in Alaskan tundra. Holarct Ecol 12: 324–334

Peterjohn WT, Correll DL (1984) Nutrient dynamics in an agricultural watershed: observations on the role of a riparian forest. Ecology 65: 1466–1475

Pomeroy LR (1970) The strategy of mineral cycling. Annu Rev Ecol Syst 1: 171–190

Risser PG (1988) General concepts for measuring cumulative impacts on wetland ecosystems. Environ Manage 12: 585–589

Risser PG (1990) Landscape pattern and its effects on energy and nutrient distribution. In: Zonneveld IS, Forman RT (eds) Changing landscapes: an ecological perspective. Springer, Berlin Heidelberg New York, pp 45–56

Robinson GR, Holt RD, Gaines MS, Hamburg SP, Johnson ML, Fitch HS, Martinko EA (1992) Diverse and contrasting effects of habitat fragmentation. Science 257: 524–526

Scherer D, Parlow E (1994) Terrain as an important controlling factor for climatological, meteorological and hydrological processes in NW-Spitzbergen. Z Geomorphol N F Suppl 97: 175–193

Shaver GR, Nadelhoffer KJ, Giblin AE (1991) Biogeochemical diversity and element transport in a heterogeneous landscape, the North Slope of Alaska. In: Turner MG, Gardner RH (eds) Quantitative methods in landscape ecology. The analysis and interpretation of landscape heterogeneity. Ecological Studies 82. Springer, Berlin Heidelberg New York, pp 105–125

Siegel DI (1988) Evaluating cumulative effects of disturbance on the hydrological function of bogs, fens, and mires. Environ Manage 12: 621–626

Slocombe DS (1993) Implementing ecosystem-based management. BioScience 43: 612–622

Stakhiv EZ (1988) An evaluation paradigm for cumulative impact analysis. Environ Manage 12: 725–748

Tenhunen JD, Lange OL, Hahn S, Siegwolf R, Oberbauer SF (1992) The ecosystem role of poikilohydric tundra plants. In: Chapin FS III, Jefferies RL, Reynolds JF, Shaver GR, Svoboda J (eds) Arctic ecosystems in a changing climate: an ecophysiological perspective. Academic Press, San Diego, pp 213–238

Walker DA, Acevedo DA, Everett KR, Gaydoes KR, Brown J, Webber PJ (1982) Landsat-assisted environmental mapping in the Arctic National Wildlife Refuge, Alaska. CRREL Rep 82-27, US Army Cold Regions Res Eng Lab, Hanover, New Hampshire

Walker DA, Webber PJ, Binnian EF, Everett KR, Lederer ND, Nordstrand EA, Walker MD (1987) Cumulative impacts of oil fields on northern Alaskan Landscapes. Science 238: 757–761

Washburn AL, Weller G (1986) Arctic research in the national interest. Science 233: 633–639

Webber PJ (1978) Spatial and temporal variation of the vegetation and its productivity. In: Tieszen LL (ed) Vegetation and production ecology of an Alaskan arctic tundra. Ecological Studies 29. Springer, Berlin Heidelberg New York, pp 37–112

Webber PJ, Ives JD (1978) Damage and recovery of tundra vegetation. Environ Conserv 5: 1–12

Webster JR, Waide JB, Patten BC (1974) Nutrient recycling and the stability of ecosystems. In: Howell FG, Gentry JB, Smith MH (eds) Mineral cycling in southeastern ecosystems. ERDA Symp Ser (CONF-740513), TIC and ERDA, Springfield, Virginia, p 898
Winter TC (1988) A conceptual framework for assessing cumulative impacts on the hydrology of nontidal wetlands. Environ Manage 12: 605–620

Subject Index

Acidophilic plants 47, 388
Advanced very-high resolution radiometer 38, 157
Advection
  role in snowmelt 136, 152
Air temperature see Temperature
Albedo 131–132, 155, 160–163, 421
Amchitka Island 27
Amino acid uptake 209
Anaktuvuk glaciation 96
Anaktuvuk Pass 51
Animal habitat 56–57, 60–61, 422
Anoxia see Waterlogged soils
ANWR see Arctic National Wildlife Refuge
ARC see US Arctic Research Commission
ARCO 46
Arctic LTER 19, 25, 330, 420, 422
Arctic National Wildlife Refuge 4–5, 12, 52–54, 425–426
Arctic System Science Program 27
ARCUS see Arctic System Science program
Atigun Pass 330
Atkasook 22
Atmospheric deposition 187, 192–195
  see also Disturbance
Atmospheric transport 333–335, 420
AVHRR see Advanced very-high resolution radiometer

Barrow 4, 19, 21, 23, 25, 48, 50–51, 125, 133, 228, 244, 297, 302, 311, 422, 424
Beaded stream 12, 79, 85, 87, 93, 275, 283, 420
Beaufort Sea 3
Bedrock 74, 97, 109
Beringia 80
Biomass
  aboveground, along toposequence 118–119
  and spectral indices 157

and respiration 234
in water tracks 118–119, 211
live aboveground in heath 114–116
of aboveground litter in heath 115, 117
of lichen species in heath 117
photosynthetic in heath 116
turnover 117
see also Plant growth, Productivity, Model
Blading see Disturbance
Bonanza Creek LTER 258, 267, 422
Boundaries
  transport across 14, 198, 295
  see also Disturbance
Boundary layer atmospheric, planetary 330–333
Brine see Disturbance

Canol Pipeline 43, 57
Canopy
  leaf angle 300
  leaf area index 223, 226, 247, 250, 299–302, 370–371, 374
  shading effects 115
  stem angle 300
  stem area index 299–300, 374
  structure 126, 300, 371
  see also Model
Cape Simpson 45, 49
Cape Thompson 24, 27, 43, 48, 51
Carbon balance 258, 275
  sink 247–252, 380–383
  source 247–252, 380–383
  see also Climate Change, Model, Net CO$_2$ Exchange
Carbon isotope ratio
  of leaves 229
Carbon storage 223, 275, 369, 384
Cassiope Dwarf Shrub heath
  definition 110–111
  see also Model, Vegetation

CDS heath see *Cassiope* Dwarf Shrub heath
Cephalodia   120–121
Climate change   3, 5, 24, 36, 38, 257, 275, 347, 369, 422
  and elevated carbon dioxide   223, 347
  and nutrient availability   217–218, 305, 359–360
  see also Model
Cloud index   161–162
Colluvial basin   78, 79
  see also Vegetation
Colonization   39, 43–44, 46, 55
Community see Vegetation
Community CO_2 uptake
  contribution by growth forms   125
  see also Net CO_2 exchange
Competition
  among growth forms   115, 125
  microbiota and plants   208–210, 350
Cumulative impacts see Disturbance

Dalton Highway   3–4, 12, 43, 57–58, 267, 325–326, 342, 360, 388
Darcy equation   313, 371, 391
Deadhorse see Prudhoe Bay
Decomposition
  and blading   42, 49
  and lemming populations   20
  controls on   206–208, 347, 350
  rates   347, 356–358, 420
  see also Model
DEM see Digital elevation model
Dempster Highway   57
Denali National Park   57
Department of Energy   6–8, 12, 419
Devon Island   261, 424
Diesel spill see Disturbance
Digital elevation model   374, 390–391
Discharge
  and thermokarst erosion   396
  effects of topography   311–315, 372–373
  from water tracks   186, 188, 196
  from watershed   144, 150–151, 186, 188, 277–278, 280–283, 377–378
  see also Model
Dissolved CO_2 in lakes and streams   26, 275
Dissolved organic carbon   191–192, 196–197, 252, 275, 278–287, 420
  measurement methods   277, 279

Disturbance
  and albedo   161
  and effects on material transfer across boundaries   295
  and effects on community spatial patterns   407–408
  animal digging   50
  atmospheric   5, 36
  blading   36, 38, 42, 48–49
  brine   36, 45–47
  bulldozing see blading
  cumulative impacts   35, 36, 43, 61–62, 421
  definition   15, 35
  deposition see atmospheric
  diesel spill see hydrocarbon spill
  dust   5, 35–36, 38, 58–59, 216–217, 325–326, 339–343
  effects on nutrients   215–218
  energy development   3, 6, 24
  fire   5, 36, 38, 47–48, 193–198
  gasoline spill see hydrocarbon spill
  gravel pads see road construction
  gravel pits   5, 36, 38, 42, 55–56
  gray water   5, 55, 217
  hydrocarbon spill   5, 35–36, 38, 43–45
  ice road see snowpad
  impoundment   5, 36, 42–43, 54, 59–61
  manipulations   7, 421
  material site see gravel pit
  off-road vehicle   35–36, 38, 48, 50–52, 215
  oil spill see hydrocarbon spill
  road construction   5, 36, 38, 42–43, 54–55, 59, 61, 420
  scale   15, 38
  seawater   38, 45–47
  seismic trail see snowpad
  snowpad   5, 36, 38, 42, 52–54
  solid waste see trash
  storm surge see seawater
  trash   36, 38–39, 42, 52
  see also Model, Herbivory
Diurnal rhythms
  and gas exchange   226
DOC see Dissolved organic carbon
DOE see Department of Energy
Dry deposition see Atmospheric deposition
Dryas heath
  definition   110–111

Dryfall see Atmospheric deposition
Dust
    and community composition 246
    and depth of thaw 245–247
    and ecosystem CO_2 exchange 224,
        245–247
    and soil characteristics 247
    and soil water content 245–247
    chemical composition 194, 217, 333–
        335
    deposition into vegetation 194, 325,
        335–342, 396
    effects on extracellular enzyme activi-
        ty 360–361
    effects on plants 58, 216–217, 388,
        406
    effects on CO_2 efflux 240
    mass transfer through atmosphere
        333–335
    particle size spectra 340–343
    see also Model, Disturbance

Eagle Creek 23–25, 126
Eagle Summit 21, 23, 424
Ecosystem carbon balance see Net CO_2
    exchange
Electron transport capacity 227–228,
    371
Energy balance 131, 133, 136, 141, 149,
    421–422
Eriophorum vaginatum 110, 224, 241,
    303–305, 310, 383, 389
    and methane oxidation 270
    effects of dust 246
    effects of water tracks 211
    effects on water table and depth of
        thaw 237, 259
    net photosynthesis 227–228, 231,
        233, 301
    nitrogen limitations to growth 305,
        411
    nutrient uptake 213–214, 306–307
    phosphatase activities of roots 361–
        362
    respiration 234–236
    retranslocation of nutrients 214–215
    rooting strategy 212
    spatial pattern of tussocks 13
    see also Model
Evaporation 132, 135, 143, 145–146
Evapotranspiration 131–132, 135–136,
    143, 145, 174, 371–372, 376–379, 391,
    420–421

Fairbanks 44
Fire see Disturbance
Fish Creek 39, 43, 48–49, 51
Floodplain 79
Flora
    at Imnavait Creek 79–80, 98
Flow path 14, 294–295
Franklin Bluffs 44
Frost scars 36, 78
Functional groups see Models

Gas exchange
    see Leaf photosynthesis, Net CO_2 ex-
        change, Soil CO_2 efflux
Gasoline spill see Disturbance
Geostatistics 13, 73, 313–315
    see also Semivariance
Glacial till 74–75, 78, 85, 109
Glacier National Park 408
Global climate model 257
Gravel pads see Disturbance
Gray water see Disturbance
Greenland 3
Growth form
    composition at sites 24
    of shrubs in heath 112
    see also Models (functional groups)

Heat flux see Soil heat flux
Herbivory 20, 23, 38
Heterogeneity see Spatial heterogeneity
Hickel Highway 48
Hydraulic conductivity 141, 371, 377
Hydrograph 141, 188, 278, 281–282
    see also Discharge
Hydrologic response units see Spatial si-
    mulation units
Hydrology
    transport 12, 14, 26, 185, 252, 281–
        283, 285–287, 369, 420
    water saturation 185–186, 189, 373,
        382, 420
    see also Model

IBP see International Biological Program
Ice
    on Imnavait Creek 278, 280–281
Ice road see Disturbance
Immobilization
    Carbon 209, 307–311, 359–360
    Nitrogen 204, 307–311, 359–360
Imnavait Creek 4, 10, 419, 424–426
    characteristics 275–277, 283

Imnavait Creek nutrients see Water
Imnavait Creek temperature see Temperature
Impoundment see Disturbance
Integrated flow systems see Spatial simulation units
Intensive research site   4, 12, 74–77, 80–85, 88–93, 110
International Biological Program   22, 125, 302, 296, 349–350
Invertebrates
    in Imnavait Creek   277
IRS see Intensive research site
Isotopes
    analysis in watersheds   165
    as tracers   165–179
    Beryllium-7   166–167, 174–175, 178
    Cesium-137   166, 174–178
    characteristics   166
    Cs uptake in plants   176–178
    Deuterium   165, 171–174
    evaporation and soil moisture   172–173
    hydrological flow paths   168–172
    Lead-210   166, 174, 178
    Nitrogen-15   209
    Oxygen-18   165–174, 179
    Potassium-40   177
    sulfur cycling   168
    Sulfur-35   166–167, 174
    transpiration   173–174
    unit definition   166
Itkillik glaciation   13, 97

Kevo, Finland   126
Kokolik River   47
Kuparuk oil field   3, 5, 46, 61
Kuparuk River   4, 10, 79, 92, 141, 258, 267, 285

Landscape   7, 9, 12–15, 26, 63, 96–97, 158, 160, 223–224, 293, 311–312, 318–319, 383, 387, 419–420, 422–426
    and soil CO_2 efflux   233–234, 240–246
    and net CO_2 exchange   247–250
    defined as Spatial simulation units   294–295
    fragmentation   422
    indices   399–400
    patterns in leaf photosynthesis   230–233
    see also Model, Toposequence, Net CO_2 exchange

Latent heat flux see Evapotranspiration
Lateral flow   144, 185, 286, 419, 423
Leaching in active layer   196
Leaf area
    determination of   225
Leaf area index see Canopy
Leaf photosynthesis
    and landscape gradients   230–233
    and light   226–228
    and micronutrients   230
    and nutrition   229–230, 232
    and phenology   227–228
    and temperature   226–227
    and water availability   229
    measurement methods   224–225
Leaf pigmentation
    and phenology   228
Leaf respiration   234
Lemmings   20, 23
Lichen heath
    definition   110–111
Lichens
    carbon content   118
    community sampling   113
    on rocks   84–85
Loess   36, 38
Long-term research   7, 19, 25, 330, 421–422

Mackenzie River   80
Management   7, 12, 14–15, 61–63, 293, 422, 425–426
Meade River   22–25, 373
Meteorological station   74
Methane
    and climate change   251, 257–258, 270–271
    and soil temperature   258, 262
    and water table   258, 262–264, 268–269
    controls on net flux   258, 260–265
    oxidation   258
    sampling procedures   258–260
Microbes
    and disturbance   215–218
    and methane oxidation   259–260, 265–267
    and mineralization   207
    and nutrient availability   205–206, 350, 354
    and nutrient availability   209
    and root competition   208
    see also Competition

Mineralization
   carbon 307–311
   controls on 205–208
   nitrogen 204–205, 307–311, 409–410
   phosphorus 205
Model
   ABISKO 350–353, 357–358
   ARTUS 24, 296, 298, 350–353, 357–358
   BARK 350–351, 353–354, 357–358, 362
   biomass 305, 395
   bottom-up 297
   canopy 299–302
   carbon balance 380–382
   Cassiope Dwarf Shrub heath
      (moist) 248–249, 299–302, 370–371
   climate change scenarios 26, 304–306, 311, 359–360
   decomposition 307–311, 348–349
   disturbance 293, 304, 396–398, 402–408
   dust 325–342, 360–361
   effects of elevated CO_2 305, 360
   effects of road-building 420
   Eriophorum vaginatum 301, 303–307, 362, 389
   extrapolation 15, 411, 419
   functional groups 299–301, 316–317, 370–371, 373–375, 388, 393, 407
   GAS-FLUX 224, 247–250, 296–302, 312, 369–371, 374–375, 378–384
   GENDEC 296, 308–311, 350–351, 354–362
   HBV 150–152, 296
   hydrology 150–152, 311–315
   International Biological Program 22, 296, 349–350
   landscape 12, 109, 294–297, 311–319, 342, 369–370, 382–383, 387–388, 419, 422
   NECS 24
   nitrogen availability 304–305, 359–360, 393–394, 399–401, 409–411
   nitrogen uptake 296, 306–308
   of road construction and effects of
      dust 325–326, 335–342, 360–361, 396–397, 406
   of road construction and effects on
      discharge 396–397, 402–406
   photosynthesis 299–302
   plant growth 302–306

   productivity 394–395
   riparian 248–249, 299–302
   sensitivity analysis 307
   SHE 371
   snowmelt 148–150
   soil CO_2 efflux 299–302, 308, 371, 375, 379, 381–382
   Sphagnum 299–302, 394–396, 398, 404–406
   summary of R4D products 296–298
   T-HYDRO 296–297, 312–315, 390–391
   T-MAP 296–297, 312, 387–388, 390, 397
   T-NUT 296–297, 390, 393–394
   T-PLT 296–297, 390–391, 394–396
   T-VEG 296–297, 390, 393
   TDHC 146, 296
   top-down 297, 311–317
   TOPMODEL 249–250, 296–297, 312, 316, 370–384
   topographically derived 296, 315–317, 369, 383, 419–420
   toposequence 300–302
   tussock tundra 248–249, 299–302, 370–371
   TVM 296–297, 315–317
   types 296, 302, 348–349
   uncertainty 15, 308, 318, 343, 380, 398
   validation 304, 318, 355–356, 398
   vegetation 315–317, 374–375
   see also Road construction
Moss see *Sphagnum*
MSS see Multispectral scanner
Multispectral scanner 38, 156
Mycorrhizal fungi 209, 214, 360

NARL see Naval Arctic Research Laboratory
National Petroleum Reserve 39, 43, 52, 55, 57
National Research Council 6–8, 14–15, 419, 425–426
Naval Arctic Research Laboratory 19
Naval Petroleum Reserve see National Petroleum Reserve
NDVI see Normalized difference vegetation index
Net CO_2 exchange
   and landscape gradients 247–250, 369, 380–384
   and site hydrology 248–252

in communities 223–224, 250–252, 369, 380–384
measurement methods 225
Net radiation 132, 136, 141
Nitrification 204, 208–210
Nitrogen
availability 359, 410
availability index 393–394
content of leaves 232
content of lichens 118, 120
control over growth 203, 302–303
fixation 121
in soil 203–204
pools along toposequence 121–124
uptake 203, 210, 306–307, 389
see also Model
Nitrophilic plants 50
Niwot Ridge, Colorado 21
Noatak 47–48
Normalized difference vegetation index 157–160, 313–315, 391, 398–399, 401–402, 409–413
and dust transect 159–160
relation to discharge 313–314, 410
relation to productivity 157, 313, 410
Northwest Territories 3
NPR-A see National Petroleum Reserve
NRC see National Research Council
Nutrient
absorption 211–214
analyses 112
recovery hypothesis 19–21
retranslocation 214
sequestering 422
turnover 9, 20, 25
Nutrient availability
and blading 42, 49
and lemming populations 20
and off-road vehicles 50
and organic matter quality 208
and vegetation composition 24–25, 232, 423

Off-road vehicle see Disturbance
Oil seeps 36, 38, 45
Oil spill see Disturbance
ORV see Disturbance
Oumalik 39, 43, 48–49, 51
Overland flow 185–186, 189, 373, 382

Palsa 92
Particulate organic carbon 252, 275, 278–287, 420

measurement methods 277, 279
Patch ecosystems 14, 294–295
see also Spatial simulation units
Percent cover
by litter 113
by rocks and bare soil 114
see also Vegetation
Percolation 140
Permanent plots
at Imnavait Creek 74
Persistence 9
Phenology
and landscape gradients 232
in GAS-FLUX 300–302, 370–371
of tundra species 223
Phosphatase activity
in tussocks 361–362
Phosphorus
availability 360
content of leaves 232
content of lichens 118, 120
control over growth 203
pools along toposequence 121–124
pools in heath 121
Photobiont
blue-green algal 120
green algal 118, 120
Photosynthesis see Leaf photosynthesis, Net CO_2 exchange
Pixel 312–314, 316, 372–373, 391–392, 399, 404–405
Plankton 283, 285
POC see Particulate organic carbon
Polygon 20, 43, 53, 60, 79
Porosity
soil profile 375
Precipitation
at Imnavait Creek 131, 136, 145–146, 151, 241, 243, 371–372, 382, 390–391, 420
chemistry 193–194
intercepted by plants 379
Productivity
controls on 203, 301, 304–306, 369–370, 389–390
of mosses 115
plant 20, 25, 313, 420, 423
see also Model
Productivity: biomass ratio 389, 396
Prudhoe Bay 3–5, 21, 35, 43–46, 50–52, 54–55, 57–59, 61, 421

R4D program 6, 8, 12, 19, 28, 419, 425–426

R4D region   4, 10–12, 74–76, 79, 81
RATE see Research on Arctic Tundra Ecosystems
Recovery
   of ecosystems   8–9, 15, 21, 35–37, 43–44, 47–49, 421–422
Reflectance   132, 155–160
   see also Albedo, Normalized difference vegetation index
Regional perspectives   61, 424
Rehabilitation see Restoration
Remote sensing   12, 38
Research on Arctic Tundra Ecosystems   22–24
Resilience   8–9, 15, 37, 50, 53, 423–424
Resistance   3, 8–9, 423–424
Restoration
   of ecosystems   5, 21, 35, 37, 44–45, 55–57
Revegetation see Restoration
Rime
   chemistry   195
Riparian
   nutrient pools   109
   see also Model
Road construction see Disturbance, Models
Rooting zone   212, 420
Roots
   and nutrient uptake   212–214, 306–307
   uptake kinetics   306–307
Rubisco   300, 371
RuBP see Rubisco
Runoff   131, 136, 141, 143, 150, 188
Runoff phases
   during snowmelt   189
   during summer   280, 284–285
Runoff plots   78, 139, 142–143, 187

Sagavanirktok glaciation   13, 75, 95
Sagavanirktok River   97, 294, 310
Scaling   15, 28, 257, 411, 419
Scandinavia   3
Seawater see Disturbance
Seed bank   39, 57
Seedling   39, 57
Semivariance   13, 313–315, 407–408
Sensible heat   132–133
Sensitivity see Terrain sensitivity
Seward Peninsula   47–48, 52
Shortwave radiation   132
Siberia   3, 62

Signy Island, Antarctica   355–356
Snow ablation   132, 141–142, 149
Snow accumulation   131–132, 137–139, 419
Snow distribution   36, 137–139, 143
Snow gauge   74, 78, 186–187
Snowbank see Snowbed
Snowbed   23, 36, 38
   see also Vegetation, Snowbed
Snowmelt   131–132, 141–142, 148, 151, 165, 187, 280
Snowmelt phases
   defined   189
Snowpack   135–139, 142–143, 187, 198
Soil
   active layer   96, 132, 141, 149–150
   aeration   235, 240–244, 419
   bulk density   235
   C:N ratio   235, 310, 355
   Ca^{2+} content   96
   carbon   235, 306–311
   extracellular enzyme activity   206, 360–362
   freezing and thawing   140, 207, 419
   heat flux   132, 135, 421
   heterotrophs   209
   insulation   20, 421
   moisture   144, 296, 308–309, 353, 371, 390–394
   nitrogen   235, 306–311
   NO_3^- content   96
   organic horizon   96, 98, 347
   organic matter   308–309, 347, 350
   pH   96, 241
   solution   306–307
   thermal conductivity   135
   water content variation   238
Soil basal respiration   235–236
Soil CO_2 efflux
   and belowground biomass   233
   and landscape gradients   240–244
   and soil moisture   234, 238–243
   and soil temperature   234–236, 239–245, 420
   and water table   234–237, 240–244, 248, 301, 369
   and water table   301, 379
   measurement methods   225
   see also Model
Soil temperature see Temperature
Solar radiation   131–132, 230
Solid waste see Disturbance
Spatial heterogeneity   13, 387

and importance in models   294, 318,
    382–383, 387–388
and landscapes   9, 293, 318–319, 387
Spatial simulation units
    defined   294–295
    hydrologic response units   292–295,
        372, 420
    integrated flow systems   295
    see also Flow paths
Spectral vegetation indices see Normali-
    zed difference vegetation index
Sphagnum   174, 243, 286, 299–300, 389,
    397–398, 404, 406
    effects of dust   217
    photosynthesis and water availabili-
        ty   229, 247, 302
    water content   238
    see also Model
SPOT see Systeme Probatoire
    d'observation de la Terre
Stability
    ecosystem   8–9, 421, 423–424
Stigstuv, Norway   126
Stomatal conductance
    and cold soils   228
    and humidity   229
Stone stripes   75, 78
Stream gauge see Weir
Subpermafrost water   141
Succession   21, 30, 37, 57, 60, 421–422
Surficial geology
    at Imnavait Creek   76
Surficial geomorphology at IRS   77
Systeme Probatoire d'observation de la
    Terre   38, 157

TAPS see Trans-Atlantic Pipeline System
TDR see Time domain reflectometry
Temperature
    air   131, 133, 138, 151, 382
    of Imnavait Creek   277–278, 280
    soil   131, 299–300, 308, 405
Temperature acclimation   226
Terrain sensitivity   3, 97–98, 421
Thaw depth
    and blading   147
    and dust   245–247
    and hydrocarbon spills   44–45
    and off-road vehicles   50
    and seawater spill   46
    and snowpads   54
    natural   20, 88, 96, 147, 237, 241–243,
        420

Thaw lake   20–21, 36, 38, 43, 60
Thaw lake cycle   20–21
Thematic mapper   38
Thermal gradient   138, 140, 146–148
Thermokarst   3, 5, 36, 42, 47–49, 52, 59–
    60, 97, 148, 396, 421
Time domain reflectometry   375–376,
    379, 381
TM see Thematic mapper
Toolik Lake   4, 24–25, 97, 325, 330, 342,
    355–356, 419, 422, 424
Toolik River   10
Topographic index   372, 383
Topographically derived model see Model
Topography
    effect on snow distribution   137–139
    and community distribution see Vege-
        tation
Toposequence
    and NDVI   159
    and nitrification   196, 204
    and nitrogen mineralization   196,
        204
    and soil water   237–238
    and thaw depth   237
    at Imnavait Creek   14, 74, 93–98, 187,
        223–224, 230–231, 294, 420
    at Sagavanirktok River   202, 294
    role in future research   422–423
    see also Model
Trace gases   3
    see also Methane
Trans-Atlantic Pipeline System   3, 13,
    52–53, 55, 57
Transpiration   135, 376–377
Trash see Disturbance
Trophic structure   26
Tundra Biome program   21–22
Tussock tundra
    heterogeneity   13–14, 73
    nutrient pools   109
    see also Model, Vegetation

US Arctic Research Commission   6
UTM grid   4, 73

Valdez   3–4
Vapor pressure   135
Vegetation
    and spectral reflectance   158
    and topography   374–375, 388–389
    at Imnavait Creek   82–84
    classification   80, 109

colluvial basin   84–85, 87, 92
composition   96
exposed   81, 84, 88
heath   84–86, 110–111, 125–126
in marshes (see mire)
in mire   81, 84, 92–93
in snowbeds   81, 84–86, 88–89,
    126
in tussock tundra   84, 89–90, 110
in water tracks   84–85, 87, 90–92,
    110
map unit   80, 98
moist upland   84–86, 90
mosaic   13, 109
percent cover by growth forms   96,
    112–114
physiognomy   81
riparian   81, 90–92, 110
sedge meadow   110
see also Model

Water
    and nutrient availability   389
    balance   131, 136, 372, 373, 376–379,
        421
    control on plant structure/function
        211, 311–312, 369, 383, 419–420
    moss storage   379

post-snowmelt solution chemistry
    185, 195–199
snowmelt solution chemistry   185,
    189–192, 198
see also Discharge, Hydrology
Water table
    depth and spatial variability   234–
        237, 240–244, 248, 301, 371–373,
        375–376, 379–381, 419
Water track
    definition   75, 78
    nutrient pools   109
    with *Carex aquatilus*   92, 241
    with *Eriophorum angustifolium*   92,
        241
    see also *Eriophorum vaginatum*, Vege-
        tation
Waterlogged soil   89, 196
Weir
    on Imnavait Creek   74, 186–187
Wetfall see Atmospheric deposition
Wind
    and snow distribution   137
    at Imnavait Creek   134–135
    at Toolik Lake   331
Wrangel Island   228
Yukon   3
Yukon-Tanana uplands   47

Ecological Studies
Volumes published since 1990

Volume 80
Plant Biology of the Basin and Range
(1990)
B. Osmond, G.M. Hidy, and L. Pitelka (Eds.)

Volume 81
Nitrogen in Terrestrial Ecosystems:
Questions of Productivity, Vegetational
Changes, and Ecosystem Stability (1991)
C.O. Tamm

Volume 82
Quantitative Methods in Landscape
Ecology: The Analysis and Interpretation
of Landscape Heterogeneity (1990)
M.G. Turner and R.H. Gardner (Eds.)

Volume 83
The Rivers of Florida (1990)
R.J. Livingston (Ed.)

Volume 84
Fire in the Tropical Biota: Ecosystem
Processes and Global Challenges (1990)
J.G. Goldammer (Ed.)

Volume 85
The Mosaic-Cycle Concept of Ecosystems
(1991)
H. Remmert (Ed.)

Volume 86
Ecological Heterogeneity (1991)
J. Kolasa and S.T.A. Pickett (Eds.)

Volume 87
Horses and Grasses: The Nutritional
Ecology of Equids and Their Impact
on the Camargue (1992)
P. Duncan

Volume 88
Pinnipeds and El Niño: Responses
to Environmental Stress (1992)
F. Trillmich and K.A. Ono (Eds.)

Volume 89
Plantago: A Multidisciplinary Study (1992)
P.J.C. Kuiper and M. Bos (Eds.)

Volume 90
Biogeochemistry of a Subalpine Ecosystem:
Loch Vale Watershed (1992)
J. Baron (Ed.)

Volume 91
Atmospheric Deposition and Forest
Nutrient Cycling (1992)
D.W. Johnson and S.E. Lindberg (Eds.)

Volume 92
Landscape Boundaries: Consequences
for Biotic Diversity and Ecological Flows
(1992)
A.J. Hansen and F. di Castri (Eds.)

Volume 93
Fire in South African Mountain Fynbos:
Ecosystem, Community, and Species
Response at Swartboskloof (1992)
B.W. van Wilgen et al. (Eds.)

Volume 94
The Ecology of Aquatic Hyphomycetes
(1992)
F. Bärlocher (Ed.)

Volume 95
Palms in Forest Ecosystems of Amazonia
(1992)
F. Kahn and J.-J. DeGranville

Volume 96
Ecology and Decline of Red Spruce,
in the Eastern United States (1992)
C. Eagar and M.B. Adams (Eds.)

Volume 97
The Response of Western Forests
to Air Pollution (1992)
R.K. Olson, D. Binkley, and M. Böhm (Eds.)

Volume 98
Plankton Regulation Dynamics (1993)
N. Walz (Ed.)

Volume 99
Biodiversity and Ecosystem Function
(1993)
E.-D. Schulze and H.A. Mooney (Eds.)

Volume 100
Ecophysiology of Photosynthesis (1994)
E.-D. Schulze and M.M. Caldwell (Eds.)

Ecological Studies

Volumes published since 1990

Volume 101
Effects of Land Use Change on Atmo-
spheric CO₂ Concentrations: South and
South East Asia as a Case Study (1993)
V.H. Dale (Ed.)

Volume 102
Coral Reef Ecology (1993)
Y.I. Sorokin

Volume 103
Rocky Shores: Exploitation in Chile
and South Africa (1993)
W.R. Siegfried (Ed.)

Volume 104
Long-Term Experiments With Acid Rain
in Norwegian Forest Ecosystems (1993)
G. Abrahamsen et al. (Eds.)

Volume 105
Microbial Ecology of Lake Plußsee (1993)
J. Overbeck and R.J. Chrost (Eds.)

Volume 106
Minimum Animal Populations (1994)
H. Remmert (Ed.)

Volume 107
The Role of Fire in Mediterranean-Type
Ecosystems (1994)
J.M. Moreno and W.C. Oechel

Volume 108
Ecology and Biogeography of Mediterranean Ecosystems in Chile, California
and Australia (1994)
M.T.K. Arroyo, P.H. Zedler,
and M.D. Fox (Eds.)

Volume 109
Mediterranean-Type Ecosystems.
The Function of Biodiversity (1995)
G.W. Davis and D.M. Richardson (Eds.)

Volume 110
Tropical Montane Cloud Forests (1995).
L.S. Hamilton, J.O. Juvik,
and F.N. Scatena (Eds.)

Volume 111
Peatland Forestry. Ecology and Principles
(1995)
E. Paavilainen and J. Päivänen

Volume 112
Tropical Forests: Management and Ecology
(1995)
A.E. Lugo and C. Lowe (Eds.)

Volume 113
Arctic and Alpine Biodiversity. Patterns,
Causes and Ecosystem Consequences (1995)
F.S. Chapin III and C. Körner (Eds.)

Volume 114
Crassulacean Acid Metabolism. Biochemis-
try, Ecophysiology and Evolution (1995)
K. Winter and J.A.C. Smith (Eds.)

Volume 115
Islands. Biological Diversity and Ecosystem
Function (1995)
P.M. Vitousek, L.L. Loope, and H. Adsersen
(Eds.)

Volume 116
High Latitude Rainforests and Associated
Ecosystems of the West Coast
of the Americas: Climate, Hydrology,
Ecology and Conservation (1995)
R.G. Lawford, P. Alaback,
and E.R. Fuentes (Eds.)

Volume 117
Anticipated Effects of a Changing Global
Environment on Mediterranean-Type
Ecosystems (1995)
J. Moreno and W.C. Oechel (Eds.)

Volume 118
Impact of Air Pollutants
on Southern Pine Forests (1995)
S. Fox and R.A. Mickler (Eds.)

Volume 119
Freshwater Ecosystems of Alaska (1996)
A.M. Milner and M.W. Oswood (Eds.)

Volume 120
Landscape Function and Disturbance
in Arctic Tundra (1996)
J.F. Reynolds and J.D. Tenhunen (Eds.)

Volume 121
Biodiversity and Savanna Ecosystem
Processes. A Global Perspective (1996)
O.T. Solbrig, E. Medina, and J.F. Silva (Eds.)